TRADE, GROWTH AND DEVELOPMENT
ESSAYS IN HONOR OF PROFESSOR T.N. SRINIVASAN

CONTRIBUTIONS
TO
ECONOMIC ANALYSIS

242

ELSEVIER
Amsterdam – Lausanne – New York – Oxford – Shannon – Singapore – Tokyo

TRADE, GROWTH AND DEVELOPMENT
ESSAYS IN HONOR OF PROFESSOR T.N. SRINIVASAN

Edited by

Gustav RANIS
Yale University, New Haven, U.S.A.

and

Lakshmi K. RAUT
University of Chicago, Chicago, U.S.A.

1999

ELSEVIER
Amsterdam – Lausanne – New York – Oxford – Shannon – Singapore – Tokyo

ELSEVIER SCIENCE B.V.
Sara Burgerhartstraat 25
P.O. Box 211, 1000 AE Amsterdam, The Netherlands

First edition 1999

Library of Congress Cataloging in Publication Data
A catalog record from the Library of Congress has been applied for.

ISBN: 0 444 50071 5

The paper used in this publication meets the requirements of ANSI/NISO Z39.48-1992 (Permanence of Paper).

Printed in The Netherlands.

INTRODUCTION TO THE SERIES

This series consists of a number of hitherto unpublished studies, which are introduced by the editors in the belief that they represent fresh contributions to economic science.

The term 'economic analysis' as used in the title of the series has been adopted because it covers both the activities of the theoretical economist and the research worker.

Although the analytical methods used by the various contributors are not the same, they are nevertheless conditioned by the common origin of their studies, namely theoretical problems encountered in practical research. Since for this reason, business cycle research and national accounting, research work on behalf of economic policy, and problems of planning are the main sources of the subjects dealt with, they necessarily determine the manner of approach adopted by the authors. Their methods tend to be 'practical' in the sense of not being too far remote from application to actual economic conditions. In addition they are quantitave.

It is the hope of the editors that the publication of these studies will help to stimulate the exchange of scientific information and to reinforce international cooperation in the field of economics.

The Editors

Preface

T.N. Srinivasan began his academic career with training in India in mathematics and statistics before coming to Yale, with the encouragement of Tjalling Koopmans, to earn his M.A. and Ph.D. degrees. T.N. subsequently returned to India and worked for many years in the Indian Statistical Institute. After a three-year sojourn in the World Bank, he rejoined the Yale faculty in 1980 as the Samuel C. Park, Jr. Professor of Economics.

T.N. has been a strong adherent of the neo-classical tenets of free markets and free trade and can feel justifiably proud of the fact that the profession has moved in his direction in recent decades. His use of mathematical and statistical tools, a rarity in the development profession at the time he began his professional career, have stood him in good stead over time. As an Indian and with a life-long interest in the application of development economics to his home country, he is fond of quoting Nehru and other early contributors to post-independence Indian economic policy. He himself was a relatively early advocate of theoretical conceptions and policy applications which have now become part of the conventional wisdom in India. He is truly one prophet who is appreciated in his own country.

On March 27 and 28, 1998, T.N.'s professional colleagues and students joined in honoring him at a Festschrift Conference at Yale leading to this volume. Its title, *Trade, Growth and Development*, appropriately spans T.N.'s professional interests which are, and have continued to be, remarkably broad and deep. Not all of the fields to which he contributed, e.g., on the role of government and on poverty, nutrition and income distribution, are represented here. But the nineteen contributions presented do reflect four main areas of his scholarly activities.

Part I deals with various dimensions of the issue of growth. Indeed T.N.'s dissertation, as well as his first published papers, focussed on the choice of technique problem within a two-sector growth model. This was a hot topic at a time when central planning was considered a prerequisite for successful growth, with the Soviet model popular and Mahalanobis' influence dominant. The problem comes down to choosing an optimal capital intensity for a developing country. The standard assumptions are two sectors, a machine making sector and a consumer goods sector, fixed proportions, and constant returns to scale. Srinivasan showed that, given government's objective to reach the maximum sustainable consumption level as soon as possible, one should invest heavily in the capital goods sector and let the consumer goods sector be neglected for a time. T.N.'s contribution is evaluated eloquently by both Robert Solow, in Chapter 1, and by Ali Khan, in Chapter 2. Solow points out how T.N. finessed the Ramsey-type problem, i.e., how to find the largest discounted value of future utility streams achievable via the right technology choice, in the absence of Pontryagin techniques and without having to choose a unique social rate of time preference. In Chapter 2, Ali Khan traces later developments in the choice of technique literature and points out that the above Ramsey problem is still with us. He briefly traces the evolution of T.N.'s own work from project choice and growth to the criteria for project evaluation under distortions.

In Chapter 3, the late Rolf Mantel raises the question of how the assumption about the social rate of time preference affects the capital accumulation path in an optimal growth model of the neo-classical variety. Since one can assume either an increasing or decreasing marginal rate of time preference, Mantel shows that, given the hybrid assumption, i.e., initially decreasing, then increasing impatience, one can obtain two locally stable steady states. Normally, the dynamics in Ramsey-type models are simple in the sense that they exhibit monotonicity and turnpike properties, and thus do not produce macroeconomic fluctuations. Chapter 4, by Nishimura and Raut, deals with another dimension of the same model, i.e., what happens when fertility is endogenous in the neo-classical type of growth model. They show that the tunpike property no longer holds in such a case and go on to discuss a different version of the neo-classical model, one with over-lapping generations. With endogenous fertility, one can now obtain quite striking differences in the dynamics, including the possibility of macroeconomic fluctuations.

Part II of this volume is concerned with international trade issues, an area in which T.N., during his long and highly productive scholarly career, has undoubtedly invested the largest portion of his prodigious energies. Indeed his work, much of it joint with his long time partner and close friend, Jagdish Bhagwati, on trade under various domestic market imperfections can justifiably be called a landmark in the literature. His more recent contributions in this arena have been concerned with such contemporary issues as trade and the environment, trade and labor conditions, child labor, etc. as well as on the GATT/WTO world trade structure. Ever faithful to the neo-classical teachings of free trade, rightly suspicious of fair trade as a code word for protectionism, he has also criticized the "new trade theory" as "old wine in new bottles."

Chapter 5 in this volume, by Bhagwati, mirrors one of Srinivasan's major concerns, that of analytical support for a multilateral world trading system. There has been widespread suspicion, no doubt fuelled by vested interests, that, with the opening of trade to developing countries, industrial country jobs will be increasingly exported and wages, especially of the unskilled, will fall. Bhagwati's paper seeks to allay such fears by showing that, under a variety of more realistic assumptions, the Stolper-Samuelson theorem may not hold. He indeed argues that, if anything, trade between North and South has actually moderated the fall of wages in the North. The adjoining paper by Krueger (Chapter 6) discusses the various policy regimes for exchange rates as used by a majority of developing countries. This, in the postwar era, has been called a "stop and go" policy paradigm, with fixed exchange rates followed by periodic devaluations, as real exchange rates appreciate as a consequence of relatively high domestic inflation. The 80s saw the rise of the NIC's in Asia, which mostly kept the real exchange rates fixed or in a variable fixed peg regime. Following this example in the 90s, Latin American countries have tried to use the exchange rate to control inflation. Krueger concludes that they are bound to fail for the same reason that the fixed exchange rate system of the early 70s failed.

Keyzer and van Wesenbeeck (Chapter 7) consider in more detail the neoclassical model with imperfect competition and derive the optimal tariff argument, discussing various cases that may arise in reality, plus the additional complexities of such a modeling exercise. The Bhattarai, Ghosh, and Whalley contribution (Chapter 8), an application of general equilibrium analysis to policy issues, includes a specification of the import and export demand of the rest of the world and shows how alternative closure assumptions affect the general equilibrium conclusions. The most common type of problem in such models is that they define trade balance as a model feature rather than an equilibrium condition. The authors actually refer to two points made by Srinivasan earlier, i.e., that the exchange rate becomes superfluous if all three equations, for demand, supply and the trade balance, are included in the model — moreover that a country cannot be a price taker in import markets and simultaneously a price maker in export markets. The chapter illustrates the importance of closure by simulating the offer curves in a product differentiation model and showing that one can get perverse offer curves for reasonable parameter values, in which case the Nash equilibrium in a tariff game may not exist.

Part II concludes with two papers in applied trade theory. The contribution by Levy (Chapter 9), presents the arguments for and against preferential trade agreements when contrasted with multi-lateral trade negotiations. While it is widely believed that preferential trade agreements, by at least moving in the direction of a free trade regime, lead to an improvement in welfare for all participants, Levy examines the arguments pro and con of both North-South trade agreements like NAFTA, and South-South arrangements like ASEAN. The paper by Goto (Chapter 10) analyzes the impact of regional integration on the exports of a new member. It applies the differentiated products model of international trade, but relates this application of the "new trade theory" to agriculture instead of to manufacturing as is standard practice. The question posed has particular relevance for rice markets in Japan. With the help of a formal model, Goto shows that the rise in exports will depend directly on the prior level of protection and inversely on the amount of product differentiation. He shows that the homogeneous model of trade liberalization tends to exaggerate the likely impact of tariff reductions and that the actual impact may be insignificant when products are widely differentiated.

Part III of the volume focuses on general equilibrium and development policies, mainly as applied to India. Some years ago, surveying the literature on social cost-benefit analysis, Srinivasan prophetically wrote that ". . . consensus has yet to come on the analytical basis for many of the project-appraisal techniques. Nor is there agreement on the utility of computable general equilibrium models either in understanding the development process, or in devising development policy. . . ." Together with his collaborators, Bell and, more recently, Dawkins and Whalley, T.N. has consistently pointed out several deficiencies in the existing approaches to project appraisal and the reasons for the declining interest in CGE methods for development planning and policy analysis. He has equally consistently argued for the use of general equilibrium prices to evaluate relatively large projects and criticized CGE models as lacking dynamics, expectations, and flexibility. The papers in Part III address some of these concerns.

The contribution by Narayana and Parikh (Chapter 11), for example, traces the history of Indian planning over the decades, emphasizing both theoretical justifications and pitfalls, reflecting very much the change in T.N.'s own views on the subject over time. While Indian planners were initially emboldened by the apparent success of the Soviet model and tried to imitate it in some respects, the authors provide several reasons for ultimate policy reversal, including inadequate information and implementation, and, most basically, the rigidity of the behavioral assumptions. They conclude that the reforms of recent decades have made central planning both obsolete and unnecessary. Go and Mitra (Chapter 12) proceed to focus on the likely impact of these policy reforms by means of a set of simulation exercises. Their approach is somewhat unique in the sense that it combines an aggregate as well as disaggregated model approach; a 6-sector model is used to calculate parameters which are then used as inputs into a disaggregated 72-sector model. The authors consider the effects of different levels of trade liberalization and of fiscal adjustment in the absence and presence of foreign finance, and claim validation of their model when results turn out to be roughly similar to *ex post* data.

The paper by Majumdar and Ossella (Chapter 13), is reminiscent of the theories of balanced and unbalanced growth popular in the early development literature. They seek to address an important policy issue, i.e., given India's overall limited resources, in which "leading sector" of the economy should the government invest in order to maximize social welfare. Deploying Leontief's input–output matrix, they conclude that investment in infrastructure increases long run growth more than the equivalent investment applied to any other sector.

The final contribution in this part, by Heckman, Lochner, and Taber (Chapter 14), is very much in sympathy with T.N.'s strongly held view against the prevalent practice of using partial equilibrium analysis to perform policy evaluations. As T.N. has pointed out in recent papers with Shoven and Bell, on the uses and abuses of economy-wide planning models, he much prefers to apply the more precise general equilibrium techniques, especially in the Indian context. While the Heckman et al. chapter focuses on education and tax policies for the U.S., the techniques developed would have T.N.'s endorsement for application to policy analysis in the developing world. Heckman and his associates construct a model which generates the phenomenon of a widening wage gap between skilled and unskilled labor as recently observed in most industrialized countries. The result, generated within an overlapping generations framework, is driven by skill-biased technical change.

The volume's final part, on market failures and economic structure, leads off with a chapter by Chander (Chapter 15) which considers whether local coalitions facilitate or inhibit the formation of global coalitions, applied, for example, to a treaty on the environment. Chander shows that under specified conditions coalitions can actually inhibit cooperation by raising the reservation price of participants since it tends to generate positive externalities for other coalitions.

A widespread assumption in economic theory generally and in development economics, in particular, is one of complete markets, an assumption which, if risky, of course, makes life much easier. The paper by Udry (Chapter 16) attempts to examine this issue in the context of African development. Complete markets imply separation, i.e. that households' production decisions are made independently of their consumption decisions. The argument that supports the assumption of separation is that some developing country markets may be efficient even as others may be missing or inefficient. Udry works with data from Burkina Faso and Kenya and finds that the null hypothesis of separation is rejected everywhere. In Burkina Faso this may be due to the absence of credit and insurance markets and in Kenya due to the inefficiency of land and labor markets.

Bell's paper (Chapter 17), also focusses on the agricultural sector and examines the existing literature on the interlinking of rural markets. In the absence of uncertainty and unlimited liability he finds some instruments to be redundant, e.g., the landlord money-lender can maximize profits by offering either a no-usury interest rate loan or a no-rent or fixed rent contract. Under different assumptions about uncertainty and monitoring costs, moral hazard problems can arise. In a very different context, i.e., examining the dynamics of the Kibbutz in Israel, Byalsky, Keren, and Levhari (Chapter 18) note that this institution serves not only production but also collective consumption decisions, implying that consumers have bargaining power they would not have as individuals. In the case of a non-cooperative Nash equilibrium with asymmetric information, the problem of moral hazard can be overcome by peer pressure in a close-knit community. However, given the changing technology of consumption, i.e., with private goods becoming more important than public goods, the Kibbutzim are losing their natural advantage and becoming less important in Israel.

The final chapter (Chapter 19) of Part IV, by Scarf, deals with another market failure, that due to non-convexity related to the so-called social planner problem. In an activity analysis context increasing returns to scale can cause non-convexities which nullify the basic welfare theorems, the foundations of modern microeconomics, and lead to the inevitable demise of competition. Scarf demonstrates that in such cases the pricing test of optimality for a program may fail and goes on to suggest an alternative, i.e., what he refers to as the quantity test of optimality.

The editors would like to thank the Economic Growth Center and the Department of Economics at Yale as well as the University of Hawaii for their support of this enterprise, Professor Mukul Majumdar for his advice and Mainak Sarkar for his research assistance, as well as Glena Ames and Lora LeMosy for their valuable help at various crucial administrative and typing stages.

Gustav Ranis
Yale University
New Haven, Connecticut

Lakshmi K. Raut
University of Chicago
Chicago, Illinois

April, 1999

About the Authors

Clive Bell is Professor of Economics at Heidelberg University. His current interests include rural economic organization, inequality, growth and fiscal policy, all with reference to India. He has taught at Johns Hopkins, Regensburg, Vanderbilt and Sussex, and has held a position as senior economist at the World Bank.

Jagdish Bhagwati is Arthur Lehman Professor of Economics and Professor of Political Science at Columbia University. Until 1980, he was Ford International Professor of Economics at MIT. He served as the Economic Policy Adviser to the Director General of GATT (1991–93). Five volumes of his scientific essays have been published by The MIT Press, and three Festschrift volumes have been published in his honor in the United States, the UK and the Netherlands. He has received several prizes, among them the Bernhard Harms Price (Germany), the Mahalanobis Memorial Medal (India), the Kenan Enterprise Award (USA), the Seidman Distinguished Award in Political Economy (USA) and the Freedom Prize (Switzerland).

Keshab Bhattarai is a Research Associate who has been working as a general equilibrium modeller with John Whalley. He is interested in using static and dynamic general equilibrium models to analyze micro and macro policy issues of developed, developing and global economies. He has constructed static and dynamic applied general equilibrium tax models of the UK economy. He received his Ph.D. from Northeastern University.

Michael A. Byalsky is an Adjunct Lecturer of Mathematical Methods at the Department of Economics of the Hebrew University of Jerusalem. He worked for a long period at some leading research institutes of the former USSR in various branches of mathematical economics, as well as participating in the government advisory group on perestroika problems. Since his repatriation to Israel in 1990, he has worked as Lecturer at Bar-Ilan, Tel-Aviv and Hebrew Universities.

Parkash Chander is a Professor of Economics at the Indian Statistical Institute. His main areas of research are environmental economics, public economics, and tax evasion. He has also done work on mechanism design and comparative economic systems. He has held visiting positions at Johns Hopkins, Caltech, and CORE. He received his Ph.D. from the Indian Statistical Institute.

Madanmohan Ghosh is a Post-Doctoral Fellow at the University of Western Ontario. His principal area of research is applied general equilibrium modelling in the fields of international trade and public finance. He has been providing technical support to an IDRC-funded project in Vietnam since November, 1996. He has worked in various capacities at India's National Institute of Public Finance and the Indian Statistical Institute. He received his Ph.D. from the Jawaharlal Nehru University, New Delhi, India.

Delfin S. Go is a Country Economist at the World Bank specializing in Macroeconomics and Southern Africa. He was formerly with the Development Economics Research Group of the World Bank where his principal areas of research were taxation, investment and growth. He received his Ph.D. in Political Economy and Government from Harvard University.

Junichi Goto is a Professor of Economics at Kobe University and a Visiting Professor at the Inter-American Development Bank. His principal area of research is international trade. He has also done work on international migration and other aspects of labor economics. He has held positions as an economist at the World Bank, as deputy director of the Labor Economics Division of the Japanese Ministry of Labor, as a visiting scholar at MIT and Yale University. He received his Ph.D. from Yale.

James Heckman is Henry Schultz Distinguished Service Professor of Economics at the University of Chicago and Director of the Center for Social Program Evaluation at the Harris School of Public Policy at the University of Chicago. He is a Senior Fellow of the American Bar Foundation. Heckman has done work in econometrics, labor economics, public finance and program evaluation. He received his Ph.D. from Princeton and has taught at Chicago for most of his professional life, with short stays at Columbia and Yale. Heckman is a member of several scholarly bodies: Fellow of the Econometric Society, Member of the National Academy of Sciences and Fellow of the American Academy of Arts and Sciences. He is the recipient of the J.B. Clark award of the American Economic Association.

Michael Keren is a Professor of Economics at The Hebrew University of Jerusalem. His principal area of research is economic systems, and until the present decade he has focused on the economies of Eastern Europe and their reforms. He is now concerned with the transition of these economies to market systems. He received his Ph.D. from Yale University.

Michiel Keyzer has been associated with the Centre for World Food Studies of the Vrije Universiteit (SOW–VU) since 1980. He received his Ph.D. at the Vrije Universiteit, Amsterdam. In 1995 he became professor of economics and director of the SOW–VU. His main research activities are in the area of mathematical economics and economic model building. He has published in the field of general equilibrium modelling, welfare economics, and economic policy analysis, with special reference to food and agriculture. He has been the leader of research projects in South and Southeast Asia and West Africa and was involved in various studies on the reform of Europe's Common Agricultural Policy.

M. Ali Khan is Abram Hutzler Professor of Economics at The Johns Hopkins University. His principal areas of research are mathematical economics and the history of economic thought. He has also done work in international trade and economic development. He received his Ph.D. from Yale University.

Anne O. Krueger is Herald L. and Caroline L. Ritch Professor of Sciences and Humanities in the Economics Department, Director of the Center for Research on Economic Development and Policy Reform, and a Senior Fellow at the Hoover Institution at Stanford University. She is also a Research Associate of the National Bureau of Economic Research. Her primary fields of interest are international trade and economic development and she has published extensively in those areas. Professor Krueger also taught at the University of Minnesota and Duke University, was Vice President, Economics and Research at the World Bank from 1982 through 1986. She received her Ph.D. from the University of Wisconsin.

David Levhari is a Professor of Economics at The Hebrew University of Jerusalem. His research is in economic theory, and he is interested in the economics of uncertainty and organization. He received his Ph.D. from MIT.

Philip I. Levy is an Associate Professor of Economics at Yale University. His principal area of research is international trade and political economy. He has also done work on economic development. He has held positions as a staff economist for the President's Council of Economic Advisers as well as for the GATT. He received his Ph.D. from Stanford University.

Lance Lochner is an Assistant Professor of Economics at the University of Rochester. His research focuses on issues in labor markets, education, and crime. In particular, he is currently studying the determinants of rising wage inequality and policies designed to improve national skill levels and combat inequality. He received his Ph.D. from the University of Chicago.

Mukul Majumdar is H.T. and R.I. Warshow Professor of Economics at Cornell University. He received his Ph.D. from the University of California at Berkeley. His primary area of research has been the economic theory of intertemporal allocation of resources. He is also a member of the Center for Applied Mathematics at Cornell and has contributed to the mathematical theory of chaotic and stochastic processes.

Rolf R. Mantel, whose recent sudden demise was a shock to all, was Professor of Economics and Director of the Department of Economics of the University of San Andres in Argentina. He received his Ph.D. from Yale University in 1966. During the sixties Rolf became interested in the problem of the calculation of general equilibrium. Later on his interests shifted to the characterization of optimal growth paths in the case of endogenous preferences. One of his best known contributions was the analysis of aggregate excess demand functions, together with H. Sonnenshein and G. Debreu. He was a Fellow of the Econometric Society and President of its Latin American Standing Committee. He received an honorary degree from the University of Tucuman, a Guggenheim Fellowship, and was President of the Argentine Association of Political Economy. He was Visiting Professor at Yale, Harvard and Northwestern Universities, and very recently, Visiting Fellow at the Weizman Institute in Israel.

Pradeep Mitra is Director of the Poverty Reduction and Economic Management Unit in the Europe and Central Asia Region of the World Bank. He was educated at Presidency College, Calcutta, the Delhi School of Economics and Balliol College in Oxford, where he received a D. Phil. in Economics and was a Rhodes Scholar. He has taught at the Universities of London and Delhi and has published widely in the areas of public economics and development economics.

N.S.S. Narayana is a Professor of Economics at the Bangalore Centre of the Indian Statistical Institute. His research interests include computable general equilibrium models, agricultural economics and econometric applications. He has earlier held positions as Consultant in the Planning Commission (New Delhi), Reader in the University of Madras, and Research Scholar at the International Institute for Applied Systems Analysis (Austria). He is on the Editorial Board of the *Indian Journal of Agricultural Economics*. He has a B.E. from Andhra University and a Ph.D. from the Indian Statistical Institute.

Kazuo Nishimura is a Professor of Economics at the Institute of Economic Research, Kyoto University. His principal area of research is nonlinear dynamics and intertemporal equilibrium theory. He previously held positions at the State University of New York at Buffalo and the University of Southern California. He received his Ph.D. from the University of Rochester.

Ilaria Ossella expects to complete her Ph.D. in Economics from Cornell University in June 1999. Beginning August 1999, she will hold the position of Assistant Professor of Economics at Illinois Wesleyan University. Her principal area of research is in economic development, growth and trade.

Kirit S. Parikh is Founder/Director and Vice-Chancellor of the Indira Gandhi Institute of Development Research (IGIDR), Mumbai. He received his Sc.D. in Civil Engineering and M.S. in Economics from MIT. He has served as Senior Economic Advisor to the Administrator of UNDP (1997–98), Program Leader of the Food and Agricultural Program of the International Institute for Applied Systems Analysis (IIASA) (1980–86), and Professor of Economics (and sometimes Head) of the Indian Statistical Institute, New Delhi (1967–80). He is a member of the Economic Advisory Council of the Prime Minister of India and has been a member of the Economic Advisory Councils of four Prime Ministers. He has authored and co-authored 16 books in the areas of planning, water resource management, appropriate technology for housing, optimum requirement for fertilizers, energy systems, national and international policies for poverty reduction, trade policies, economic reforms, sustainable human development, environmental accounting and general equilibrium modeling.

Gustav Ranis is the Director of the Center for International and Area Studies and Frank Altschul Professor of International Economics at Yale University. He has served as Director of the Pakistan Institute of Development Economics, Director of the Economic Growth Center at Yale, and as Assistant Administrator for Program and Policy in the U.S. Agency for International Development. He has been a consultant to The World Bank, ADB, AID, OECD, UNIDO, FAO, UNDP, the Ford and Rockefeller Foundations, among others, and currently serves as a member of the advisory board of a number of third world research institutions. He has worked extensively in and on Korea, Taiwan and Japan as well as on Indonesia, the Philippines, the Caribbean, Mexico and Ghana. He has been a Distinguished Visitor to China under the Advisory Panel on Chinese Economics Education. In 1982 he was awarded an Honorary Degree by Brandeis University. He received his Ph.D. from Yale University.

Lakshmi K. Raut is a Research Associate at the University of Chicago. His principal areas of research are growth and development, human resources, social security, social mobility, and Industrial R&D. He has held positions as Associate Professor at the University of Hawaii-Manoa, as Assistant Professor at the University of California-San Diego, as Hewlett Post-Doctoral Fellow at the University of Chicago, and as Consultant at the World Bank. He received his Ph.D. from Yale University.

Herbert Scarf is Sterling Professor of Economics at Yale University. His research interests have included non-cooperative game theory inventory theory, the stability and computation of economic equilibria, convergence of the core to the Walrasian equilibrium, the existence of a nonempty core for general cooperative games, economies of scale in production, and the study of production sets arising from activity analysis models with discrete activity levels. He has been at the Cowles Foundation and the Department of Economics at Yale since 1963. He received his Ph.D. from Princeton University.

Robert M. Solow is Institute Professor Emeritus at the Massachusetts Institute of Technology, where he has been a professor of economics since 1949. He taught macroeconomics and other subjects to undergraduate and graduate students until January, 1996. Professor Solow studied at Harvard and received the Nobel Prize in Economics in 1987 for his theory of growth. For a number of years he served as member of the Board of Directors of the Federal Reserve Bank of Boston and was Chairman of that Board for three years. He is currently a member of the National Science Board. He is past president of the American Economic Association and of the Econometric Society, a member of the National Academy of Sciences and a fellow of the British Academy.

Christopher Taber is currently an assistant professor in the Department of Economics and the Institute for Policy Research at Northwestern University. He has been at Northwestern since receiving his Ph.D. from the University of Chicago in 1995. Much of Professor Taber's work is on human capital accumulation, but he has also written on labor economics, econometrics and public economics.

Christopher Udry is a Professor of Economics at Yale University. His principal area of research is development economics. Most of his work concerns microeconomic aspects of development in Africa. He was formerly an Associate Professor of Economics at Northwestern and has been a visiting scholar at Ahmadu Bello University in Nigeria and at the University of Ghana. His Ph.D. is from Yale.

Lia van Wesenbeeck has been associated with the Centre for World Food Studies of the Vrije Universiteit (SOW–VU) since 1993. She started her Ph.D. research on the topic of imperfect competition and international trade in a general equilibrium setting in 1995, under the supervision of Michiel Keyzer. At the moment, this thesis is nearing completion.

John Whalley holds appointments at the Universities of Western Ontario in Canada and Warwick in the UK. He works on policy issues in trade, public finance, and environmental issues, using numerical simulation methods as a central part of his research methodology. He is a fellow of the Econometric Society and a Fellow of the Royal Society of Canada. He is a joint managing editor of the trade policy journal *The World Economy.*

Contents

Preface vii
About the Authors xiii

PART I. Economic Growth

Chapter 1 Srinivasan on Choice of Technique 3
Robert M. Solow
Chapter 2 Srinivasan on Choice of Technique Once Again 9
Ali Khan
Chapter 3 The Effects of a Decreasing Rate of Time Preference on the Accumulation of Capital 21
Rolf R. Mantel
Chapter 4 Endogenous Fertility and Growth Dynamics 39
Kazuo Nishimura and Lakshmi K. Raut

PART II. Trade Theory and Policy

Chapter 5 Play it Again, Sam: A New Look at Trade and Wages 57
Jagdish Bhagwati
Chapter 6 Exchange Rate Policies for Developing Countries: What Has Changed? 71
Anne O. Krueger
Chapter 7 Trade Models of Imperfect Competition 89
Michiel Keyzer and Lia van Wesenbeeck
Chapter 8 More on Trade Closure 125
Keshab Bhattarai, Madanmohan Ghosh and John Whalley
Chapter 9 Developing Countries and the Lure of Preferential Trade Agreements: Beware of the Hook 137
Philip I. Levy
Chapter 10 The Impact of Regionalism on Agricultural Trade: APEC and Japanese Rice Imports 161
Junichi Goto

PART III. General Equilibrium and Development Policy

Chapter 11 Virtual Reality in Economy-wide Models: Some Reflections on Hope and Despair with Reference to India ... 201
N.S.S. Narayana and Kirit S. Parikh

Chapter 12 Trade Liberalization, Fiscal Adjustment and Exchange Rate Policy in India ... 229
Delfin S. Go and Pradeep Mitra

Chapter 13 Identifying Leading Sectors that Accelerate the Optimal Growth Rate: A Computational Approach ... 273
Mukul Majumdar and Ilaria Ossella

Chapter 14 General-Equilibrium Cost-Benefit Analysis of Education and Tax Policies ... 291
James J. Heckman, Lance Lochner and Christopher Taber

PART IV: Market Failures and Economic Structure

Chapter 15 International Treaties on Global Pollution: A Dynamic Time-Path Analysis ... 353
Parkash Chander

Chapter 16 Efficiency and Market Structure: Testing for Profit Maximization in African Agriculture ... 363
Christopher Udry

Chapter 17 Explaining Interlinking ... 395
Clive Bell

Chapter 18 The Kibbutz as a Labor-Managed Club: Public and Private Goods, Incentives, and Social Control ... 419
Michael Byalsky, Michael Keren and David Levhari

Chapter 19 What Do We Know about Production Sets Arising from an Activity Analysis Model with Integral Activity Levels? ... 443
Herbert E. Scarf

Index ... 459

Part I

Economic Growth

Trade, Growth, and Development
G. Ranis and L.K. Raut

CHAPTER 1

Srinivasan on Choice of Technique

Robert Solow

Department of Economics, MIT, Cambridge, MA 02139

One day, probably around 1960, I got a letter from Yale. It pointed out that there was a slip in the way I had written down the solution of a differential equation in my 1956 *Quarterly Journal* article. The writer also said that I had seemed to use the words "uncorrelated" and "independent" as if they meant the same thing, and he included a standard textbook counter-example. My correspondent was one T.N. Srinivasan, then a graduate student. I have not been able to find the letter or my reply in my files. I do remember that the slip in question had no consequences for the theory of economic growth, and I corrected it — with acknowledgment, I hope — the next time the article was reprinted. I also reminded Mr. Srinivasan that I had taught probability and statistics for years and was well aware of the difference between statistical independence and lack of correlation, but thought that the slight abuse of language had been useful in the context.

This is how friendships begin. I can not remember when I first actually met T.N., but I do know that he sent me the manuscript of his second published paper ("Optimal Savings in a Two-Sector Model of Growth," *Econometrica*, 1964, 358–74) and I sent back a page or so of comments. On this occasion, I want to comment on T.N.'s first published paper. It was called "Investment Criteria and Choice of Techniques of Production," and one knows how important it must have been to him, because it is signed Thirokodikaval N. Srinivasan. In fact, this paper represents T.N.'s Ph.D. thesis, or a large part of it. It was published in *Yale Economic Essays*, a journal devoted to Yale Ph.D. theses, in 1962. It is interesting not only for its content, but because it sends a clear signal of T.N.'s lifelong commitment to rigorous thought on questions of economic development.

The precise connotation of the phrase "choice of techniques" may not be clear to contemporary readers. In the decade before Srinivasan's thesis, many of the most important students of economic development were concerned with this question: how should a developing country identify the best capital-intensity or labor-intensity as it chooses among alternative techniques of production for an investment project to be initiated? Srinivasan mentions solutions proposed by Alfred Kahn, Hollis Chenery, Walter Galenson and Harvey Leibenstein, Otto Eckstein, Maurice Dobb, Amartya Sen and Francis Bator, among others.

All are aware that a poor country with — if it is lucky — decades of development ahead of it will want to look beyond immediate profitability. Equally profitable investments can, in practice, make unequal contributions to the society's longer-run investible surplus, especially because governments in poor countries are no more or even less able than those in rich countries to make lump-sum taxes and subsidies. Somehow a good criterion of choice has to find a way to capture an investment's contribution to longer-run growth. So Kahn and Chenery defined a notion of Social Marginal Productivity, Galenson and Leibenstein a notion of Marginal Per Capita Investment Quotient, Eckstein a Marginal Growth Criterion, and Dobb and Sen chose versions of a Reinvestible Surplus.

All of these criteria were getting at something important, but all suffered from a common deficiency. Srinivasan puts it this way after his review of the literature: ". . . most of the authors analyze the problem of growth in a static framework." But I think that this does not quite name the gap that the thesis will try to fill. The trouble with these earlier attempts is that they were not embedded in a coherent model of long-run growth, and so each of them was forced to draw economy-wide implications from project-by-project characteristics. In practice this approach might be necessary in the end, but it is not the way to gain clarity about the nature of the problem. Srinivasan's thesis, or at least the part of it reported in this article, does not culminate in a formula that could be handed to an agency allocating credit or foreign exchange in a developing country; but it is a step toward a coherent theory of "choice of technique." There are no acknowledgments in the printed article; but if I had to guess who was young Mr. Srinivasan's thesis advisor, I would suspect the sober influence of Tjalling Koopmans.

Here is a quick sketch of the model T.N. uses. There is a capital-goods sector that employs labor and one fixed type of machine to produce many kinds of machines, including the type used in that sector. This production takes place with fixed proportions and constant returns to scale. (The input requirements are the same, no matter what type of machine is being produced. There is some loss of generality here, but the payoff in simplicity is very great.)

The consumption-goods sector uses machines of various kinds together with labor to produce a single, homogeneous consumer good. There are as many fixed-proportions, constant-returns-to-scale activities in this sector as there are kinds of machines, in principle countably many. Once inefficient machine-types are weeded out, there is a piecewise-linear convex unit isoquant for the consumer good, with each vertex corresponding to a machine type. The model is completed by a geometrically growing labor force and a common depreciation rate for each type of machine.

The only significant resource-allocation decisions in this model are: (a) the assignment of available labor to the capital-goods sector and to the various machine-types in the consumption-goods sector and (b) the choice as to which types of machines to build in the capital-goods sector. The "choice of technique" issue boils down to the choice of machine-types (and can indeed be represented as a choice of capital-intensity in the consumption-goods sector). But the question is now about programming a growth path, not about choosing among investment projects. Young Srinivasan does not dwell on this transformation, but it is what he has done. There is a gain in clarity and generality, as noted earlier, but the "planning entity" — whatever it is — is unlikely to have any simple, easily comprehensible rule of thumb at its disposal.

Now Srinivasan tries to win back some simplicity in an interesting way. (This is my interpretation; T.N. never stands back to describe an overall strategy.) Any economist familiar with growth theory will know that this sort of model can at best achieve a long-run steady state with constant per-capita consumption. (Remember there is no technological progress. If there were labor-augmenting technological progress, one would do the usual trick and shift to consumption per efficiency-unit of labor.) So there will be a maximal sustainable rate of consumption per worker or MSCP. (Ned Phelps was a graduate student at Yale at about this time, so the idea of the "Golden Rule" must have been in the coffee-room air. T.N. does not use that phrase, but the MSCP is his more complicated model's version of the Golden Rule consumption level. In the conventional one-sector model, the Golden Rule plays off the marginal product of capital against the extra saving needed to equip a growing labor force with a slightly higher capital intensity. Srinivasan's two-sector model must also take account of the fact that increasing capital intensity decreases the direct labor required to produce a unit of consumption, but increases the indirect labor required in the investment-goods sector to support the greater capital intensity.)

This reasoning picks out the consumption-sector technique that maximizes the steady-state level of consumption per head or MSCP. (Two adjacent techniques may be tied.) It is then straightforward to calculate the corresponding steady-state allocation of available labor between machine-building and machine-using. Srinivasan more or less takes it for granted that the goal of development planning is to reach and maintain the maximum per capita level of consumption. The (eventual) choice of technique is just subsumed under this version of development planning.

Of course, each poor country starts with historically given initial conditions. Its consumer-goods sector will not be equipped with the right sort of machines to achieve MSCP, and the machine-building industry may be initially too small either to achieve the MSCP configuration in a hurry or to maintain it if it were achieved. In practice, even at this level of abstraction, there remains the problem of getting to MSCP, and this involves an ongoing choice-of-technique question too.

The thesis resolves this problem by studying feasible consumption paths that get to the MSCP as quickly as possible. It is obvious that this fastest-approach policy will want to compress consumption in the early stages, while the appropriate stock of capital goods is being built up. Srinivasan points out that this effect could be softened by redefining feasibility to include a floor under consumption per head. Naturally, the higher the floor, the longer it will take the shortest path to achieve the MSCP configuration. He finds it necessary to ignore this complication, however.

The fastest-approach criterion leads, as one would expect, to an emphasis on heavy industry. Initially the focus is entirely on building up the investment-goods sector. Only the type of machine used in that sector is produced; the initial stock of machines in the consumer-goods sector is allowed to depreciate. In fact, the capital-goods sector is temporarily overbuilt. At a certain time, the capital-goods sector switches to producing machines of just the type used in the consumer-goods sector when it has reached the MSCP configuration. The timing is such that when the capital-goods industry depreciates to just the size it should be when it is sustaining maximal consumption, the consumption-goods sector has also reached the appropriate size, and the whole economy clicks into the MSCP configuration.

This sort of heavy-industry-first pattern of development is reminiscent of the ideas of P.C. Mahalanobis, the famous statistician who turned his mathematical talent to development planning after Indian independence. There is no reference to those proposals in the paper; but an even younger T.N. had been a student at the Indian Statistical Institute in Calcutta in the mid-1950s, so it is quite likely that he had been exposed to Mahalanobis's ideas at the source. (My own recollection is that Mahalanobis had come to this point of view using much more primitive models of a growing economy.)

Now I can summarize T.N.'s solution to the choice-of-technique problem. Find the capital-intensity (i.e., machine-type) that corresponds to the maximum sustainable consumption per head. This defines a steady-state configuration with a machine-building sector just big enough to maintain the stock of (appropriately designed) machines required in the consumption sector, with the two sectors together just capable of employing the growing labor force. Then, starting from the historically given initial state of the economy, follow the path that leads to the achievement of the terminal steady state in the shortest possible time. This will involve building up the machine-building sector first, while consumption makes do with the depreciating stock of "old" machines. After that phase, the economy makes a run for the terminal steady-state.

So far I have steadfastly avoided asking the question that anyone coming to this problem in the 1990s will be clamoring to ask: why not solve the Ramsey problem for this model? That is to say, why not define a strictly concave social utility function for current per capital consumption, and a social rate of time preference for discounting utility, and then find the largest discounted value of the stream of future instantaneous utilities achievable with this technology? We all know, for example, that such a society, even if it found itself in the steady-state with maximum sustainable consumption, would choose to move to a lower level of consumption, enjoying a temporary splurge in the meanwhile.

That is for T.N. to tell us in his memoirs. I can only speculate from internal evidence. Toward the end of the published thesis, T.N. considers two alternative criteria for development planning and choice of technique. One, the Galenson–Leibenstein suggestion, is to aim at maximizing consumption per head at some single specified date in the future. That approach is found wanting for two reasons: it is unreasonably sensitive to the particular terminal date chosen; and it can impose an unreasonable burden on those who come along after the arbitrary terminal date.

The second alternative is the maximization of the present value of the stream of future consumption per head, with discounting at the social rate of time preference. (That is, Ramsey with linear utility and positive time preference.) He rejects this again on two grounds. He pretty clearly shares Ramsey's belief that the only ethically defensible social rate of time preference is zero. Far future generations should not be discriminated against simply because they come later. And he finds that adoption of this criterion can indeed lead to unjustifiable neglect of early consumption. The modern reader should keep in mind that there is no technological progress in the Srinivasan model. So consumption per head cannot grow forever. With zero pure time preference, therefore, eventually the discount rate on goods must fall to zero. (And MSCP comes back into its own.)

There is still a loose end. A sufficiently sharply-concave utility function would enforce a closer approach to intergenerational equality. That was Ramsey's own way out of the difficulty. It is interesting that T.N. never really considers that possibility. Early on, when he first considers the approach to MSCP, he imagines that there might be a social welfare function defined on *paths* of per capita consumption. But, he says, "(S)uch a formulation is much too general." A time-additive concave social welfare function defined on current consumption is much less general. T.N. must surely have known of this option, but he never mentions it.

I can imagine a couple of reasons. Here I am merely expressing my own reservations about this widely accepted and used device; I have no warrant to impute them to T.N. in 1962 except for the possibility that if such doubts seem reasonable to me they might have seemed reasonable to him. The conventional sum-of-strictly-concave-instantaneous-utilities is in some ways too general and in some ways not general enough. It is too general in that the range of permissible degrees of concavity (e.g., the range of permissible elasticities in the convenient constant-elasticity form) is so wide that the implications for the optimal saving rate are too broad to be useful. It would be different if there were any way of narrowing the relevant range. (This is for the social welfare function, mind you.) We have more experience now than T.N. had in 1962; but I am not sure that our conventional choices are any more than arbitrary conventions. Someone seriously interested in development planning, then and there, could easily have concluded that fussing about the elasticity of the marginal social value of consumption per head was not a profitable use of time.

In another and perhaps more important respect, the time-additive convention is not general enough. It makes the marginal rate of substitution between consumption at time s and time t depend only on the level of consumption at those two moments. There is no room for learning, habit-formation, continuity, complementarity or any other consideration that could make the social value of consumption depend on its context, or on anything besides its current level. It is not hard to imagine that these gaps and difficulties could lead a clever, and impatient 29-year-old Indian economist to think: why not just decide where we are going, and go there quickly? The other extreme, to focus entirely on sustainability and intergenerational equality, leads to freezing consumption at an unreasonably low level, and that would never have attracted T.N.

Trade, Growth, and Development
G. Ranis and L.K. Raut

CHAPTER 2

Srinivasan on Choice of Technique Once Again

M. Ali Khan

Department of Economics, The Johns Hopkins University, Baltimore, MD 21218

In his 1964 article for Abba Lerner's sixtieth birthday, Paul Samuelson has a section entitled *Portrait of the Artist as a Young Blade,* and in which he writes "I believe it was Wasily Leontief who once said you can gauge the quality of a scientist by his first paper." In his article for T.N. Srinivasan's sixty-fifth birthday, Robert Solow returns to T.N.'s first paper — I say return because already in his 1962 article on substitution and fixed proportions in the theory of capital, Solow had sighted in the context of "discussions of planned economic development . . . an interesting Yale Ph.D. thesis by Mr. T.N. Srinivasan [on] investment criteria and choice of techniques of production."[1]

As a graduate student at Yale during the early seventies, I was of course well aware of Solow [31]. For one thing, it began with the unforgettable line, "I have long since abandoned the illusion that participants in this debate actually communicate with each other. So I omit the standard polemical introduction, and get down to business at once." It also has no separate bibliographic section: in addition to T.N.'s thesis, the last two paragraphs mention the work of Dobb, Sen, Salter, Butt and Samuelson. In the paragraph before these two, we read "The model I have used in this paper has long played a part in what might be called conversational economics. Mrs. Robinson and I have discussed it in correspondence.[2] But it has only lately been subjected to careful analysis, and never to my knowledge in the way I have treated it here."

I then read Solow [34] as showing how Solow [31] goes along with Srinivasan [35], and I read it, in addition, as a song of celebration (in the sequel I refer to Solow [34] simply as Solow). And when an *ustad* celebrates another *ustad* in

[1] In his 1962 article [31], Solow dates the thesis to 1961; it was subsequently published in 1962 as Srinivasan [35]).

[2] Some of this correspondence is tracked down in Turner [41].

song, the subcontinental tradition is for the *shagird* to limit themselves to that nodding of the head that has by now become obligatory for the portrayals of Indians and Pakistanis in TV sitcoms. But Lakshmi Raut insists that I add my voice as a complimentary/complementary attempt, and the Evenson–Peck–Ranis letter of invitation is also interested in my association with Professor Srinivasan. So I'll take the two tasks in order, and try to be brief.

I

Solow reminds us that the "precise connotation of the phrase 'choice of technique' may not be clear to contemporary readers [but that] in the decade before Srinivasan's thesis many of the most important students of economic development were concerned with this question."[3] One may add that the phrase continued to color the theory of dynamic economics more generally at least for a decade after T.N.'s thesis (see, for example, Bardhan [2], Cass–Stiglitz [7], Weitzman [43, 44], and Stiglitz [38, 39, 40]). At any rate, it is difficult to date precisely when a particular intellectual conversation begins, but Dobb's 1951 comment[4] on Strumilin's 1946 proposals clearly sets out the issues: to determine (a) the total amount, (b) the distribution, and (c) the technical form of investment to be made out of current income. The last consideration, of course, pertains to the "choice of technique," and the phrase is used, in particular, as chapter headings in Dobb [11] and as the title of Sen's 1956 Cambridge dissertation.[5] The synonymous phrase "degree of 'capital intensity' of investment" was also initially used by Dobb [10] and Sen, but never really gained currency.[6]

There is, of course, another piece of Solow's published in 1962 (henceforth Solow [32]); it is a review of Dobb [11], and in it Solow puts the issues this way:

> "Society" must lay down some consistent preferences among alternative time-profiles of consumption; only some among these are feasible, given circumstances, resource endowments, and expected technology and labor supply; thus constrained, "society" chooses the most preferred time-profile.
>
> The existence of a machine-tool sector opens up the possibility of postponing an increase in consumption and in the investible surplus, and

[3] The fact that the phrase does not even make it to the index in an important recent text in development economics allows one to appreciate the shift in research interest; see Ray [22].

[4] See "A Note on the discussion of of the problem of choice between alternative investment projects;" subsequently published as [9, Ch. XV].

[5] See Sen [28, 29]; it is of some (tangential) interest that Sen's Appendix B is on "Strumilin on time preference".

[6] It was in terms of this phrase that Dobb referred to (c) above, and Sen [28] used it to title his Chapter III.

> concentrating instead on producing more and more efficient machine-tools so that, at a later date, large-scale production of tractors and of consumables can begin. This is the kind of problem that can be solved only by Ramsey–Fisher considerations and not by any simple growth-rate comparison.

When we use a phrase to understand and represent a particular intellectual conversation, it is important not to miss what went with "the sense of the phrase." Thus, when Solow asks T.N. what "anyone coming to this problem in the 1990's will be clamoring to ask: why not solve the Ramsey problem for this model," it might be worthwhile to give to the 1998 reader some inkling of the range of attitudes that went with the phrase at the time that T.N. was working on his dissertation.

What we now take for granted as the mainstream was, at least in some prominent circles, only a particular school then. Thus, after disposing of a positive rate of discount, and of the "so-called 'diminishing marginal utility of income' introduced by Economists of the Utility School," Dobb is adamant in that "Utility can provide us with no adequate criterion [and indeed that] no 'automatic criterion can afford an answer, to be read off as from a slide rule'." In a parallel discussion, Sen writes [28, Ch. VIII], "Another objection to Professor Tinbergen's approach, or for that matter to any approach with mathematical precision, is that there is no reason why the particular utility function should be the relevant one. The problem of choosing the *correct* utility function is an insoluble one." To continue with Dobb [9, pp. 260–61],

> Some would say that the notion cannot be given any precise meaning; and that, even if it could be given a meaning, the relation would not be independent of historical change (with changing social relations and social standards, changed products and changed wants), and hence could not be deduced for any future period by extrapolation from the past.

It goes without saying that there were several inter-related issues involved, and they continued to dominate part of the conversation. Koopmans' [17] exposition is fully informed on arguments against a positive rate of time preference, and refers to von Weiszäcker and Gale on the then recently-developed, "overtaking criteria" (see Brock [4] for details and references). On the question of the utility function itself, Sen had already emphasized in 1956 the inevitability of "choosing between alternative sets of time series" [10, Ch. VIII], and followed it up a decade later with a discussion of variational problems involving utility functions ([29] and also the Appendix to Chapter where the relevance is drawn to the prisoner's dilemma problem). Even in 1962, and in Cambridge itself, Mirrlees [19] had already added his voice to Solow's in the context of an "inter-economist dialogue":

> Better to use, like Ramsey, explicit but sensible utility functions and get concrete results, than to stop at the point when we say 'it is all very complicated'. [However,] the economist who uses utility functions in a dialogue with Nehru or the Indian ryot, is daft. Utility functions are for the dialogue with himself."

Solow [33, esp. ftn 4] is a subsequent example of such a dialogue — one in which the author is *plus Rawlsien que le Rawls.* However, it is clear that I have devoted enough time (and space) to this part of the conversation; I refer the (still interested) reader to Solow [32] and move on.

II

Solow gives a verbal exposition of T.N.'s model, and I suggest that the reader complement his description by looking at the tabular presentation in Srinivasan [35]. It illustrates the care with which T.N. formalizes his two-sector, putty–clay vision, and also brings out clearly the differences between his model and that of Solow [31]. In terms of the problem at hand, and particularly in terms of the qualitative properties of the solution obtained, the differences in technological specification seem to be inessential: one-hoss shays of fixed life versus radioactive decay; and machines built by unassisted labor but with varying input requirements, versus a single identical technique involving capital for all machines.[7]

Solow furnishes a verbal exposition of T.N.'s solution and draws attention to the "heavy-industry-first pattern of development, reminiscent of the ideas of P.C. Mahalanobis." One is also reminded of Dobb's "accelerator in reverse, a process that leads to the creation of capital goods well in advance of any market demand for them."[8] However, what I would like to underscore is that "only one of infinitely many alternative techniques is used for producing consumer goods."[9] T.N. explained his characterization of the choice of technique in two possible ways: in terms of maximizing the sustainable rate of consumption, and in terms of minimizing the labor cost of a unit of sustainable rate of consumption per worker, SCP.

> If we consider two economies which differ only in their rates of growth of the labor force, the economy with the faster growing labor force will

[7]Referring to the assumption of common input requirements for the different types of machines, Solow writes, "There is some loss of generality here, but the payoff in simplicity is very great." The payoff from this assumption is fully exploited in Shell–Stiglitz [30].

[8]See Halevi [14]. Of course, in an optimal growth exercise, the relevant consideration relates to the possibilities of decentralization, and to the existence of decentralizing prices. Perhaps this is why Pontryagin's principle was a tailor-made tool.

[9]See Srinivasan [35, p. 69]; this point is also singled out in the summary of the thesis (p. 58) and in Section 7 devoted to "summary and conclusions."

> use a technique which is at least as labor intensive as [that of] the other economy.[10] [In terms of the second interpretation, it may be noted that] the ratio of direct and indirect labor services needed per unit of SCP seems to correspond to Marx's concept of the organic composition of capital.[11]

Solow points out why T.N. rejected the option of "maximization of the present value of the stream of future consumption per head, with discounting at the social rate of time preference. (That is, Ramsey with linear utility and positive time preference.)" This precise problem was taken up, of course, in Stiglitz [37] in another rendition of what one may legitimately call the Solow-Srinivasan model.[12] In Stiglitz's version, the man–machine ratio is identical for each type of machine, but different types have different productivities and require differing amounts of labor to be built.[13] The choice of technique issue remains as to the kind of machine that is to be built. What is fascinating to see is the extent to which Stiglitz [37] is anticipated in Srinivasan [35]. Stiglitz introduces his work as follows:

> We shall show that no matter what the initial endowments and no matter how many techniques are available to the economy, the optimum requires that only one type of machine ever be constructed: the type which minimizes the (correctly calculated) labor costs, [and at which the] net social marginal product is equal to the rate of growth of population."

In his 1970 reply to Joan Robinson, Stiglitz again emphasized that "a striking feature of this economy was that only one type of machine was ever constructed." Like T.N., Stiglitz also characterized the optimum machine both in terms of the "labour theory of value" and of the "neo-neo-classical theorem, or the golden rule." Of course, Srinivasan [35] is pre-Pontryagin and in discrete time, while Stiglitz [37] is post-Pontryagin and in continuous time.[14]

---

[10] It is this interpretation that Solow focusses on: "T.N. does not use the phrase, but the MSCP is his, more complicated, model's version of the Golden-Rule consumption level."

[11] See Srinivasan [35, pp. 69–70]. Students of intellectual history will surely be amused to see Dobb's [9, p. 260, ftn 2] on Marx's 'organic composition of capital' acknowledged, so to speak, in Srinivasan [35, ftn 2]. Of course, Dobb [10] discusses this connection at more length and in his other writings.

[12] Stiglitz [38] refers to the "usual Robinson-Solow vintage model," and contrasts it with the "alternative technological assumptions of the Johansen vintage model." In Stiglitz [40, p. 138], it is all subsumed under the somewhat mischievous heading of the "Wicksell–Solow Robinson neo-classical story." In addition to Robinson [23], Okishio [21] is another relevant reference.

[13] Stiglitz follows T.N. on the question of depreciation but assumes no population growth.

[14] My elders may know better, but in so far as I can tell, Cass [5] seems to be the relevant index of this periodic division.

In concluding this section, I would like to give some idea of the intellectual activity that was generated once the relevance of the Ramsey–Fisher considerations was clearly seen. Solow's 1962 explicit admission that "raw materials were [being] ignored" was answered in Weitzman [43] with a correspondingly extended three-sector version of the Fel'dman–Mahalanobis model. Weitzman [44] provides a rigorous analysis in response, at least in as-if hindsight, to Dobb's [11, pp. 11–12] identification of the importance of substantial indivisibilities in infrastructure. The issue of a suitable response to stagnant export earnings was already implicit in Solow [31], and Bardhan begins his 1971 piece by quoting him as follows:

> I am told that there has been discussion in India of whether to import fertilizer, or import fertilizer-making machines or import machinery for making fertilizer-making machinery."

In sum, it is a most interesting trajectory, and I leave it to you to explore it,[15] and also possibly think about the research objective it maximized.

III

What I would like to still take up, however, is the "loose end" identified by Solow: the question as to why T.N. did not consider the Ramsey problem with a "time-additive concave social welfare function. [He] must surely have known of this option, but he never mentions it." Solow answers his own question in terms of marginal rates of substitution between periods and the inability of different social welfare functions to give a sharp-enough answer.[16] I would also like to add the not unimportant consideration that it is a difficult mathematical problem that still remains open.

It is easiest to approach the question through Joan Robinson's [24] criticism of Stiglitz' characterization of the choice of technique in a socialist economy. Referring to the Stiglitz' optimal path, she wrote:

> This would involve (when the initial stock of equipment is all of a low technique) ceasing to consume and living on air during the first phase of the plan. Alternatively, we might take literally the assumption of a zero gestation period for equipment so that the final position was reached before breakfast on the first day of the plan.

It is difficult to imagine the Arrow–Debreu assumption of markets opening only at one time figuring prominently in conversations then going on at

[15] I do not mention, for example, the Austrian turn, as in Yasui–Uzawa [46] and Findlay [13]; also see Cass [6].

[16] In terms of the first, I refer the reader to Beals-Koopmans [3], Wan [42], Samuelson [27] and their followers. Koopmans [18] still remains a useful survey. I see Solow's second consideration as pertaining to the discussions on utility functions in Section II above.

Cambridge, but Stiglitz [38] responds to Robinson's concerns about the difficulties in getting on to the turnpike.[17] In the context of a version of the 1968 problem discussed above, but now with a minimum consumption constraint, Stiglitz observes the following:

> Although I noted that introducing a non-linear utility function, for example, would alter my results, the extent to which it would necessitate modification of the analysis has only recently become clear: the process of optimal capital accumulation may be characterized by discontinuities in the choice of technique, temporary retrogressions, i.e., going from a more to a less-capital intensive technique, or even reswitching, i.e., building a machine of one type, then of another and finally returning to the first type.[18]

Whereas we now do have a good idea of the pathologies and anomalies involved, we do not yet have a formulation of the problem in the generality that Solow would like, much less a complete characterization of the optimal trajectory in a world of a set of social discount rates that includes zero. The loose end remains loose.

IV

The main point, of course, is the "gap that [T.N.'s] thesis will try to fill," and Solow's criticism of earlier work on the grounds that it was not "embedded in a coherent model of long-run growth." In T.N.'s model, "the 'choice of technique' issue boils down to the choice of machine types. But the question is now about programming a growth path, not about choosing among investment projects." What is interesting is the extent to which T.N.'s own subsequent research moves away from growth-theoretic considerations towards contexts that are static. In Srinivasan–Bhagwati [36], a useful point of comparison with the work done fifteen years earlier, the problem of the choice of technique remains well within the orbit of his interest, but it is now phrased in terms of the determination of shadow prices of foreign exchange and of labor, and the relation of these to effective protection rates and to domestic resource costs. Without making too much of it, let me also note here that T.N.'s sole Palgrave entry is on *distortions.*

Given that the essentials of the action in the earlier work revolved around the labor market, this was in some sense inevitable. (Sen's movement in the

[17] And let me say, in parenthesis, as one whose car broke down on the George Washington Bridge on the way to this conference, getting off the turnpike can be sometimes as difficult as getting on it!

[18] See Stiglitz [38]; also Stiglitz [40], Cass–Stiglitz [7] and the useful Figure 7.8 in the Cowles Foundation Discussion Paper No. 303 underlying Stiglitz [39, 40]. It may also be worth pointing out here that Robinson's difficulties with zero gestation periods was never taken seriously in this conversation.

1968 introduction is also relevant here.) The more articulated and detailed the distortion, the less the need to complicate it by intertemporal considerations. To put the matter in a more technical way, one simply shifts to the first stage of the maximization of the Hamiltonian in a dynamic exercise, but now conducts it in the context of Samuelson's [25] GNP function or Chipman's [8] production function for foreign exchange (in addition to Sen [29], Weitzman [45] is also relevant in this connection). This trade-off between statics and dynamics is already explicit in Solow [32]:

> Whether the private rate of return to any concrete investment will approximate its social rate of return is a matter for case by case analysis. There are many causes for the two to diverge: externalities or "neighborhood effects"; the extra burden of uncertainty to the private entrepreneur, which society may reduce by pooling risks or by coordination; and the possibility that wages may, for one reason or another, exceed the true opportunity cost of labor. In such cases, private profitability calculations will go astray, unless appropriately modified.

It was precisely in the context of determining such appropriate modifications that I first received written comments from T.N. As I recall, T.N. was visiting Hopkins that semester, and I was working on development policy in an economy with different and easily-identifiable ethnic groups. It was an unwieldy model,[19] and I have T.N.'s sharp one-page comment — that, too, in green ink. As any reader of Srinivasan [35] would have expected, the comment directly focused on the simple motivating intuition that was missing in the formidable-looking formulae that I had derived. I also remember receiving it within a week after putting the draft in his box, but then, the speed of response has already been singled out in other presentations at this conference.

V

This was not my first encounter with T.N. — I recall my first meeting vividly. In early 1972, I got a call from Tjalling Koopmans' secretary informing me that a distinguished Indian economist — a Yale Ph.D. — was visiting New Haven, and would I join him and Salim Rashid[20] for tea to discuss the turbulent situation in the subcontinent. I remember enough of the conversation to say that it was T.N. who mostly talked, and mostly about "business" in the sense of what I understand Solow [31] to be calling "business" — the need for good data and careful measurement, for the identification of empirical

[19] See Khan [15]; a fuller articulation is in Khan–Datta-Chaudhuri [16] and in subsequent work.

[20] Professor Salim Rashid of Illinois was then my East Pakistani classmate, and was about to assume, in a month or so, his current Bangladeshi identity.

regularities, for insightful theorems — and also, more than once, for openness and civility. Koopmans looked on, as the cliché goes, with his "mild and magnificent eye," with what Salim and I clearly saw to be approval, mixed with pride.

Happy Birthday, T.N.

References

1 Bardhan, P.K., Economic Growth, Development, and Foreign Trade: A Study in Pure Theory, Wiley-Interscience, New York, 1970, Ch. 8.

2 Bardhan, P.K., Optimum growth and allocation of foreign exchange, *Econometrica* 39 (1971), 955–91.

3 Beals, R. and T.C. Koopmans, Maximizing stationary utility in a constant technology, SIAM Journal of Applied Mathematics, 17 (1969) 1001–15.

4 Brock, W.A., Optimal control and economic dynamics, in: J. Eatwell, P.K. Newman and M. Milgate (eds.), The New Palgrave, Macmillan, London, 1987.

5 Cass, D., Optimal growth in an aggregative model of capital accumulation, Review of Economic Studies, 32 (1965), 233–40.

6 Cass, D., On the Wicksellian point-input, point-output model of capital accumulation: a modern view (or, neoclassicism slightly vindicated), Journal of Political Economy, 81 (1973) 71–97.

7 Cass, D. and J.E. Stiglitz, The implications of alternative savings and expectations hypotheses for choices of technique and patterns of growth, Journal of Political Economy, 77 (1969), 586–627.

8 Chipman, J.E., The theory of exploitative trade and investment policies: a reformulation and synthesis, in: L.E.D.Di Marco (ed.), International Economics and Investment: Essays in Honour of R. Prebisch, Academic Press, New York, 1972.

9 Dobb, M., On Economic Theory and Socialism, Routledge-Kegan Paul, London, 1954.

10 Dobb, M., Second thoughts on capital-intensity of investment, Review of Economic Studies, 24 (1956) 33–42.

11 Dobb, M., An Essay on Economic Growth and Planning, Monthly Review Press, New York, 1960.

12 Dobb, M., The question of "Investment-priority for heavy industry," in: Capitalism, Development and Planning, International Publishers, New York, 1967, Ch. 4.

13 Findlay, R., An "Austrian" model of international trade and interest rate equalization, Journal of Political Economy, 86 (1978) 989–1007.

14 Halevi, J., Investment planning, in: J. Eatwell, P.K. Newman and M. Milgate (eds.), The New Palgrave, Macmillan, London, 1987.

15 Khan, M. Ali, A multisectoral model of a small open economy with non-shiftable capital and imperfect labor mobility, Economics Letters, 2 (1979), 369-75.

16 Khan, M. Ali and T. Datta-Chaudhuri, Development policies in LDCs with several ethnic groups a theoretical analysis, Zeitschrift für Nationaloekonomie, 45 (1985), 1–19.

17 Koopmans, T.C., Intertemporal distribution and optimal aggregate economic growth, in: W. Fellner et al. (eds.), Ten Economic Studies in the Tradition of Irving Fisher, Johns Wiley and Sons, New York, 1967, Ch. 5.

18 Koopmans, T.C., Objectives, constraints and outcomes in optimal growth models, Econometrica, 35 (1967), 1–24.

19 Mirrlees, J.A., Choice of techniques, Indian Economic Review, 6 (1962), 93–102.

20 Mirrlees, J.A. and N.H. Stern (eds.), Models of Economic Growth, John Wiley and Sons, New York, 1973.

21 Okishio, N., Technical choice under full employment in a socialist economy, Economic Journal, 76 (1966) 585–92.

22 Ray, D., Development Economics, Princeton University Press, Princeton, 1998.

23 Robinson, J., Exercises in Economic Analysis, 1960, pp. 38–56.

24 Robinson, J., A model for accumulation proposed by J. E. Stiglitz, Economic Journal, 79 (1969) 412–13.

25 Samuelson, P.A., Prices of factors and goods in general equilibrium, Review of Economic Studies, 21 (1953), 1–20.

26 Samuelson, P.A., A.P. Lerner at sixty, Review of Economic Studies, 31 (1964), 169–78.

27 Samuelson, P.A., Turnpike theorems even though tastes are intertemporally dependent, Western Economic Journal, 9 (1971), 21–26.

28 Sen, A.K., Choice of Techniques, Basil Blackwell, Oxford, 1960.

29 Sen, A.K., Optimal savings, technical choice and the shadow price of labour, Introduction in: A.K. Sen, Choice of Techniques, 3rd edition, Basil Blackwell, Oxford, 1968.

30 Shell, K. and J.E. Stiglitz, The allocation of investment in a dynamic economy, Quarterly Journal of Economics, 81 (1969) 592–609.

31 Solow, R.M., Substitution and fixed proportions in the theory of capital, Review of Economic Studies, 29 (1962), 207–18.

32 Solow, R.M., Some problems in the theory and practice of economic planning, Economic Development and Cultural Change (1962) 216–22.

33 Solow, R.M., Intergenerational equity and exhaustible resources, Review of Economic Studies, 41 (1974), S29–S35.

34 Solow, R.M., Srinivasan on choice of technique, this volume, 1999.

35 Srinivasan, T.N., Investment criteria and choice of techniques of production, Yale Economic Essays, 1 (1962), 58–115.

36 Srinivasan, T.N. and J.N. Bhagwati, Shadow prices for project selection in the presence of distortions: Effective rates of protection and domestic resource costs, Journal of Political Economy 86 (1978).

37 Stiglitz, J.E., A note on technical choice under full employment in a socialist economy, Economic Journal, 78 (1968) 603–609.

38 Stiglitz, J.E., Reply to Mrs. Robinson on the choice of technique, Economic Journal, 80 (1970) 420–22.

39 Stiglitz, J.E., The badly behaved economy with the well-behaved production function, in: J.A. Mirrlees and N.H. Stern (eds.), Models of Economic Growth, John Wiley and Sons, New York, 1973, Ch. 6. (Originally circulated as Cowles Foundation Discussion Paper No. 303, 1970.)

40 Stiglitz, J.E., Recurrence of techniques in a dynamic economy, in: J.A. Mirrlees and N.H. Stern (eds.), Models of Economic Growth, John Wiley and Sons, New York, 1973, Ch. 7. (Originally circulated as Cowles Foundation Discussion Paper No. 303, 1970.)

41 Turner, M.S., Joan Robinson and the Americans, M.E. Sharpe, London, 1989.

42 Wan, H.W., Jr., Optimal savings program under intertemporally dependent preferences, International Economic Review, 11 (1970) 521–49.

43 Weitzman, M.L., Shiftable versus non-shiftable capital: A synthesis, Econometrica, 39 (1970), 511–29.

44 Weitzman, M.L., Optimal growth with scale economies in the creation of overhead capital, Review of Economic Studies, 37 (1970) 555–70.

45 Weitzman, M.L., On the welfare significance of national product in a dynamic economy, MIT Working Paper No. 125, 1974.

46 Yasui, T. and H. Uzawa, On an Åkerman–Wicksellian model of capital accumulation, Economic Studies Quarterly, 14 (1964) 1–10.

Trade, Growth, and Development
G. Ranis and L.K. Raut

CHAPTER 3

The Effects of a Decreasing Rate of Time Preference on the Accumulation of Capital[1]

Rolf R. Mantel

Department of Economics, Universidad de San Andrés, Vito Dumas 284, (1644) Victoria, Prov. Bs. As., Argentina

3.1 Introduction

The main theme of this investigation is the structure of intertemporal preferences and its consequences for the theory of optimal economic growth. The interpretation given to the model to be presented is that of the head of a family who has to decide on how future consumption is to evolve, taking into account not only the present generation — the members of the consumption unit living today — but also their descendants — the future generations. For that reason it will be accepted that the family has an unbounded existence, a situation which is intuitively clear if the present generation care for their children and all future generations equally cares for their children. Hence, each member cares directly or indirectly for all their descendants. An alternative interpretation of the same problem is that of the central planner, a government which has to decide the future evolution of consumption for a country, where again it is necessary to care not only for the present generation but also the future ones.

The main conclusion will be that the assumption of increasing marginal impatience, frequently used in the literature, is not very representative of real world behavior, and also lacks the richness of results offered by the contrary assumption, decreasing marginal impatience, which was proposed by Irving Fisher [9] 90 years ago.

[1] The underlying investigation for this paper was finished while the author was Visiting Scientist at the Faculty of Mathematical Sciences of the Weizmann Institute of Science, Israel, funded by the Fundación Campomar, Argentina.

3.2 The Heritage of Irving Fisher

As was formalized later by Fisher [10] himself in his book *The Theory of Interest*, he assumes that a typical family, confronted with an election between a constant — i.e., stationary — flow of commodities on the one hand, and a flow which differs from the first one only in the first two consecutive periods on the other, will tend to choose the second one only if it offers a higher level of consumption in the future than in the present period.

Figure 3.1 reproduces Fisher's graph 31 [10, p. 246]. The horizontal axis measures present consumption and the vertical axis measures future consumption of a certain household. The 45° line corresponds to equal consumption levels in all periods, that is, a stationary consumption program. The curves representing equal satisfaction were called "Willingness Lines" by Fisher; today they are called indifference curves. As Fisher drew them, higher incomes — as measured by higher stationary consumption levels — correspond to curves which become flatter as they intersect the 45° line. Fisher argued that if one compares two situations of a household corresponding to different levels of stationary consumption — with an initial situation in which it consumes the same amount in each period — the consumption unit will be more willing to give up present consumption in exchange for future consumption. In more modern terminology, it will be more patient: the higher the level of its stationary consumption is, the richer it is.

The argument is presented in Fisher's book (p. 247). In the sequel, the intuition contained on that page will be presented.

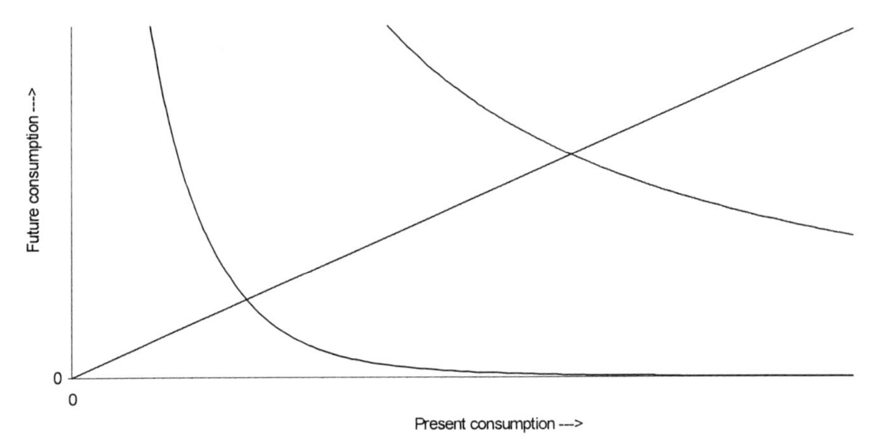

Figure 3.1: Willingness lines

3.3 Present-day Implementations of Fisher's Ideas

Besides some heated discussions between Eugen von Boehm Bawerk and Irving Fisher, the latter's work relating to preference over time did not have a major impact on economic analysis in his time. This, in spite of the modern and advanced character of his ideas.

The earliest work utilizing the apparatus created by Fisher (that the present author could find) is that of Leontief [16, 17], who in 1958 gave the conditions for a stationary equilibrium: the equality between the marginal product of capital (MPK) and the marginal rate of time preference (MRTP). Even though his model does not exhibit maximizing behavior, his graph — reproduced later on as Figure 3.8 — is applicable in that case, since the Euler equations, necessary for an optimal trajectory, imply that there will be capital accumulation if and only if the MRTP exceeds the MPK. Additional references to this subject will be made in the sequel.

A ground-breaking article by Tjalling C. Koopmans [13] in *Econometrica* revived the subject. Koopmans showed that if the preferences of the consumer — concerning intertemporal consumption programs — satisfy certain postulates, then they can be represented by means of an aggregator function, and that preferences are, according to the terminology introduced by Lucas and Stokey, recursive. Intuitively, the prospective utility of a consumption program can be evaluated by combining the first period consumption level with the prospective utility of the tail of the program, starting during the second period.

This structure of preferences has later been utilized by many authors. Two are the usual treatments of time, either as a continuous variable represented by intervals of real numbers — it is intuitive to consider that time evolves continuously, that there is no interval, however small, during which nothing happens — or else as a discrete variable, represented by equally spaced time periods (days, months or years, with no concern about what happens within each of these periods). The discrete time case was used by Fisher [9, 10] in his work, and later more fully developed by Koopmans [13]. The continuous case originates with the articles of Uzawa [30] and the present author [19, 20, 21]. An interesting interpretation in terms of generations is due to Robert Barro, where the consumption of each period corresponds to the consumption of one generation and the prospective utility corresponds to the utility of the consumption program for all future generations. In that case, the aggregator function is the utility function of the present generation, taking into account its concern for the well-being of its descendants.

There are two main lines of thought relating to this subject in present-day economic thinking. One of them corresponds to the idea of increasing marginal impatience, a term coined by Lucas and Stokey [18]. It is the more usual assumption, and has been introduced by Uzawa [30]. The other one is that of decreasing marginal impatience, introduced by Fisher, with consequences rigorously analyzed by Beals and Koopmans [3] in the discrete time case, and by the present author [21] in the continuous time case. Iwai [12] analyzed additional situations for the discrete time case not covered by the analysis of Beals and Koopmans.

There is a third hybrid line of thought. It consists in assuming marginal impatience to be decreasing for low levels of consumption, and increasing for high levels. This assumption has been used by the present author [21] and Fukao and Hamada [11].

In the literature, one can find arguments in favor of both increasing and decreasing marginal impatience.

Justifications of increasing impatience have been given, for example, by Stokey, Lucas and Prescott [29]. These authors require the stability of their dynamic trajectories. Consequently, they say (p. 161) that

> the required condition is that . . . the consumer become more impatient (in the margin) for higher levels of stationary consumption. We shall call this condition . . . increasing marginal impatience.

A footnote adds to this statement:

> A similar condition was used by Uzawa [30], and by many other subsequent formulations in continuous time.

But if one goes to the original source, Uzawa [30, p. 489] is not too explicit. One reads,

> The . . . assumption . . . [of an increasing rate of time preference] requires that an increase in the level of consumption in a certain future date will increase the rate of discount for all consumption realized after that moment.

The reader will observe that he is describing what is happening, but not why it should happen.

Obstfeld [26, p. 47] presents a survey on the subject and dedicates more space for his thoughts on it. He says that

> Readers of the literature in this field frequently remain with the impression that the models analyzed are not only analytically intractable but are based on restrictive assumptions.

To this he adds in a footnote:

> Thus, Blanchard and Fischer warn students that models of time preference of the Uzawa variety "are not recommended for general use."

But he finally concedes (p. 52) that

> . . . once the horizon of the consumer extends indefinitely . . . the assumption . . . is necessary for the convergence; . . . the condition . . . will be imposed from this moment on.

In spite of these words uttered by Obstfeld, the truth is that Blanchard and Fischer [4] presented an argument against increasing impatience. In their book, these two authors write (pp. 74–75) that

> The Uzawa function, with its assumption [of an increasing rate of time preference], is not particularly attractive as a description of preferences and is not recommended, in general, for its utilization.

It is quite obvious that their negative comment refers to the increasing rate of time preference and not to the model itself.

The interpretation of the present author differs from that of Obstfeld. The latter suggests that the authors he mentions say that models with a variable rate of time preference are intractable, whereas the present author interprets that they say that an increasing rate of time preference is not a reasonable assumption. The readers may draw their own conclusions.

Ronald Findlay [8, p. 48] says that the assumption of increasing impatience

> seems to contradict the widespread notion that the poor are more impatient than the rich.

He arrives at the result that more impatience leads to a lower standard of living assuming different rates of time preference for different countries at all levels of consumption, so that preferences of different countries are different. This seems to be a different problem than that of a changing rate of time preference at different levels of consumption for the preferences of the same country.

Justification for the assumption of decreasing impatience have been provided by Fisher, as mentioned at the beginning of this investigation.

3.4 Consequences of a Variable Rate of Time Preference

Beals and Koopmans [3] showed the consequences of the assumption of decreasing impatience for the shape of optimal growth trajectories, as has the author [21] and by Iwai [12].

The research of the author during the end of the 60s produced several papers. One of them was published in *Económica* (Universidad Nacional de La Plata) [21], whereas the last of that period was presented at the Second World Congress of the Econometric Society, Cambridge, England [22].

Some years ago, in June, 1992, Robert Barro visited Argentina and gave a very interesting conference on international comparisons between rates of growth which differed due to the effects of the accumulation of human capital (Barro and Xala-i-Martin [2]). There the author made some comments referring to research carried out during the 60s, when such differences were attributed to the interaction of the preferences with existing resources rather than to technological factors. Barro asked how the present author felt, given the recent revival in the interest for growth theory. The answer was that "it feels wonderful. As my grandmother used to say, when a dress is no longer fashionable, the only thing one has to do is to keep it in a trunk. After 20 years it will be surely be fit to be used again."

Consequently, the author took out his notes of 30 years ago to rewrite and actualize them. The results of that effort will be presented very succinctly. For the details, the reader should consult with the bibliographic references listed at the end.

3.5 Recent Results

The heritage of Irving Fisher — the material of those two pages of his book mentioned above — has been shown. Now a jump of several years will be undertaken in order to make reference to research evolving from those pages: Koopmans [13], Beals and Koopmans [3], Diamond [7], Koopmans, Diamond and Williamson [14], and many others.

In the first place, some reference will be made to the notation to be used. For that purpose Figure 3.2 will prove useful. There, along the horizontal axis, time is measured, while the vertical axis represents aggregate consumption. Even though the theory is valid for consumption bundles in more general spaces, for illustration a single commodity will suffice. Perhaps the simplest interpretation of the aggregate neoclassical model to be utilized in the sequel is to imagine that our economy produces a single consumption good — in Argentina we would immediately think of beef — produced by a single capital good — cattle. The net increase in the herd due to new births, the number of which depend on the stock of capital (the herd of cattle) can than be used for investment, increasing the herd, or for consumption (by slaughtering the animals for the purpose of consuming the beef). Thus the same commodity — cattle — can be used for consumption and for investment.

Simplifying, the cows that are not eaten produce offspring which turn into more cows. Of course, this is just a caricature for the sake of the argument. In Figure 3.2, the horizontal axis represents time, t, and the vertical axis the consumption rate, x. A consumption program is a function of time, represented by the curve. Time is divided into three intervals, corresponding to three sections of the consumption program. The first one is a section of the consumption program extending from the initial moment 0 to the instant t, and is denoted by ${}_0x_t$, where the subindex zero preceding x indicates the moment in which the consumption program begins. The subindex t following x indicates the moment in which the program ends. To the right there follows another section, indicated by ${}_tx_s$, to show the starting point t as a subindex to the left and the ending time s as the subindex to the right. The third section of the program is labelled ${}_sx$ so that it starts at time s; since there is no terminating right subindex, this section is supposed to extend forever.

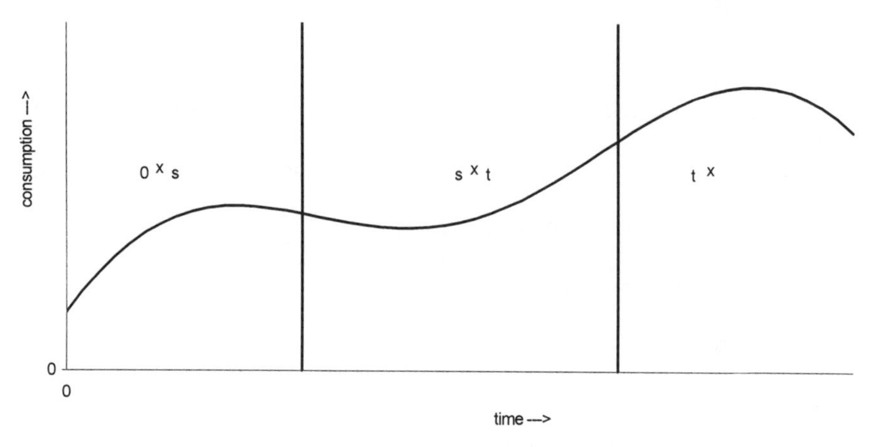

Figure 3.2: Sections of a consumption path

With this notation one can proceed to present the main characteristics of the preferences over consumption programs. In this, the ideas set forth by Koopmans for the case of discrete time will be presented and adapted to a formulation in continuous time.

The key step in the translation of Koopmans' work from his discrete-time formulation to the continuous-time case resides precisely in his postulate of limited non-complementarity over time. The continuous-time case is illustrated in Figure 3.3. The corresponding formulae are presented in the following exhibit.

Postulate of limited non-complementarity over time

For all programs x, y, for all constant programs b, c, and for all instances s,

$$W(b_{s,s}x) \geq W(c_{s,s}x)$$

implies

$$W(b_{s,s}x) \geq W(c_{s,s}y).$$

This, together with the next postulate, is the most important assumption Koopmans introduced in order to simplify the analysis of the structure of time preference. It says that if W indicates the welfare — that is to say, the utility of the consumer, the welfare for the country — and if one compares two consumption programs which differ only in the initial constant section, a modification of the tail of the programs by a section which is the same in both cases will not affect the ordering restricted to these two programs in the preferences of the agent.

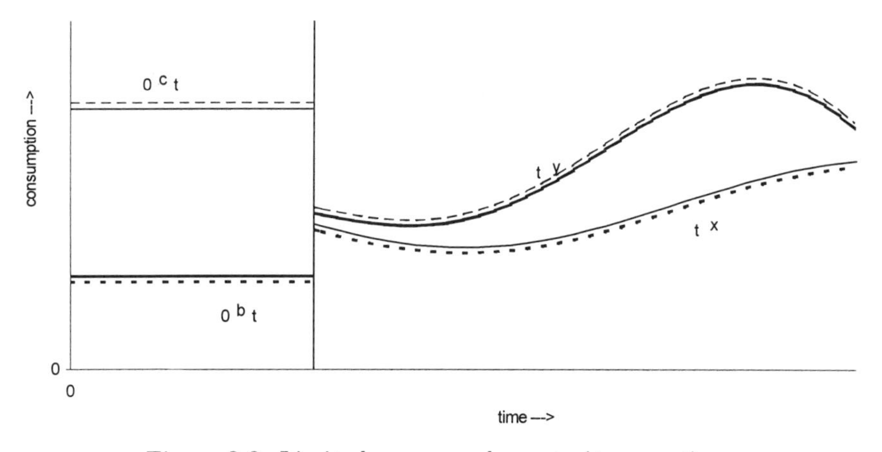

Figure 3.3: Limited non-complementarity over time

In the next exhibit, the postulate of stationarity is presented analytically (illustrated in Figure 3.4). It says that if there are two programs which coincide in some initial section where the two programs have a common initial section ${}_0c_t$ — one can advance the timing of both programs by discarding the common initial section and consuming the quantities planned for the future earlier by the same time interval t, using only the rest — the tail — of the programs. That operation will not affect the ordering restricted to those two programs. In fact, for this postulate the initial section need not be constant.

Postulate of stationarity

For all programs x, y, and for all instances s,

$$W(x_{s,s}x) \geq W(x_{s,s}y)$$

if and only if

$$W({}_sx) \geq W({}_sy)$$

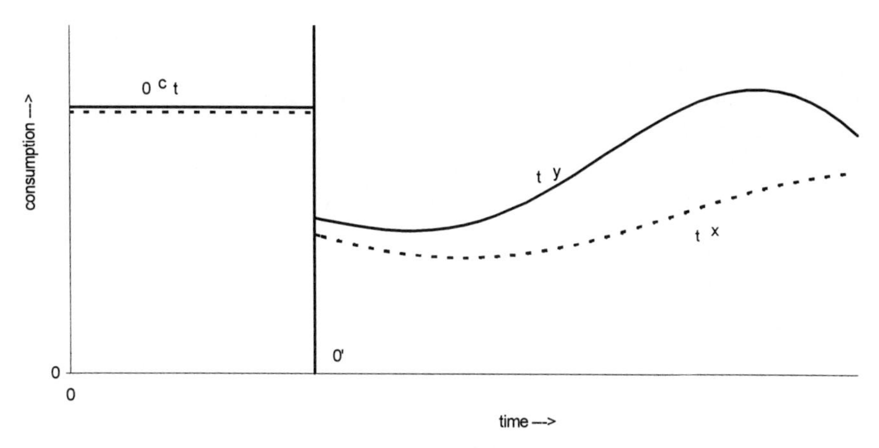

Figure 3.4: Stationarity of preference

With these assumptions, Koopmans — for the case of discrete time, with annual, weekly, or some other equally spaced time periods — proved that the prospective utility function, as Koopmans called it, can be expressed as a function of the first period consumption and the prospective utility of the rest — the tail of the program. A similar analysis for the case of continuous time allows the evaluation of the utility of consumption programs to be carried out by solving a differential equation.

Example: The linear aggregator

$$\frac{dW}{dt} = \rho(u)[W - u]; \; u \equiv u[x(t)], W \in [\underline{W}, \overline{W}] \; \forall t \geq 0$$

imply

$$W({}_0x) = \int_0^\infty e^{-\int_0^t \rho[u(x(\sigma))]d\sigma} v[(x(s))]ds$$

where

$$v(u) \equiv \rho(u)u, \; \rho(u) \geq \varepsilon > 0 \; \forall u$$

The simplest example of the continuous time case can be seen in the previous exhibit. Even though the rate of time preference is not constant, the *aggregator function* — that is, the name the right hand side of the differential equation receives, as a function of the instantaneous utility u and the level of welfare W — is linear in the prospective utility or welfare W. The differential equation expresses that the change in welfare — or prospective utility — due to a marginal advancement of the consumption program is a proportion ρ — the rate of discount or rate of time preference — multiplied by the difference between the prospective utility and the instantaneous utility of the present rate of consumption. In the case of the linear aggregator, it is possible to provide a closed form solution of the differential equation in the form of an integral which provides the value of the prospective utility. Here, as in the sequel, the instantaneous utility function $u(\cdot)$ is assumed to be continuous and strictly increasing.

The aggregator function in the general case

The differential equation

$$\frac{dW({}_t x)}{dt} = F[u(x(t)), W({}_t x)];\; W({}_t x) \in [\underline{W}, \overline{W}]\ \forall t \geq 0$$

has a unique solution with initial value

$$W({}_0 x)$$

if

$$\frac{\partial F}{\partial W} \geq \varepsilon > 0$$

The more general case, corresponding to Koopmans' analysis for discrete time, is exhibited above. Again, one has a differential equation showing how the evaluation of the welfare $W({}_t x)$ of the program initiating at instant t changes as the program is advanced. That change depends on the function $F(\cdot,\cdot)$ on the utility $u(x(t))$ of the present consumption rate, and on the level of the prospective utility of the future consumption program, $W({}_t x)$.

Figure 3.5 shows a diagram with time on the horizontal axis and welfare W on the vertical. The values attained by the instantaneous utility of current consumption along a certain consumption program are shown as a line oscillating with a relatively high frequency. The dotted lines correspond to different solutions W of the differential equation, for different initial values for welfare. If one starts with a value which is too high, the trajectory tends to exceed the upper boundary of the graph, thus exceeding the maximum admissible welfare level. If, on the other hand, one starts with a value which is too low, the trajectory tends to escape sooner or later toward the lower boundary on the horizontal axis, which represents the minimum admissible level of welfare.

Applying the fundamental theorem of existence and unicity of solutions to differential equations — we assume that the aggregator function $F(u, W)$ is continuous, decreasing in u, strictly increasing in W, and satisfies a Lipschitz condition with respect to welfare W — it can be shown that there is exactly one initial value which maintains the trajectory forever within the band given by the two extreme welfare levels. In the figure, this trajectory has been also drawn in. That unique initial value is the evaluation of the welfare corresponding to the given consumption program. The corresponding trajectory stays forever within the bounds of the picture, and so provides the prospective utility corresponding to the instantaneous utility of the consumption program shown as a continuous line.

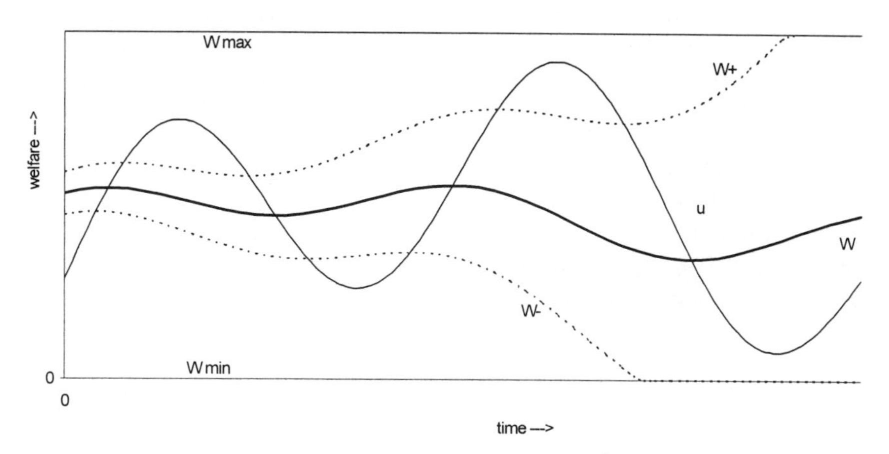

Figure 3.5: The process of evaluating welfare

All this, of course, can be shown rigorously, as is done in many of the references at the end.

It was mentioned before that there is a hybrid form for the rate of time preference, used by the author [21] and by Fukao and Hamada [11]. In Figure 3.6, the rate of consumption is represented on the horizontal axis and the rate of time preference ρ on the vertical axis. The form the curve shows is initially decreasing (Fisher's branch) and later increasing (Uzawa's branch).

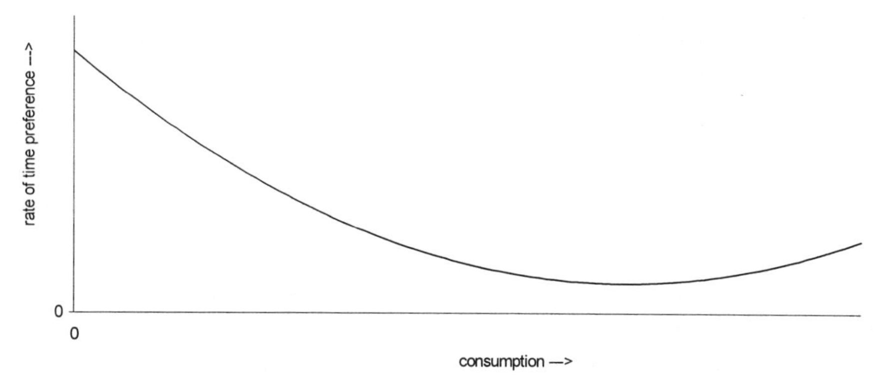

Figure 3.6: Hybrid rate of time preference

The taste of the pudding is in the eating. After a logical jump one reaches Figure 3.7. Here one measures the capital stock $k(t)$ — heads of cattle — on the horizontal axis and the consumption rate $x(t)$ on the vertical axis.

One wishes to draw the optimal capital-consumption trajectories in the phase plane when one maximizes welfare, $W(x)$, subject to the usual neoclassical restriction $dk(t)/dt = f(k(t)) - x(t)$ relating net investment to net output and consumption. To that effect three curves — the isoclines — are shown. The one marked $f(k)$ corresponds to the production function. It satisfies the usual assumptions of being concave and initially increasing. It marks the points at which investment becomes zero, so that there is no change in the stock of capital. Hence, nonoptimal trajectories become vertical there. If less is consumed than what society produces, the capital stock increases and one moves toward the right. If one consumes more than the output, so that the point of the trajectory is above the curve, consumption exceeds the amount which would allow maintaining the capital stock, and the latter declines; the movement is toward the left.

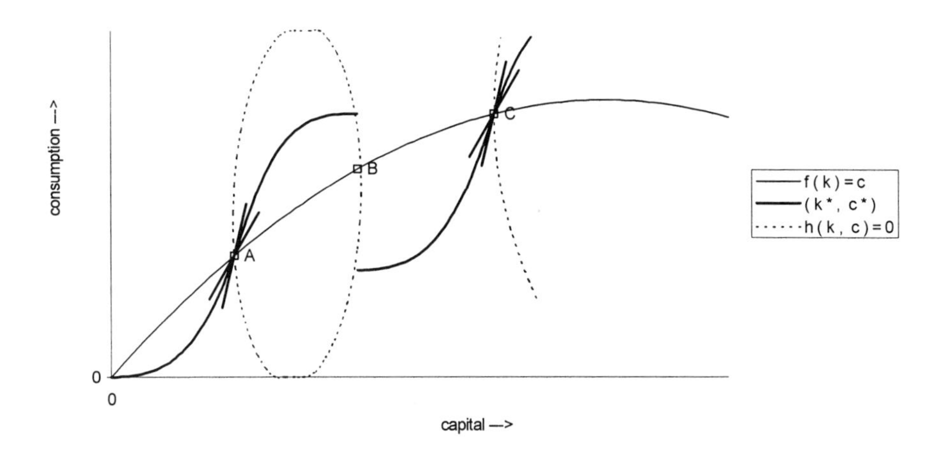

Figure 3.7: Optimal capital-consumption trajectory with a neoclassical technology

The other two curves represent situations in which optimal consumption is stationary; the consumption rate does not change on points of these curves. Outside of the closed region of a quasi-elliptical shape enclosed by the curve through the points A and B, optimal consumption will tend to increase. The candidates for optimal trajectories can be obtained by solving the usual Euler equations of the calculus of variations, or equivalently, from Pontryagin's maximum principle. Given the initial capital stock $k(0)$, there will then be only one solution to the state and Euler equations which will remain within the picture forever and satisfy the transversality condition.

The optimal trajectories of the social planner or family head planning future consumption are marked in solid lines. One trajectory starts from the origin, marked with 0, and reaches point A, which is a stationary equilibrium: if the initial capital stock were to coincide with the abscissa of that point, there would be no incentive to move away from there; society would be content with that capital stock forever, consuming exactly what is produced. If, on the other hand, the initial capital stock were lower, there would be a desire to consume less in the short run in order to grow and accumulate capital, tending toward point A.

If the initial capital stock is above A but below B, the situation reverses: capital will be decumulated by consuming more than output, tending in the long run to the capital stock represented by point A.

To the right of B the tendency is again to accumulate, so that the economy moves toward C, a point qualitatively similar, from the local point of view, to point A. An initial capital stock in excess of that of point C will induce a decumulation, so that again the economy tends toward C.

It has to be remarked that the results of this analysis differ drastically from the usual results obtained from the neoclassical model of growth with no technical progress, where the economy always reaches the same long-run capital stock no matter what its initial capital endowment is. There is no path dependence, in the long run there is no memory about what happened in the beginning. No matter how rich society is in the beginning, no matter how poor, it will always aim at the same long-run capital stock.

This is the usual result for a constant rate of time preference, as has been demonstrated by many researchers since the time of Ramsey in 1928, including the important work by Koopmans [15] and Cass [5] in the mid-1960s. It turns out that the same result is true when impatience is increasing *à la* Uzawa.

On the other hand, if impatience decreases with the level of consumption, the long-run state will depend on the initial conditions. If society is rich enough to start out to the right of point B, it will have the incentive to save, accumulate and grow, reducing if necessary its present consumption for the sake of an increased future consumption reaching point C.

But below that critical level of wealth, society will desire to decumulate, eat more steaks, let the stock of cows diminish, and end up at point A, which for the future generations is decidedly inferior to point C. The present-day head of the family — or the social planner — will sacrifice the future generations in order to increase the standard of living of the present generation. The satisfaction of the children of our children is not enough to offset today's loss in satisfaction if consumption has to be reduced to invest in increases in wealth.

In order to interpret what has been said, if one is confronted at the beginning with a relatively rich family or society — one that is to the right of critical point B in terms of initial capital — that family will be willing to sacrifice its consumption for a higher consumption for its descendants, so as to end up in the long run at point C. If, on the other hand, the present generation owns a level of initial capital below the critical level, its descendants will end up being even poorer, since the present generation will not be willing to undertake the sacrifice that a reduction in consumption entails.

In Figure 3.8, which has been adapted from a similar figure presented by Leontief [16], one can see the relation between the marginal product of capital and the rate of time preference. On the horizontal axis the capital stock is measured, with the two concepts just mentioned on the vertical axis. The key, long-run equilibrium points of the previous graph can be read off from this graph from the intersections of the two curves.

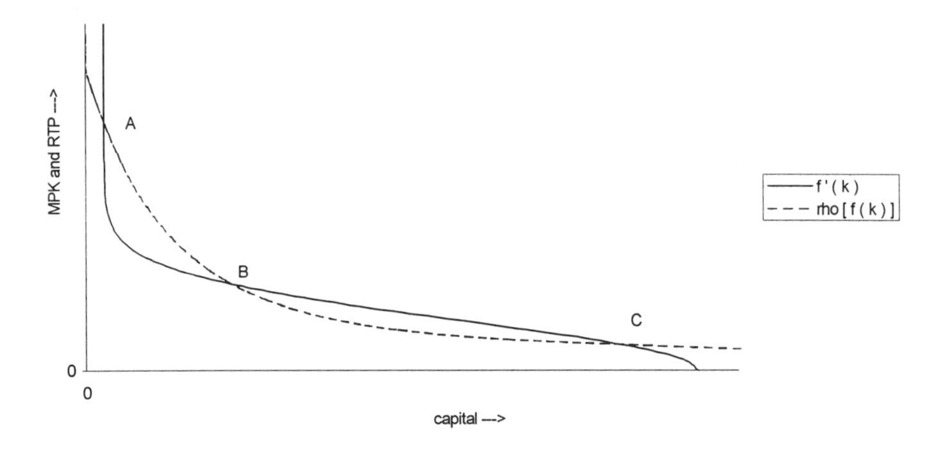

Figure 3.8: Leontief's graph

Another way of seeing the form the optimal trajectories adopt is shown in Figure 3.9, where time is measured along the horizontal axis and capital along the vertical. There it is shown how the capital stock develops over time, given different initial conditions. The same points A, B, C are marked along the vertical axis as they were marked along the horizontal axis in Figure 3.7. Two trajectories are marked $A+$ and $A-$, corresponding to the stationary state marked with A. Of course, the same story can be repeated. If one starts off at a point below A one grows, but in the long run one tends to the relatively poor equilibrium A. If the initial capital exceeds A but not B, society will tend to eat up its capital stock, reaching the same state of relative poverty.

The stories behind trajectories $C+$ and $C-$ are different, because if initial capital exceeds B and if it is below C, society will decide to sacrifice consumption, consuming less output, so that future generations may reach the point of relatively high wealth C in Figure 3.7.

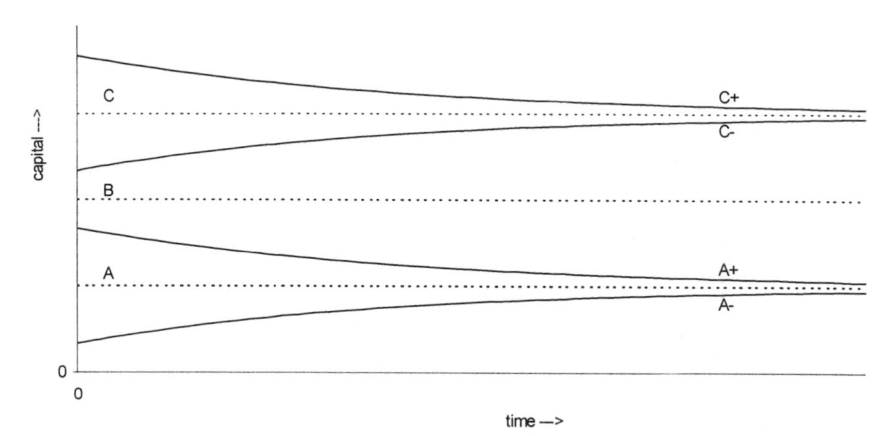

Figure 3.9: Optimal trajectories in time-capital space

With this rather schematic description it is hoped that the reader will acquire an approximate idea of how the trajectories depend on the past history. Of course one can imagine much more complicated situations, with any number of stationary solutions in which the system may turn up in the long run. The purpose of the present article is just to whet the appetite of the reader for this fascinating subject.

References

1 Arrow, Kenneth J., Samuel Karlin and Patrick Suppes (eds.), Mathematical Methods in the Social Sciences, 1959, Stanford University Press, Stanford, CA, 1960.

2 Barro, Robert J. and Xavier Sala-i-Martin, Convergence, Journal of Political Economy, 100 (1992) 223–51.

3 Beals, Richard and Tjalling C. Koopmans, Maximizing stationary utility in a constant technology, SIAM Journal of Applied Mathematics, 17 (1969) 1001–15.

4 Blanchard, Olivier and Stanley Fischer, Lectures on Macroeconomics, MIT Press, Cambridge, MA, 1991.

5 Cass, D., Optimum savings in an aggregative model of capital accumulation, Review of Economic Studies, 32 (1965) 233–40.

6 Debreu, Gerard, Topological methods in cardinal utility theory, in: Kenneth J. Arrow, Samuel Karlin and Patrick Suppes (eds.), Mathematical Methods in the Social Sciences, 1959, Stanford University Press, Stanford, CA, 1960, pp. 16–26.

7 Diamond, Peter A., The evaluation of infinite utility streams, Econometrica, 33 (1965) 170–74.

8 Findlay, Ronald, Factor Proportions, Trade, and Growth, MIT Press, Cambridge, Massachusetts, 1995.

9 Fisher, Irving, The Rate of Interest, New York, 1907.

10 Fisher, Irving, The Theory of Interest, Macmillan, Macmillan, New York, 1930.

11 Fukao, Kyoji and Koichi Hamada, The Fisherian time preference and the evolution of capital ownership patterns in a global economy, Yale University: Economic Growth Center Discussion Paper No. 579, August 1989.

12 Iwai, Katsuhito, Optimal economic growth and stationary ordinal utility: A Fisherian approach, Journal of Economic Theory, 5 (1972) 121–51.

13 Koopmans, Tjalling C., Stationary ordinal utility and impatience, Econometrica, 28 (1960) 287-309.

14 Koopmans, Tjalling C., On the concept of optimal economic growth, in: P. Salviucci (ed.), The Econometric Approach to Development Planning, North-Holland, Pontificiae Academiae Scientiarum Scripta Varia No.28, Amsterdam, 1965, pp. 224–87.

15 Koopmans, T.C., P.A. Diamond and R.E. Williamson, Stationary utility and time perspective, Econometrica, 32 (1964) 82–100.

16 Leontief, Wassilly, Theoretical note on time-preference, productivity of capital, stagnation and economic growth, American Economic Review, 48 (1958) 105–11.

17 Leontief, Wassilly, Time-preference and economic growth: Reply, American Economic Review, 49 (1959) 1041–43.

18 Lucas, Robert and Nancy Stokey, Optimal growth with many consumers, Journal of Economic Theory, 32 (1984) 139–71.

19 Mantel, Rolf R., Tentative list of postulates for a utility function for an infinite future with continuous time, Cowles Foundation for Research in Economics, CF–70525(1), May 25, 1967.

20 Mantel, Rolf R., Maximization of utility over time with a variable rate of time-preference, Cowles Foundation for Research in Economics, CF-70525(2), May 25, 1967.

21 Mantel, Rolf R., Criteria for Optimal Economic Development, Instituto Torcuato Di Tella Documento de Trabajo No. 38, June (Spanish) and No. 38b, December (English). Published as Criterios de desarrollo económico óptimo, Económica (La Plata) 3, No. 3, 1968.

22 Mantel, Rolf R., On the utility of infinite programs when time is continuous, paper presented at the Second World Congress of The Econometric Society, Cambridge, United Kingdom, 1970; mimeo, Instituto Torcuato Di Tella, August in Spanish, September in English; Abstract in Econometrica, 38.

23 Mantel, Rolf R., Grandma's dress, or what's new for optimal growth, Revista de Análisis Económico, 8(1) (1993) 61–81.

24 Mantel, Rolf R., Why the rich get richer and the poor get poorer, Estudios de Economía 22(2)2 (1995) 177–205.

25 Mantel, Rolf R., Optimal growth with a variable rate of time preference: discrete time, Conference on Economic Growth, Technology, and Human Resources, Tucumán, Argentina, December 19–20, 1996.

26 Obstfeld, Maurice, Intertemporal dependence, impatience, and dynamics, Journal of Monetary Economics, 26 (1990) 45–75.

27 Pontryagin, L.S. et al., The Mathematical Theory of Optimal Processes, John Wiley & Sons. Inc., New York, 1962.

28 Ramsey, Frank P., A mathematical theory of saving, Economic Journal, 38 (1928) 543–59.

29 Stokey, Nancy L., Robert E. Lucas and Edward E. Prescott, Recursive Methods in Economic Dynamics, MIT Press, Cambridge, MA, 1989.

30 Uzawa, Hirofumi, Time preference, the consumption function, and optimum asset holdings, in: J.N. Wolfe (ed.), Value, Capital, and Growth. Papers in Honor of Sir John Hicks, University Press, Edinburgh, 1968.

31 Westfield, Fred M., Time-preference and economic growth: Comment, American Economic Review, 49 (1959) 1037–41.

32 Wolfe, J.N. (ed.), Value, Capital, and Growth. Papers in Honor of Sir John Hicks, University Press, Edinburgh, 1968.

Trade, Growth, and Development
G. Ranis and L.K. Raut

CHAPTER 4

Endogenous Fertility and Growth Dynamics[1]

Kazuo Nishimura[a] and Lakshmi K. Raut[b]

[a]Institute of Economic Research, Kyoto University, Kyoto, Japan

[b]Department of Economics, University of Chicago, Chicago, IL 60637

4.1 Introduction

The main issue we address in this paper is what happens when we endogenize fertility in a neoclassical equilibrium growth model. It changes a wide spectrum of standard neoclassical growth results, ranging from the Ricardian equivalence theorem to the turnpike theorem. In this paper, however, we focus only on growth dynamics.

In recent years, there has been a lot of interest in endogenizing economic fluctuations of the long-waves type, as observed by Kondratieff [11], Kuznets [12] and Easterlin [8], and of the short-waves type, as observed in business cycles and stock market volatilities. Much of this research has been conducted in the neoclassical Ramsey growth framework, and in the overlapping generations general equilibrium framework. Two-period lived overlapping generations models generally exhibit very complex dynamics. When parental altruism is introduced in a two-period lived overlapping generations model (as in Barro [1]), the equilibrium dynamics of the model, under certain assumptions, can be characterized by an optimal growth path of a neoclassical one-sector Ramsey model, which does not exhibit dynamic complexities for its growth path. Subtle differences, however, occur in the equilibrium dynamics of these models when fertility is endogenized.

In Section 4.2, we first summarize the widely studied dynamics of the standard neoclassical growth model with exogenous fertility, and then report the striking differences in dynamics that arise when fertility is endogenized. We also point out the limitations of the literature in dealing with the dynamics of the Ramsey growth model when fertility is endogenized with two-period

[1]We like to thank Jonathan Conning and T.N. Srinivasan for their comments on an earlier draft.

lived overlapping generations of agents who are altruistic towards their children (i.e., when the Barro model is extended to endogenize fertility). We then set out in Section 4.3 to formulate such a model and in Section 4.4, we contrast its dynamics with the dynamics of existing equilibrium growth models of endogenous fertility.

4.2 Backdrop

4.2.1 Growth dynamics in the Ramsey model with exogenous fertility

There are several reasons for the recent revival of interests in the Ramsey model. To endogenize the savings rate in the Solow growth model, it is often assumed that consumption and savings decisions are made by an infinitely lived agent or by one-period lived agents having an altruistic utility function. In either case, competitive equilibrium can be characterized as the solution of a Ramsey type growth model. More specifically, suppose agents are born as adults and live only for one period. We denote by c_t^t, the consumption in period t of an adult of generation t and by n_t, the number of children she has. Let the utility V_t of an adult of generation t, be given by

$$V_t = U(c_t^t) + \gamma(n_t) \cdot V_{t+1}, \quad t = 0, 1, \ldots \tag{4.1}$$

where $U(c_t^t)$ is the felicity index of one's life time consumption, and $\gamma(n_t)$ is the discount factor which may depend on the number of children. Suppose the child-rearing cost in period t is θ_t, which includes the opportunity cost of parents' child-rearing time cost. Then all the above problems can be cast in the form

$$\max_{\{k_t\}_1^\infty} V_0 = \sum_{t=0}^{\infty} \left[\prod_{\tau=o}^{t} \gamma(n_\tau) \right] U(c_t^t) \text{ subject to}$$
$$c_t^t + n_t[k_{t+1} + \theta_t] = f(k_t) + (1-\delta)k_t, k_0 \text{ given} \tag{4.2}$$

When fertility is exogenous, the general practice is to assume $\gamma(n_t) = \gamma$ a constant, and thus, $[\prod_{\tau=o}^{t} \gamma(n_\tau)] = \gamma^t$, and $\theta_t = 0$, for all $t \geq 0$.

In the one-sector Ramsey growth model, if the felicity index U is concave and the production function is concave, the optimal capital–labor ratio path, $\{k_t\}_1^\infty$, of the problem in (4.2) is monotonic in the sense that a capital–labor ratio path will be higher or lower in all periods if it starts from a higher level of k_0). Furthermore, the optimal capital-ratio path exhibits a turnpike property that as $t \to \infty$, k_t tends to a locally stable steady-state. Thus, it is apparent that the complex dynamics in optimal growth framework that are shown to occur in multi-sector growth framework by Benhabib and Nishimura [4] and

Boldrin and Montrucchio [6] cannot occur in standard neoclassical one-sector Ramsey growth framework.

In the above Ramsey framework, however, there is no overlap of generations, so the framework is not suitable for analyzing many economic issues that involve economic decisions over the life-cycle, such as the old-age pension motive for savings, pay-as-you-go social security transfers, or inter-vivos transfers. Two-period[2] Samuelson [20]–Diamond [7] overlapping generations (OLG) framework is more suitable for this. To describe it briefly, suppose that each agent lives for two periods: adult and old. Let us denote by c_t^t the adult age consumption of an agent of period t, and by c_{t+1}^t her consumption in period $t+1$ (she is old in period $t+1$). Suppose the felicity index of her consumption over the life-cycle is $U(c_t^t, c_{t+1}^t)$. Grandmont [9] and others have shown that the competitive equilibrium of this type of OLG economy produces very complex dynamics, including periodic fluctuations and chaos. Under the assumption that the life-cycle felicity index U is separable, it can be easily established (also see below for an exposition) that a social planner's problem of optimizing the (discounted) sum of utility functions of various generations turns out to be a one-sector Ramsey growth problem, and thus no new dynamic issues arise for this class of problems.

An important extension of this Samuelson–Diamond framework is by Barro [1], who assumes that parents have altruistic utility functions of the type (4.1) in which $U(c_t^t)$ is replaced by $U(c_t^t, c_{t+1}^t)$. This adds bequest motives to the life-cycle motives of the Samuelson–Diamond model. Even in this framework, under some additional assumptions (e.g., utility functions of children are consistent with their parents' utility functions, and $U(c_t^t, c_{t+1}^t)$ is separable, and the bequest motive is operative), Weil [21] has shown that the competitive equilibrium path of the economy is characterized by the optimal solution of an one-sector Ramsey growth model. Thus, at least for this class of OLG economies, parental altruism rules out complexities in the dynamics of competitive equilibrium path. Recently, however, Michel and Venditii [14] have shown that if the life-cycle felicity index, $U(c_t^t, c_{t+1}^t)$ is not separable, then the optimal growth path in the one-sector growth model can produce cycles.

4.2.2 Endogenous fertility and growth dynamics

Fertility has been endogenized in growth framework mainly along two lines:[3] in one strand, the motive for children is parental altruism or love, and in

[2] More than two but finite period lived OLG framework can be identified with a two-period lived agent OLG model. so it is general enough to restrict our discussions to two-period lived OLG models.

[3] There is, however, another class of models which endogenize population growth by postulating that population growth rate is a function of per capita income. See Nerlove and Raut [16] for an account of the dynamic properties of this kind of model. Our focus in this paper is, however, the models that endogenize fertility as pre-natal choice.

the second strand, the motive for children is old-age pension. These two frameworks of endogenous fertility are, respectively, the analogues of the one-sector Ramsey framework and the two-period lived OLG framework. More specifically, Barro and Becker [2] and Becker and Barro [3] consider an one-period lived OLG framework with dynastic utility function of the form (4.1), and the corresponding optimal growth problem is then characterized by (4.2), with n_t as a choice variable. Kemp and Kondo [10], Lapan and Enders [13] and Nishimura and Kunaponagkul [17] also make fertility endogenous in the Ramsey framework, but they assume that $\gamma(n_t)$ is constant. In such models, to incorporate a motive for children in the utility function, the tradition has been to assume that the felicity index, U, depends on c_t^t as well as on n_t. The two-period lived OLG growth models that incorporate the life-cycle motive for savings and the old-age pension motive for children, include Neher [15] and more recently Raut [18], and Raut and Srinivasan [19].

While there are subtle differences in the dynamics of the above two frameworks of endogenous fertility, they also share some common dynamics which are strikingly different from the dynamics of the neoclassical growth model with exogenous fertility. We point out a few important ones here.

- When fertility is endogenous, the nature of equilibrium dynamics in both types of models depends crucially on the form of the child-rearing cost. For instance, if the child-rearing cost involves resources other than parental time, the capital–labor ratio in both type of models converge to steady-state in two periods (see Barro and Becker [2] for this in their model, and see Raut [18] for the second type of model).
- As we mentioned earlier, within the precinct of the one-sector growth framework, when the fertility is exogenous, it is necessary to have two-period lived overlapping generations of agents with a nonseparable life-cycle felicity index in order to generate cycles in the optimal growth path (Michel and Venditti [14]). When the fertility is endogenous, however, the optimal growth path of the one-sector Ramsey growth model can generate cycles (see Barro and Becker [2] with Cobb–Douglas form for $\gamma(n_t)$, Benhabib and Nishimura [5] with any concave $\gamma(n_t)$, and also Lapan and Enders [13]).
- In a two-period OLG framework with an endogenous fertility, in which population density creates external effect on the total factor productivity of the economy, it is shown in Raut and Srinivasan [19] that depending on the nature of externality, the child-rearing cost function, the values of the parameters, and the initial conditions of the economy, the model can produce complex nonlinear dynamics, which "not only include neoclassical steady-state with exponential growth of population with constant per capita income and consumption, but also growth paths which do not converge to a steady-state and are even chaotic." There are also

equilibrium paths in which capital–labor ratio attains steady-state in finite period, but the fertility rate fluctuates over time.

It is important to integrate the old-age pension motive with an altruistic motive for childbearing and saving in the endogenous fertility case for the same reasons Barro introduced it in the exogenous fertility case and to examine its implications for equilibrium dynamics. Could it eliminate some of the complex dynamics of the two-period lived overlapping generations model with endogenous fertility as in Raut and Srinivasan [19]? Could it preserve the dynamics of the one-period lived Barro–Becker growth model with endogenous fertility? The existing results from the literature are not adequate to reflect on these issues. The reason is that, unlike in the exogenous fertility case mentioned above, even with separable life-cycle felicity index $U(c_t^t, c_{t+1}^t)$, we cannot convert this model into a Ramsey growth model of the type that has been studied so far. To see this, suppose that $U(c_t^t, c_{t+1}^t) = u(c_t^t) + v(c_{t+1}^t)$. Then the optimal growth problem of Barro model with endogenous fertility becomes

$$\max_{\{n_t, k_{t+1}, c_t^t, c_{t-1}^t\}_0^\infty} V_0 = \sum_{t=0}^{\infty} \left[\prod_{\tau=o}^{t} \gamma(n_\tau) \right] U(c_t^t, c_{t+1}^t) \text{ subject to}$$

$$c_t^t + c_t^{t-1}/n_{t-1} + n_t \left[\theta_t + k_{t+1}\right] = f(k_t) + (1-\delta)\, k_t, k_0 \text{ given} \tag{4.3}$$

Notice that we can rewrite V_0 as

$$V_0 = \gamma(n_0)u(c_0^0) + \sum_{t=1}^{\infty} \left[\prod_{\tau=o}^{t-1} \gamma(n_\tau) \right] \cdot \left[\gamma(n_t)u(c_t^t) + v(c_t^{t-1})\right]$$

For $t \geq 1$, let us denote by

$$\tilde{U}(c_t, n_t, n_{t-1}) \equiv \max_{c_t^t, c_t^{t-1}} \gamma(n_t)u(c_t^t) + v(c_t^{t-1}) \text{ subject to}$$

$$c_t^t + c_t^{t-1}/n_{t-1} = c_t$$

Then we have the following Ramsey model with endogenous fertility

$$\max_{\{n_t, k_{t+1}\}_0^\infty} V_0 = \gamma(n_0)u(c_0^0) + \sum_{t=1}^{\infty} \left[\prod_{\tau=o}^{t} \gamma(n_\tau) \right] \tilde{U}(c_t, n_t, n_{t-1}) \text{ subject to}$$

$$c_t + n_t \left[\theta_t + k_{t+1}\right] = f(k_t) + (1-\delta)\, k_t, k_0 \text{ given} \tag{4.4}$$

As pointed out earlier, the previous literature has either assumed $\tilde{U}$ to be independent of n_t's or $\gamma(n_t)$ to be independent of n_t, or both. In this paper, we study the dynamics of the above type of model. We do not impose separability of U from the beginning; we impose it only to derive some specific results. We do, however, impose some reasonable restrictions on the nature of the bequest and division of current consumption between living adult and old generations in each period to make our analysis possible within an one-sector optimal growth framework.

4.3 Basic Framework

4.3.1 Production sector

We assume that the productive sector has a constant returns to scale production function $Y_t = F(K_t, L_t)$, which uses capital K_t and labor L_t to produce output Y_t in each period t, $t \geq 0$. Capital takes one period to gestate. Old members of the households own capital. We adopt the convention that the producer borrows from the old members of the households the stock of capital K_t at the beginning of period t and pays them $(\partial F/\partial K)K_t$ amount of rental income during the period t and stock of depreciated capital $(1-\delta)K_t$. This depreciated capital, $(1-\delta)K_t$, is bequeathed to the L_t children by the L_{t-1} old parents at the end of period t before they die. Thus, at the beginning of period $t+1$ the stock of capital available for production is:

$$K_{t+1} = (1-\delta)K_t + L_t s_t \tag{4.5}$$

On the right-hand side of the above, the first term is the inherited capital and the second term is the new capital added by the adults of period t. We assume that $s_t \geq 0$, which is equivalent to the assumption that capital is irreversible.

>From (4.5) we have the following relationship:

$$k_{t+1} = \frac{(1-\delta)k_t + s_t}{n_t} \tag{4.6}$$

where n_t is the number of children chosen by an adult of period t.

4.3.2 Households

At the beginning of time, $t = 0$, assume that there is only one adult agent who has at her disposal an initial endowment of capital $k_0 > 0$. Each person lives for three periods: young, adult, and old. While young she is dependent on her parent for all decisions, including childhood consumption. As an adult, she earns income w_t in the labor market, out of which she decides the amount of savings s_t and the number of children $n_t \geq 0$. In the next period, she inherits $(1-\delta)k_t$ amount of physical capital assets from her deceased parents, and lives off the income $\rho_{t+1}[(1-\delta)k_t + s_t]$ from her assets, where ρ_{t+1} is the rental rate of capital in period $t+1$.

We assume that utility of agent t, V_t depends on her own life-cycle consumption and the discounted sum of the utilities of her identical children V_{t+1} as follows:

$$V_t = U(c_t^t, c_{t+1}^t) + \delta(n_t)n_t \cdot V_{t+1} \tag{4.7}$$

where $\delta(n_t)$ is the weight given to each child's utility. We assume that $\delta(n_t)$ is the decreasing function of the number of children, n_t. We denote by $\gamma(n_t) = \delta(n_t)n_t$.

The recursive equation (4.7) leads to the following welfare for agent $t = 0$ as a function of the stream of lifetime consumptions and fertility levels of her own and future generations:

$$V_0 = \sum_{t=0}^{\infty} \left(\prod_{\tau=0}^{t} \gamma(n_\tau) \right) U(c_t^t, c_{t+1}^t). \tag{4.8}$$

Assuming perfect foresight and complete enforceability of her decisions $\{n_t, k_{t+1}\}_0^\infty$ on subsequent generations, and for a given stream of future social security benefits $\{b_{t+1}\}_0^\infty$, the problem of the adult of generation $t = 0$ could be formally stated as follows:

$$\max_{\{(n_t, k_{t+1}\}_0^\infty} V_0 = \sum_{t=0}^{\infty} \left(\prod_{\tau=0}^{t} \gamma(n_\tau) \right) U(c_t^t, c_{t+1}^t) \text{ subject to} \tag{4.9}$$

$$c_t^t = w_t - s_t - \theta_t n_t$$

$$c_{t+1}^t = \rho_{t+1}[(1-\delta)k_t + s_t]$$

$$t \geq 0, w_0 \text{ is given} \tag{9}$$

where we have

$$\left. \begin{array}{l} k_{t+1} = \dfrac{(1-\delta)k_t + s_t}{n_t} \\ \rho_{t+1} = f'(k_{t+1}) \\ w_{t+1} = f(k_{t+1}) - k_{t+1}f'(k_{t+1}) \end{array} \right\} , \ t \geq 0 \tag{4.10}$$

Let us denote by $w(k) \equiv f(k) - kf'(k)$. Assume that the utility function, production function, and the degree of altruism are all concave and increasing; there exists a positive value $\bar{\gamma} < 1$, $\gamma(0) = 0$ and $\gamma(n) \leq \bar{\gamma}$ for all n. Under these conditions, the solution of the above problem in (4.9) is equivalent to the solution of the following Bellman equation of the dynamic programming problem:

$$V(k_t) = \max_{n_t, k_{t+1}} U(c_t^t, c_{t+1}^t) + \gamma(n_t)V(k_{t+1}) \text{ subject to}$$

$$c_t^t = w\,(k_t) + (1-\delta)k_t - (\theta + k_{t+1})n_t$$

$$c_{t+1}^t = f'(k_{t+1})k_{t+1}n_t, \ t \geq 0, k_0 \text{ is given} \tag{4.11}$$

4.4 Dynamic Properties of Competitive Equilibrium Paths

In order to study the dynamic properties of the competitive equilibrium path, we assume that the depreciation rate $\delta = 1$. We introduce the following notation. Define

$$W(k_t, k_{t+1}, n_t) \stackrel{\text{def}}{\equiv} U(w(k_t) - (\theta + k_{t+1})n_t, n_t R(k_{t+1}) + \gamma(n_t)V(k_{t+1}) \quad (4.12)$$

where, $R(k_{t+1}) = f'(k_{t+1})\, k_{t+1}$. Define

$$n(k_t, k_{t+1}) \stackrel{\text{def}}{\equiv} \arg\max_{n_t \geq 0} W(k_t, k_{t+1}, n_t). \quad (4.13)$$

Denote by

$$\bar{W}(k_t, k_{t+1}) \stackrel{\text{def}}{\equiv} W(k_t, k_{t+1}, n(k_t, k_{t+1})) \quad (4.14)$$

The original problem in (4.11) of finding $\{n_t, k_{t+1}\}_0^\infty$ is now equivalent to solving (4.13) and the following:

$$V(k_t) = \max_{k_{t+1}} \overline{W}(k_t, k_{t+1}). \quad (4.15)$$

Suppose the above problem has a unique solution, denoted as $k_{t+1} = P(k_t)$. This function is known as the *policy function*. The dynamic behavior of the optimal solution path is determined from the properties of the first order difference equation given by the policy function, which we study now.

Let the partial derivative of $W(x_1, x_2, x_3)$ with respect to x_i be denoted by $W_i(x_1, x_2, x_3)$, and the second order partial derivatives of W by W_{ij}. Then, we have

$$W_1 = -k_t f''(k_t) U_1 \quad (4.16)$$

$$W_2 = -n_t[U_1 - R'(k_{t+1})U_2] + \gamma(n_t)V'(k_{t+1}) \quad (4.17)$$

$$W_3 = -(\theta + k_{t+1})U_1 + R(k_{t+1})U_2 + \gamma'(n_t)V(k_{t+1}). \quad (4.18)$$

We assume that there exists an interior solution so that $W_2 = W_3 = 0$ are satisfied. We further assume that $W_{33} \neq 0$ and apply the implicit function theorem to $W_3(k_t, k_{t+1}, n_t) = 0$ to obtain:

$$\frac{\partial n(k_t, k_{t+1})}{\partial k_t} = -\frac{W_{31}}{W_{33}} \quad \text{and} \quad \frac{\partial n(k_t, k_{t+1})}{\partial k_{t+1}} = -\frac{W_{32}}{W_{33}}. \quad (4.19)$$

Using the above relationships, we can study the short- and long-run effects of an exogenous increase in capital–labor ratio on fertility. In particular, we can find conditions for the Easterlin hypothesis to hold. Notice that

$$\frac{\partial n_t}{\partial k_t} = \frac{\partial n(\cdot,\cdot)}{\partial k_t} + \frac{dk_{t+1}(\cdot)}{dk_t}\frac{\partial n(\cdot,\cdot)}{\partial k_{t+1}} \quad (4.20)$$

where $k_{t+1}(\cdot)$ is the solution of the problem in (4.14). To determine the sign of the partial defined in (4.20), and also for later use, we need to determine the signs of the partial derivatives in (4.19). For that we apply the implicit function theorem on equations (4.18) and (4.17) and obtain the following:

$$\mathrm{W}_{33} = (\theta+k_{t+1})^2 U_{11} - 2(\theta+k_{t+1})R(k_{t+1})U_{21} + R(k_{t+1})^2 U_{22} + \gamma''(n_t)\mathrm{V}(k_{t+1}) \tag{4.21}$$

$$\begin{aligned} W_{32} &= n_t[(\theta+k_{t+1})U_{11} - \{(\theta+k_{t+1})R'(k_{t+1}) + R_t\}U_{21} \\ &\quad + R(k_{t+1})R'(k_{t+1})U_{22}]R'(k_{t+1})U_2 - U_1 + \gamma'(n_t)\mathrm{V}'(k_{t+1}) \end{aligned} \tag{4.22}$$

$$W_{31} = f''(k_t)k_t[(\theta+k_{t+1})U_{11} - R(k_{t+1})U_{21}] \tag{4.23}$$

$$W_{21} = f''(k_t)k_t[n_t(U_{11} - R'(k_{t+1})U_{21})]. \tag{4.24}$$

If we assume that U is strictly concave, then it follows immediately that W_{33} in (4.21) is strictly negative. When we assume that U is separable, it easily follows that $W_{31} > 0$. However, even for nonseparable utility functions, $W_{31} > 0$, as we will see in the following three specific class of economies. We cannot, in general, determine the sign of W_{32} in equation (4.22). Assuming that $W_{32} < 0$, which is true for each of the following three examples, we note that dn_t/dk_t and dk_{t+1}/dk_t are inversely related. Thus in cases when $dk_{t+1}/dk_t < 0$, i.e., when the optimal $\{k_t\}_0^\infty$ is *oscillatory*, we have a theoretical basis for the well-known Easterlin [8] hypothesis, stated in the introduction.

In the following theorem we find conditions that characterize the dynamic properties of the optimal path. Let us assume that $\bar{W}$ in (4.15) is twice continuously differentiable at each point of an interior solution path $\{k_t\}_0^\infty$. By differentiating the first order condition of the problem (4.15), we note that dk_{t+1}/dk_t $(\equiv dP(k_t)/dk_t) = -\bar{W}_{21}/\bar{W}_{22}$. Since $\{k_t\}_0^\infty$ is optimal, $\bar{W}$ is locally concave with respect to the second argument at each k_t. Hence, we have $\bar{W}_{22} < 0$. Thus we note that the sign of $dk_{t+1}/dk_t >= $ or < 0 according to whether $\bar{W}_{21} >=$ or < 0. More precisely, we have the following theorem (also see Benhabib and Nishimura [5] for an alternative proof).

Theorem 1 *Let $\{k_t\}_0^\infty$ be an interior optimal solution of the problem* (4.8) *with $k_0 \neq k^*$, then the following are true:*
(i) $\bar{\mathrm{W}}_{21} < 0 \Rightarrow (k_t - k_{t+1})(k_{t+1} - k_{t+2}) < 0$,
(ii) $\bar{\mathrm{W}}_{21} > 0 \Rightarrow (k_t - k_{t+1})(k_{t+1} - k_{t+2}) > 0$,
(iii) $\bar{\mathrm{W}}_{21} = 0 \Rightarrow k_{t+2} = k^*$, *for all* $t \geq 0$.

Let us determine the sign of this crucial cross partial derivative.

The cross partial of $\bar{W}$ is related to the second derivatives of $W(k_t, k_{t+1}, n_t)$ as follows:

$$\bar{W}_{21} = W_{21} + W_{31}\frac{\partial n_t}{\partial k_{t+1}} + W_{23}\frac{\partial n_t}{\partial k_t} + W_{33}\frac{\partial n_t}{\partial k_t}\frac{\partial n_t}{\partial k_{t+1}}. \tag{4.25}$$

By substituting (4.19) into (4.25) we have

$$\bar{W}_{21} = \frac{W_{21}W_{33} - W_{31}W_{32}}{W_{33}} \tag{4.26}$$

Substituting (4.21)–(4.24) in (4.26) we have the following:

$$\begin{aligned}\bar{W}_{21} &= W_{33}^{-1} k_t f''(k_t)[\{n_t R(k_{t+1})[R(k_{t+1}) - (\theta + k_{t+1})R'(k_t)]\} \\ &\quad (U_{22}U_{11} - U_{12}^2) - \{(\theta + k_{t+1})U_{11} - R(k_{t+1})U_{21}\}(R'(k_{t+1})U_2 - U_1) \\ &\quad (1 - \gamma'(n_t)n_t/\gamma(n_t)) + ((U_{11} - R'(k_{t+1})U_{21}))\,\gamma''(n_t)\mathrm{V}(k_{t+1})n_t] \\ &\quad W_{33}^{-1}\,\triangle\,, \text{ say.}\end{aligned} \tag{4.27}$$

It is not possible in general to determine the sign of the above cross partial derivative. We impose restrictions on the forms of the utility function $U(\cdot,\cdot)$ and the degree of altruism function $\gamma(n)$ along the lines of the available results in the literature to determine the sign of the above partial derivative and hence the dynamics of the equilibrium path.

4.4.1 Constant marginal utility of young age consumption

Let the utility function be given by $U(c_t^t, c_{t+1}^t) = c_t^t + v(c_{t+1}^t)$, that is, the marginal utility of the first period consumption is constant. In this case, $U_1 = 1$, $U_{11} = U_{12} = 0$ imply that $\bar{W}_{21} = 0$ and thus we have the following result:

Theorem 2 *In economies with special types of separable utility functions of the form $U(c_t^t, c_{t+1}^t) = c_t^t + v(c_{t+1}^t)$, the optimal sequence of capital–labor ratio, $\{k_t\}_0^\infty$ reaches steady-state at $t = 1$.*

It follows from the above theorem that the optimal fertility level, n_t, also reaches steady-state at $t = 1$. From equations (4.19) and (4.20), it follows that fertility and income are positively related in such economies. Since a steady-state is attained in finite time period, there is a unique steady-state.

In Barro and Becker [2] one-period lived agent framework, the above result is true for Cobb–Douglas $\gamma(n)$. In our two-period lived agent framework, the result is true for any general functional form for $\gamma(n)$, provided we restrict the felicity index to be separable and a linear function of the first period consumption.

4.4.2 Constant discount rate for progeny's welfare

In this section, we consider the case when $\gamma(n) = \gamma$, where $1 > \gamma > 0$, and characterize dynamic properties of optimal paths in terms of properties of a felicity index function, $U(c_t^t, c_{t+1}^t)$. We extend the one-period lived agent framework of Nishimura–Kunapongkul, Kemp–Kondo, and Lapan–Enders to two-period lived agents framework. By assuming $\gamma(n_t) = \gamma$, agents in these models are assumed to care about the welfare of a representative child; in one-period lived agent framework, previous models incorporate motives for children by assuming the utility function, U, to depend on c_t^t and n_t. We further assume that

$$U_{12} > 0 \tag{4.28}$$

$$f'(k) + f''(k)k > 0. \tag{4.29}$$

Notice that since $\gamma(n_t) =$ constant, we have

$$W_{33} = (\theta + k_{t+1})^2 U_{11} - 2(\theta + k_{t+1})R(k_{t+1})U_{21} + R(k_{t+1})^2 U_{22} \tag{4.30}$$

$\bar{W}_{21}$ from (4.27) can be expressed as

$$\begin{aligned} \triangle = \; & k_t f''(k_t)U_2^{-1}(\theta + k_{t+1}) \left[(\theta + k_{t+1})n_t[U_1 - R'(k_{t+1})U_2](U_{22}U_{11} - U_{12}^2) \right. \\ & - \left. [U_2U_{11} - U_1U_{21}][R'(k_{t+1})U_2 - U_1]\right] \\ & k_t f''(k_t)U_2^{-1}(\theta + k_{t+1})(U_1 - R'(k_{t+1})U_2) \\ & \times \left[n_t(\theta + k_{t+1})\left(U_{22}U_{11} - U_{12}^2\right) + (U_2U_{11} - U_1U_{21})\right]. \end{aligned} \tag{4.31}$$

For further simplification of the above, let us consider the following life-cycle utility maximization problem by a representative agent who takes I, $(\theta + k)$, and μ as given to solve:

$$\begin{aligned} & \max_{\{n\}} U(c_1, c_2) \text{ subject to} \\ & c_1 + (\theta + k)n = I \\ & c_2 = n\mu \end{aligned}$$

Substituting the expression for c_2 in the utility function U, and denoting the Lagrange multiplier corresponding to the first constraint as λ, we have the following first order necessary conditions:

$$U_1 - \lambda = 0, \tag{4.32}$$

$$\mu U_2 - (\theta + k)\lambda = 0, \tag{4.33}$$

$$I - c_1 - n(\theta + k) = 0. \tag{4.34}$$

From (4.32)–(4.34), we get the following well known results from the demand theory:

$$\frac{dn}{dI} = \frac{(\theta + k)U_{11} - \mu U_{21}}{W_{33}}, \tag{4.35}$$

$$\frac{d\lambda}{dI} = \frac{\mu^2 \left(U_{11}U_{22} - U_{12}^2\right)}{W_{33}}. \tag{4.36}$$

Let us denote the income elasticity of the demand for children, $e_n \equiv (I/n)(dn/dI)$, and the income elasticity of the marginal utility of income, $e_\lambda \equiv -(I/\lambda)(d\lambda/dI)$, for the above utility maximization problem. Using the facts that $\lambda = U_1$ from equation (4.32) and $\mu = (\theta + k_{t+1})U_1/U_2$ from equation (4.33), after simplifications we arrive at the following:

$$e_n - e_\lambda = \frac{I}{n}\frac{(\theta + k_{t+1})U_1}{W_{33}U_2^2}(n_t(\theta + k_{t+1})(U_{22}U_{11} - U_{12}^2) + (U_2U_{11} - U_1U_{21})). \tag{4.37}$$

Since $W_2 = 0$ is satisfied by the optimal path, we have

$$U_1 - R'(k_{t+1})U_2 = \gamma(n_t)\mathrm{V}'(k_{t+1}) > 0 \tag{4.38}$$

Substituting equation (4.37) in equation (4.31) and using equation (4.38), we have the following theorem:

Theorem 3. *Let $k_0 \neq k^*$, then we have*
(i) $e_\lambda > e_n \Rightarrow \{k_t\}$ *is monotonic,*
(ii) $e_\lambda = e_n \Rightarrow k_t = k^*$ *for all* $t \geq 2$,
(iii) $e_\lambda < e_n \Rightarrow \{k_t\}$ *is oscillatory.*

Corollary: *The economies for which $e_\lambda = e_n$, we also have that $n_t = n^*$ for all $t \geq 1$, and when $e_\lambda < e_n$, we also have oscillatory $\{n_t\}$. We cannot tell how $\{n_t\}$ will behave if $e_\lambda > e_n$.*

The above theorem is an extension of the Lapan–Enders characterization of competitive equilibrium dynamics for the one-period lived Ramsey model. It should be noted that the results (i) and (iii) in the above theorem remain true even when γ depends on n, but it is close to a constant function.

4.4.3 Dynasty of one-period lived agents

We have pointed out earlier that most growth models of endogenous fertility and savings in the dynastic framework assume that agents live one period (Barro and Becker [2], Becker and Barro [3], Benhabib and Nishimura [5], and others). We can nest those models and derive their dynamic properties from our extended framework as follows: In the optimization problem (4.9) view saving s_t is for the purpose of bequest as opposed to old-age pension that we have maintained so far. In the notation of problem (4.9), these assumptions are equivalent to the following:

$$\mathrm{U}_2 = \mathrm{U}_{22} = \mathrm{U}_{21} = 0. \tag{4.39}$$

Using (4.39) in (4.27) and noting that $V'(k_{t+1}) = (\theta + k_{t+1})U_1/\gamma'(n_t)$ in this special case, we have the following:

$$\begin{aligned}\triangle &= k_t f''(k_t)(\theta + k_{t+1})U_{11}U_1 \left[1 + n_t \left(\frac{\gamma''(n_t)\gamma(n_t) - [\gamma'(n_t)]^2}{\gamma'(n_t)\gamma(n_t)}\right)\right] \\ &= k_t f''(k_t)(\theta + k_{t+1})U_{11}U_1[1 - e(n_t)] \end{aligned} \tag{4.40}$$

where $e(n)$ is given by

$$e(n) \equiv \frac{-n}{\left(\frac{\gamma'(n)}{\gamma(n)}\right)} \frac{d\left(\frac{\gamma'(n)}{\gamma(n)}\right)}{dn} \tag{4.41}$$

Thus, the sign of $\overline{W}_{21}$ depends on the sign of $1 - e$, which depends only on the degree of altruism function $\gamma(n)$, but not on the utility function U or production function. Thus, we have proved the following result:

Theorem 4

(i) *if* $e < 1$, $\{k_t\}_0^\infty$ *is monotone,*
(ii) *if* $e = 1$, $\{k_t\}_0^\infty$ *reaches steady-state at* $t = 1$,
(iii) *if* $e > 1$, $\{k_t\}_0^\infty$ *oscillates.*

Corollary: *Let* $k_0 \neq k^*$. *If* $e = 1$, $\{n_t\}$ *reaches steady-state at* $t = 1$. *If* $e < 1$, $\{n_t\}$*is oscillatory.*

Barro and Becker [2] assumed Cobb–Douglas form for $\gamma(n)$ which forces $e = 1$.

It is important to note that Theorem 4(i) and (ii) are true even when U_2, U_{22}, and U_{21} are not zero, but are very close to zero. This extends Benhabib and Nishimura [5] to the dynastic model with two-period lived agents.

4.5 Conclusion

The one-sector neoclassical Ramsey growth model with exogenous fertility exhibits simple dynamics: The optimal path exhibits monotonicity and turnpike property. In this paper, we have briefly reviewed a host of one-sector equilibrium growth models with exogenous fertility, including the two-period lived overlapping generations model with parental altruism (i.e., the Barro model) which unifies the bequest motive and life-cycle motive for savings. Under general conditions, the dynamics of competitive equilibrium path and social optimal path in all these models share the above dynamic properties of the one-sector neoclassical Ramsey growth model with exogenous fertility.

The existing growth models with parental altruism and endogenous fertility study dynamics of the optimal path by restricting their analyses to one-period lived overlapping generations framework. Surveying the findings of various papers, we find strikingly different dynamics in one-sector equilibrium growth models when fertility is endogenous. We also show that unlike the exogenous fertility case, the existing Ramsey growth models of endogenous fertility are not able to embed the two-period lived overlapping generations model with parental altruism and endogenous fertility. We have formulated a two-period lived overlapping generations model of endogenous fertility with parental altruism and extended some of the existing results from the one-period lived framework to our general framework.

References

1 Barro, Robert J., Are government bonds net wealth?, Journal of Political Economy, 82 (1974).

2 Barro, Robert J. and Gary S. Becker, Fertility choice in a model of economic growth, Econometrica, 57(2) (1989) 481–501.

3 Becker, G.S. and R.J. Barro, A reformulation of the economic theory of fertility, Quarterly Journal of Economics, 53(1) (1988) 1–25.

4 Benhabib, Jess and Kazuo Nishimura, Competitive equilibrium cycles, Journal of Economic Theory, 35 (1985) 284–306.

5 Benhabib, Jess and Kazuo Nishimura, Endogenous fluctuations in the Barro–Becker theory of fertility, in: A. Wenig, K.F. Zimmermann (eds.), Demographic Change and Economic Development, Springer-Verlag, Berlin, Heidelberg, 1989.

6 Boldrin, M. and L. Montrucchio, On the indeterminacy of capital accumulation paths, Journal of Economic Theory, 40 (1986) 26–39.

7 Diamond, Peter, National debt in neoclassical growth models, American Economic Review (1965).

8 Easterlin, Richard, Population, Labor Force, and Long Swings in Economic Growth, New York: Columbia University Press, New York, 1968.

9 Grandmont, J.M., On endogenous competitive business cycles, Econometrica, 53 (1985) 995–1045.

10 Kemp, M.C. and Hitoshi Kondo, Overlapping generations, competitive efficiency and optimal population, Journal of Public Economics, 30 (1986) 237–47.

11 Kondratieff, N.D., The long waves in economic life, Review of Economics and Statistics, 17 (1935) 105–15.

12 Kuznets, S., Long swings in the growth of population and in related economic variables, Proceedings of the American Philosophical Society, 102 (1958) 25–52.

13 Lapan Harvey E. and Walter Enders, Endogenous fertility, Ricardian equivalence, and debt management policy," Journal Public Economics, 41 (1990) 227–48.

14 Michel, Philippe and Alain Venditti, Optimal growth and altruism in overlapping generations models, Economic Theory (1994).

15 Neher, P.A., Peasants, procreation and pensions, American Economic Review, 61 (1971) 380–89.

16 Nerlove, M. and L.K. Raut, Growth models with endogenous population: A general framework, in: Mark R. Rosenzweig and Oded Stark (eds.), Handbook of Population and Family Economics, Vol. 1B, Ch. 20., Elsevier Pub. Co., New York, 1997.

17 Nishimura, Kazuo and Vijit Kunaponkul, Economic growth and fertility rate cycles, Chaos, Solutions and Fractals, 1 (1991) 475–84.

18 Raut, L.K., Capital accumulation, income distribution and endogenous fertility in an overlapping generations general equilibrium model, Journal of Development Economics, 34(1/2) (1991) 123–50.

19 Raut, L.K. and T.N. Srinivasan, Dynamics of endogenous growth, Economic Theory, 4 (1994) 777–90.

20 Samuelson, P., An exact consumption-loan model of interest with or without the social contrivance of money, Journal of Political Economy, 66 (1958), 6.

21 Weil, Philippe, Love thy children: Reflections on the Barro neutrality theorem, Journal of Monetary Economics, 19 (1987) 377–91.

Part II

Trade Theory and Policy

Trade, Growth, and Development
G. Ranis and L.K. Raut

CHAPTER 5

Play it Again Sam: A New Look at Trade and Wages[1]

Jagdish Bhagwati

Department of Economics, Columbia University, 420 West 118th Street, New York, NY 10027

5.1 Introduction

T.N. Srinivasan has written profusely, and profoundly, on both development economics and on international trade. So, I could offer him my tribute by writing on either. But T.N. will appreciate, as the no-nonsense economist of integrity he has always been, that I must follow my comparative advantage and choose to write on trade. That I do today, tackling a question of immense topicality and great policy concern: to wit, and to put it strikingly, the effect of trade with poor countries on the poor in the rich countries.

Indeed, the prolonged decline in real wages of our unskilled workers, and the widely-shared sense that crystallized during the national debate over NAFTA that trade with Mexico would harm our workers have produced arguably the most animated and politically salient debate among economists on the question: *does trade with poor countries immiserize our unskilled workers?*[2]

[1]My thanks go to Susan Collins, Don Davis, Alan Deardorff, Vivek Dehejia, Elias Dinopoulos, Robert Feenstra, Richard Freeman, Pravin Krishna, Paul Krugman, Robert Lawrence, Ed Leamer, Dani Rodrik, T.N. Srinivasan, and Martin Wolf for helpful conversations over the years on the issues discussed in this paper.

[2]In my view, this issue was created by NAFTA because bilateral trade agreements inevitably lead to a focus on the characteristics of products, endowments, governance, etc., of the specific country with whom you are negotiating. With Mexico being impoverished, with illegal workers streaming across the Rio Grande, it was inevitable that objections would arise as to how freer trade would indirectly hurt our workers the way illegal immigration from Mexico was allegedly doing. By contrast, there were no such questions raised vis-á-vis the Uruguay Round because the multilateral trade negotiations were with several countries, both rich and poor, and it would have therefore been simply absurd for anyone to object to them by raising the red flag over the implications of trade with countries such as India where there are even more poor than in Mexico! I have discussed this downside of regional-

Yet, despite the immense number of academic analyses, confusion reigns and general pessimism prevails. I propose here to remove the confusion and to reach a more optimistic conclusion.

5.2 Two Different Questions

In doing so, I first note that as Deardorff and Hakura [9] and Bhagwati and Dehejia [6] noted earlier, the question posed is ambiguous (these essays appeared in the volume edited by Bhagwati and Kosters [7]). Different questions must be distinguished, each appearing at first blush to be like the other, while being quite distinct with different answers. In particular, I distinguish between two questions, both of importance and each corresponding in some way to what seems to agitate policymakers in some vague, if not inchoate, fashion. As it happens, I argue that the answer to each question, for different reasons, is not as alarming as in the popular perception of the threat from trade (with poor countries) to our workers.

Question 1. *If the rich countries (the North hereafter) were to liberalize their trade with the poor countries (the South hereafter), or if the South were to liberalize its trade with the North, e.g., as NAFTA did for the US and Mexico, then would this reduce the real wages of workers in the North?*[3]

Question 2. *Can the observed changes in real wages in the North be explained by changes in trade (opportunities) coming from the South rather than by factors internal to the North?*

As I will presently argue, Question 1 focuses exclusively on trade liberalization in the South and/or in the North and its consequences for real wages in the North. By contrast, Question 2 contrasts the effects on real wages in the North as a result of all factors (that would not be confined to trade liberalization alone but extend also to technical change and factor accumulation): these factors would then be grouped into those coming from the South and those coming from within the North, and interacting via trade.

In both cases, however, my answers are comforting rather than pessimistic. In each case, there are two steps involved in linking trade with real wages. The first step is to assert that the (relative) prices of labor-intensive goods have fallen within the North because of trade. The second step is then to argue

ism, and more generally of PTAs (preferential trade agreements) in the course of discussing President Clinton' s failure to secure fast-track authority from Congress in "Think big, Mr. Clinton," *The Financial Times*, Tuesday, November 25, 1997.

[3] If wages are inflexible downwards, then unemployment would increase instead of wages falling. The former is assumed generally to happen in the US, the latter in Europe.

that therefore, as in the Stolper–Samuelson (SS) theorem, the real wages of labor have fallen in the North. For Question 1, I show below (in the stylized 2×2 model) that step 1 certainly holds. But step 2, involving the empirical applicability of the SS theorem, is open to serious objections and the effect on real wages of all factors, including labor, could well be favorable. Therefore, one may well be optimistic on the impact of trade on real wages if the empirical relevance of the SS theorem is denied. As for the answer to Question 2, the answer I give below is decidedly optimistic in the sense that the first step itself cannot be taken: changes emanating from the South, in their totality (as distinct from merely trade liberalization in the South), will likely raise rather than reduce the prices of labor-intensive goods, *ceteris paribus.*[4]

5.3 "Exclusively Trade Liberalization" Question

This question refers to the effects of trade liberalization, whether in the South or in the North (vis-à-vis each other), on real wages of workers in the North. Thus, in a stylized 2×2 model where the South exports the (unskilled)- labor-intensive good Y while the North exports the capital-intensive good X, the answer to this question is straightforward.

In Figure 5.1, depicting the offer curves of the South and the North, with tariffs in each region leading to the tariff-ridden offer curves intersecting at Q, consider trade liberalization by the South. This shifts its offer curve to OS' and the new trade equilibrium to R. Clearly, the trade liberalization will increase the supply of exports from the South at every relative goods price (i.e., terms of trade) and will reduce the world price of the labor-intensive good Y and hence, given any tariff in the North, also the *domestic* price of Y in the North.

If, however, the North liberalizes its trade, ON shifts to ON' while OS is unchanged, leading to trade equilibrium at Z. But while this raises the world price of the labor-intensive good Y, its domestic price will fall in the North (unless the Metzler paradox obtains, so we must rule it out).

Thus, whether the trade liberalization occurs in the South or the North, we can expect it to lead to a fall in the domestic price of the labor-intensive good Y in the North. Hence, it inevitably sets the stage for the Stolper–Samuelson (SS) theorem.

If SS reigns, the fall in the domestic price of the labor-intensive good Y will lead to a fall in the real wage of labor in the North. Thus, while the first step

[4] Of course, if you believe that the SS theorem does not apply, so that the terms of trade improvement implied by falling world prices of labor-intensive goods will improve both national income and the real wage of labor, then a rise in the price of these goods is not a cause for celebration! The conclusion in the text is therefore comforting only if you believe in the stranglehold of the SS theorem on reality.

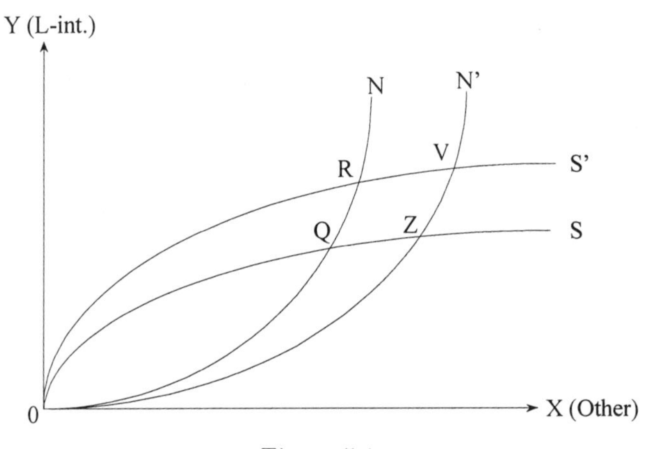

Figure 5.1

in the argument linking trade liberalization to decline in our real wages, via a fall in the price of labor-intensive goods, is *theoretically* satisfied (though, as argued in the next section, the stylized *facts* show that the prices of labor-intensive goods, whose behavior must reflect not just the trade liberalization we are discussing presently, actually rose slightly in the period when real wages declined), we still have to ensure that the second step, the applicability of the SS theorem, can also be taken.

But then we must recall that the SS theorem cannot be regarded as necessarily defining the empirical reality. In fact, the theorem became well-known precisely because it simply established a *possibility* when no one thought it possible to do so. In particular, these distinguished authors managed to show that under certain restrictive conditions, one could indeed infer an unambiguous effect on real wages from a change in the goods prices. Until SS did so, it was generally believed that while nominal wages would fall for workers intensively employed in the good whose price had fallen, the effect on *real* wages was ambiguous: it would depend on the consumption pattern of the workers, since a fall in the nominal wage could be offset by a preference in consuming the good whose price had fallen. The SS result, by showing that (given their model and its assumptions) we did not need to know what consumer preferences were in order to infer the impact on real wages, became a theoretical curiosum, as it were; few regarded it as an inevitable empirical reality or even as capturing a central tendency. Today, however, in a supreme irony, it seems as if it is regarded as our inescapable fate.

And that is a singular mistake. For, as discussed extensively in Bhagwati and Dehejia [6], we must recognize that specialization in production will mean that instead of one factor being hurt while the other benefits from the change in the goods price, as in the SS case, *both* (of the two) factors will benefit from the price fall. Scale economies can also do this. Improvement in overall efficiency following trade competition can do it too. In fact, these "lift-all-boats" effects can kill the SS "redistributive" effect. As it happens, the calculations of Brown, Deardorff and Stern [8] (with the aid of their well-known computable Michigan model during the NAFTA debate), allowing for the restrictive SS conditions not to be fully met, showed a real wage *improvement* for American workers from NAFTA. So, the asserted link between trade and real wage decline, as precisely postulated here, breaks down; the SS theorem, whose applicability is not inevitable or in my judgment even likely, is then not the dagger aimed surely at our workers' jugulars!

I might add that there is nothing in what I have said above about the Factor Price Equalization (FPE) theorem. The FPE theorem requires a great deal of added baggage: structure must be put on the South so as to make, for instance, its production functions identical to those of the North, to rule out factor intensity reversals, to assume identity of tastes across counties. Indeed, many of these assumptions are unrealistic (e.g., we know from the work in the 1960s by Minhas and Arrow–Chenery–Minhas–Solow that factor intensity reversals are not merely possible, since estimated CES production functions have different cross-sector elasticities, but also likely because endowments lie on different sides of the factor-intensity-crossover point). But that is no cause for concern, of course, unless we also wish explain what is happening in the South as a result of trade liberalization. All we need to do, in explaining the past and future link between trade (with the South) on the real wages of the unskilled in the North is to start from the fact that the South is a net exporter of labor-intensive goods and then to examine the effects of trade liberalization, as we have done, on goods prices in the North and therefrom on the real wages in the North. That is just what I have done here.

5.4 "Total Trade" Question

But then let me ask the altogether different Question 2, distinguished above, which relates to whether a shift in the offer curve of the South, arising from the *totality* of all relevant factors such as factor accumulation and not merely trade liberalization, can explain the decline in real wages in the North. Again, we would have to argue that this shift leads to a decline in the average world prices of labor-intensive goods, by augmenting their supplies (i.e., the offer curve shifts outwards), and then again via the SS theorem to a decline in real wages. The second SS step runs into the same difficulties as with Question 1 in the preceding section. But so does the first step, because we must now reckon with factors such as capital accumulation and technical change as well, as I demonstrate presently.

The analysis of what happens to the offer curve of the South, as a result of several factors distinguished below, explains why the offer curve will not necessarily shift outwards so as to push down, *ceteris paribus*, the prices of labor-intensive goods in world trade (and hence be the cause of the declining wages in the North by triggering the SS theorem). In fact, it can be expected to exhibit the opposite tendency, reducing the overall excess supplies of labor-intensive goods and hence leading to a rise in their prices instead, *as seems to have happened.* It also explains a number of other stylized facts. I show this now, first by stating the stylized facts, and then developing the shift-of-the-offer-curve explanation.

5.4.1 Stylized facts

A number of stylized facts have emerged in the empirical studies spawned by the trade-and-wages debate:

(a) The most important fact is evidently the behavior of the prices of labor-intensive goods in world trade, which in the US, fell in the 1970s but (slightly) rose in the 1980s and early 1990s. The latter phenomenon is now conceded by all serious scholars, including early skeptics such as the world class econometrician and trade economist Ed Leamer, who has done a considerable amount of careful empirical work on the subject.[5]

(b) However, in an eyescan "refutation" of the SS theorem, US real wages (of unskilled workers), defined first as "compensation per worker" and next as the less satisfactory "average hourly earnings," continued to rise during the 1970s while they fell by the latter measure and their rise was seriously moderated by the former measure, during the 1980s and early 1990s (see Figure 5.2).

(c) The wage differential between unskilled and skilled workers has risen not just in the US and other OECD countries, but also in some other countries, e.g., in Chile, Uruguay, Colombia, Costa Rica and Mexico in the last decade (see Robbins [12]).

Trade in labor-intensive manufactures of the poor countries has not only been a story of all these countries becoming larger exporters of such manufactures. Over time, per capita incomes grow more rapidly in some (e.g., East Asia in the 1970s and 1980s) as compared to others. The former subset of poor

[5] The only exception is provided by Sachs and Shatz [13] in the in-house journal of the Brookings Institution. However, their evidence to the contrary is not compelling, in view of their regression failing to meet the requisite standards of statistical significance. Even then, these authors get their insignificant regression to show only a slight fall in the prices of labor-intensive goods, and that too by excluding computers without plausible justification.

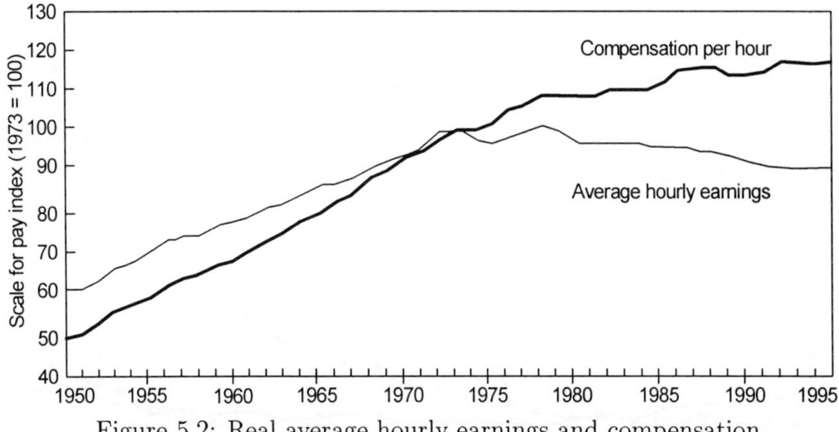

Figure 5.2: Real average hourly earnings and compensation

Note: Compensation per hour includes wages and salaries of employees plus benefits (employers' contributions for social insurance and private benefit plans). It covers the nonfarm business sector. Average Hourly Earnings does not include nonwage benefits. It covers production and nonsupervisory workers in the private nonfarm sector of the economy. Both measures are adjusted for inflation using CPI-U-X1.

Sources: Bureau of Labor Statistics, Economic Report of the President, 1996; M.Kosters, AEI, May 1996.

countries then become net importers of labor-intensive manufactures themselves so that the net exports of the poor-countries group (constituted by the two subsets of countries taken together) to the group of rich countries grow less dramatically than many fear. The fear comes from an erroneous assumption that each poor country will become an increasing supplier of labor-intensive manufactures to the rich countries, leading to an avalanche of exports. International economists, among whom the late Bela Balassa deserves pride of place, have long understood this phenomenon empirically, calling it the phenomenon of "ladders of comparative advantage."

This more comforting picture is exactly what Ross Garnaut [10] of ANU has shown in Figure 5.3. There, the 1970s witness East Asia steadily increasing net exports of labor-intensive manufactures while Japan reduces them. The same pattern repeats itself in the 1980–94 period when East Asian (NIE) net exports decline from over 10 percent of world trade in labor-intensive manufactures to nearly zero while China goes almost in a crossing diagonal from around 2 to over 14 percent, the difference between the two leaving greatly reduced the net impact on what Garnaut calls the "old industrial countries" on the average. This is, of course, what I just recalled as the "ladder of comparative advantage" and countries climb up on it with growing per capita incomes.

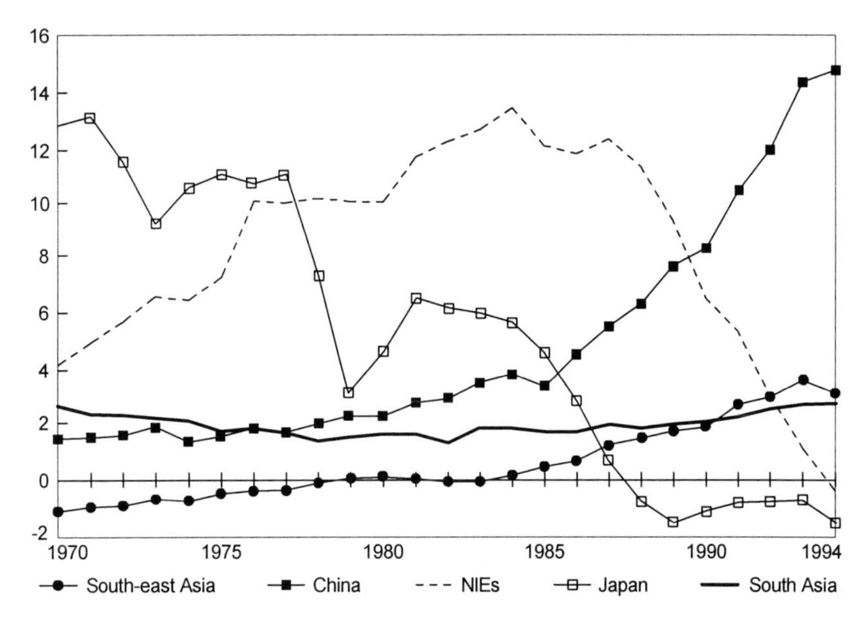

Figure 5.3. Ratio of net exports to world imports of labor-intensive manufactures, East and South Asia 1970–94 (%)

Notes: South-east Asia includes ASEAN (excluding Singapore) and Vietnam; NIEs include Taiwan, Hong Kong, Korea and Singapore; and South Asia includes India, Pakistan, Bangladesh and Sri Lanka.

Source: UN trade data, International Economic DataBank, The Australian National University.

5.4.2 Analysis

These stylized facts can be explained, and their underlying causes understood, by returning to the offer curve analysis. Essentially, I plan to answer Question 2 at the outset by analyzing immediately how the South's offer curve would shift as a result of various reasons such as capital accumulation. (In the Section 5.4, I will then go on to discuss the corresponding shifts in the North's offer curve as well, for identical reasons, seeking to enact the whole story of what happened in the recent period.)

The underlying changes that shift each offer curve are clearly: (1) capital and labor accumulation; (2) technical change; and (3) trade liberalization. (I say "trade liberalization" instead of the more generic "trade policy change" because we have been witnessing liberalization rather than growth of trade barriers in the last few decades.) Consider each of these three factors in turn.

(1) When capital and labor accumulate at the same rate (say, $x\%$), the offer curve will obviously *expand* outwards by an identical rate. But if capital accumulates more rapidly (say, at $y\%$), then we have to account for the effect of that extra non-uniform expansion of capital ($(y - x)\%$).

This latter effect, which international economists call the Rybczynski effect, reduces the excess supply of the labor-intensive good Y. It will thus contract or shrink the offer curve.

The net effect of factor accumulation will then depend on the relative strength of these two effects. But evidently, if capital accumulation is considerable, as it has been in East Asia for over two decades, that could well be a cause of their offer curve exhibiting a shrinking of their production of labor-intensive goods and their withdrawal from exports of such goods in world markets. (In fact, the East Asian "miracle" has been precisely in the "miraculous" investment rates that these countries have chalked up, as I have argued in Bhagwati [5].)

(2) If technical change is occurring and is contributing to the expanding per capita incomes as well, we can generally expect it to be occurring faster in the modern industries that use human and conventional capital intensively. In that case, one can expect again a pull of resources away from the labor-intensive industries towards the production of the progressive industries, thus contributing to a *decline* or shrinking of the South's offer curve rather than to its further outward expansion. (This tendency is conclusive when the technical progress is Hicks-neutral but may not be so decisive if it is biased.)

(3) Trade liberalization, on the other hand, will *expand* the offer curve, as already discussed in analyzing Question 1. (I might add that import protection could, as emphasized by Paul Krugman in his classic work, leads to export promotion eventually. This possibility is being ignored.)

Hence, there are two factors (trade liberalization and uniform expansion of all factors of production) which push the South's offer curve out, and two factors (beyond-uniform accumulation of capital and technical change) which pull it in. Very likely, the former two factors were more important than the latter two in the 1970s. In that era, the oil shock had generally depressed growth rates of per capita incomes in many developing countries while trade liberalization continued only at a moderate pace. From the early to mid-1980s through 1990s, however, the growth rates in developing countries were generally more robust, with the huge East Asian growth rates continuing more or less, while the "miracle" spread to other countries in Asia westwards. I would expect therefore that in the 1970s, compared to the later period, the expansion of the South's offer curve would be greater, and hence the downward pressure on the prices of labor-intensive goods, *ceteris paribus*, would be less. In fact, my stylized view, which I develop shortly, is that the later period, due to these effects, actually saw a *shrinking* of the net group offer curve of the South leading to a *price rise* of labor-intensive goods. Furthermore, the moderately *expanded quantities* of trade, despite that, are to be attributed to a simultaneous outward shift in the group offer curve of the North, as its demand for imports of labor-intensive goods rose for the same reasons operating in the North.

Given the asymmetries I have just argued for the South between the two periods, the 1970s and later, it is not surprising that the stylized facts on prices of labor-intensive goods show that the 1970s witnessed a fall in them, with opposite behavior in the later period.

Furthermore, if capital accumulation is a major factor in some Southern countries, one should expect the "ladder" phenomenon that Garnaut has documented (Figure 5.3). And if conventional capital is a complement to skilled labor but a substitute for unskilled labor, the accumulation of capital would generally tend to widen the differential in reward in favor of skilled labor, as has happened in some of the better-performing countries.[6]

To recapitulate the main conclusion, therefore, the analysis of the factors that shift the South's offer curve shows that changes in trade with the poor countries, which arise from the *totality of changes* coming from them, can be plausibly argued to have been, in the recent period and on balance, benign as far as the fear of falling (average) world prices of labor-intensive goods is concerned. And if we then expect these forces to continue operating, with rapid growth diffusing through the developing world as broad economic reforms take root, then we can well expect that the future will also be benign.

Thus, the answer to Question 2, posed above, is essentially benign. The picture is dramatically different from the one we get (as when we discuss Question 1) if we focus exclusively on trade liberalization by developing countries:[7] a process which, in any case, is spread out over time in most cases and therefore is not likely to outweigh, at any time, the effect of rapid growth rates.

5.5 The Full Story

To grasp fully what happened in the recent period, however, we need to bring into the analysis the shift in the North's offer curve, a shift arising from the same constellation of causes that were discussed in relation to the South's offer curve.

[6] The statement about the wage differential means, of course, that we depart from the 2×2 structure on factors and goods. But nothing qualitative that was derived within that framework needs to be modified.

[7] The difference is in the first step of the two-step argumentation outlined above: the prices of labor-intensive goods may now be expected to rise, rather than fall, once factors other than trade liberalization are taken into account.

When this is done, we are confronted with an interesting contrast: the factors that worked to reduce the supply of labor-intensive goods from the South work in reverse for the North's offer curve, since it *imports* labor-intensive goods. Hence, all factors tend to increase the North's demand for labor-intensive goods, reinforcing the upward pressure on the prices of labor-intensive goods that come from the South itself.

(1) Thus, while uniform expansion of factor supply will push out the North's offer curve, further capital accumulation will reduce the output of labor-intensive goods and thus reduce their net supply and increase import demand. This will reinforce the outward shift in the offer curve.
(2) A similar result would follow from technical change concentrated in the capital-intensive industries, disallowing complexities that can follow from biased technical change.
(3) And trade liberalization, of course, will also shift out the offer curve.

So, we have a situation where all factors tend to reinforce one another, raising the demand for imports of labor-intensive goods and hence, *ceteris paribus*, raising their world prices. Associated with this, there would be expanded trade volumes. We would thus observe increasing "import penetration ratios" in the import-competing industries of the North as, in fact, we have. (Note that this outcome is a result of purely domestic factors, and is not to be attributed to an exogenous increase in export supplies from the South. In fact, as I argued above, my informed guess is that the export supplies from the South shrank, not rose, in the post-1980s period.) The world prices of labor-intensive goods would then be expected to rise with the North's increased demand for them, as in fact they have done.[8]

Insofar as these shifts in the two offer curves — one (for the South) shrinking and the other (for the North) expanding — translate into increased domestic prices for the labor-intensive goods and in the quantities traded, we are left with the question: what can we say about the accompanying effects on the real wages of unskilled labor in the North?

Clearly, if we ask a *ceteris paribus* question, namely, what is the effect of the shift in the South's offer curve on real wages in the North, then it has to be (if the SS theorem holds) positive. For it leads to a rise in the prices of labor-intensive goods, not a fall. If we bring both shifts into the picture, and look for a total answer, then clearly the answer has to be as follows: the factors underlying the North's expanding demand for labor-intensive imports may reduce the real wages of labor, *ceteris paribus*, but that fall will be *moderated* by the effect of the exogenous shift in the South's offer curve. In short, I

[8]So would the domestic prices of the labor-intensive goods in the North, except when the cause of the change is trade liberalization by the North (as discussed in the analysis of Question 1).

would maintain this answer to Question 2 posed above: *trade with the South has moderated the adverse impact, such as may be, from technical change, on real wages in the North caused by technical change.*

To *recapitulate* the substance of my argument, if I was asked to put the most plausible story together from the previous analyses of the shifts in the South's and North's offer curves, it would be as follows. Based on a simultaneous shift in the offer curves of the South and the North (as in Figure 5.1, where we go from an initial trade equilibrium at Q to V where both shifted curves ON' and OS' now intersect):

- Ongoing changes in capital accumulation and technical change, working alongside and offsetting the effects of trade liberalization, are likely to have been predominant in the world economy. These caused a mildly upward, instead of a substantial downward, shift in the average world prices of labor-intensive manufactures.

- The net effect of these forces has also been to raise the domestic prices of labor-intensive manufactures in the North as well.

- Insofar as the factors operating within the South and affecting its offer curve (i.e., the "trade opportunities" the South offers us or, in popular imagination, threatens us with) are considered, my conclusion is that they likely have been, on balance, increasing the average prices of labor-intensive goods in world trade during the years when real wages have fallen in the North.

- If, there, the SS theorem is invoked, the changes exogenously emanating from the South cannot be responsible for the decline of the real wages in the North: they push the goods prices in the wrong direction.

- But the overall increase in the world prices of labor-intensive manufactures also reflects a shift away from the production of labor-intensive goods in the North, a shift due to endogenous factors such as capital accumulation and technical change. By adding (as argued above) to the deterioration of the North's terms of trade — as the exogenous shrinking of the South's supply of labor-intensive exports entails — they further reduce the primary gain in income that these per capita income-augmenting fundamentals imply.

- Whatever the effect on real wages in the North, caused by the fundamental factors underlying the shift in import demands for labor-intensive goods in the North itself, there is no way we could argue that the forces shifting the export supplies from the South have had an adverse effect on them. Rather, they have made the real wages better than they would have been, if the SS theorem holds, since the *ceteris paribus* effect of trade with the South will be to improve the real wages in the North. Thus, it will have raised, and not lowered, the traded prices of labor-intensive goods.

I think that this conclusion is pretty plausible. It puts me on the side of those who deplore the usual declamations against globalization on the grounds that trade with poor countries hurts our workers. But it puts me right at the edge of that group since the most that they have said, in ways that I am not enthusiastic about analytically, is that the adverse effect is small or even negligible. I actually say that it is favorable, not adverse! And I expect it to remain so in the foreseeable future.

So, I claim the distinction of counting myself out of the "consensus" often asserted in Washington (especially in the think tanks distinguished by their armor rather than their grey cells, and even in the Bretton Woods institutions that seek amiably-agreed views) that economists "believe" that the adverse effect of trade on real wages is around 10–20 percent or 15–20 percent.

This was the range that Dani Rodrik recently concluded in his alarmist pamphlet on globalization for the Institute for International Economics; it is also to be found in an IMF pamphlet reported on in *The Economist.* The former was based on negligible work; the latter simply averaged, under instructions, all empirical studies on the subject and ignored the fact that, in science, the average of good and bad is bad. If I am wrong, it will not be because of these forgettable contributions but because of the fault lines in my own argumentation. However, I hope to stand, alone for now, but not lonely for long.

References

1 Arrow, Kenneth, Hollis Chenery, Bagicha Minhas and Robert Solow, Capital-labor substitution and economic efficiency, Review of Economic Studies, 43 (1961) 225–50.

2 Bhagwati, Jagdish, Challenges to free trade: Old and new, 1993 Harry Johnson Lecture, Economic Journal (1994), March.

3 Bhagwati, Jagdish, Trade and wages: Choosing among alternative explanations, Federal Reserve Bank of New York Economic Policy Review (1995), January.

4 Bhagwati, Jagdish, Trade and wages: A malign relationship?, Paper presented at a Brookings Conference in February 1995 (forthcoming in a volume edited by Susan Collins).

5 Bhagwati, Jagdish, The "miracle" that did happen: Understanding East Asia in comparative perspective, Keynote Speech at Cornell University Conference; mimeo, Columbia University, May 3, 1996.

6 Bhagwati, Jagdish and Vivek Dehejia, Freer trade and wages of the unskilled, in: Jagdish Bhagwati and Marvin Kosters (eds.), Trade and Wages: Leveling Wages Down?, The AEI Press, Washington, DC, 1994.

7 Bhagwati, Jagdish and Marvin Kosters (eds.), Trade and Wages: Leveling Wages Down?, The AEI Press, Washington, DC, 1994.

8 Brown, Drusilla, Alan Deardorff and Robert Stern, Protection and real wages: Old and new trade theories and their empirical counterparts, mimeo, Michigan, Ann Arbor, 1994.

9 Deardorff, Alan and Dalia Hakura, Trade and wages: What are the questions?, in: Jagdish Bhagwati and Marvin Kosters (eds.), Trade and Wages: Leveling Wages Down?, The AEI Press, Washington, DC, 1994.

10 Garnaut, Ross, Open Regionalism and Trade Liberalization, ISEAS, Singapore, 1996.

11 Krugman, Paul, Import protection as export promotion, in: H. Kierzkowski (ed.), Monopolistic Competition and International Trade, Blackwell, 1989.

12 Robbins, Donald, HOS hits facts: Facts win; evidence on trade and wages in the developing world, Harvard Institute for International Development, Cambridge, MA, 1996.

13 Sachs, Jeffrey and Howard Shatz, Trade and jobs in U.S. manufacturing, Brookings Papers in Economic Activity, 1994.

14 Samuelson, P.A., Factor price equalization, Economic Journal (1948), June.

15 Samuelson, P.A., Factor price equalization once again, Economic Journal (1949), June.

16 Stolper, Wolfgang and P.A. Samuelson, Protection and real wages, Review of Economic Studies (1941).

Trade, Growth, and Development
G. Ranis and L.K. Raut

CHAPTER 6

Exchange Rate Policies for Developing Countries: What Has Changes?[1]

Anne O. Krueger

Center for Research on Economic Development and Policy Reform,
Stanford University, Corner of Galvez & Serra Sts., Stanford, CA 94305

6.1 Introduction

Casual perusal of any economics journal, or even the newspaper, has provided evidence of the lack of consensus with regard to appropriate exchange rate policies for developing countries, at least since the onset of the "Asian crisis" in the summer of 1997.[2] The lack is remarkable in view of the importance of the topic and recent experience, and in light of the earlier — and probably correct — consensus of the 1960s.

One of T.N. Srinivasan's many important contributions to understanding development and development policy focused on the role of trade and exchange rate policy for economic development. In the 1950s and early 1960s, India was following a fixed exchange-rate policy, along with restrictive trade policies to encourage import substitution. Those policies led to severe balance of payments difficulties, quantitative restrictions on imports, and other measures that were inimical to efficient resource allocation and growth. T.N.'s contributions to understanding how and why these policies were undesirable were important, not only at an analytical level, but also empirically for all developing economies that had adopted fixed exchange rates while undergoing domestic inflation rates above world levels.

In this essay, therefore, I ask how our understanding of these issues may be altered by the Asian crisis. I start with the understanding of the 1960s with

[1] I am indebted to Aaron Tornell for helpful comments on an earlier draft of this paper and to Evren Ergin for valuable research assistance

[2] There is also disarray with regard to developed countries' exchange rate policies, as is evidenced by debate over the EMU. There are several aspects of that discussion, however, that differ significantly from those of developing countries, so analysis here is limited to the latter.

regard to exchange-rate policy for developing countries. That is the topic of Section 6.1. In a second section, the evolution of exchange rate policies in the 1980s and 1990s is addressed. A final section then asks how different the crises of the 1990s are from those of earlier decades.

6.2 Policy Prescriptions of the 1960s

During the 1950s and 1960s, of course, the developed countries were almost exclusively on the Bretton Woods fixed-but-adjustable exchange rate system, which increasingly became a dollar standard. The US dollar was the strongest currency in the world, and most prices of internationally traded goods were set in US dollars (see, for example, McKinnon [17]). The American rate of inflation was relatively low, and most indices of dollar prices of internationally traded goods showed them to be stable over the first fifteen years after the Korean war boom. At the outset of the 1950s, the dollar was one of only four fully convertible currencies. Other developed countries were attempting to shift from barter trading arrangements to multilateral clearing (the European Payments Union), and then to current account convertibility and Article VIII status with the International Monetary Fund (IMF). But they all maintained fixed exchange rates, and adjusted them, only reluctantly, in the event of a foreign exchange crisis.

Heavily influenced by thinking stimulated by the Great Depression, most approaches to economic policy followed James Meade in seeing the twin objectives of economic policy as being internal and external balance.

Conventional wisdom with regard to exchange rate policy was, in very broad outline, along the following lines. In the short run, it was of course possible to cover any foreign exchange shortfall by running down foreign exchange reserves. But in the longer run, there were believed to be three broad approaches for assuring external balance. First, policy makers might adopt quantitative restrictions (import licensing) to constrain actual imports to a value consistent with available foreign exchange. Second, policy makers could tighten monetary and fiscal policy[3] to achieve a reduction in the demand for imports (which was regarded as a function predominantly of income with some — perhaps small — responsiveness to price) and an increase in the supply of exports, thus tending to restore external balance. Third, policy makers could change the exchange rate, thus inducing domestic residents to reduce their desired level of imports and making exports more profitable.[4]

[3] At that time, it was generally thought that there was very little private capital mobility for developing countries, so monetary policy was not seen as an instrument for inducing capital flows.

[4] It will be recalled that during the 1950s, there was a debate between the "elasticities" approach, under which imports and exports were regarded as functions of their relative

In the understanding of that time, QRs (quantitative restrictions on imports) were deemed unacceptable — their economic costs were thought to be far too high (and GATT signatories had largely committed themselves to rely on tariffs and eschew QRs, in any event). It was generally thought that monetary and fiscal policy should be used to achieve internal balance, while the exchange rate could be changed or, preferably, be permitted to float (see below) in order to achieve external balance. The classic article was, of course, Friedman [7]. See also Johnson [10].

Clearly, when the current account was in surplus and domestic demand was "too low," monetary and fiscal policy could be relaxed and move the economy toward both internal and external balance. Likewise, with inflationary pressures and a current account deficit, restrictive monetary and fiscal policies could serve to move toward both targets. But there were bound to be times when there was an inadequate level of domestic demand and current account deficits, or excess domestic demand and current account surplus. In these situations, monetary and/or fiscal policy alone could achieve only one target (recall that capital mobility in response to interest rate differentials was thought to be unimportant). The case for flexible exchange rates rested on the proposition that monetary and fiscal policy should (or would) be geared toward internal balance, leaving the exchange rate as the policy instrument for external balance.

Although the developed countries were on an almost-dollar standard with fixed-but-adjustable exchange rates, the majority of economists were convinced of the desirability of flexible exchange rates. Given that there *would* be conflicts between internal and external balance, two alternatives were seen: (1) let foreign exchange reserves dwindle and permit speculators a one-sided bet to determine when the authorities would be forced to alter the nominal exchange rate (when their reserves become inadequate to defend the exchange rate, sacrificing "internal balance") or (2) adopt a floating exchange rate. The goal of full employment was regarded as too important to be sacrificed to the balance of payments, and hence, economists advocated flexible exchange rates. See Friedman [7] and Johnson [10]. Still, policy makers in developed countries opted for fixed nominal exchange rates and for giving speculators the one-sided bet.

In most developing countries, by contrast, policy makers believed they had three objectives: internal balance, external balance, and economic growth. In most cases, developing countries also fixed their nominal exchange rates in those years. However, they were experiencing inflation while world prices were constant, and simultaneously, greatly increasing demand for internationally

---

prices and the "absorption" approach which started from the identity that any current account deficit (surplus) was by definition an excess (shortfall) of expenditures relative to income.

traded goods as efforts were made to raise investment levels, levels which had a relatively high import content. With appreciating real exchange rates, and high protection through tariffs and import restrictions pulling resources into import-competing activities, export earnings failed to grow rapidly (despite the healthy growth of the international economy). Both of these factors contributed to a growing excess demand for foreign exchange at fixed nominal exchange rates.[5]

As with the developed countries, there were three policy choices: (1) the exchange rate could be altered; (2) domestic macroeconomic policy instruments could be adjusted to levels designed to eliminate excess demand for foreign exchange; or (3) quantitative restrictions (QRs) on imports could be adopted. As foreign exchange reserves were diminishing, policy makers in most developing countries rejected the devaluation alternative. Likewise, because of the conventional wisdom of the time, inherited from Keynes, that restrictive monetary and fiscal policy could only result in recession, and because of policy makers' determination that development plans should not be cut back, the second alternative was also ruled out.

That left QRs on imports as the only way[6] to manage external accounts. Since it was, at that time, widely believed that import-substitution would in any event be desirable as the means to achieve industrialization (rationalized by the infant industry argument) and rapid growth, there was little resistance to QRs. (See Krueger [13] for a description of the factors contributing to an emphasis on industrialization through import substitution in the 1950s and 1960s.)

However, throughout the 1950s and 1960s, evidence mounted of the high costs of closed economies, import licensing, and the necessary accompanying exchange controls. Among economists, it was increasingly recognized that those costs could be avoided *only* with a floating exchange rate or an exchange rate regime that kept the real exchange rate at realistic levels for a sustainable current account position.

Many countries continued to maintain a fixed exchange rate system, and hence found their currencies increasingly overvalued with respect to exports and imports. While inflation in developing countries was low in the 1950s and 1960s as contrasted with their inflation rates in the late 1970s and 1980s, it was nonetheless high enough so that the degree of real exchange rate appreciation

[5] In some instances, exchange rates had been fixed at levels that were influenced by high primary commodity prices associated with the Korean war boom. When commodity prices fell, the impact on foreign exchange was similar to that resulting from upward shifts in demand.

[6] Of course, policy makers tried to maximize foreign aid and other resources they could obtain. But once those measures were taken, excess demand for foreign exchange continued to mount, and eventually, quantitative restrictions became unavoidable in light of a reluctance to make the alternative adjustments.

mounted over time.

A "stop-go" pattern of growth[7] gradually became recognizable in many developing countries. As import demand outstripped export supply growth, authorities at first restricted "unnecessary" imports, tightened (and delayed) import licensing, and otherwise attempted to restrict the quantity of imports. Over time, however, the restrictive effect on economic activity caused by the lack of spare parts, raw materials, and capital goods imports mounted. (See Bhagwati [2] and Krueger [11] for analyses of the sequence and causes of foreign exchange crises.) Imports would exceed planned levels as the authorities attempted to avoid the worst consequences of reduced imports. But countries would eventually encounter a severe balance-of-payments crisis. Thus, their policy makers would approach the IMF and agree on a stabilization program under which the exchange rate would be devalued, credit and other ceilings to bring down the rate of inflation would be negotiated, and a Stabilization Program would begin.

The rate of growth of output would slow markedly or, in some cases, be followed by a recession and a drop in output.[8] Government investment plans would be cut back, and private domestic spending would drop. In response to the slowdown in domestic economic activity and to the restrictive effects of the monetary and fiscal contraction associated with devaluation, exports would increase sharply while imports would fall.[9] Hence, the early response to devaluation would be a significant improvement in the trade balance. Policy makers would therefore begin increasing public investment and expenditures, and growth of output would resume or accelerate.

As that happened, imports would increase more rapidly, while increased domestic demand would siphon off export growth. As inflation picked up, appreciation of the real exchange rate would also affect the exports, imports, and the current account balance. Foreign exchange reserves would once again be run down. Countries would attempt to persuade donors to increase foreign aid, and delay issuing foreign exchange to domestic importers. But as the costs of "foreign exchange shortage" mounted, increasing restrictiveness of the import licensing regime would not be able to keep pace with flagging export receipts. In addition, importers would attempt to "speculate" against devaluation, despite capital controls, by trying to delay payments on imports

[7] The term was first applied to developing countries, as far as I know, by Carlos Diaz-Alejandro [5] in relation to Colombia's experience under the fixed-but-adjustable exchange rate system.

[8] In fact, while fiscal and monetary tightening were restrictive, import liberalization was expansive. Not all stabilization episodes were therefore contractionary. See Behrman [1] for an analysis of the relative impacts of these two effects in the Chilean case.

[9] Part of the decline in imports was the result of inventory decumulation as inventories, built up by speculators in anticipation of devaluation, were then sold off. A parallel phenomenon took place with respect to exports, as exporters who had withheld exports, pending devaluation, reduced their inventories. See Diaz-Alejandro [6] for a discussion.

and build inventories, while would exporters delay exporting and try to keep their foreign exchange receipts abroad. Ultimately, another "foreign exchange crisis" would ensue, and the stop-go pattern would repeat itself.

Though there are obvious differences between developing countries, this sequence took place in many of them. Almost invariably, the pressure to go to the IMF became irresistible when countries' foreign exchange obligations exceeded their ability to pay, and restrictions on imports resulting from "foreign exchange shortage" were seen to be sufficiently harmful to the growth-and-development objective. When longer term credits dried up and letters of credit could no longer be issued, some developing countries even resorted to use of high-cost suppliers' credits to finance imports (which were in such sufficiently short supply that it was profitable to do so).

IMF programs usually involved not only the measures mentioned above, but also debt rescheduling and additional capital inflows, usually financed by official creditors. Debt rescheduling took place under the auspices of the unofficial "Paris Club" of official lenders, and the "London Club" for private creditors. In many instances, a major problem was to ascertain the size of the outstanding debt.

Balance of payments crises and stabilization programs with these broad features took place, for example, in Turkey in 1958, 1960 and 1970; in Brazil in 1957, 1961, 1964, 1974 and many times in the 1980s; and in Colombia in 1954, 1957 and 1966. (See Krueger [12] for data on the timing of these episodes in ten countries, and for data on the behavior of the real exchange rate over the cycle.)

In Turkey, for example, a "crisis" in 1958 finally led to an IMF stabilization program after the country's access even to suppliers' credits had been lost and the country — which has no oil of its own — could not even import oil with a harvest approaching. The level of economic activity was in fact falling because of the absence of imports, and inflation was accelerating. The IMF program included an adjustment of the exchange rate from TL 2.80 per US dollar to TL 9.00 per US dollar.[10] Whereas real GDP had been declining at an annual rate in excess of 5 percent prior to the August policy reforms, it began increasing soon afterward, and reached a positive rate of 5 percent for 1959. It should also be noted that debt rescheduling was an important part of the 1958 Turkish stabilization program.[11]

The Turkish exchange rate remained fixed in nominal terms until 1970, while the Turkish rate of inflation was 5–6 percent above the rate of Turkey's

[10] There were a large number of "add-ons" for various categories of transactions, so that most exporters received somewhere between 2.8 and 5 TL per dollar, while importers paid between 2.8 and 6 TL per dollar; so the de facto devaluation, while large, was less than the de jure devaluation. See Krueger [11] for particulars.

[11] In fact, advertisements had to be placed in European newspapers, asking holders of Turkish debt to inform the authorities so that debt rescheduling could proceed.

major trading partners. By the late 1960s, export earnings had again stopped growing, economic growth was again decelerating, and arrears were mounting. A second 1970 devaluation marked the beginning of the next cycle, with attendant debt rescheduling, an altered exchange rate, and tightened monetary and fiscal policy. In each instance in Turkey (as well as in most other countries), the trigger point came when financing for imports deemed essential could no longer be found because creditors were aware of the levels of outstanding obligations and were unwilling to extend further loans.

These stop-go cycles masked the underlying slowdown in growth rates that appears to have been occurring among inner-oriented developing countries, at least until the 1980s. For purposes of understanding exchange rate policy, however, four features are noteworthy. First, the stop-go cycles were themselves seen as the cause of slower growth. Second, real exchange rates were abruptly depreciated, then gradually appreciated until the next stabilization episode. Third, in the types of situations described here, there appears to have been no intention to change the nature of development strategy; rather, the intent was to remove the constraint that was perceived as binding to growth — that is, the foreign exchange constraint.[12] Fourth, although almost no developing country had capital account convertibility, the build-up of debt and arrears, and the increasing restrictiveness of the import licensing regime were the classic hallmarks of fixed nominal and increasingly overvalued real exchange rate regimes.[13]

As the costs of the stop-go cycle and QRs became evident, some developing countries adopted a crawling peg exchange rate, under which the exchange rate was frequently adjusted by small amounts in order to maintain its value in real terms. The crawling peg experience of Colombia was widely applauded as a successful model of exchange rate management, for example (see Diaz-Alejandro [5] for an account of the Colombian experience). Korea, from 1964 until the end of the decade, permitted the exchange rate to float so that the real exchange rate was relatively constant (see Frank, Kim and Westphal [8] for an account).

But the important point is that the very high (and rising over time) cost of QRs implied that any mechanism which prevented their imposition or intensification was viewed as preferable. Hence, economists (and the staff of

[12] The "two-gap" model developed by Chenery and his associates was based on the premise that foreign exchange was a binding constraint for developing countries.

[13] There are a number of definitions of "overvalued" exchange rates. One could mean an exchange rate at which there is no excess demand for foreign exchange given the structure of tariffs (but not quantitative restrictions). Alternatively, one could mean an exchange rate at which there would be no excess demand for foreign exchange when quantitative restrictions were removed. These definitions sufficed in an era when there were almost no private capital flows apart from trade finance. More recently, of course, a definition has to include both a specification as to the height of protection and as to the present value of the current account (presumably equaling zero).

the World Bank and the IMF) advocated exchange rate policies that avoided their imposition. However, the GATT did permit "balance of payments" exceptions for developing countries, which enabled them to impose QRs. As experience with them mounted, the extent to which they were a high-cost alternative became increasingly evident.

At the same time, however, the East Asian "tigers" — Hong Kong, Singapore, South Korea and Taiwan — shifted away from the inner-oriented trade strategies, and began focusing on export-led growth. To achieve this, they rapidly found that a realistic exchange rate was an essential prerequisite to an outer-oriented trade strategy. All four found mechanisms that maintained the real exchange rate at a realistic level; indeed, for a period of time, the South Koreans adjusted the exchange rate in response to changes in the rate of growth of exports.[14]

While the negative experience with QRs did not substantially alter views regarding the desirability of import substitution in developing countries, the success of the East Asian "tigers" began raising questions about the overall desirability of an inner oriented trade strategy. As policy makers became convinced that they could achieve more under a less restrictive trade and payments regime, with its accompanying IS orientation, more and more countries began changing their exchange rates and their exchange rate regimes, in an effort to open their economies.

To state the matter in a way that will be useful in contrasting the 1960s and the 1990s, we see that in the earlier era, developing countries maintained unrealistic nominal (and real) exchange rates for extended periods of time. They were enabled to do so by imposing QRs as a means of maintaining external balance. The costs of the exchange-rate policy were the high costs of QRs. Effective rates of protection were very high; a lack of competition resulted in very low productivity growth, and exports failed to grow rapidly.

One other characteristic of the early import-substitution-exchange-control regimes should be noted for later reference: despite the absence of current- (much less capital-) account convertibility, debt buildups were an almost universal accompaniment of the build-ups to balance of payments crises and IMF stabilization programs.

For although it is true that there was considerably less private capital mobility in the 1950s and 1960s than there has been in subsequent decades, there was normal trade financing, as well as official capital flows. Debt rescheduling, as was seen in the Turkish case, was a normal accompaniment to an IMF

[14]For later purposes, it is highly relevant that the Koreans began borrowing substantially from international banks to finance investment which exceeded domestic savings by 8, 9 and even 10 percent annually. The very high rate of return on capital in the early years meant that this borrowing did not result in an increase in the debt-service ratio. As Korean growth continued, the domestic savings rate rose and borrowing from abroad decreased.

Stabilization Program.[15]

It was often only when no additional credit — official assistance, normal import financing, or even suppliers' credits — was available that measures would be taken to correct the underlying payments imbalance.

6.3 The First Instrument Reassignment

It is difficult and probably unnecessary to judge whether it was the high cost of QR regimes or the experience of the East Asian tigers that led to the shift, by one developing country after another, to a policy regime in which the nominal exchange rate was, in one way or another, adjusted to maintain a degree of constancy of the real exchange rate.

The East Asian NICs, of course, had altered their policies in the 1960s. Korea, for example, at first used and altered export subsidies, tax rebates, and other incentives to maintain exporters' real won receipts per dollar of exports fairly constant while the nominal exchange rate was maintained constant (see Frank, Kim and Westphal [8] for a detailed account). Gradually, these instruments diminished in importance. At times, the nominal exchange rate was allowed to float, while at other times, it was fixed but altered at fairly frequent intervals.[16]

With the oil price increases of the 1970s and the lessons of QR regimes learned, exchange rates were increasingly seen as the policy instrument used to provide incentives for exporting. With the debt crisis of the early 1980s, this trend accelerated.[17] In some instances, QRs were removed and tariffs reduced, in an effort to switch to an outer-oriented trade regime. In other instances, the exchange rate became more realistic, although protection and import substitution policies still prevailed. For example, Turkey in 1980 went to a floating rate regime, eliminating QRs and reducing tariffs significantly over the following five years; by contrast, Ghana devalued the cedi in the early 1980s from 2.75 new cedis per US dollar in 1982 to 30 in 1983 and to 50 in 1984. In the latter case, the black market exchange rate had been about ten times the official rate, and the devaluation considerably reduced excess demand for foreign exchange without fundamentally altering the inner-oriented stance of trade policy.

---

[15] The World Bank [20, p. 23] lists 24 debt reschedulings between 1974 and 1979; these took place under the Paris Club, Aid Consortia, and the London Club.

[16] The won was permitted to appreciate significantly in real terms in the late 1970s, when the Korean authorities were attempting to shift the economy to heavy and chemical industries. Export earnings growth slowed dramatically and in fact, in dollar terms, export earnings were falling in 1979 when the authorities began reversing policies. Thereafter, greater attention was paid to maintaining a realistic exchange rate for exports.

[17] To give one example: after three periods of fixed nominal exchange rates (in the 1950s, after 1958 and 1970), the Turkish authorities switched to a floating rate regime in 1980.

Table 6.1 gives some data on the exchange-rate regimes of developing countries.[18] In 1982, most were still on a fixed exchange rate system, with the occasional devaluations described above. As the 1980s progressed, more and more developing countries switched towards a regime in which their exchange rates were adjusted more frequently or were permitted to float freely. To be sure, frequent adjustment does not imply constancy of the real exchange rate: Mexico, for example, preannounced the rate of devaluation of the peso at a rate below the rate of inflation. As can be seen in Table 6.3, Mexico's real appreciation between 1987 and 1994 was substantial. But the underlying fact remains: many developing countries had realized the disadvantages of an increasingly overvalued real exchange rate between nominal devaluations, and had adopted exchange rate regimes which permitted less abrupt changes in the time path of the real exchange rate.

Table 6.1: Developing countries' exchange rate regimes, various years (number of countries)

	Pegged exchange rates		Limited	More flexible systems	
				Managed	Independently
Year	Single	Composite	flexibility	floating	floating
1982	65 (55)	30 (25)	(a)	23 (20)	
1988	57 (52)	20 (18)	0	22 (20)	11 (10)
1996	38 (27)	18 (13)	3 (2)	37 (26)	46 (32)

Notes: Figures in parentheses are percentage of developing countries in the indicated category.

a. For 1982, categories differed from those of later years. There were at that time still 44 developing countries listed as having multiple exchange rates (in the sense of having more than one exchange rate for imports, exports, or other transactions, or for some combination thereof) in addition to the basic mechanism tabulated above. Also, for 1982, there was no breakdown of types of float.

Source: International Monetary Fund, Exchange Arrangements and Exchange Restrictions, various issues, Table 2, p. 11 (1988), pp. 531-46 (1996), pp. 496–501 (1982).

As the switch in policy instruments took place, officials still adhered to the goal of rapid economic growth. However, in many instances, past policy inefficiencies cumulated and resulted in sizable government expenditures, including deficits of state-owned enterprises (SOEs). Furthermore, the mandate for growth led to efforts to increase government investment in the face of falling real rates of return on investments, especially in countries where import substitution policies persisted.

[18] In the 1950s and early 1960s, there were still several developing countries operating a multiple exchange rate system. Brazil and Thailand may have been the most visible of these. The systems, however, proved administratively chaotic, and by the late 1960s, almost all countries had moved to a single exchange rate, possibly providing a special rate for one or two categories (such as a major primary commodity export or tourism).

6.4 The Third Stage: The Exchange Rate as an Anti-inflation Instrument

Hence, monetary and fiscal policy continued to be targeted towards growth, while exchange rate policy was permitted to operate, at least to some extent, to maintain external balance (or at the very least, to reduce the negative allocative effects of QRs). Although external balances improved in many of these countries, and the reliance on quantitative restrictions was greatly diminished, there was one significant negative side effect. Namely, in many countries, there was a rapid increase in the rate of inflation. Table 6.2 gives data on the average rate of inflation among developing countries for five-year intervals since 1965. As can be seen, the average rate was below 10 percent in the 1960s — well above that in developed countries, but significantly below the rate of later periods. Although the averages are influenced by extremely high rates in a few countries — such as Brazil's rate of 2,700 percent in 1990 and Argentina's 3,430 percent in 1989 — a sharply accelerated rate of inflation was a fact of life in many countries.

Table 6.2: Rates of inflation in developing countries (average annual inflation rate for wholesale prices)

	1965–70	1970–75	1975–80	1980–85	1985–90	1990-95
Inflation rate	9.6	18.3	25.7	41.4	70.4	50.2

Source: IMF, International Financial Statistics, Yearbook 1994, pp. 104–105 (1965–93 and January 1998), pp. 57, 59 (1994-95). Note that data for years prior to 1994 are for wholesale prices; data for 1994 and 1995 are consumer prices. Numbers are simple averages of country rates.

Indeed, whereas the characteristic distinguishing the "typical" developing country from developed countries in the 1960s had been the difference in trade and exchange regimes, differences in price level behavior became perhaps an even bigger difference by the 1980s. Moreover, as inflation rates accelerated, the economic costs of these high rates became more evident to policy makers in developing countries.

Whereas Stabilization Programs in both the first and second phases were motivated in large part by balance of payments difficulties, many of the reforms of the 1980s were motivated by the desire to reduce inflation (see Little et al. [16] for an analysis of the triggering events for "crises"). To name just a few well-known cases: the second stage of Mexico's reforms started in 1987 when the rate of inflation reached 100 percent; in Brazil, the real plan was introduced as an anti-inflationary program (after several earlier unsuccessful starts); and in Argentina, reform was undertaken when the rate of inflation exceeded 3,000 percent.

In these reform efforts, the centrality of the inflation-reducing objective led policy makers to adopt exchange rate policies which would, it was thought, bring down the rate of inflation. These policies, known as nominal anchor exchange rate policies, were designed on the underlying premise that without a nominal anchor, inflation or inflation-induced inertia would persist. The central idea was that if the nominal exchange rate was fixed (or preannounced with anticipated depreciations at rates below the prevailing rate of inflation), the law of one price would become operative and constitute a significant drag on the inflation rate. It was, of course, generally assumed that governments would balance their fiscal accounts and adopt appropriate monetary policies simultaneously with the introduction of the new exchange rate policies. But it was equally assumed that without a nominal anchor exchange rate policy, inflation rates would be reduced significantly more slowly and with a significantly lower cost in terms of foregone output (see Calvo, Reinhart and Vegh [3] and Calvo and Vegh [4] for an analysis of the theory and effects of exchange-rate targeting).

A nominal anchor exchange rate policy in the 1990s operated little differently from a fixed-exchange rate with above-average inflation policy in the 1950s and 1960s. That is, the real exchange rate was appreciating sufficiently enough so that the current account balance incipiently turned less positive or more negative. However, in the 1950s and 1960s, that incipient movement was thwarted by the imposition of QRs; in the late 1980s and 1990s, the incipient movement was financed by private capital flows.

It was recognized that in the 1990s, the mobility of private capital flows to developing countries had increased markedly. At the same time, most people advocating nominal anchor exchange rate policies also recognized that the rate of exchange rate depreciation would be below the rate of inflation during the transition period. Thus, there would be a period during which the real exchange rate would appreciate. Moreover, during that period, capital inflows would be likely to increase in response to the attractiveness of holding the nominal-anchor currency during the period when foreign investors could achieve the real rate of return in domestic currency plus the rate of real currency appreciation (see Krueger [14] for an elaboration of this argument).

6.5 What Is Different in the 1990s?

The theoretical literature regarding exchange-rate targeting has focused on examining the mechanisms by which real currency appreciation affects the behavior of the key macroeconomic variables (see, for example, McKinnon and Pill [18]). But, regardless of the mechanism, the broad outline of the evolution of the key variables under a nominal anchor exchange rate policy and real currency appreciation is much the same as that under a fixed nominal exchange rate with a higher rate of inflation than that in the trading-partner countries. In essence, using the nominal exchange rate as an anchor involves a period during which there is deliberate real appreciation of the currency. Authorities, knowing this, may initially "overdevalue" the currency. And it is recognized that real appreciation of the currency cannot indefinitely continue. "Exit" policy is clearly crucial, and most proponents of nominal anchor exchange rate policies recognize that an otherwise successful policy can ultimately be costly if exit is not achieved in a timely fashion.[19] While it is certainly possible in principle that an initial "overdevaluation" could be followed by a period during which the currency appreciated at a decreasing rate (in real terms) as inflation slowed to the world level, it is difficult to imagine how the parameters of the structure of the economy (especially during a time period in which inflation was greatly slowing down) could be sufficiently well known to policy makers to enable them to set the initial exchange rate correctly. Moreover, if those parameters were known, the authorities could adopt appropriate monetary and fiscal policy in order to reduce the inflation rate to the desired extent, without resort to a period of nominal anchor policies and their attendant costs in terms of shifting resource allocation, etc.

For present purposes, however, the more important consideration is that during the period of real currency appreciation, resources will be pulled out of tradable goods and into home goods, thus resulting in an increase in the size of current account deficits (or a reduction in the size of surpluses). While capital inflows may enable the financing of these deficits for a period of time, the cumulative current account deficits (and their time trend) will, in the absence of a positive external shock, certainly invite a speculative attack on the currency at some point. See Krugman [15] for the classic analysis, and Tornell [19] for an application to protectionist regimes. See also Garber and Svensson [9] for a recent survey of the literature. Fixing an inappropriate nominal exchange rate for a country with an inflation rate above that in its major trading partners is little different from adopting a nominal anchor regime, in which the nominal exchange rate is deliberately adjusted by less than the inflation differential.

A key difference between the 1950s and 1960s and the 1990s lies in the locus of the economic costs of an inappropriate exchange rate policy: in the 1990s, there was excessive accumulation of debt at interest rates well above the real returns achievable in traded goods, and the costs were borne through the excessive indebtedness. Elsewhere, I have calculated that the excess costs of Mexican borrowing to the Mexican economy during the 1987–94 period were on the order of 5 percent of GDP. A low real interest rate domestically

[19] The two most frequently cited cases of successful nominal anchor exchange rate policies are Israel in the period immediately after 1985 and Poland in the first year after the transition began. In both instances, the time period during which the policy continued to be followed was fairly short.

discourages saving and encourages investment (in the nontraded goods sector) while the higher interest rate received by foreigners induces a capital inflow. The result can either be overconsumption (as analyzed by McKinnon and Pill [18]) or investment in low-yielding nontradable goods sectors (such as construction).

We can, therefore, turn to the Asian crisis and ask how these considerations apply. Table 6.3 gives data on real effective exchange rates for ten developing countries from 1985 to 1997. As can be seen, Argentina's policies resulted in sharp real appreciation of the currency from 1985 to 1989, fluctuations between 1989 and 1991, and then real appreciation until 1995, after the introduction of the quasi-currency board. Brazil adopted the real as an anti-inflation device in the early 1990s and has experienced significant real appreciation since that time. Chile, by contrast, has discouraged capital inflows and has experienced real depreciation of the currency. Indonesia, Korea (to a lesser degree), Malaysia, and Thailand all permitted their currencies to appreciate in real terms from the mid 1980s, as did Mexico from 1987 to 1994. By contrast, Turkey (which has had inflation rates between 60 and 100 percent annually) has permitted the nominal exchange rate to depreciate by more than enough to maintain the real exchange rate.

Table 6.3: Real broad effective exchange rate indices, 1985–97 (1990 = 100)

	Argentina	Brazil	Chile	Indonesia	Korea	Malaysia	Mexico	Taiwan	Thailand	Turkey
1985	121.78	71.94	129.13	170.56	105.11	154.43	126.50	100.68	121.28	81.83
1986	104.48	62.97	111.28	132.57	89.37	126.36	90.17	91.71	103.09	84.08
1987	92.58	61.32	105.78	103.69	88.42	188.74	92.92	96.90	96.88	84.28
1988	103.27	66.99	98.64	101.83	96.35	106.06	112.30	100.46	97.40	87.18
1989	87.97	82.73	101.88	102.83	107.78	103.48	107.57	107.03	100.37	95.43
1990	100.01	100.01	100.00	100.01	100.02	100.00	100.00	100.00	99.99	99.99
1991	115.43	80.39	106.10	100.91	97.46	98.79	106.16	97.37	102.33	97.13
1992	113.41	73.08	113.80	99.58	88.37	106.42	107.74	95.96	98.67	89.12
1993	114.94	82.19	113.86	101.53	85.80	109.51	116.58	92.80	100.13	92.59
1994	111.40	94.16	113.93	100.28	84.10	106.30	112.19	91.12	99.44	72.82
1995	108.86	100.19	120.22	98.69	85.60	106.10	78.88	91.54	97.71	75.63
1996	112.76	98.45	126.38	103.22	87.97	110.98	89.77	89.93	105.33	74.23
1997	120.29	104.65	134.80	96.30	83.19	108.26	102.67	91.23	96.59	78.29

Note: Data are averages of monthly figures.
Source: JP Morgan website.

Table 6.4 provides data on the relative magnitude of the current account balance in each of these countries. As can be seen, in most instances, these balances tracked the behavior of the real exchange rate fairly closely: Malaysia, Mexico and Thailand reached a current account deficit of around 8 percent;

Table 6.4: Current account deficit (as a percentage of GDP), 1985–96

	Argentina	Brazil	Chile	Indonesia	Korea	Malaysia	Mexico	Thailand	Turkey
1985	−1.08	−0.13	−8.58	−2.21	−0.80	−1.92	0.44	−3.95	−1.51
1986	−2.58	−1.98	−6.73	−4.90	4.37	−0.37	−1.07	0.57	−1.93
1987	−3.82	−0.49	−3.55	−2.77	7.40	8.15	3.03	−0.73	−0.92
1988	−1.28	1.26	−0.97	−1.57	7.99	5.38	−1.37	−2.68	1.76
1989	−1.70	0.22	−2.50	−1.09	2.42	0.83	−2.80	−3.46	0.88
1990	1.16	−0.80	−1.76	−2.61	−0.69	−2.03	−3.02	−8.50	−1.74
1991	−1.51	−0.38	0.32	−3.32	−2.82	−8.89	−5.12	−7.71	0.17
1992	−3.65	1.62	−1.64	−2.00	−1.28	−3.76	−7.31	−5.66	−0.61
1993	−2.89	0.00	−4.54	−1.33	0.31	−4.47	−6.37	−5.09	−3.57
1994	−3.67	−0.21	−1.24	−1.59	−1.01	−5.86	−7.80	−5.65	2.01
1995	−1.43	−2.64	0.23	−3.55	−1.81	−8.62	−0.26	−8.11	−1.42
1996	−1.41	−3.20	−4.06	−3.30	−4.75	−5.20	−0.60	−7.90	−0.82

Source: World Development Indicators Database.

Note: Malaysia 1995 figure calculated using current account data from IFS, GDP at market prices from World Data. All 1996 figures are calculated using IFS current account and GDP data, or they come from IMF Annual Report 1997. The 1996 figure for Indonesia covers the IMF fiscal year 1995 96. The remaining are those reported in World Development Indicators.

Argentina and Korea reached around 4 percent. Only in Brazil, Chile[20] and Turkey were large current account deficits avoided in the 1990s.

While many additional factors need to be taken into account when analyzing the experience of the East Asian (and other) countries,[21] the evidence is strong that overly appreciated real exchange rates were indeed a common factor in all of the crises. To be sure, the crises might have been postponed (and indeed, several countries did raise interest rates to attempt to maintain their exchange rate regimes) by tightening monetary policy still further and thus inducing further capital inflows. But the economic costs of such high payments on debt to foreigners mount rapidly. Just as in the 1950s and 1960s, the economic costs of exchange controls and quantitative restrictions on imports ultimately mounted to such unacceptable levels that in the 1990s, the height of interest rates that would have been necessary to induce additional capital inflows would have resulted in high economic costs to the borrowers. Once it became evident that the authorities were no longer willing to pay such costs, a "balance of payments" crisis ensued.

[20] In the Chilean case, efforts were made to restrain capital inflows, and, as can be seen, the real exchange rate was permitted to depreciate in response to current account deficits. In the mid-1980s, Chile's current account deficit was very large (8.58 percent of GDP in 1985), which is what prompted the initial real depreciation.

[21] Among the important factors that differentiate the various countries are the impact of the appreciation of the dollar relative to the yen, the declining price of oil (especially for Indonesia), and the authorities' responses to their difficulties.

There are, to be sure, several differences. Most of the Asian countries in crisis in the 1990s are considerably more advanced, and have more responsive economic structures, than did the developing countries experiencing balance of payments crises in the 1950s and 1960s. And there seems to be an increased willingness to learn. In a world of private capital mobility, there is probably little choice for countries unwilling to adopt a currency-board-like system: a freely floating exchange rate that determines the appropriate incentives for tradable and nontradable goods production and consumption is the appropriate policy instrument for achieving external balance. Monetary and fiscal policy must be used to achieve domestic price-level stability targets. If those instruments are appropriate, the resulting macroeconomic balance can be expected to achieve a rate of capital inflows that is sustainable.

It is to be hoped, that in light of the Asian experience, policy makers revert to the conventional wisdom of the 1950s and 1960s: the exchange rate should be used to achieve external balance, in conjunction with monetary and fiscal policies geared toward domestic economic objectives. The economic costs of persisting in maintaining unrealistic nominal exchange rates in the 1950s and 1960s manifested themselves in high-cost import substitution activities and slow growth. In the 1990s, the economic costs of persisting in maintaining an exchange rate policy geared to inflation targets are manifest in excessive and extremely costly, debt.

References

1 Behrman, Jere R., Foreign Trade Regimes and Economic Development: Chile, Columbia University Press for the National Bureau of Economic Research, New York, 1975.

2 Bhagwati, Jagdish N., Foreign Trade Regimes and Economic Development: Anatomy and Consequences of Exchange Control Regimes, Ballinger Press for the National Bureau of Economic Research, Lexington MA, 1978.

3 Calvo, Guillermo, Carmen Reinhart and Carlos Vegh, Targeting the real exchange rate: Theory and evidence, Journal of Development Economics, 47 (1995) 97–133.

4 Calvo, Guillermo and Carlos Vegh, Inflation stabilization and BOP crises in developing countries, in: John Taylor and Michael Woodford (eds.), Handbook of Macroeconomics, North Holland, 1999.

5 Diaz-Alejandro, Carlos, Foreign Trade Regimes and Economic Development: Colombia, Columbia University Press, New York, 1975.

6 Diaz-Alejandro, Carlos, Southern cone stabilization plans, in: William R. Cline and Sidney Weintraub (eds.), Economic Stabilization in De-

veloping Countries, Brookings Institution, Washington, DC, 1981, pp. 119–41.

7 Friedman, Milton, The case for flexible exchange rates, in: Essays in Positive Economics, University of Chicago Press, Chicago, 1953, pp. 153–207.

8 Frank, Charles R., Jr., Kwang Suk Kim and Larry E. Westphal, Foreign Trade Regimes and Economic Development: Korea, Columbia University Press, New York, 1975.

9 Garber, Peter and Lars E.O. Svensson, The operation and collapse of fixed exchange rate systems, in: Gene M. Grossman and Kenneth Rogoff (eds.), Handbook of International Economics, Vol. 3, North-Holland-Elsevier, Amsterdam, 1995, pp. 1865–1911.

10 Johnson, Harry G., The case for flexible exchange rates, Federal Reserve Bank of St. Louis Review, 5l (1969) 12–24.

11 Krueger, Anne O., Foreign Trade Regimes and Economic Development: Turkey, Columbia University Press, New York, 1975.

12 Krueger, Anne O., Foreign Trade Regimes and Economic Development: Liberalization Attempts and Consequences, Ballinger Press, Lexington, MA, 1978.

13 Krueger, Anne O., Trade policy and economic development: How we learn, American Economic Review (1997).

14 Krueger, Anne O., Nominal anchor exchange rate policies, forthcoming in Gary R. Saxonhouse and T.N. Srinivasan (eds.), Development, Duality and the International Economic Regime, University of Michigan Press, Ann Arbor, 1999, pp. 375–97.

15 Krugman, Paul, A model of balance of payments crises, Journal of Money Credit and Banking, 51 (1979) 311–25.

16 Little, I.M.D., Richard N. Cooper, W. Max Corden and Sarath Rajapatirana, Boom, Crisis and Adjustment. The Macroeconomic Experience of Developing Countries, Oxford University Press, Oxford, 1993.

17 McKinnon, Ronald, The rules of the game: international money in historical perspective, Journal of Economic Literature 31 (1993) 1–44.

18 McKinnon, Ronald and Huw Pill, Credible liberalizations and international capital flows: The 'overborrowing syndrome', in: Takatoshi Ito and Anne O. Krueger (eds.), Financial Deregulation and Integration in East Asia, University of Chicago Press, NBER–East Asia Seminar on Economics, Vol. 5, Chicago, 1996, pp. 7–48.

19 Tornell, Aaron Time inconsistency of protectionist programs, Quarterly Journal of Economics (1991).

20 World Bank, World Development Report, Oxford University Press, London, 1983.

Trade, Growth, and Development
G. Ranis and L.K. Raut

CHAPTER 7

Trade Models of Imperfect Competition

Michiel Keyzer[a] and Lia van Wesenbeeck[b]

[a,b]Centre for World Food Studies, Vrije Universiteit, De Boelelaan 1105, 1081 HV Amsterdam, The Netherlands

7.1 Introduction

For many years Heckscher–Ohlin models have dominated the international trade literature. Their ability to capture the complexity of international trade relations through diagrammatic expositions has contributed to their popularity. T.N. Srinivasan has always been a master at drawing complex figures of this kind on any paper and blackboard. He presumably owes this skill to his early experience in writing the Sanskrit character "OM," the sound of silence and sign of reflection, the figure that should speak for itself, but only to the initiated. As shown in Figure 7.1,[1] this symbol contains the major elements of modern trade theory: a bliss point A, an iso-utility curve B and a highly non-convex production possibility set, with boundary C.

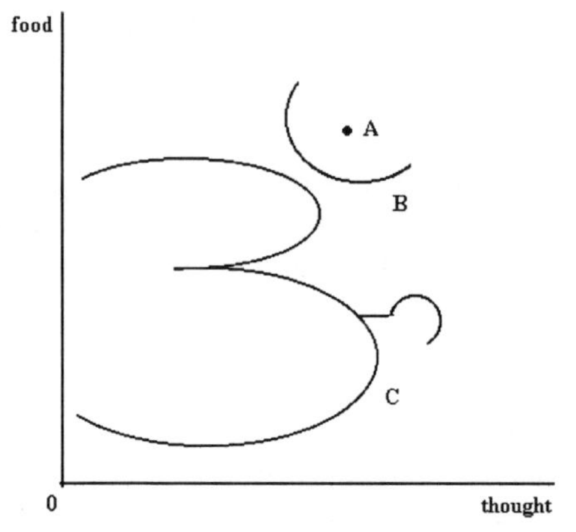

Figure 7.1: The letter "OM"

[1] This is a European interpretation of the production possibility frontier; T.N. Srinivasan would, no doubt, reverse the axes.

The early exercises with trade models mainly sought to explain how specialization patterns, terms of trade, and consumer welfare react to opening to trade, changes in tariffs, and technical change. Typical questions are whether the home country will specialize on the capital intensive good, and whether wages and wage/rental ratios will tend towards equalization between the home and the foreign country, in spite of the immobility of labor and capital. However, in the seventies, the assumptions of constant returns to scale (convex technologies) and perfect competition of the Heckscher–Ohlin model were increasingly exposed as unrealistic, because they made it difficult to account for intra-industry trade and trade between countries with similar relative factor endowments (Tharakan and Kerstens [82] and Greenaway Hine and Milner [41] present empirical evidence to this effect). The model also failed to explain that, in practice, governments are reluctant to apply the free trade principle to themselves, in spite of the welfare gains from liberalization predicted by the Heckscher–Ohlin model. This led to the development of the so-called "new trade" theory, which has sought to incorporate increasing returns to scale and imperfect competition, and more generally, possible strategic behavior of agents. The proper treatment of these aspects within models that deal with trade liberalization is important not only to obtain a more accurate understanding of the welfare gains that are at stake, but also to analyze the hypothesis that trade liberalization can promote concentration. However, it proved difficult to realize this project under the common assumptions of the general equilibrium model that trade theorists had relied upon so far, and simplifying assumptions had to be introduced, especially with respect to the structure of preferences and technology. A significant strand of literature introduced approaches from industrial organization and became partial. In his Comment on Helpman, T.N. Srinivasan [81] criticizes this development, arguing that the earlier theory of optimal tariffs was able to account for many of the phenomena in a more general framework (see De Graaff [38], Johnson [49], Kemp [51]).

This paper builds on this idea. It reviews various models of the new trade theory and indicates how their main concepts can be captured within a numerically tractable, general equilibrium model of optimal tariffs, arguing that this allows us to avoid many of the *ad hoc* assumptions in the new trade literature, while maintaining a welfare theoretic perspective. In this exercise our interest is not to generalize the specification for its own sake, but we believe that the analysis of the effect of trade reforms, in particular the GATT/WTO agreement, calls for models that are firmly rooted in welfare theory and do not take a lenient view on imperfect competition. More specifically, it might happen that a country reduces its tariffs but re-introduces them through the back-door, by hiding them in the mark-up of its oligopolistic firms, while benefitting from the fact that the firms in other countries have smaller market shares and are no longer protected by tariffs. Clearly, such a country can hardly be expected to champion an international competition policy.

However, going back to Harry Johnson's theory of optimal tariffs does not resolve all problems. In fact, this model, like new trade theory, side-steps the difficult issue of modeling imperfect competition within a general equilibrium framework. This motivates the second part of the paper (Sections 3–5). The relationship between international trade and general equilibrium theory has always been very special, since the literature on international trade was applying general equilibrium theory long before applied general equilibrium modelers became active in the footsteps of Johansen [48] and Scarf and Hansen [77]. General equilibrium theorists have always remained somewhat suspicious of the general equilibrium applications in trade theory, wondering how "traditional" trade theorists could manage to neglect intermediate inputs as they arrived at conclusions on the basis of their skillfully drawn diagrams. The situation repeated itself when these trade theorists claimed they were able to compute optimal tariffs in general equilibrium, and subsequently when new trade theorists extended this claim to the whole field of imperfect competition, without worrying much about issues such as the choice of numeraire, non-existence of equilibrium, rationality of anticipations, and communication of information between agents. The paper highlights the assumptions that are commonly invoked to facilitate the task of modeling rational behavior, and proposes an approach that allows us to relax these assumptions, making it possible to represent agents whose anticipations vary in depth and can range from very naive to highly sophisticated.

The paper is structured as follows. In Section 7.2, we specify the classical tariff game within a general equilibrium setting. Section 7.3 describes the oligopolistic equilibrium among producers and discusses reasons for welfare maximizing governments to tolerate such oligopolies. It also reviews the simplifying assumptions in the new trade models, which in our view, limit their usefulness in applied policy analysis of trade reforms and trade agreements. Section 7.4 introduces a general equilibrium model with more flexible assumptions on preferences and technology, multinational firms, and taxes and subsidies on profit remittances, by extending the model of Brander and Spencer [17]. However, this general model has the limitation that it cannot give unambiguous policy guidance beyond the prescriptions of free trade and perfect competition of classical welfare theory, precisely because it is a true general equilibrium model. Hence there is a definite need for numerical simulation exercises. In Section 7.5 we address the difficulty that this requires computing every agent's monopoly power (the elasticity of his objective demand) and propose an easy and flexible technique that computes the elasticity as the dual variable of a mathematical program. This makes it possible to avoid the cumbersome and error-prone calculation of these elasticities via analytical derivatives or numerical approximation, without compromising on the rationality of the strategic agents. Section 7.6 concludes.

7.2 A Tariff Game

7.2.1 Optimal tariffs

Until the emergence of its "new trade" branch, the interest of international trade theory in strategic behavior was limited to national governments setting their tariffs on imports and exports so as to affect international prices in their favor. The early literature referred to these as optimal tariffs, set by a single country treated as a "monopolist," who faced competitive agents in the rest of the world. Soon it was realized that this simplification might be misleading and tariff games were formulated where every government explicitly takes the tariffs of other countries as given to arrive at an "optimal tariff" in a "Nash" sense (see, e.g., Berthélémy and Bourguignon [8]). This game yields results that will serve as the general framework for our discussion, as they carry through under a wide range of assumptions, including increasing returns to scale in production. We will specify a general equilibrium model of international trade in which countries set tariffs on their net imports. We will assume that every country has perfect insight in the market clearing process and takes only the tariffs of other countries as given, while anticipating all effects on consumers and producers. For the time being, we disregard all restrictions on tariffs that would result from GATT/WTO type of agreements.[2]

We thus consider a world economy with n countries $(i = 1, ..., n)$, a single consumer per country, and r commodities indexed $k \in K = \{k|k = 1, ..., r\}$. In every country the consumer maximizes a utility function $u_i(x_i)$ that satisfies standard assumptions.[3] This consumer behaves non-strategically and maximizes his utility, subject to a budget constraint, taking the prices p_i within country i as given. He receives an income H_i, obtained from (i) sale of endowments ω_i at a price p_i (where the endowment ω_{ik} is taken to be positive for at least one commodity) and (ii) transfers from the national government, T_i. The consumer therefore solves the usual utility maximization problem:

$$\max_{x_i \geq 0}\{u_i(x_i)|p_i x_i \leq H_i, \text{ given } H_i = p_i\omega_i + T_i\} \tag{7.1}$$

Every country (government) imposes a tariff rate τ_i on net imports z_i, whose proceeds are returned to the consumer as a transfer T_i. Due to this, domestic

[2] In the application currently under construction, we explicitly take into account possible GATT/WTO restrictions.

[3] **Assumption U** (utility functions). *The utility function* $u_i : R^r_+ \to R$, $u_i(x_i)$ *is* (i) *concave,* (ii) *continuously differentiable on the positive orthant, and* (iii) *increasing in commodities* $k \in K_i \subset K$ *and stationary in other commodities.*

prices relate to international prices p^w according to:

$$p_{ik} = p_k^w(1 + \tau_{ik}), \text{ for all } k \tag{7.2}$$

We start with the case of pure exchange. The net imports z_i of the country are then equal to consumption minus endowments. Every country sets its tariffs so as to maximize consumer utility, subject to a balance of payments constraint $P_i^w(z_i, \tau_{-i})z_i = 0$, where $P_i^w(\cdot)$ denotes the price correspondence for country i, which here, for convenience, is assumed to be a differentiable function of country i's net imports and of the given tariffs τ_{-i} (this requirement will be relaxed in the sequel). Hence optimal net imports z_i follow from the program:

$$\max_{z_i}\{U_i(z_i)|P_i^w(z_i, \tau_{-i})z_i = 0\} \tag{7.3a}$$

where $U_i(z_i)$ is the trade welfare function, defined as:

$$U_i(z_i) = \max_{x_i \geq 0}\{u_i(x_i)|x_i - \omega_i = z_i\}$$

Here all goods are treated as internationally tradable, but it is also possible to deal with non-tradable goods by dropping them from the commodity list of the vector z_i, while keeping them in the domestic commodity list. In the optimum, the first-order conditions of this program include:

$$\partial U_i/\partial z_{ik} = \lambda_i P_{ik}^w(1 + \tau_{ik}) \text{ for } \tau_{ik} = (\partial P_i^w/\partial z_{ik})z_i/P_{ik}^w \tag{7.3b}$$

where λ_i denotes the consumer's marginal utility of income and the variable τ_{ik} is the optimal tariff rate. We can interpret this tariff as the flexibility of the net import schedule that the country faces and view (7.3b) as a representation similar to the Amoroso–Robinson condition [75]. Together with the tariff setting rule (7.3b), program (7.3a), defined for every country i, specifies the tariff game as a Nash equilibrium with every country setting its optimal tariff taking those of other countries as given. Equation (7.3b) shows that this game has the following important property (see, e.g., Bhagwati and Srinivasan [10, Ch. 17]) which we formulate as a proposition, for later reference:

Proposition 1. *The optimal tariff τ_{ik} is approximately zero for all commodities k in which the country lacks monopoly power, i.e., has partial $\partial P_i^w/\partial z_{ik}z_i \cong 0$.*

In particular, this proposition suggests that for small developing countries, import substitution policies tend to be welfare deteriorating, since they specifically apply to commodities for which the country lacks monopoly power.

Next, let us check whether this result holds in an economy with production, and increasing returns in particular. For this we introduce an index set J for producers, with subscript j. Let J_i denote the subset for producers of country i. Producer (firm) j supplies a net output y_j. The firm has a technology that can be represented[4] through a transformation function $F_j(y_j, \delta_j)$. In this transformation function, the binary variable δ_j switches production on and off and gives the firm a possibility of inaction (zero inputs, zero outputs). When the switch is on, a firm might incur losses due to setup costs ($f_j(0) < 0$) and lacking demand.[5]

We notice that rather than treating the setup cost as a technological datum, we can assume it to be chosen from a continuum. For example, we can assume a trade-off to exist between setup costs and variable costs. If the market that can be captured is large, high setup costs will be preferred. (The strategic use of alternative technologies is discussed in more detail in Gershenberg [36], Davies et al. [23] and Huizinga [47].) Let $f_j(v_{1j}, s_j)$ now denote the variable cost part of the production function, where v_{1j} are the variable inputs (with $f_j(0, s_j) = 0$), s_j is the reduction factor in setup costs, and $v_{2j} = a_0 s_j$ the setup inputs. The cost minimization becomes:

$$\begin{aligned} C_j(p_j, q_j) \;=\; & \min p_j(v_{1j} + v_{2j}) \\ & \quad v_{1j}, v_{2j} \geq 0, s_j \geq 0 \\ & \text{subject to} \\ & \quad f_j(v_{1j}, s_j) \geq q_j \\ & \quad v_{2j} = a_0 s_j \\ & \quad s_j \leq 1 \end{aligned} \tag{7.4}$$

For $s_j = 0$, setup costs will vanish and the firm has the possibility of inaction. This makes every firm's technology convex, and since in the optimum every active firm will be profitable, the switch δ_j can be dispensed of when convenient. In the sequel we will interchangeably use transformation functions with and without δ_j, and when we drop this switch, we implicitly assume setup costs to be chosen endogenously, as in (7.4).

Since we have assumed, for the time being, that every country is the sole owner of the firms located within its borders, we can now define the trade

---

[4] In accordance with the practice of the international trade literature, we restrict attention to a single-output firm with setup costs. The transformation function is specified as follows:

Assumption F (Transformation function with setup costs). *For every producer j, the transformation function $F_j : \mathbb{R}^r \times \{0,1\} \to \mathbb{R}$, $F_j((y_{j1}, ..., y_{jr}), \delta_j) = \max_k\{q_j - f_j(v_{j1}, ..., v_{jr})\delta_j | v_{jk} \geq 0,\ q_{jk} \geq 0,\ b_{jk}q_j - v_{jk} = y_{jk}\}$, where b_{jk} is unity for the commodity k that is produced by firm j and zero otherwise; and the production function $f_j(v_j)$ is* (i) *concave, continuously differentiable and* (ii) *uses directly or indirectly a commodity that is not produced in the economy.*

[5] Associated with the production function $f_j(v_j)$, there is a cost function $C_j(p_i, q_j) = \min_{v_j \geq 0}(p_i v_j | f_j(v_j) \geq q_j)$. The setup costs are $C_j(p_i, 0)$ and the associated demand for inputs can be obtained as the derivatives with respect to prices.

welfare function for the model with production as:

$$U_i(z_i) = \max_{x_i \geq 0, y_j \text{ all } j \in J_i, \delta_j} \left\{ u_i(x_i) | F_i(y_j, \delta_j) \leq 0, z_i = x_i - \omega_i - \sum_{j \in J_i} y_i, \delta_j = 0, 1 \right\}$$

Taking this function to be differentiable and assuming the price function $P_i^w(z_i, \tau_{-i})$ to have its earlier properties preserved, we obtain program (7.3a) with tariff setting rule (7.3b) as before, and Proposition 1 holds. We also note that for given optimal tariffs τ_i, the Nash equilibrium of this model reduces to an ordinary general equilibrium of an international trade model with fixed tariff rates (Shoven and Whalley [78]). Moreover, we can as before decentralize the welfare optimum as follows: firms without setup costs can freely determine their net supply y_j so as to maximize their profits, while taking prices as given. Other firms will have to apply with the government for a license, that will be granted on the basis of the welfare optimum (which sets δ_j to 0 or 1). Some of these firms might incur a loss, in which case government will provide a transfer payment linked to δ_j being set to unity.[6] The transfer is financed from direct taxes: hence the consumer will receive the balance of net profits and transfers. All licensed firms act as independent profit maximizers, taking prices as given. There is a strict ban on oligopolies. The following proposition summarizes this as:

Proposition 2. *Under the Assumptions* U *and* F *and if every country faces the price function* $P_i^w(z_i, \tau_{-i})$ *and takes the tariffs of the other countries as given, an optimal policy will impose tariffs at the border, but will keep domestic markets perfectly competitive, with the qualification that it will pay lump sum transfers to cover the possible losses of producers who should be active according to the government plan.*

Clearly, this tariff game has the essential limitation that although each country acts with perfect foresight on all markets, the outcome could be most unfavorable, as every country could end up entrenched behind its tariff walls. Indeed, Proposition 2 has straightforward corollaries in terms of cooperation between countries. If countries cooperate by coordinating their tariff policies and are willing to provide compensating payments to the participants who are worse off, they cannot lose with respect to utility levels U_{i0} achieved in the absence of such cooperation. Cooperation will mean that they maximize

[6] For example, if the cost function is linear in q_j (constant returns to scale in variable inputs) and written as $C_j(p_i, q_j) = c_j(p_i)q_j + C_{0j}(p_i)$, profit maximization will imply that the price of the output becomes equal to $c_j(p_i)$ and that the firm will need full compensation for its setup costs. All licensed firms will make independent profit maximizing decisions.

joint utility of coalition c according to some positive welfare weights α_i:

$$\begin{aligned}
&\max \sum_{i \in I_c} \alpha_i U_i(z_i) \\
&\quad z_i,\ \text{all } i \in I_c, z_c \\
&\text{subject to} \\
&\quad P_c^w(z_c, \tau_{-c}) z_c = 0 \\
&\quad z_c = \sum_{i \in I_c} z_i \\
&\quad U_i(z_i) \geq U_{i0}
\end{aligned} \tag{7.5}$$

where $P_c^w(z_c, \tau_{-c})$ is the price function faced by the coalition. Obviously, the potential gains are due to the improved coordination in the tariff setting. Moreover, since the model of the coalition (say, a regional grouping of countries) has precisely the same structure as country model (7.3a), Proposition 2 holds for the coalition: free trade conditions will prevail internally, including zero tax on profit outflows, and eventually, if all countries join, the world economy will as a Grand Coalition have free trade as well. We do not discuss the problem of agents free riding on an existing cartel. It may be argued, however, that this problem is more severe in the case of a private cartel between firms than in the situation of a coalition between countries. A nation that stays out of, say, a customs union loses its open access to the consumers of the union, whereas a private producer only frees himself of the obligation to restrict net supply. Another point to be emphasized here is that we did not take into account any set-up costs in the services provided by governments. The economies of scale which could be realized here are an additional argument in favor of coalition formation by countries.

7.2.2 Ownership of firms abroad

The possibility of owning firms abroad makes it necessary to represent profit remittances and to address the issue of profit taxes. Allowing for the consumer of country i to own shares in firms located in other countries will cause profit remittances to appear on the balance of payments. We choose to represent this by attributing all profits to a residual (scalar) production factor e_j that is equal to unity. More specifically, we rewrite the profit maximizing decision of the competitive firm j as:

$$\begin{aligned}
&\max p_j q_j - C_j(p_j, q_j/e_j) e_j \\
&\quad q_j \geq 0, e_j \\
&\text{subject to} \\
&\quad e_j = 1 \qquad\qquad (\pi_j)
\end{aligned} \tag{7.6}$$

where π_j denotes the shadow price of the constraint. Since this objective function has constant returns to scale in (q_j, e_j), by Euler's rule, the value of the objective will be equal to this shadow price, even if the cost function is not convex in q_j. Hence we can treat the ownership of firms abroad as the ownership of an "endowment" e_{ij}, that adds up to unity, carrying a "price" π_j, and treat these factors as part of the commodity list. Proposition 2 will now hold in terms of this extended list, which implies that countries will have the power to tax the profits. Since the ownership of shares in firms is assumed to be fixed, countries cannot affect the net demand, but they can influence the profit itself. Notice that the "world price" for this endowment is equal to the profit earned domestically, minus the "import tariff," i.e., the tax on profit outflow. Hence every country is a monopsonist with respect to the capacity use of the firms located within its border and can in principle tax all profits away, which amounts to expropriating all foreigners.

To interpret such an outcome, let us move to a dynamic setting, as in Lucas and Stokey [61], where every consumer is an infinitely lived dynasty and all commodities carry a time subscript. The profit will then measure the discounted returns evaluated at time $t = 0$ and be equal to the market value of the shares, while the commodity endowments will refer to initial capital stocks, and to a stream of labor supplied at $t = 0, 1, \dots$, until infinity. The expropriation would amount to a redistribution of initial stocks. The country would not be penalized for this action, since its firms would always be able to borrow to finance future investments. In short, in this model with no constraints on foreign borrowing, no externalities associated to foreign investments, no uncertainty, and no fear of retaliation by foreign governments, the common objections against nationalization do not apply. The only motive for a country to abstain from it would be that when foreign countries have no stake in the profitability of the home firm, they will have less incentive to promote the profitability of this firm by favorable pricing.

7.2.3 Replacing the price function by explicit constraints

The theory of optimal tariffs usually assumes the price function $P_i^w(z_i, \tau_{-i})$ in model (7.3a) to exist and to be known. As pointed out by Marschak and Selten [63], among others, the concept of a price function is problematic within a general equilibrium context, because the strategic agents (and the modeler) can only have knowledge of constraints relating to excess demand and budgetary equilibrium, and only under highly restrictive assumptions will these constraints generate a differentiable price function. If the non-negativity restrictions on excess demand define multiple equilibria, they define a price correspondence rather than a function. Furthermore, even if there exists such a price function, it will not be given in explicit form, nor will it be possible to characterize its derivatives in any detail. Consequently, models that contain this price function cannot produce any general conclusion as to the magnitude or even the sign of the optimal tariff. In a full general equilibrium setting, we must replace the price function by explicit excess demand constraints, and we have to rely on numerical exercises where we can compute optimal tariffs for specific case studies.

We will now formulate these constraints explicitly. This will also provide us with the setup required in Section 7.3 for the computation of demand elasticities.. By Proposition 2, we can at given tariffs represent the economy of other countries through the tariff ridden Walrasian net import correspondence:

$$z_i(p_i, H_i) = x_i(p_i, H_i) - \omega_i - \sum_{j \in J_i} y_j(p_i)$$

where $x_i(p_i, H_i)$ and $y_j(p_i)$ are the correspondences that result from utility and profit maximization.[7] The net demand vectors are taken to include demand and supply for fixed factors. Defining p_i^w as the international price expected by country i, the excess demand constraint is written as:[8]

$$Z_i(p_i^w, \tau_{-i}, H_{-i}) + z_i = 0$$

where $Z_i(p_i^w, \tau_{-i}, H_{-i}) = \Sigma_{h \neq i} z_h(p_i^w + \tau_h, H_h)$. ($z_i$ is the strategic part of excess demand, here the full net imports.) We also specify the balance of payment constraint of these countries as:

$$B_i(p_i^w, \tau_{-i}, H_{-i}) = 0$$

with elements $B_{ih}(p_i^w, \tau_{-i}, H_{-i}) = p_{ik}^w z_h([p_{ik}^w(1 + \tau_{hk})], H_h)$ for every country $h \neq i$ (the square brackets are used to denote a vector). Finally, we must introduce a price normalization. As pointed out by Gabszewicz and Vial [35], under imperfect competition, "price normalization matters." This is essentially because (i) under imperfect competition any strategic profit maximizing producer suffers from money illusion and (ii) normalizing on a subset of prices may restrict the instrument space available to the agent who would seek to manipulate this price. As yet, our model does not contain strategic producers and we can avoid restrictions on the instrument space by enabling the strategic agent to anticipate effects on all prices. For convenience we will normalize anticipated prices on the simplex: $||p_i^w|| = 1$.

We can now write the model for the strategic government of country i as maximizing national welfare subject to a world excess demand constraint and

[7] Net imports may not be single valued, since in order to cover the cases with constant returns, we did not impose strict quasi-concavity on utility function, and production could be set valued since we did not require strict concavity of the production function. Differences (and sums) of correspondences refer to the differences (and sums) of all pairs, for example: $Z(p) = X(p) - Y(p) = \{z | z = x - y, x \in X(p), y \in Y(p)\}$.

[8] In this function τ_{-i} is a column vector of length $(m-1)r$ consisting of the vectors $\tau_1, ..., \tau_{i-1}, \tau_{i+1}, ..., \tau_m$ of country specific tariffs for all $h \neq i$. The vector τ_j denotes the tariff in the country where firm j is located.

country specific balance of payments constraints, with every country taking the tariffs imposed by other countries as given:[9]

$$\begin{aligned}
&\max U_i(z_i) \\
&\quad p_i^w \geq 0, z_i, H_{-i} \\
&\text{subject to} \\
&\quad p_i^w z_i = 0 && (\lambda_i) \\
&\quad Z_i(p_i^w, \tau_{-i}, H_{-i}) + z_i = 0 && (\nu_i) \\
&\quad B_i(p_i^w, \tau_{-i}, H_{-i}) = 0 && (\zeta_i) \\
&\quad ||p_i^w|| = 1 && (\xi_i)
\end{aligned} \tag{7.7a}$$

with

$$\tau_{ik} = (\partial U_i^w / \partial z_{ik})/(\lambda_i p_{ik}^w) - 1 \tag{7.7b}$$

where p_i^w, z_i and the dual variable λ_i are evaluated in the optimum. A Nash equilibrium of programs (7.7a) and (7.7b) will define the solution of a classical tariff game within a general equilibrium setting. At given tariffs, the excess demand constraint and the balance of payment equations jointly determine a market clearing price p_i^w and a vector of incomes. Under relatively weak assumptions, the solution will be locally unique after price normalization (Dierker [25]), in which case all prices can be made to coincide in the Nash equilibrium: $p_i^w = p^w$ for all i.

7.3 Imperfect Competition between Firms

7.3.1 Reasons for tolerating monopolies

The upshot of the discussion so far is that the setup costs emphasized in the new trade theory do not alter the basic precepts of the theory of optimal tariffs, that they highlight the incentives for cooperation between countries and do not provide a basis for tolerating monopolies. However, there are various reasons for concluding differently.

First, countries will be inclined to tolerate oligopolies that purely operate "offshore," i.e., buy and sell goods that hardly affect the country's trade prices. (As in Brander and Spencer [17] and Eaton and Grossman [31]. Echeverri-Carroll [32], among others, studies the positive effect the use of increasingly advanced technology in parent firms has on starting offshore production facilities in developing countries.)

Second, decentralization of the cooperative agreement in program (7.5) will as a rule require income transfers among consumers (member countries)

---

[9] We neglect the nonnegativity constraint on tariff ridden prices and take net import to be infinite whenever p_{ik} is negative.

to compensate the losers. Apart from the fact that it might prove difficult in practice to mobilize such funds on a lasting basis, there is the theoretical problem that individuals are supposed to take the transfers as given. This implies that they will have an incentive to form a cartel that adjusts its supply in order to affect prices in its favor, obviously without having to resort to tariffs. In fact, such an incentive towards rent-seeking behavior (Bhagwati and Srinivasan [10]) was already present before the conclusion of a tariff agreement, but at that stage a national welfare maximizing government would still have a clear interest in fighting such oligopolies. Once this government has given up its power of conducting an independent trade policy, it has to weigh the gains from oligopolistic rents earned abroad against consumer losses due to oligopolistic pricing on the domestic market. This may induce such a government to accept oligopolies so as to allow its earlier tariffs to be hidden within the oligopolistic firms owned by its people. The same argument applies outside economic unions, when countries reduce tariffs under GATT/WTO agreements,[10] with the important difference that at the international level there is no recognized authority to monitor competition practices so as to protect vulnerable parties. This is precisely why the issue deserves attention.

Third, the tariff game itself is a form of imperfect competition between governments that may leave consumers trapped at a heavily distorted equilibrium. In this case, consumers in different countries might find it profitable to cooperate via the joint ownership of a multinational firm. Such a firm will often use both quantity and price signals to manage its internal operations. Consequently, it does not have to price competitively the many specialized intermediate products it moves across various borders. In fact, the outside market exists only for few of these products, such as for specialized managerial knowledge. (The importance of such specific assets is emphasized in, among others, Buckley and Casson [18, 19], Porter [72], Purkayastha [74], and Markusen [62].) Therefore, the multinational has virtually every liberty to price its internal deliveries so as to minimize its outlays on tariffs and profit

[10]The issue is by no means an academic one: Sleuwaegen and Yamawaki [79] study the effects of the formation of the European Common Market on seller concentration; Canals [20] concentrates on the recent trend of strategic internationalization of European banks; Bishop and Kay [13] raise issues for industrial policy makers and business strategists related to the sharp increase in mergers across borders between European-based firms; and Linda [59] uses the European Commission's verdict in the Nestle–Perrier cartel case (July 22, 1992) to illustrate the increased concentration of industry and the importance of anti-cartel policies at a European level. The influence of multinational enterprises in an international context is studied by Green [39, 40] and Newberry [70], who concentrate on the importance of preventing market power of privatized electricity companies in Wales and England. Olkinger and Fernandez-Cornejo [71] focus on the impact of pesticide product regulations on the number of firms and multinational presence in the US; UNTCT, 1988 provides a review of the extent of multinational activity in biotechnology; and Lofgren [60] stresses the increasing bargaining power of multinational enterprises, which dominated the pharmaceutical industry in Australia vis-à-vis the Australian government.

taxes. (Chan and Chow [21] give an empirical assessment of tax auditing to detect transfer pricing in China; Becker [7], among others, studies the fiscal and legal aspects of transfer pricing.) This policy of "transfer pricing" often leaves national governments with no other choice but to offer tax exemptions for such intermediate deliveries.[11]

Finally, uncertainty plays an important role in research and development intensive sectors. This makes it virtually impossible for a central agent to select all R&D-projects eligible for public funding. At the same time, the marginal costs of production of such high-tech products often show so little increase with scale that perfect competition will never yield a price that can cover the setup costs. This is the type of argument that is usually given in support of patents, trademarks, and other private licensing regulations (as in Brander [15], Curtis [22], Krugman [53], Anis and Ross [4], and Motta and Thisse [67]), and also more specifically of multinationals who can take the risks of developing R&D intensive new products through their large markets and diversified asset portfolios (see e.g., Buckley and Casson [18], Alam and Langrish [3], and Markusen [62] for empirical support of this argument and UNTCT [84] for case studies on biotechnology and new energy technologies).

For optimal government policy, accepting monopolies means that in modeling terms, rather than treating the technology of firms as a datum, the behavioral response of the oligopolists becomes part of the constraint set. Hence, if all oligopolists are represented, the government will anticipate the resulting oligopolistic equilibrium. We will now describe such an equilibrium.

7.3.2 Cournot and Chamberlinian competition; cartels

The theory of imperfect competition was originally developed in a partial equilibrium setting and neglects the conflict that might exist between shareholders of a firm, or between different countries that own one firm. It treats producers as independent agents who maximize profits without taking the interests of the shareholders of the firm into account. The theory distinguishes various kinds of competition (e.g., Cournot, Chamberlin and Stackelberg) which refer to a particular structure of anticipations and choice variables of the participating agents. Cournot competition has been used most widely and considers a producer who takes the net supply y_{-j} of his competitors as given and anticipates the depressing effect on the price that would result from an expansion in his own production, which leads him to restrict his net supply. In a general equilibrium setting, Chamberlin competition is a special case where the competitors do not supply perfect substitutes (therefore, each producer has a monopoly). Under Stackelberg competition there is the asymmetry that

[11] In modeling terms this means that the goods are represented as internal to the firm and do not enter the commodity list of net imports.

one producer (the leader) has the ability of anticipating all actions y_{-j} of the other producers (the followers). Hence, in a model where strategic producers anticipate the behavior of non-strategic producers, the three types of competition coexist. With a transformation function satisfying Assumption F, the profit maximization problem of strategic producer j is:

$$\begin{aligned} &\max_{y_j} P_j(y_j, y_{-j}, \tau) y_j \\ &\text{subject to} \\ &\quad F_j(y_j) \leq 0 \qquad (\rho_j) \end{aligned} \tag{7.8a}$$

where ρ_j is the associated dual variable and $P_j(\cdot)$ denotes the price function for producer j, which is assumed to be a differentiable function of producer j's net supply, the net supply of competing producers, and the tariffs τ set by governments. Under Cournot competition, there will be a single output and perfect substitutability between the outputs of the various oligopolists; Chamberlinian competition treats these outputs as imperfect substitutes. Hence both cases are covered by this notation for the price function

$$P_j - \mu_j = \rho_j \partial F_j(y_j)/\partial y_j \tag{7.8b}$$

for

$$\mu_{jk} = -(1/P_{jk})\partial P_j/\partial y_{jk} y_j$$

where μ_{jk}, usually a positive number, is the optimal mark-up rate, which can be interpreted as the negative of the flexibility of the net demand curve perceived by the producer. In the tariff game we made a distinction between the monopolistic behavior of the utility maximizing government and the price-taking behavior of the utility maximizing consumer. Here, we can apply a similar separation by letting the "manager" of the firm pursue competitive profit maximizing behavior taking the mark-up ridden prices $P_j - \mu_j$ as given, where μ_j is the mark-up set by the "master program." The resulting profits are to be distributed to consumers: this calls for a modification of the valuation of the fixed factor in (7.6) to account for the monopolistic rent. Therefore, we include an additional fixed factor m_j, associated to market power (for example, patents):

$$\begin{aligned} &\max_{\delta_j \in \{0,1\}, e_j, m_j, y_j} P_j(y_j/m_j, y_{-j}, \tau) y_j \\ &\text{subject to} \\ &\quad F_j(y_j/e_j, \delta_j) e_j \leq 0 \qquad (\rho_j) \\ &\quad e_j = 1 \qquad (\psi_j) \\ &\quad m_j = 1 \qquad (\varphi_j) \end{aligned} \tag{7.9}$$

where similar to (7.6), the scalar e_j is the fixed factor associated to non-homogeneity of the production function. Profits π_j will be equal to the sum of the associated shadow prices, ψ_j plus ϕ_j. This formulation allows us to proceed as before, treating the fixed factors as part of the commodity vector, and to distribute profits as revenue from the sale of endowments.

Together with the mark-up setting rule (7.8b), program (7.8a), defined for every strategic producer j, specifies the tariff game as a Nash equilibrium with every producer setting his optimal net supply while taking the net supply of others as well as the government tariffs as given. Equation (7.8b) shows that this game has a property equivalent to Proposition 1: the optimal mark-up μ_{jk} is approximately zero for all commodities k in which the firm lacks monopoly power, i.e., has $\partial P_j / \partial y_{jk} \cong 0$. We also note that for fixed tariffs τ and mark-ups μ, the Nash equilibrium reduces to an ordinary general equilibrium of an international trade model with fixed tariffs and mark-ups, as was often used in applications (Harris [44], de Melo and Tarr [64], Francois and Shiells [34]).

Notice that the mark-up is perfectly equivalent to a producer tax (or subsidy) that is redistributed to the owners of the firm. This reiterates the point that protectionism can have various incarnations. For example, the European Union in 1992 started a process of replacing export subsidies on cereals with producer subsidies and supply restrictions that maintain income protection in an indirect way. For sugar, an "official" producer cartel has been in existence for decades, that achieves the same objectives without government subsidies (see, e.g., Folmer et al. [33]).

Models (7.8) and (7.9) relate to a single firm. To represent a multinational, we need a model of cooperation among various firms. For this, we define a set C of cartels, indexed c, which maximize the joint profits of their members and distribute these according to some given sharing rule.[12] Let the subset J_c of J denote the firms j that belong to cartel c, and subset J_{c_i} denote the strategic firms located in country i. Every firm is located in a single country. All firms not fully owned by the home consumer are strategic (belong to some cartel). Therefore, a cartel could be a multinational operating in several countries.[13] We will now reformulate program (7.8a) in the case of excess demand and balance of payments constraints. Cartel c solves:

[12] In the case of "small players," the share in profits can be determined according to the marginal contribution. In the case of large players, the Shapley value of cooperative game theory provides a possible sharing rule (see Myerson [68]) but many other rules can be envisaged.

[13] Notice that since this is a general equilibrium model, both horizontal and vertical integration can be represented as cartel formation of firms.

$$\begin{aligned}
&\max_{\substack{p_c^w \geq 0, y_j \text{ all } j \in J_c, H}} \sum_{j \in J_c} p_c^w y_j \\
&\text{subject to} \\
&\quad F_j(y_j) \leq 0 \qquad (\rho_j) \\
&\quad Z_c(p_c^w, \tau, H, y_{-c}) - \sum_{j \in J_c} y_j = 0 \quad (\nu_c) \\
&\quad B_c(p_c^w, \tau, H, y_c, y_{-c}) = 0 \qquad (\zeta_c) \\
&\quad ||p_c^w|| = 1 \qquad (\xi_c)
\end{aligned} \tag{7.10a}$$

where symbols in brackets denote Lagrange multipliers (that are supposed to exist) and $y_c = \{y_j\}_{j \in J_c}$ is the vector of net supplies of the firms belonging to the cartel, y_{-c} is the vector of net supplies of firms not belonging to the cartel, and $p_{jk} = p_{ik}^w(1 + \tau_{jk})$ for $j \in J_i$ is the price applicable to firm j. The excess demand and the balance of payment function for the cartel are defined by:

$$Z_c(p_c^w, \tau, H, y_{-c}) = \Sigma_i \hat{z}_i([p_{ck}^w(1 + \tau_{ik})], H_i) - \sum_{j \in J_{-c}} y_j$$

$$B_{ci}(p_c^w, \tau, H, y_c, y_{-c}) = p_c^w(\hat{z}_i([p_{ck}^w(1 + \tau_{ik})], H_i) - \sum_{j \in J_i} y_j$$

and $\hat{z}_i(\circ)$ is the net demand correspondence, excluding strategic production, but including the profit maximizing net supply by non-strategic firms. It must be emphasized that the choice of numeraire matters in this case. The mark-up rate can be computed from the optimal solution as:

$$\mu_{jk} = 1 - (\rho_j \partial F_j / \partial y_{jk}) / p_{ck}^w \tag{7.10b}$$

When solved for every cartel c, equations (7.10a) and (7.10b) jointly define a Nash equilibrium among Cournot oligopolists.

7.3.3 Simplifying assumptions and modifications

The models of the new trade theory usually impose requirements that are far more specific and thus more restrictive than our Assumptions U and F on preferences and technology. We will now briefly review these requirements, arranged under the following five conditions:

Condition 1 (Utility). Utility functions are of the form $u_i(x_{1i}, x_{2i}) = u_{1i}(x_{1i}) + x_{2i}\iota_2$, and H_i exceeds $p_1 x_{1i}$ for x_{1i} solving $\partial u_{1i}(x_{1i})/\partial x_{1i} = p_1$, at relevant values of income and prices.

Condition 2 (Inputs). Strategic producers only use numeraire goods as inputs. These goods are not produced by any strategic producer and are used in production with constant marginal productivity.

Condition 3 (Ownership). The home consumer is the sole owner of the strategic firms located in the home country.

Condition 4 (Entry–exit). No firm switches off in a discrete fashion.

Condition 5 (Non-rival inputs). Whenever present, non-rival inputs go unpriced.

Utility. Condition 1 makes it possible to neglect income effects in consumer demand. In addition, the price of the numeraire good will be equal to unity (we can dispense of the price normalization constraints) and the excess demand for non-numeraire goods will not depend on income. For numeraire goods, however, it will still depend on income and, since welfare maximization is the government objective, the balance of payments constraint of the home country should be maintained to compute the residual income available for consumption of the numeraire goods. It must be emphasized that this condition is non-constructive since it is an *ex post* requirement that depends on equilibrium tariffs and prices.

Dixit and Stiglitz [28] do not make this assumption. Their paper started a line of research that seeks to explain intra-industry trade between countries by arguing that consumers like a variety of products. The preference for a variety is expressed via an additive utility function $u_{i1}(x_i) = \sum_k v(x_{ik})$ where k denotes the variety; $v(x_{ik})$ is concave and increasing in every k. Every variety is produced with the same setup costs C_0 and variable costs c and satisfies Condition 2. The concavity of the utility function reflects the preference for a large number of varieties, but the number of varieties every consumer faces is given. The Dixit–Stiglitz model is generally considered one of the standard models of monopolistic competition. Yet, Yang and Heijdra [87] and Heijdra and Yang [45] criticize two underlying assumptions: (1) each producer may ignore the cross-price elasticity of demand for a variety of a good and (2) the influence of an individual price change on the overall price level can be ignored because the number of varieties is large. This is questionable since this number is to be determined endogenously and may therefore, in principle, be small. Heijdra and Yang even show that in the case that the number of goods produced in equilibrium is large, the influence of individual prices on the overall price level may not be trivial either. They also object to the constant elasticity of substitution between varieties that is implicit in the model and thus independent of the number of goods available.

Notice that there is an alternative formulation which does not suffer from these deficiencies. We assume that the utility function $v(x_{ik})$ is not only concave and increasing in every k but also nonnegative and having a bounded slope at the origin.[14] Hence, as the utility is the same across varieties, the individual consumer actually has the utility function $u_i(x_i; n) = v(x_i/n)n$ where n is the number of varieties, that is, it is treated as a real-valued number rather than as an integer (Dixit and Stiglitz [28] use the form: $U = [\Sigma_k D_k^{\alpha}]^{1/\alpha}$, $0 < \alpha < 1$). This also makes it possible to nest the utility function for varieties of one commodity within a general utility function over several commodities. It is even possible to define a perfectly general utility and merely label some of the commodities as "number of varieties of product k." As in Dixit–Stiglitz, the number of varieties can be determined via a zero-profit condition which stipulates that:

$$(p - c)\Sigma_i x_i(p; n) = nC_0 \tag{7.11}$$

Since there is only one producer for each variety, it is natural to use this construction in the context of Chamberlinian monopolistic competition, where every producer of a variety acts as a profit maximizer imposing a mark-up rate $\mu_j \cong -v''x/v'$, which depends on p and n.[15] This model should not have any good x_2 entering the utility function through an additive term and consumed in positive quantity, because this makes it impossible to solve for n in (7.11).[16,17] The model will generate intra-sectoral international trade because every country is a monopolist for its own varieties, and since the number n of varieties essentially is a good, a rising income will increase the number of varieties.

[14] In the papers surveyed, no explicit reference is made to the nonnegativity requirement. This requirement is essential. If it is not met, a larger number of varieties might lead to a fall in utility, even if the utility function is increasing. The assumption of a bounded slope keeps the problem well defined: varieties that are not supplied should not be allowed to have an infinite marginal utility, for otherwise equilibrium might not exist.

[15] This neglects the effect on marginal utility of income; the motivation given in the papers which use the Dixit–Stiglitz approach is that the producer of one variety can hardly affect this value, since he cannot coordinate his actions with the producers of other varieties. Yet due to this approximation, the model belongs to the class of models with subjective demand (Negishi [69]), as opposed to those with objective demand considered so far.

[16] For example, Krugman [52] and Dixit [27] use this type of model to show that there are gains from trade, even in the case of similar or identical countries. By contrast, Driffil and van der Ploeg [29] show that international coordination of trade unions undermines the beneficial effects of trade liberalization.

[17] Preference for variety has also been represented via Lancaster's [57, 58] characteristics approach, but there are only few applications in the new trade literature. Lancaster [58] studies the effects of tariffs and protection on product differentiation, Helpman and Razin [46] allow for foreign direct investment and argue that these might harm a recipient country, and Eaton and Kierzkowski [31] show that trade may reduce product variety through the elimination of producers serving small markets with idiosyncratic tastes.

Inputs. If Conditions 1 and 2 are imposed simultaneously, strategic producers need not be concerned with the market supply and demand of numeraire goods (e.g., with their monopsonistic influence on the labor market). By Condition 1, the price of these goods will be unity and by Condition 2 the terms of trade with respect to the numeraire cannot be influenced via an adjustment of supply or demand for numeraire goods. This enables strategic producers to disregard balance of payments constraints and excess demands for numeraire goods. The price normalization constraint could already be dispensed of by virtue of Condition 1.

Ownership. Condition 3 avoids the intricacies of profit taxes but it makes it more difficult to justify the acceptance of oligopolistic behavior.

Entry–exit. Condition 4 eliminates the discontinuities in net supply that arise when a firm is switched off, but we have seen that the variable setup cost formulation enables us to circumvent this problem. The entry–exit conditions lead us to a wide range of models in the international trade literature, for example in relation to the determination of the number of varieties. Potential entrants and leavers often appear as non-strategic producers with whom the strategic producer plays a leader–follower game. This actually means that the supply correspondences of the entrants appear as part of the non-strategic net demand. Several authors deal with the situation where active producers, because of asymmetries in information, are able to mislead potential entrants through wrong signals about entry costs. Milgrom and Roberts [66] is a well-known reference on this subject, for a more recent treatment, see, for example, Salonen [76].

If one assumes the entrant's information is accurate, there are basically two options in representing entry and exit. One is to treat entrants as infinitesimally small firms with, say, a cost function $C_j(p_i, q_j)$. Aggregate output will then be denoted by $Q_j = q_j n_j$ and $C_j(p_i/n_j)n_j$ will be the aggregate cost function, where n_j is the (real-valued) number of firms. This creates a constant returns technology, with n_j as a free input. Therefore, profits will be zero in equilibrium for the entrants. If the inputs for entry exclusively involve competitively priced goods (numeraire goods), this establishes an upper bound on the price the monopolist can charge.

The alternative approach is to consider a finite number of potential entrants as part of the set of producers j, all being present from the outset but only becoming active when prices are favorable. The single-output firm with decreasing returns, for example, will only be active if the output price exceeds the slope of the cost function $C_j(p_i, q_j)$ at $q_j = 0$. Since in this setting all entrants are in fact ordinary producers with entry costs appearing as part of their technology, it is also possible to treat some of the entrants as strategic producers.

The issue of entry has played an important role in the new trade and industrial organization literature on innovation.[18] The successful innovator enjoys a temporary monopoly that is gradually eroded by the entry of competitors who come in once their investments have come to maturity and the protection from patents has faded away. The follower benefits from lower costs of R&D but enjoys less monopoly power, which explains the falling trend in the life cycles of many high-tech products (a classic reference on the product life cycle is Vernon [85]; an empirical appraisal of product life-cycles in high-tech industries is given in Agarwal [1]). The Brander and Spencer type of general equilibrium model, in its dynamic interpretation, can represent this process if one treats first-generation products as research intensive inputs that generate specialized knowledge as a by-product that can be used by the firms that produce second generation commodities. Since this second generation use of knowledge will, in many cases, be possible without licenses and patents, it has to be treated as a non-rival input that can benefit many producers simultaneously.

Trade models (Grossman and Helpman [42, 43]) often follow endogenous growth theory in treating knowledge as a free non-rival good. This is commonly referred to as a spillover effect. However, such an unpriced use causes inefficiencies. The general equilibrium model can represent this by "forcing" the second-generation producer to use this input, but it can also treat the case where the non-rival input is priced, competitively or not. See for example Verspagen [86] on the empirical relevance of spill-over effects.

7.4 A Tariff Game with Governments Anticipating the Imperfect Competition among Firms

7.4.1 Formulating the game

Once governments treat the monopolistic behavior of firms as a reality, and are for some reason committed to abstention from any price–quantity regulation on the domestic market, all they can do is anticipate the interactions among producers and the response to changes in tariffs. This is the situation dealt with in the celebrated article of Brander and Spencer [17], who analyze a situation without setup costs and show that even in this case, it may be optimal for the government to impose an import tariff because this raises the profit earned abroad by the home monopolist. They define a two-country

[18] For empirical studies on entry behaviour and factors influencing the entry decision, see e.g., Agarwal [2], who considers the survival of firms and finds that early entrants enjoy a higher survival rate than entrants in later stages; and Porter [73], who provides a bibliography of recent empirical research in industrial organization.

model where a two-level game is played: two strategic producers operate at the lower level, while at the upper level the respective national governments impose a tariff on trade (not on profits). The strategic producers compete in a Cournot fashion: they supply on both markets and determine their output level so as to maximize profits, taking as given the tariffs in both countries. Both governments maximize national welfare taking the tariff of the other country as given, but anticipating the effect of their own tariff on the supply of the strategic producers and on the equilibrium price on the market for the associated good. Hence, they play a Nash tariff game among each other and a Stackelberg leader–follower game with the producers. The paper marked the beginning of a large body of literature on managed trade which shows that a tariff may be welfare improving for the country that imposes it because it can "bring home" part of the monopolistic rent. Conversely, the model makes it clear that a country can experience a situation where it benefits from monopolistic practices by its own producers because they capture rents abroad, even if their technology has non-increasing returns. Surprisingly, despite the wide attention the paper has received, the model has so far not been generalized into a full-fledged general equilibrium model. Our generalization will:

(i) allow for an arbitrary number of commodities and (strategic and non-strategic) agents,
(ii) give up the simplifying assumptions summarized under Conditions 1–5.

Several vintages of applied general equilibrium models of imperfect competition are available which also relax these assumptions. In the early versions (Harris [44], de Melo and Tarr [64]), oligopolists assume fixed price elasticity of demand, which leads to fixed mark-up rates, or specify elasticities according to some known function; the more recent versions (Mercenier and Schmitt [65]) represent more rational anticipations by deriving their elasticity functions from optimizing behavior of rational agents. The disadvantage of the elasticity approach is that it lacks transparency. For every application the modeler has to engage in new painstaking effort to obtain explicit forms of elasticity functions, and there is no clear way to detect errors in the process. Building on Ginsburgh and Keyzer [37, Ch. 11], we propose an alternative route that obtains mark-up rates as dual variables rather than through user-specified functions.

For this purpose, we will modify the tariff game (7.7) so as to incorporate the anticipation of the outcome of the Cournot competition (7.10) among strategic producers. First, however, we redefine the excess demand function in (7.7) to isolate the net supply by strategic producers (non-strategic producers remain, as before, part of the net demand function $z_h(\circ)$)

$$Z_i(p_i^w, \tau_{-i}, H_{-i}, y_{-i}) = \sum_{h \neq i} (z_h([p_{ik}^w(1+\tau_{ik})], H_h) - \sum_{j \in J_h} y_j \qquad (7.12a)$$

$$B_{ih}(p_i^w, \tau_{-i}, H_{-i}, y_{-i}) = p_i^w\left(z_h([p_{ik}^w(1+\tau_{hk})], H_h) - \sum_{j\in J_h} y_j\right) \tag{7.12b}$$

Next, we determine the strategic net supply y_j within the tariff setting model of the government. For this we include the first-order conditions of producer problem (7.10). To avoid cumbersome notation we denote the first-order conditions of program (7.10a) together with condition (7.10b) by:

$$K_c(p_c^w, z_i, \tau_{-i}, H_{-i}, y_c, y_{-c}, \rho_c, \nu_c, \zeta_c, \xi_c) = 0$$

where ρ_c, ν_c, ζ_c and ξ_c are the dual variables of (7.10a). The program for the government of country i that plays the tariff game reads as:

$$\begin{aligned}
&\max U_i(z_i)\\
&\quad p_i^w \geq 0,\ z_i, H_{-i}, y_j \text{ all } j \in J_c, \rho_c, \nu_c, \zeta_c, \xi_c \text{ all } c\\
&\text{subject to}\\
&\quad p_i^w\left(z_i - \sum_{j\in J_i} y_j\right) = 0 \qquad (\lambda_i)\\
&\quad Z_i(p_i^w, \tau_{-i}, H_{-i}, y_{-i}) + z_i - \sum_{j\in J_i} y_j = 0\\
&\quad B_i(p_i^w, \tau_{-i}, H_{-i}, y_{-i}) = 0\\
&\quad ||p_i^w|| = 1\\
&\quad K_c(p_i^w, z_i, \tau_{-i}, H_{-i}, y_c, y_{-c}, \rho_c, \nu_c, \zeta_c, \xi_c) = 0
\end{aligned} \tag{7.13a}$$

with in the optimum

$$\tau_{ik} = (\partial U_i / \partial z_{ik})/(\lambda_i p_{ik}^w) - 1 \tag{7.13b}$$

In the Nash-equilibrium of the model the following conditions should hold:

(i) Every strategic government solves (7.13a) and (7.13b),

(ii) Price anticipations are consistent: $p_i^w = p^w$ for all i and c.

7.4.2 Properties of the game

Though the resulting program looks intractable, we will, as mentioned earlier, propose a scheme of decomposition that makes it easy to handle numerically. First, however, we will compare the welfare properties of program (7.13) with those of the tariff game (7.3). For this we treat the net supply y_j of strategic producers as "foreign" and redefine the trade welfare function in terms of net demand z_i, that is, as consumption minus endowments (and if relevant, minus net supply by non-strategic producers). If we take both the price $P_i^w(z_i, \tau_{-i})$ and the strategic supply $y_j(z_i, \tau_{-i})$ to be differentiable functions, the tariff game (7.13) can be written in compact form as:

$$\max_{z_i} \left\{ U_i(z_i) | P_i^w(z_i, \tau_{-i}) \left(z_i - \sum_{j \in J_i} y_j(z_i, \tau_{-i}) \right) \right\} = 0 \tag{7.14a}$$

with first-order conditions

$$\partial U_i / \partial z_{ik} = \lambda_i (P_{ik}^w (1 + \tau_{ik})), \text{ for}$$

$$\tau_{ik} = (1/P_{ik}^w) \left(\partial P_i^w / \partial z_{ik} \left(z_i - \sum_{j \in J_i} y_j(z_i, \tau_{-i}) \right) - P_i^w \sum_{j \in J_i} \partial y_j(z_i, \tau_{-i}) / \partial z_{ik} \right) \tag{7.14b}$$

This expression shows that Proposition 1 no longer holds: the country will in general impose tariffs even on commodities for which it has no monopoly power, because it will seek to influence the net supply of the home oligopolists. Nonetheless, it will have free trade domestically without consumer or producer taxes, as in Proposition 2, but here the abstention from the use of domestic taxes s_i on producers is purely due to the fact that this instrument was not made available. If such an instrument were at hand, the oligopolist's net supply would be $y_j(z_i, \sigma_i, \tau)$ and the government could in general reach higher welfare through it. This illustrates that in this model, tolerating monopoly and refraining from counteracting policies on the home market implies a "voluntary" loss of control, which seems in sharp contrast to the assumed Stackelberg structure of the game. If acceptance of monopoly is based on anticipated retaliation by foreign shareholders, or on the home government's lack of knowledge about the reactions of the oligopolist, it would seem more natural to take, say, the mark-up rate as given and formulate a model where both governments play a Nash game with other firms and other governments.

Furthermore, this generalized Brander–Spencer model has the limitation that it treats the composition of the coalition as fixed, as in all of non-cooperative games, and it does not impose GATT/WTO induced restrictions on tariffs. We view it as a framework for simulation exercises that makes it possible to compute the payoffs of alternative coalition structures under specified restrictions on tariffs, so as to "test" the conjecture that trade liberalization leads to concentration. Hence there is a need to devise a numerical solution scheme for this game.[19]

[19] Illustrative numerical examples in GAMS can be accessed via the Internet (Keyzer [52]). These relate to models with either tariffs or imperfect competition but do not combine both. A more elaborate application is currently under development as part of the thesis-research of the second author.

7.5 Solving the Game

7.5.1 Problems of multiplicity

We will argue in this section that multiplicity of representations is a major weakness of the models of imperfect competition. By this we mean that the theoretical and empirical information available to the model builder leave much room for choice at the stage of actual model specification and that this makes it necessary to test numerically the sensitivity with respect to various equally plausible assumptions. We will list various reasons for this problem, and propose a solution technique for our generalized version of the Brander and Spencer model. The difficulties are of two types. The first relates to choice of specification, i.e., to formulating in an unambiguous manner the theoretical model that is to be applied, and the second relates to choosing a particular stationary or equilibrium point among various possible candidates.

7.5.2 Choice of specification: Navigating between partial and general equilibrium

In the competitive setting, where consumers and producers take prices as given, the distinction between partial and general equilibrium models is relatively clear: both give a full account of supply and demand by all agents on a market, but partial models relate to a subset of goods while general equilibrium models in principle cover the whole set. In practice, it obviously is impossible to deal with all goods, since several markets, say, of environmental goods or future goods, do not exist, but one usually is satisfied with the list of goods covered by national account statistics. Furthermore, under perfect competition the choice of numeraire does not matter, but as soon as one leaves the competitive setting, there are many more ways in which the model can be made partial and numeraires can be chosen.

First, price anticipations can be generated in various ways. The distinction introduced by Marschak and Selten between the planning (or anticipation) and implementation (or realization) stage of the non cooperative game may help clarify this aspect (see Marschak and Selten [63] and d'Aspremont, Dos Santos and Gérard-Varet [6]). In model (7.13), the planning stage deals with the setting of tariffs and mark-ups, and the implementation stage deals with the clearing of markets for given levels of tariffs and mark-ups. This market clearing equilibrium may be partial or general, but even if it is general, the planning stage model could be partial. In (7.3), (7.5), (7.8), (7.10), and (7.13) we have consistently assumed that agents took account of all information and therefore worked within a general equilibrium setting. A partial approach would assume that strategic agents restrict attention to a subset of the excess demand constraints, disregarding balance of payment restrictions, and take prices on other markets and consumer incomes (or other variables such as consumption levels or marginal utilities of income) as given in their planning stage or disregard these altogether. Alternatively, strategic agents might base their decisions on subjective expectations or some simple calculations on the flexibility of demand curves. While the full general equilibrium approach is unambiguously superior for normative analysis, a more partial approach might be able to describe a real world economy more accurately.

Second, the choice of tariff rates as strategic variables of the Nash equilibrium is to a large extent arbitrary. One could also work with fixed levels or with quantity controls, and even differentiate the use of these instruments between governments and firms in various countries, though this could cause certain practical problems with respect to the existence of equilibrium at the implementation stage. Every such choice would lead to a different outcome.

Third , the game structure (e.g., Stackelberg or Nash) to be imposed in a particular case is far from obvious.

Fourth, we reiterate that under oligopolistic competition among profit maximizing producers, price normalization matters (see Gabszewicz and Vial [35], Dierker and Grodal [26] and Böhm [13]). In this respect, the partial setting offers the advantage that it naturally suggests treating the rest of the economy as numeraire; Conditions 1 and 2 of Section 7.3.2 define numeraire goods in a similar manner.

Finally, as noted by Bhagwati, Brecher and Hatta [11], and in line with the earlier Debreu–Mantel–Sonnenschein theorem (see, e.g., Debreu [24]), even the Heckscher–Ohlin model with perfect competition lacks robustness and allows, as the number of agents is increased, the generation of virtually any effect. This holds *a fortiori* for general equilibrium models of imperfect competition.

7.5.3 Multiplicity of stationary points and equilibria

When it comes to actual computation of a solution, multiplicity problems arise in connection to (i) the solution of the optimization problems of strategic agents and (ii) the Nash equilibrium between these agents.

In the case of optimization problems, the difficulty is intimately associated with indivisibility and can be dealt with in various ways. One is to treat the producers as infinitesimally small units, as in the preference for variety model. The variable setup costs make it possible to avoid switches due to the non-convexity in the technology by smoothing the supply response. In the presence of a monopoly, there is the option of avoiding non-convexity of the profit function by compensating non-convexity of the cost function with a high degree of concavity of demand (d'Aspremont, Dos Santos and Gérard-Varet [6]). These modifications eliminate the discontinuity of the supply response. In other cases, as we did to incorporate the producer in (7.13), one can replace the optimization problem by its first-order conditions, an approach that was pioneered by Bonisseau and Cornet [14] in connection with increasing returns under perfect competition. The satisfaction of these conditions is then treated as part of the equilibrium problem. However, replacing a non-convex optimization problem by its first-order conditions creates an additional problem of multiplicity, since these conditions are necessary but not sufficient, even for local optimality. Hence, there is a need for some selection of the appropriate stationary point, or at least for checking numerically whether the stationary point chosen actually is an optimum.

As regards the Nash equilibrium itself, the multiplicity issue is twofold. Common to all equilibrium problems is obviously the multiplicity of the solution itself, but specifically in relation to imperfect competition there is the consistency issue: different agents may "look" at a different equilibrium when they make their decisions. The proposed scheme will only address the second problem.

7.5.4 Decomposition into a fixed rate equilibrium and tariff and mark-up setting programs

We will not compute an equilibrium by solving program (7.13) directly but by decomposing the problem into a pair of sub-problems whose combined first-order conditions coincide in equilibrium with those of this program. These two problems will be: (i) a tariff and mark-up ridden general equilibrium model and (ii) a series of mathematical programs for every strategic government that will be used to compute new values for tariffs and mark-ups, based on the equilibrium solution of (i), until a fixed point is reached.

Fixed rate equilibrium The fixed rate (tariff and mark-up ridden) equilibrium is characterized by:

(i) consumers following (7.1)
(ii) producers maximizing mark-up ridden profits subject to their transformation constraint (the fixed factor construct will ensure that all profits accrue to factors),
(iii) the price relation (7.2),
(iv) the balance of payments restriction $p^w(z_i - \Sigma_{j\in J_i} y_j) = 0$
(v) the world market equilibrium condition: $\Sigma_i(z_i - \Sigma_{j\in J_i} y_j) = 0$
(vi) a price normalization $||p_i^w|| = 1$.

To verify that such an equilibrium exists, note that at prices p_i that are positive for the commodities in which some consumer is nonsatiated, every consumer will have a positive income from endowments. By nonsatiation of the utility function, any level of T_i such that H_i is positive will determine a unique value px_i, and this value is increasing in T_i. Under Assumption F with variable setup costs as in (7.4), the supply function of mark-up ridden producers will be upper-semicontinuous and convex-valued and the profit will be a continuous function of p_i (accruing to the fixed factor). Hence, at given positive prices p_i, there will be a unique value T_i which clears the balance of payments and $T_i(p_i)$ will be a continuous function. If some price p_{ik} drops to zero, then by Assumption U(iii), this will at a given income level either not affect consumption (and no discontinuity will be created) or bring world excess demand to infinity. Therefore, net imports depend on prices only and the aggregate excess demand goes to infinity whenever the price of a consumer good goes to zero. This enables us to invoke the excess demand theorem (see e.g., Arrow and Hahn [5]).

Tariff setting. To set the tariff and mark-up rates at the anticipation stage, we notice that in (7.14b) the tariff level $\gamma_{ik} = p_k^w \tau_{ik}$ is the marginal effect on $p_i^w(z_i - \Sigma_{j \in J_i} y_j)$ of a small change in z_{ik}. Hence it can be computed as the Lagrange multiplier $(\tilde{\gamma}_i)$ associated with a fixed value of z_{ik} in:

$$\max_{z_i, p_i^w \geq 0, H_{-i}, y_j \text{ all } j, \rho_c, \nu_c, \zeta_c, \xi_c \text{ all } c} p_i^w \left(\hat{z}_i^\circ - \sum_{j \in J_i} y_j \right)$$

subject to

$$Z_i(p_i^w, \tau_{-i}, H_{-i}, y_{-i}) + z_i - \sum_{j \in J_i} y_j = 0 \tag{7.15a}$$

$$B_i(p_i^w, \tau_{-i}, H_{-i}, y_{-i}) = 0$$

$$K_c(p_i^w, z_i, \tau_{-i}, H_{-i}, y_c, y_{-c}, \rho_c, \nu_c, \zeta_c, \xi_c) = 0$$

$$p_i^w \iota = 0 \qquad (\tilde{\rho}_i)$$

$$z_i = z_i^\circ \qquad (\tilde{\gamma}_i)$$

where the symbols in brackets denote the associated Lagrange multipliers and $z_i^\circ = \hat{z}_i^\circ$ is a value obtained from the tariff and mark-up ridden equilibrium.[20] This program has the most natural interpretation of a device which computes the marginal increase in national net expenditure that would result from an increase in net imports, given all the model constraints, except the balance of payments. The new tariff rate will be:

$$\hat{\tau}_{ik} = \tilde{\gamma}_{ik} / p_{ik}^w \tag{7.15b}$$

The critical observation is that though program (7.15a) is non-convex, solving it will usually be a trivial matter. More specifically, if the oligopolistic

[20] We distinguish between $\hat{z}_1^\circ$ and z_i° to indicate that we are interested in computing the effect of a change in the constraint while keeping the parameter in the objective fixed.

equilibrium is locally unique, there will only be one price satisfying the excess demand constraint at given levels of strategic net imports z_i°, $i = 1, ..., m$, taken from the tariff and mark-up-ridden equilibrium. In this case the mark-up equilibrium price p° will be both feasible and optimal in (7.15) and this means that this program can be initialized at its optimum and that it only serves to compute the dual variable. This dual variable is a mere marginal value which can always be computed at the optimum, for feasible z_i°. As mentioned earlier, the program can be viewed as a device to calculate the flexibility of the net import function that the agent faces. We claim it to be a major simplification, as compared to the current practice in applied modelling of analytically computing these flexibilities (see, e.g., Mercenier and Schmitt [65]).

Mark-up setting. We can proceed in a similar way to compute the producer mark-up. For this, we determine the marginal contribution to profits of a change in net supply, while taking all the constraints of the producer model (7.10a) into consideration but keeping net supply itself fixed:

$$\begin{aligned}
&\max_{\substack{p_c^w \geq 0,\ y_j \text{ all } j \in J_c, H}} \sum_{j \in J_c} p_c^w \hat{y}_j^\circ \\
&\text{subject to} \\
&\quad F_j(y_j) \leq 0 \\
&\quad Z_c(p_c^w, \tau, H, y_{-c}) - \sum_{j \in J_c} y_j = 0 \\
&\quad B_c(p_c^w, \tau, H, y_c, y_{-c}) = 0 \\
&\quad ||p_c^w|| = 1 \\
&\quad y_c = y_j^\circ \qquad\qquad (\chi_j)
\end{aligned} \tag{7.16a}$$

where $\hat{y}_j^\circ = y_j^\circ$ is the value obtained in the fixed rate equilibrium. The mark-up rate should satisfy:

$$\hat{\mu}_{jk} = \chi_{jk}/p_k \tag{7.16b}$$

Nash equilibrium. In the fixed point of this process, the equalities $\tau_{ik} = \hat{\tau}_{ik}$ and $\mu_{jk} = \hat{\mu}_{jk}$ will hold. The decomposition into a fixed rate equilibrium and tariff and mark-up generating programs is only possible if the Nash equilibrium exists and we are unable to prove this, even though we may add that the equilibrium usually can be "made to exist" by appropriate calibration of model parameters. In fact, the decomposition scheme naturally lends itself to computational purposes. Starting from an initial level of tariff and mark-up rates, one calculates an equilibrium, and subsequently uses the resulting values to compute new targets and so on, until hopefully, convergence is reached.

Convergence is uncertain both because the equilibrium may not vary continuously under changes in the targets and because the dual variables may not vary continuously under changes in strategic net imports and production. Yet the scheme is of some interest even if it fails to converge, since it will always describe a sequence of equilibria (thus meeting commodity balances, budget constraints, and the behavior rules for price taking consumers) which may be interpreted as a dynamic adjustment process that describes how agents "learn" to adjust their optimal policy. Such a dynamic approach may be motivated by arguing that strategic agents can only explore the consequences of changes in their actions (here z_i) for small variations around the prevailing equilibrium. Moreover, since the programs to compute the tariffs and mark-ups merely repeat equations of the model itself, the computer programming is straightforward and coding errors can be avoided, errors which are bound to occur if elasticities have to be computed analytically by taking first derivatives of the model.

We also notice that it is easy to develop and implement variants of this scheme. Indeed, hyper-rational agents can, within the same model, coexist with naive predictors. For example, more partial and naive planning models, which consist of dropping some equations and treating as fixed some variables originating from the equilibrium model, lead to obvious changes in (7.15a) and (7.15b). This will, of course, not affect the optimal prices, imports, and production of the fixed rate equilibrium model, but it will change the dual variables and hence the tariffs and mark-ups. Change of normalization is another modification that can be implemented without effort. For example, if a single commodity, say, commodity n, is to be used as numeraire,[21] one can simply replace the normalization constraint by the restriction $p_n = 1$.

7.5.5 Checking for local optimality and for relevance of the numeraire

Once a fixed point has been reached, we can test the local optimality of the optimal solution for every country and strategic producer by solving (7.10) and (7.13) at the equilibrium. Standard optimization packages such as GAMS/MINOS recognize whether a given stationary point is a local optimum. In case these conditions are not met, it is possible to eliminate a closed neighboring set around these points from the set of acceptable tariffs and start again. More formal procedures to deal with multiplicity of equilibria are discussed in Kehoe [50], but in practice, it often proves more effective to restrict the choice sets if this type of problem arises. The relevance of the numeraire is also readily checked. If the numeraire does not matter, it will carry a zero shadow price in the original programs [but not in (7.15a) and (7.16a)].

[21] Such a normalization is possible if some consumer with positive income is known to be nonsatiated with respect to this commodity.

7.6 Conclusion

T.N. Srinivasan [82] has forcefully argued that the "new" theory never was so new. By treating imperfect competition as a fact of life, this theory basically accepts the imperfections that lead to it as purely technological characteristics rather than as distortions that should be eliminated. New trade theorists do appreciate this limitation but they defend their approach on grounds of realism. For example, Brander and Spencer [16] argue that in the presence of increasing returns in their domestic firms, national governments have an incentive to act as oligopolists in their setting of tariffs. Krugman [55, p. 253] clearly acknowledges the danger in treating free and strategic trade even-handedly:

> Strategic trade policy is, without doubt, a clever insight. From the beginning, however, it has been clear that the attention received by that insight has been driven by forces beyond the idea's intellectual importance. The simple fact is that there is a huge external market for challenges to the orthodoxy of free trade. Any intellectually respectable case for interventionist trade policies . . . will quickly find support for the wrong reasons. At the same time, the profession of international economics has a well developed immune system designed precisely to cope with these outside pressures. This immune system takes the form of an immediate intensely critical scrutiny of any idea that seems to favor protectionism.

This paper finds support for T.N. Srinivasan's view that the theory of optimal tariffs offers a suitable framework for such an investigation. However, the optimal tariff model needs extensions, because it lacks an explicit representation of anticipations. Once these extensions have been introduced, the resulting model will be able to represent the main concepts from new trade theory, but precisely because of its general equilibrium nature, it will no longer produce clear-cut analytical answers that reach beyond classical welfare theory. Yet it offers a flexible framework for simulation exercises, under different assumptions with respect to the depth of the anticipations of the strategic agents. We have proposed a solution technique for such a simulation model, that uses a dual variable instead of an elasticity to represent the agents' anticipations, thus avoiding cumbersome and error-prone calculations. It is our intention to apply this framework to compare the payoffs of alternative coalition structures of firms as well as governments under given GATT/WTO inspired restrictions on trade protection.

References

1 Agarwal, R., Technological activity and survival of firms, Economics Letters, 52(1) (1996) 101–108.

2 Agarwal, R., Survival of firms over the product life cycle, Southern Economic Journal, 63(3) (1997) 571–84.

3 Alam, G. and J. Langrish, Non-multinational firms and the transfer of technology to less developed countries, World development 9 (1981) 383–87.

4 Anis, A.H. and T.W. Ross (1992) Imperfect competition and Pareto improving strategic trade policy, Journal of International Economics, 33: 363–72.

5 Arrow, K.J. and F.H. Hahn, General Competitive Analysis, Holden-Day, San Francisco, Oliver and Boyd, Edinburgh, 1971.

6 d'Aspremont, C., R. Dos Santos and L.-A. Gérard-Varet, General equilibrium concepts under imperfect competition: A Cournotian approach, CORE discussion paper 9257, Université Catholique de Louvain, Louvain-la-Neuve, 1992.

7 Becker, H., Transfer pricing: The future of transfer pricing, Bulletin for International Fiscal Documentation, 50(11–12) (1996) 535–37.

8 Berthélémy, J.C. and F. Bourguignon, North–South–OPEC trade relations in an intertemporal applied general equilibrium model, in: J. Mercenier and T.N. Srinivasan (eds.), Applied General Equilibrium Analysis and Economic Development, The University of Michigan Press, Ann Arbor, 1994, pp. 317–40.

9 Bhagwati, J.N. and T.N. Srinivasan, Revenue seeking: a generalization of the theory of tariffs, Journal of Political Economy, 88 (1980) 1069–87.

10 Bhagwati, J.N. and T.N. Srinivasan, Lectures on International Trade, MIT Press, Cambridge, MA, 1983.

11 Bhagwati, J.N., R.A. Brecher and T. Hatta, The global correspondence principle: A generalization, American Economic Review, 77(1) (1987) 124–32.

12 Bishop, M. and J. Kay (eds.), European Mergers and Merger Policy, Oxford University Press, New York, Toronto, and Melbourne, 1993.

13 Böhm, V., The foundation of the theory of monopolistic competition revisited, Journal of Economic Theory, 63(2) (1994) 208–18.

14 Bonniseau, J.-M. and B. Cornet, General equilibrium theory with increasing returns: the existence problem, in: W. Barnett, B. Cornet, C. d'Aspremont, J.J. Gabszewicz and A. Mas-Colell (eds.), Equilibrium Theory and Applications: Proceedings of the Sixth International Symposium in Economic Theory and Econometrics, Cambridge University Press, Cambridge, 1991.

15 Brander, J.A., Intra-industry trade in identical commodities, Journal of International Economics, 11 (1981) 15–31.

16 Brander, J.A. and B.J. Spencer, Tariff protection and imperfect competition, in: H. Kierzkowski (ed.), Monopolistic Competition and International Trade, Clarendon Press, Oxford, 1984.

17 Brander, J.A. and B.J. Spencer, Export subsidies and market share rivalry, Journal of International Economics, 18 (1985) 83–100.

18 Buckley, P.J. and M. Casson, The Future of the Multinational Enterprise, MacMillan, London. 1976.

19 Buckley, P.J. and M. Casson, The Economic Theory of the Multinational Enterprise, MacMillan, London, 1985.

20 Canals, J., Competitive Strategies in European Banking, Oxford University Press, Clarendon Press, Oxford, New York, Toronto and Melbourne, 1993.

21 Chan, K.H. and L. Chow, An empirical study of tax audits in China on international transfer pricing, Journal of Accounting and Economics, 23(1) (1997) 83–112.

22 Curtis, D.C.A., Trade policy to promote entry with scale economies, product variety, and export potential, Canadian Journal of Economics 16 (1983) 109–21.

23 Davies, S., B. Lyons, H. Dixon and P. Geroski, Economics of Industrial Organization, Longman Inc., New York, 1988.

24 Debreu, G., Excess demand functions, Journal of Mathematical Economics, 1 (1974) 15–23.

25 Dierker, E., Two remarks on the number of equilibria of an economy, Econometrica, 40(5) (1972) 951–53.

26 Dierker, E. and B. Grodal, Non-existence of Cournot–Walras equilibrium in a general equilibrium model with two oligopolists, in: W. Hildenbrand and A. Mas-Colell (eds.), Contributions to Mathematical Economics in Honour of Gerard Debreu, North-Holland, Amsterdam, 1986, pp. 167–85.

27 Dixit, A.K., Growth and terms of trade under imperfect competition, in: H. Kierzkowski (ed.), Monopolistic Competition and International Trade, Clarendon Press, Oxford, 1984.

28 Dixit, A.K. and J.E. Stiglitz, Monopolistic competition and optimum product diversity, American Economic Review, 67(3) (1977) 297–308.

29 Driffill, J. and F. van der Ploeg, Trade liberalization with imperfect competition in goods and labour markets, The Scandinavian Journal of Economics, 97(2) (1995) 223–44.

30 Eaton, J. and G.M. Grossman, Optimal trade and industrial policy under oligopoly, Quarterly Journal of Economics, 51(2) (1986) 383–406.

31 Eaton, J. and H. Kierzkowski, Oligopolistic competition, product variety, and international trade, in: H. Kierzkowski (ed.), Monopolistic Competition and International Trade, Clarendon Press, Oxford, 1984.

32 Echeverri-Carroll, E., Flexible linkages and offshore assembly facilities in developing countries, International Regional Science Review, 17(1) (1994) 49–73.

33 Folmer, C., M.A. Keyzer, M.D. Merbis, H.J.J. Stolwijk and P.J.J. Veenendaal, The Common Agricultural Policy Beyond the MacSharry Reform, Elsevier Science, North-Holland, Amsterdam, New York and Tokyo, 1995.

34 Francois, J.F. and C.R. Shiells, Modelling Trade Policy: Applied General Equilibrium Assessments of North American Free Trade, Cambridge University Press, Cambridge, 1994.

35 Gabszewicz, J.J. and J.P. Vial, Oligopoly a la Cournot in general equilibrium analysis, Journal of Economic Theory, 4(3) (1972) 381–400.

36 Gershenberg, I., Multinational enterprises, transfer of managerial know-how, technology choice and employment benefits: A case study of Kenya, Working paper of the ILO Multinational Enterprises Programme, No. 28, 1983.

37 Ginsburgh, V. and M.A. Keyzer, The Structure of Applied General Equilibrium Models, MIT Press, Cambridge, MA, 1997.

38 Graaff, J.V. de, On optimal tariff structures, Review of Economic Studies, 17(1) (1949 50) 47–59.

39 Green, R., The British electricity market, Pacific and Asian Journal of Energy, 6(1) (1996) 39–52.

40 Green, R., Competition in the electricity industry in England and Wales, Oxford Review of Economic Policy, 13(1) (1997) 27–46.

41 Greenaway, D., R.C. Hine and C. Milner, Vertical and horizontal intra-industry trade: A cross industry analysis for the United Kingdom, Economic Journal, 105(433) (1995) 1505–18.

42 Grossman, G.M. and E. Helpman, Innovation and Growth in the Global Economy, MIT Press, Cambridge, MA, 1991.

43 Grossman, G.M. and E. Helpman, Trade wars and trade talks, Journal of Political Economy, 103(4) (1995) 675–708.

44 Harris, R.G., Applied general equilibrium analysis of small open economies with scale economies and imperfect competition, American Economic Review, 74 (1984) 1016–32.

45 Heijdra, B.J. and X. Yang, Imperfect competition and product differentiation: some further results, Mathematical Social Sciences, 25 (1993) 157–71.

46 Helpman, E. and A. Razin, Increasing returns, monopolistic competition, and factor movements: A welfare analysis, in: H. Kierzkowski (ed.), Monopolistic Competition and International Trade, Clarendon Press, Oxford, 1984.

47 Huizinga, H., Taxation and the transfer of technology by multinational firms, Canadian Journal of Economics, 28(3) (1995) 648–55.

48 Johansen, L., A Multisectoral Study of Economic Growth, North-Holland, Amsterdam, 1960.

49 Johnson, H.G., Optimum tariffs and retaliation, in: H.G. Johnson, International Trade and Economic Growth; Studies in Pure Theory, George Allen and Unwin Ltd., London, 1958, Ch. 2.

50 Kehoe, T.J., Computation and multiplicity of equilibria, in: W. Hildenbrand and H. Sonnenschein (eds.), Handbook of Mathematical Economics, Vol. 4, North-Holland, Amsterdam, 1991.

51 Kemp, M.C., Notes on the theory of optimal tariffs, Economic Record, 43(103) (1967) 395–404.

52 Keyzer, M.A., Library of applied general equilibrium models in GAMS, located at internet site, http://mitpress.mit.edu/books/GINTH/GAMS.html, 1997.

53 Krugman, P.R., Intraindustry specialization and the gains from trade, Journal of Political Economy, 98(5) (1981) 959–74.

54 Krugman, P.R., Import protection as export promotion: International competition in the presence of oligopoly and economies of scale, in: H. Kierzkowski (ed.), Monopolistic Competition and International Trade, Clarendon Press, Oxford, 1984.

55 Krugman, P.R., Rethinking International Trade, MIT Press, Cambridge, MA, 1990.

56 Laan, G. van der and A.J.J. Talman, Adjustment processes for finding economic equilibria, in: G. van der Laan and A.J.J. Talman (eds.), The Computation and Modelling of Economic Equilibria, North-Holland, Amsterdam, 1987.

57 Lancaster, K., Variety, Equity and Efficiency, Columbia University Press, New York, 1979.

58 Lancaster, K., Protection and product differentiation, in: H. Kierzkowski (ed.), Monopolistic Competition and International Trade, Clarendon Press, Oxford, 1984.

59 Linda, R., Options, orientations et méthodes de la politique européenne de la concurrence face aux multinationales globales: Nestlé et les marchés des boissons et des produits alimentaires, Economie Appliquée, 46(1) (1993) 31–62.

60 Lofgren, H., State supremacy in decline: The pharmaceutical industry and the government, Journal of Industrial Studies, 3(1) (1996) 87–103.

61 Lucas, R.E., Jr. and N.L. Stokey, Optimal growth with many consumers, Journal of Economic Theory, 32(1) (1984) 139–71.

62 Markusen, J.R., The boundaries of the multinational enterprise and the theory of international trade, Journal of Economic Perspectives, 9(2) (1995) 169–89.

63 Marschak, T. and R. Selten, General Equilibrium with Price Making Firms, Springer-Verlag, Berlin, 1974.

64 Melo, J. de and D. Tarr, A General Equilibrium Analysis of United States Foreign Trade Policy, MIT Press, Cambridge, MA, 1992.

65 Mercenier, J. and N. Schmitt, On sunk costs and trade liberalization in applied general equilibrium, International Economic Review, 37(3) (1996) 553–72.

66 Milgrom, P. and J. Roberts, Limit pricing and entry under incomplete information: an equilibrium analysis, Econometrica 50 (1982) 443–59.

67 Motta, M. and J.-F. Thisse, Does environmental dumping lead to delocation?, European Economic Review, 38(3–4) (1994) 563–76.

68 Myerson, R.B., Conference structures and fair allocation rules, International Journal of Game Theory, 9 (1980) 169–82.

69 Negishi, T., Monopolistic competition and general equilibrium for a competitive economy, Metroeconomica, 12 (1961) 92–97.

70 Newberry, D.M., Power markets and market power, Energy Journal, 16(3) (1995) 39–66.

71 Olkinger, M. and J. Fernandez-Cornejo, Regulation and farm size, foreign-based company market presence, merger choice in the US pesticides sector, Bureau of the Census Center for Economic Studies Discussion Paper 94-6, 1994.

72 Porter, M.E., Changing patterns of international competition, California Management Review, 27(2) (1986) 9–40.

73 Porter, R.H., Recent developments in empirical industrial organization, Journal of Economic Education, 25(2) (1994) 149–61.

74 Purkayastha, D., Firm-specific advantages, multinational joint-ventures and host-country tariff policy, Southern Economic Journal, 60(1) (1993) 89–95.

75 Robinson, J., The Economics of Imperfect Competition, Macmillan, London, 1993.

76 Salonen, H., Entry deterrence and limit pricing under asymmetric information about common costs, Games and Economic Behavior, 12(2) (1994) 245–65.

77 Scarf, H.E. and T. Hansen, The Computation of Economic Equilibrium, Yale University Press, New Haven, CT, 1973.

78 Shoven, J.B. and J. Whalley, On the computation of competitive equilibrium on international markets with tariffs, Journal of International Economics, 4 (1974) 341–54.

79 Sleuwaegen, L. and H. Yamawaki, The formation of the European common market and changes in market structure and performance, European Economic Review, 32(7) (1988) 1451–75.

80 Smale, S., A convergent process of price adjustment and global Newton methods, Journal of Mathematical Economics, 3 (1976) 1–14.

81 Srinivasan, T.N., Comment on "The non-competitive theory of international trade and trade policy" by Helpman, in: S. Fischer and D. de Tray (eds.), Proceedings of the World Bank Annual Conference on Development Economics 1989, World Bank, Washington DC, 1989.

82 Tharakan, P.K.M. and B. Kerstens, Does North-South horizontal intra-industry trade really exist? An analysis of the toy industry, Weltwirtschaftliches Archiv, 131(1) (1995) 86–105.

83 United Nations Center on Transnational Corporations (UNTCT), Transnational Corporations in Biotechnology, United Nations, New York, 1988.

84 United Nations Center on Transnational Corporations (UNCTC), Transnational Corporations and the Transfer of New Technologies to Developing Countries, United Nations, New York, 1990.

85 Vernon, R., International investment and international trade in the product cycle, Quarterly Journal of Economics, 80 (1966) 190–231.

86 Verspagen, B., Measuring inter-sectoral technology spill-overs: Estimates from the European and US patent office databases, MERIT Working Paper 2/95-007, MERIT, Maastricht, 1995.

87 Yang, X. and B.J. Heijdra, Comment — Monopolistic competition and optimum product diversity, American Economic Review, 83(1) (1993) 295–301.

Trade, Growth, and Development
G. Ranis and L.K. Raut

CHAPTER 8

More on Trade Closure[1]

Keshab Bhattarai,[a] Madanmohan Ghosh[b] and John Whalley[c]

[a]Department of Economics, University of Warwick, Coventry, CV4 7AL, UK

[b,c]Department of Economics, University of Western Ontario, London, Ontario ON N6A 5C2

8.1 Introduction

Over his career, T.N. has made contributions in a wide array of areas, and not only through his written works. In the 1970s, T.N. attended meetings and conferences where numerical general equilibrium modelers presented their trade models, some of which sought to analyze the effects of trade liberalization on various economies. T.N. picked up on the issue of trade closure: how the specification of the rest of the world and its trade behavior influences the model outcome. T.N. highlighted the seeming inconsistency of having financial variables (exchange rates) in what were presented as real side models, and other points related to closure.

This present paper, as its title suggests, offers more on trade closure, and goes beyond the immediate debate that T.N. initiated. It takes as its point of departure the claim by modelers who responded to T.N. that the only exchange rate at issue in their closure treatment is the real exchange rate (the relative price of traded and non traded goods), and that in a formulation with product differentiation on both the demand and supply sides, the issues that T.N. identified no longer apply. We argue that a new set of issues arise, such as perversely sloped offer curves in small open economy trade models and ill-defined Nash equilibria in two-country trade models. In small open economy models with Cobb–Douglas preferences, trade shocks impact only on the demand side, seemingly independent of the production side specification. In two-country models with perverse offer curves, each trading country will

[1]We are grateful to Sethaput Suthiwart-Narueput and conference participants for comments.

try to move to extreme points of the partner's offer curve, and hence Nash equilibrium may not exist. T.N. has long argued that there is no shortcut to well defined theoretical structures, and the debate on model closure is a direct reflection of his earlier contributions. Our contribution is merely a further twist on the earlier development that T.N. started.

8.2 Trade Closure

The need for trade closure arises from the development of detailed numerical single economy models for the analysis of trade liberalization or other trade policy changes. As Figure 8.1 suggests, substantial modeling effort is typically put into the development of a detailed representation of the economy in question. Some treatment is then needed of the world with which the single economy trades, and in particular, its trade behavior, through the export demand and import supply functions that the country faces in world markets.

Figure 8.1: Closure in single country numerical general equilibrium trade models

The pitfalls with trade closure are well illustrated by the simple homogenous good trade structure discussed in Whalley and Yeung [13], which in turn is reflective of closure structures actually used in some of the applied models of the early 1980s. For the two-good case, they set out a then-used system of closure as representing the behavior of the rest of the world in three equations. The first is an export demand function

$$E = E_0\left(\frac{P_E}{e}\right)^{\eta}, \quad -\eta < A < 0 \tag{8.1}$$

where E_0 represents base period exports and P_E is the endogenously determined price of exportables in own country currency. As foreign buyers purchase these exports at world prices, this is divided by an exchange rate term e to yield the prices paid by buyers P_e/e. η seemingly defines the price elasticity of the export demand function, and varying the parameter η allows modelers to vary the export demand elasticity that the country in question faces.

A related equation can be written down for the import supply function of the world to the domestic market, but with a change in sign for the elasticity parameter

$$M = M_0 \left(\frac{P_M}{e} \right)^{\mu}, \quad \infty > \mu > 0 \tag{8.2}$$

where M_0 is base period imports, P_M is the own country price of importables, (P_M/e) is the world price of imports of world suppliers to the country, and μ is a further elasticity term, now interpreted as the import supply elasticity.

A trade balance condition (at either world or own country prices) completes the specification

$$P_M M = P_E E \tag{8.3}$$

As T.N. pointed out, for such structures there cannot be any meaningful exchange rate term in this closure system if the own country model structure to which the closure is appended is (as is typical) a real side barter economy. This follows directly by substituting (8.1) and (8.2) into (8.3) and solving for e. This solution can, in turn, be substituted back into (8.1) and (8.2) to give export demand and import supply functions as

$$E = E_0 P_E^{-1} \left(\frac{M_0}{E_0} \right)^{-\eta/(\mu-\eta)} \cdot P_M^{-(\mu+1)\eta/(\mu-\eta)} \cdot P_E^{\mu(\eta+1)/(\mu-\eta)} \tag{8.4}$$

$$M = M_0 P_M^{-1} \left(\frac{M_0}{E_0} \right)^{-\mu/(\mu-\eta)} \cdot P_M^{-(\mu+1)\eta/(\mu-\eta)} \cdot P_E^{\mu(\eta+1)/(\mu-\eta)} \tag{8.5}$$

No exchange rate really exists within the structure; its presence is an illusion created by the closure structure. Moreover, as Whalley and Yeung show, η and μ also define the true export demand and import supply elasticities. These come from equations (8.4) and (8.5) as

$$\mathcal{E}_E^{FD} = \frac{\partial E}{\partial P_E} \cdot \frac{P_E}{E} = \frac{\eta(1+\mu)}{(\mu-\eta)} \tag{8.6}$$

$$\mathcal{E}_M^{FS} = \frac{\partial M}{\partial P_M} \cdot \frac{P_M}{M} = \frac{-\mu(1+\eta)}{(\mu-\eta)} \tag{8.7}$$

and are in fact connected in the two-good case by the relationship between the two export demand and import supply elasticities and the elasticity of the offer surface. Thus, for this system, modelers may believe that they control two independent elasticity parameters (η and μ), while in reality they only control one parameter $\mathcal{E}^{OC}$ which implies the two jointly determined model elasticities, i.e.,

$$\mathcal{E}^{OC} = \frac{\eta}{1+\eta} \cdot \frac{1+\mu}{\mu}. \tag{8.8}$$

modelers, therefore can easily misinterpret their model specification on the elasticity front if they adopt this form of closure treatment.

Another closure issue identified by T.N. involved the claim by some modelers to be separately incorporating price-making behavior, say, on the export side and price-taking behavior on the import side through their closure specification. As Figure 8.2 makes clear, in the simple two-good case, only one relative price is involved in trade, and the domestic economy must be either a price-taker in both imports and exports (a straight line offer curve), or a price maker in both (bowness in the offer curve). Mixed price-making/price-taking rules in the simple two-good case would thus seem to be inconsistent.

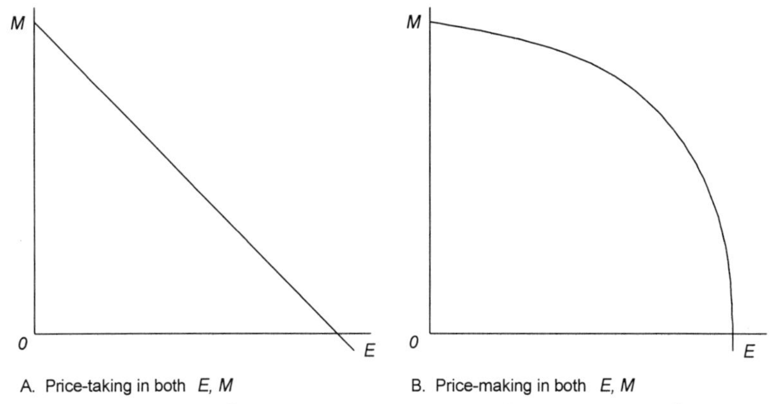

Figure 8.2: World offer curves as closure, and price-making/price-taking behavior

8.3 More on Trade Closure[2]

In more recent trade policy numerical modeling literature, it is common to use the so-called Armington assumption of product heterogeneity (see Dervis, de Melo and Robinson [4], Devarajan and Lewis [5], Robinson [12], Harrison, Rutherford and Tarr [7], Rutherford [13], and papers in the volume edited by Mercinier and Srinivasan [10]). This involves treating domestic products (which are both produced and consumed) as qualitatively different from imported or exported products. Such a specification is useful in accommodating the phenomena of cross hauling in trade data. This treatment essentially uses a constant elasticity of transformation (CET) frontier to separate each domestically produced good into exported and domestically consumed goods, and a nested CES preference function to capture imperfect substitution between domestic products and foreign imports. Price-taking behavior is commonly used for both imports and exports, but prices of domestically produced and consumed goods are delinked from world prices which now only apply to consumed imports and produced exports. Any difference between the value of

[2] This section heavily relies on our earlier paper (Bhattarai, Ghosh and Whalley [1]).

imports and exports can be met by a net foreign transfer in the trade balance equation. We argue that this widely used trade closure has some unattractive properties.

Here, we note some peculiar features of behavior that can arise from simple versions of this structure. If, for instance, preferences are Cobb–Douglas, then independent of the specification of production (transformation) elasticities, the model will generate fixed export volumes under any world price change. World price changes have no effect on either domestic production or consumption of the domestically produced good. If preferences are CES and elasticities are less than one, higher world prices for imports generate lower imports but higher exports; a perverse offer curve which asymptotically approaches the import axis and is truncated at an upper bound on exports in the two traded good case. Using a similar structure in a two-country model, it can also be shown that global equilibria may be obtained in which a country that reduces its tariffs increases its imports but reduces its exports. We also show cases where a Nash equilibrium of the two country tariff game does not exist, because of the perverse shape of offer curves.[3]

8.3.1 Model specification

For illustration we first consider a small open two-good economy which is a price taker for both exports (E) and imports (M), and also uses tariffs on imports. $\bar{P}_M$ and $\bar{P}_E$ are the fixed world prices at which the small economy trades internationally. Thus the economy faces a perfectly elastic supply function for imports and a perfectly elastic demand function for exports.

A representative consumer derives utility by consuming a domestically produced good, D and an imported good M, and hence the utility function is given by

$$U = U(D, M) \tag{8.9}$$

A commonly used functional form for the utility function in (8.9) is constant elasticity of substitution (CES).

Total production, T, is allocated between domestic sales and exports using a transformation frontier given by

$$T = T(E, D) \tag{8.10}$$

Such a frontier is often represented by a CET function.

[3] De Melo and Robinson [2, p. 62] in discussing this structure concluded that, "An external closure with symmetric product differentiation for imports and exports is theoretically well behaved, and gives rise to normally shaped offer curves." Our results suggest that this need not be the case; though some of our observations about the properties of this structure are implicit in the analytics that de Melo and Robinson present.

Optimizing behavior on the demand side implies that the ratio of the marginal rate of substitution between D and M in preferences should equal the ratio of prices of domestic goods to imported goods (inclusive of tariffs):

$$\frac{U_D}{U_M} = \frac{P_D}{\bar{P}_M(1+t)} \tag{8.11}$$

where U_D and U_M are the first derivatives of the utility function with respect to D and M, P_D is the price of the domestic good, and t is the tariff rate on imports.

Economy wide income, I, is given by the sum of the value of domestic sales plus exports and tariff revenue:

$$I = P_D + \bar{P}_E E + R \tag{8.12}$$

where R is tariff revenue. Given t, P_M, P_E, (8.11) and (8.12) determine demands for the domestically produced and consumed commodity, which we denote as D^D, and if E is known, import demands M. From (8.10), and prices P_D and $\bar{P}_D$ profit maximizing behavior gives

$$\frac{T_D}{T_E} = \frac{\bar{P}_D}{\bar{P}_D} \tag{8.13}$$

where T_D and T_E are the first derivatives of the transformation frontier with respect to D and E. Equations (8.13) and (8.10) in combination give the export supplies E, and the supply of the domestically produced and consumed good, which we denote as D^S.

To both solve for and characterize an equilibrium in this structure, external sector balance needs to be specified:

$$\bar{P}_M M - \bar{P}_E E = B \tag{8.14}$$

where B is the exogenous trade imbalance; B will equal zero if there is zero trade balance.

This treatment of trade balance is unusual since trade balance is usually not implied by Walras Law (domestically), and is a property of an international trade equilibrium. In this setup trade balance must be specified as a model feature, instead of having it emerge as a property of an equilibrium. This resolves the simultaneity between income, exports, and hence import demands.

An equilibrium for this system is given by the price of the domestic good, P_D, such that there is market clearing in the domestically produced and consumed good, i.e.,

$$D^D = D^S. \tag{8.15}$$

We now show how a range of problems with the comparative static properties of this trade structure can arise for particular functional forms and parameters often used in the trade literature. If, for example, Cobb–Douglas preferences in consumption are used, then independent of the specification of the transformation frontier, the implied offer curve for the domestic economy involves constant exports.[4] This reflects the constant expenditure shares implied by Cobb–Douglas, since in this case world prices can be accommodated by changes in M, with no further change in E or D. Such properties would not characterize a homogenous good trade model with unitary elasticities in preferences, and would not occur independently of the specification of production.

Even with CES preferences, which introduce bowness to the home country offer curve, depending upon the value of elasticity of substitution in preferences, two alternative regimes may appear, one with perversely sloped offer curves and the other with a more usual slope.

In Figure 8.3[5] and Table 8.1 we display the coordinates of offer curves for different values of σ_U (the elasticity of substitution in preferences) and for given share and shift parameters for an arbitrarily specified numerical example. Here, the perverse offer curves for the low elasticity specification imply falling E with rising M, a result opposite to that implied by conventionally

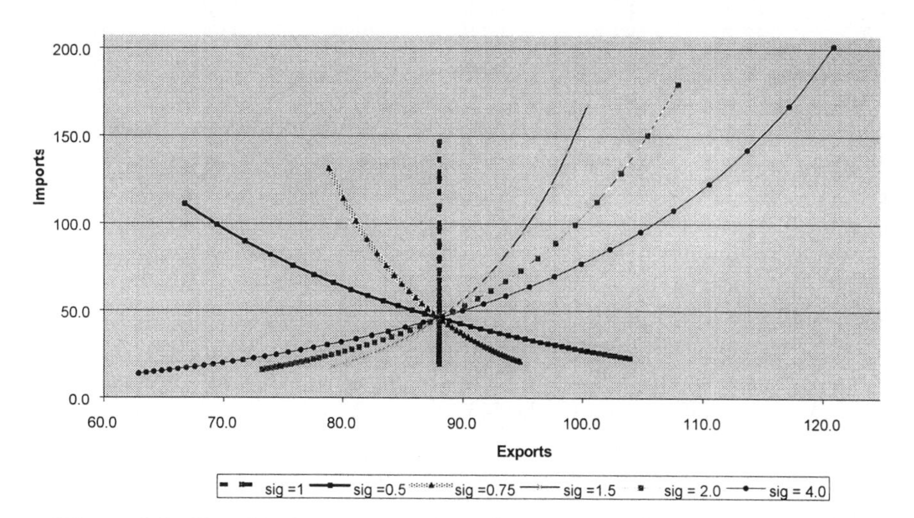

Figure 8.3: Graphical representation of perverse and non-perverse offer curves

[4] In an appendix, de Melo and Robinson [2] analytically derive the elasticity of the offer curve for this sytem and note that their formula implies an offer curve elasticity $(\partial E/\partial M) \cdot (M/E)$ of ∞. This implies the same behaviour we note above.

[5] SIG refers to a non-distorted offer curve with substitution elasticity as specified.

Table 8.1: Offer curves generated by a simple product differentiation model for alternative substitution, elasticities in preferences

A. Numerical specificaton of model	
Production specification	$T = 100$; E share = 0.4; D share = 0.6; Transformation elasticity = 0.8
Demand specification	D share = 0.7; M share = 0.3; Demand elasticity (varied as below)
World price ratio	Varied to generate offer curve

B. Model behavior

	Perverse cases					
	$\sigma_U = 0.5$		$\sigma_U = 0.75$		$\sigma_U = 1.0$	
	Imports	Exports	Imports	Exports	Imports	Exports
0.5	127.2	63.6	154.8	77.4	176.0	88.0
1.0	75.8	75.8	82.8	82.8	88.0	88.0
2.0	44.4	88.8	44.2	88.3	44.0	88.0
3.0	32.2	96.5	30.5	91.6	29.3	88.0
4.0	25.4	101.8	23.5	93.8	22.0	88.0

	Non-perverse cases					
	$\sigma_U = 1.5$		$\sigma_U = 2.0$		$\sigma_U = 4.0$	
	Imports	Exports	Imports	Exports	Imports	Exports
0.5	204.5	102.2	221.7	221.7	249.9	125.0
1.0	95.0	95.0	99.4	99.4	107.6	107.6
2.0	43.8	87.5	43.6	43.6	43.3	86.7
3.0	27.7	83.2	26.7	26.7	24.8	74.4
4.0	20.0	80.2	18.8	18.8	16.5	66.1

sloped offer curves in traditional trade models. An offer curve approaching the M axis implies an infinite number of low-priced imports for an infinitesimally small number of high-priced exports. Similarly an offer curve approaching the E axis implies an infinite number of low priced exports financing an infinitesimally small amount of high-priced imports. In reality, E will face an upper bound that reflects the maximal E value in production given finite resources and the production frontier (8.10).

Offer curves with and without tariffs are drawn in Figure 8.4[6] from the computational results in Table 8.2. In the perverse case, the imposition of a tariff shifts the offer curve to the left, as do the non perverse and perfectly inelastic offer curves.

We can also modify the above model to represent a multi-country model by allowing P_M and P_E to be endogenously determined to clear global markets (Harrison and Rutherford [7] use a multi-country structure of this form to evaluate international burden sharing under global carbon tax initiatives). Under this formulation, preferences and technology [equations (8.9) and (8.10)] for

[6] SIGT refers to a tariff distorted offer curve with substitution elasticity as specifies and SIG refers to a non-distorted offer curve with substitution elasticity as specified.

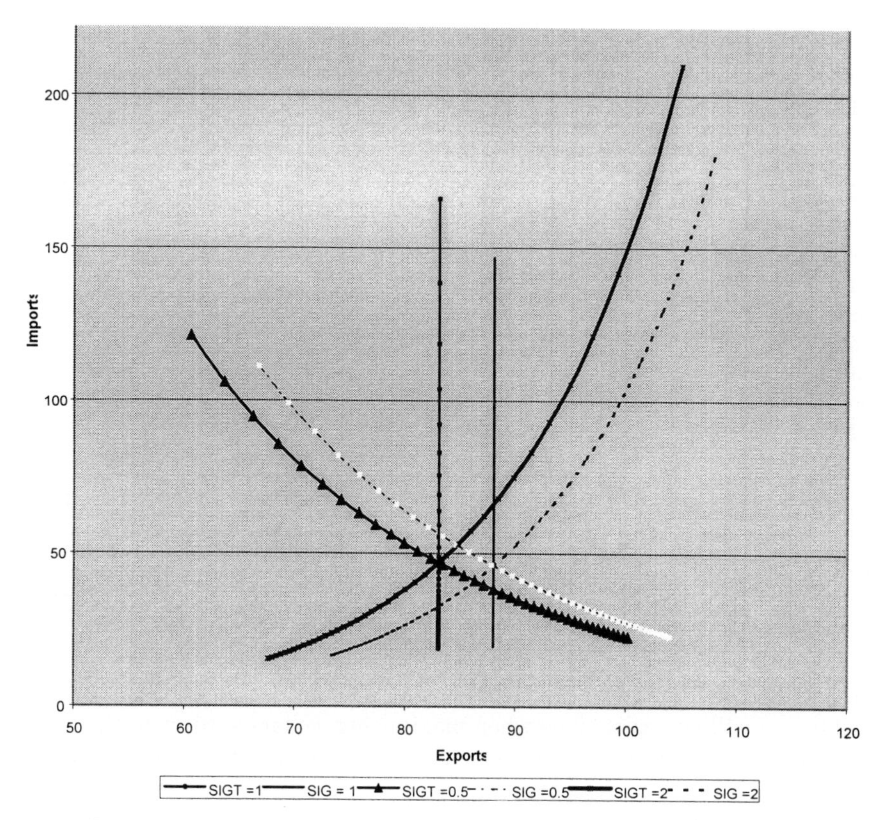

Figure 4: Tariff distorted perverse and non-perverse offer curves

each country and trade balance (8.15) are again specified for each country. The Armington assumption implies that domestically produced goods for domestic consumption are different from those produced for export sales and those imported for domestic consumption. This structure generates two separate country offer curves; and given the discussion above. three possibilities appear. These are: (1) where both offer curves are non perverse, and an intersection of two conventionally sloped offer curves characterizes equilibrium country behavior with respect to induced changes in world prices as is conventional (either due to foreign country tariff changes, or technology or demand shifts), and a Nash tariff game (as in Johnson [9] and Gorman [6]) has a well defined equilibrium; (2) either of the two countries have perverse offer curves; and (3) both countries have perverse offer curves.

In cases (2) and (3), peculiar behavior can again be exhibited by the model. In case (2) with one perverse and one non-perverse offer curve, increases in the tariff of the country with the perverse offer curve will reduce the volume of trade in both E and M, but increases in the tariff of the country with the

Table 8.2: Two-country cases of product variation trade models with perverse behavior

A. Specification				
	Shares		Elasticities	
	M	D	Perverse	Non-perverse
Preferences in 1	0.5	0.5	0.5	1.5
Preferences in 2	0.5	0.5	0.5	1.5
Technology in 1	0.5	0.5	–3.5	–1.5
Technology in 2	0.5	0.5	–3.5	–1.5
Economy-wide resources in 1	1,000			
Economy-wide resources in 2	500			

B. Model behavior								
	Perverse case				Non-perverse case			
	Country 1		Country 2		Country 1		Country 2	
	M1	X1	M2	X2	M1	X1	M2	X2
No tariff	36.5	125.1	125.1	36.5	53.4	92.9	92.9	53.4
10% tariff	36.8	122.8	122.8	36.8	53.2	89.6	89.6	53.2
20% tariff	37.0	120.7	120.7	37.0	53.0	86.6	86.6	53.0
50% tariff	37.7	115.0	115.0	37.7	52.5	78.9	78.9	52.5

non-perverse offer curve will reduce E while M increases. In addition, a Nash equilibrium will not exist since when maximizing welfare subject to the other country's offer curve, the non perverse country tries to go to an extreme point of the other country's offer curve.

We again provide some numerical simulation illustrate the possible perverse behavior that results from the two-country Armington structure. Table 8.2 reports on a case where both countries can have perverse offer curves. In this case, a reduction in tariffs by country 1 reduces imports, but increases country 1 exports, with the same being true of country 2. As noted above, no meaningful Nash equilibrium exists in case of (2) and (3).

In Armington trade models, values in excess of one are commonly used for elasticities, suggesting that the problems we highlight may not be present in most applied models. However, applied trade models are typically used with elasticities in preferences which are calibrated to import price elasticities in the neighborhood of one (the current literature consensus, though such values strike many trade economists as implausibly low), and the possible presence of perverse behavior is a source of unease (also see the discussion of these elasticities in Perroni and Whalley [11]). In general, the lesson seems to be that applied trade models need to be specified so that the offer curves generated from them are consistent with trade theory in terms of their implied model responses.

8.4 Conclusion

Closure rules used in applied Armington single country trade models involve price-taking behavior for both imports and exports, and product differentiation between domestic and traded goods. In such models, most often a CET function describes transformation between domestic sales and exports and a CES preference function represents the substitution between imports and domestically produced goods for domestic consumption. A trade balance condition is imposed to solve the model, rather than trade balance being a property of an international equilibrium via Walras Law (which is more usual). Here we argue that when the elasticity of substitution in preferences equals or is less than one, the implied offer curves can be perverse, with exports falling as imports rise with falling import prices. The Nash equilibria of tariff games may not exist in the two-country case. When domestic products and imports have unit elasticity of substitution in preferences, changes in world prices merely result in changes in import volumes, with no change in exports and consumption and production of domestic goods in the new equilibrium. Trade disturbances only affect imports.

We do not claim that all Armington trade models generate perverse offer curves; but modelers using trade closure should be aware of the somewhat unfortunate properties that can result in some situations. This develops a theme long stressed by T.N., which we re-emphasize here.

References

1 Bhattarai, K., M. Ghosh and J. Whalley, On some unusual properties of trade closure widely used in numerical modelling, Economics Letters, 62 (1999), 13–21.

2 De Melo, J. and S. Robinson, Product differentiation and the treatment of foreign trade in computable general equilibrium models of small economies, Journal of International Economics, 27 (1989) 47–67.

3 De Melo, J. and D. Roland-Holst, Tariffs and exports subsidies when domestic markets are oligopolistic: Korea, in: Jean Mercenier and T.N. Srinivasan (eds.), Applied General Equilibrium and Economic Development: Present Achievement and Future Trends, University of Michigan Press, Ann Arbor, 1994.

4 Dervis, K., J. de Melo and S. Robinson, General Equilibrium Models for Development Policy, CUP, New York, 1985.

5 Devarajan, S. and J.D. Lewis, Policy lessons from trade focussed two-sector models, Journal of Policy Modeling 12(4) (1990) 625–57.

6 Gorman, W.M., Tariffs, retaliation and the elasticity of demand for imports, Review of Economic Studies, 25(3) (1957) 133–62.

7 Harrison, G.W. and T.F. Rutherford, Burden sharing, joint implementation, and carbon coalitions, Working Paper No. 97–11, Department of Economics, University of Colorado at Boulder, 1997.

8 Harrison, G.W., T.F. Rutherford and D.G. Tarr, Trade reform in the partially liberalized economy of Turkey, The World Bank Economic Review, 7(2) (1993) 191–217.

9 Johnson, H.G., Optimal tariffs and retaliation, Review of Economic Studies, 21(2) (1953–54) 142–43.

10 Mercenier, J. and T.N. Srinivasan (eds.), Applied General Equilibrium and Economic Development: Present Achievement and Future Trends, University of Michigan Press, Ann Arbor, 1994.

11 Perroni, C. and J. Whalley, The new regionalism: Trade liberalization or insurance, NBER Working Paper No. 4626, 1994.

12 Robinson, S., (1991). Macroeconomics, Financial variables, and computable general equilibrium models, World Development, 19(11) (1991) 1509–23.

13 Rutherford, T.F., Extension of GAMS for complementary problems arising in applied economic analysis, Journal of Economic Dynamics and Control,19 (1995) 1299–1324.

14 Whalley, J. and B. Yeung, External sector closing rules in applied general equilibrium models, Journal of International Economics, 16 (1984) 123–38.

Trade, Growth, and Development
G. Ranis and L.K. Raut

CHAPTER 9

Developing Countries and the Lure of Preferential Trade Agreements: Beware of the Hook[1]

Philip I. Levy

Economic Growth Center, Yale University, P.O. Box 208269, New Haven, CT 06520–8269

9.1 Introduction

Toward the end of his recent broad overview of developing nations and the multilateral trading system, T.N. Srinivasan notes with regret the persistent popularity of Preferential Trading Arrangements (PTAs).[2] "Unfortunately," he writes, "many developing countries seem to be actively engaged in negotiating their membership in one or more such arrangement" [42, p. 118]. He concludes that it would be in the best interests of developing nations to pursue liberalization through the multilateral trading system instead. Proponents of PTAs would not accept the proposition that there is a choice to be made. They would argue that PTAs are complementary to or could even enhance prospects for multilateral liberalization.[3]

These issues have been the subject of a great deal of recent debate in light of the resurgence of interest in PTAs in the late 1980s, mostly from the perspective of developed nations.[4] In this paper I propose to survey many

[1] I am grateful to Kyle Bagwell, Jagdish Bhagwati and Anne Krueger for helpful comments. This research was supported by a grant from the Ford Foundation.

[2] Following the admonishments of Jagdish Bhagwati, I will refer to these agreements as PTAs rather than using some of the other common appelations: Regional Integration Arrangements or most commonly Free Trade Agreements, except where such usage would confuse. The agreements need not be regional (the first US agreement was with Israel) and between the slow phase-ins, exceptions made for sensitive sectors, and retention of administered protection they generally do not result in free trade, even among the member states.

[3] Again, in the terminology of Jagdish Bhagwati [6], the argument is over whether PTAs are "building blocs" or "stumbling blocs." Among the notable proponents of PTAs are Lawrence [26] and Frankel [18].

[4] There were flurries of PTAs in the 1950s and 1960s. Except for the European Community, they did not amount to much. For a recent review of post-war experience with regional trade agreements, see WTO [47].

of the recent arguments for and against PTAs with special attention to their applicability to developing nations. It will turn out, of course, that Srinivasan is right — PTAs are inadvisable for developing countries — but I hope to add arguments to his case.

Before delving into the arguments, it may be worthwhile to mention some of the ways in which developing countries' participation in the world trading system is different from that of developed countries and to recount some of the history of that participation (this section draws heavily on Srinivasan [41] and Krueger [24]). In the post-war era, developing nations saw themselves at a disadvantage in trade with respect to developed nations. The developing nations were largely producing primary products and trading for developed country manufactured products. Such exchange seemed to offer little hope of development, since it was difficult for the developing nations to compete with developed country manufactures. The predominant development philosophy, thus, was that of import-substitution industrialization. Developing countries would raise trade barriers and give their nascent manufacturing industries an opportunity to grow.

One implication of this approach was that developing nations did not participate in much of the reciprocal trade liberalization pursued under the auspices of the General Agreement on Tariffs and Trade (GATT). In fact, they were granted "special and differential treatment" by the developing nations, whereby liberalization would be extended to them without the requirement of reciprocity. This special status persists at the GATT's successor, the World Trade Organization (WTO). The developed nations accepted this unequal bargain in part because the developing nations constituted a small share of world trade, in part because they accepted the development philosophy, and in part because there was a cold war desire to appease developing countries.

In the 1980s, a number of events combined to alter the relationship between developing nations and the multilateral trading system. First, there was the end of the cold war. Second, there was the example of the Asian Tigers who had pursued an export-oriented strategy very successfully. Third, there was a debt crisis and slow growth in much of the rest of the developing world. These events helped dethrone import-substitution industrialization as the dominant philosophy and led developing nations to take an interest in liberalizing their trading regimes.[5]

Thus we come to the question at hand: what is the best way for the developing countries to integrate? They played a more active role in the Uruguay Round of GATT negotiations (1986–1993) than they had played in any of the previous rounds. They have also pursued PTAs both among

[5] Haggard [19] stresses the extent to which external pressure from international organizations helped bring about integration. Edwards [15] argues that internal motivations were important.

themselves and with developed countries.

The rest of the paper is structured as follows: Section 9.2 considers some of the theoretical arguments that are particularly applicable to developing countries. Section 9.3 returns to those arguments in the context of discussing several recent actual or proposed PTAs: APEC, Mercosur, NAFTA and FTAA.[6] Section 9.4 asks whether developing countries' stance toward preferential liberalization matters and considers potential reforms of the multilateral trading system to address the outbreak of PTAs. Section 9.5 concludes.

9.2 Theoretical Arguments for and Against Developing Country Involvement in PTAs

I begin by surveying some of the many arguments put forward for and against PTAs. I roughly divide these arguments into those that are relevant for countries at any level of development; those that are relevant primarily for PTAs involving only developing countries (South–South agreements); and those that are primarily relevant for PTAs including both developed and developing countries (see Winters [45] for an extensive survey of the work on this topic) The reader will note that the division is somewhat arbitrary, however, and there is a fair amount of overlap.

9.2.1 General arguments

Trade diversion or trade creation. The classic rebuttal to the argument that any liberalization is good liberalization comes from Jacob Viner, who analyzed the effects of customs unions.[7] Viner introduced the concepts of trade creation and trade diversion. The former is the positive effect whereby preferential trade liberalization towards a trading partner reduces the distortions associated with protection. A simple example would be when the partner receiving the preferences happens to be the worldwide low-cost supplier of the good in question. The liberalization induces new trade flows and those flows are efficient.

Trade diversion occurs when preferences are extended to a country that is not the low-cost supplier. In this case we also may induce new trade flows, but those flows will be inefficient. Not only is this damaging for world welfare, it can also decrease the welfare of the country adopting the preferences (e.g., that country could have a negligible change in domestic prices but lose the tariff revenue it obtained when it imported from the efficient supplier) (see De Melo, Panagariya and Rodrik [13] for a more thorough analysis).

---

[6] The major PTA that I omit by this approach is the European Union. There are, of course, useful lessons to be gleaned from European experience. See Winters [46].

[7] Viner's 1950 paper can be found in a collection of theoretical works on PTAs edited by Bhagwati, Krishna and Panagariya [11].

The point is, that from the viewpoint of static welfare analysis, preferential liberalization can raise or lower participant and world welfare. This is equally true for developed or developing countries, although if one assumes that developing countries are efficient producers of a more limited range of goods, a PTA between developing countries may be more likely to bring about trade diversion.

Political economy dynamics and domino effects. The Vinerian analysis is static in that it contrasts several states of the world (uniform protection, preferential liberalization, and complete liberalization) without any comment on how one moves between them. An important part of the argument in favor of PTAs is that they might *lead* to more efficient multilateral liberalization. Jagdish Bhagwati [7] has characterized this as the "dynamic time-path question." The reasoning that regional integration may be a useful means to bring developing countries into the world trading system has apparently been adopted by the Director General of the World Trade Organization, Renato Ruggiero, who stated that "in regions such as Sub-Saharan Africa (regional initiatives) may be an essential starting-point for integration of least-developed countries into the wider global economy" (quoted in Srinivasan [42, p. 342]).

Perhaps the strongest proponent of this viewpoint is Richard Baldwin.[8] He has laid out a "Domino Theory of Regionalism" whereby once regional liberalization is initiated among a group of countries, non-member countries have an incentive to join (Baldwin [2]). He finds evidence that this has happened in the desire of non-members to join in European and Western Hemisphere economic integration (Baldwin [3]). One difficulty, of course, is that the existing member countries may not wish to allow additional members.

Levy [27] makes this argument in a median-voter setting. He suggests that in a multilateral setting countries adopt trade liberalization that offers individuals within the economy some combination of benefits and costs. These benefits and costs to key individuals (e.g., the median voter) need not accord with those from the perspective of the country as a whole. PTAs then allow for key individuals to accept agreements that offer disproportionate benefits and opt out of the remaining multilateral liberalization (which would offer disproportionate costs) (see Frankel [18, Ch. 10] for a survey of political economy models of preferential trade liberalization).

Another variant on the same argument concerns the effect of PTAs on industries lobbying for protection. Baldwin [3, p. 885] describes a model in which "any liberalization, which raises imports and exports, systematically strengthens the power of pro-trade forces (exporters) and weakens that of

[8] Other prominent recent works on the dynamic effects of preferential liberalization include Bagwell and Staiger [1] and Bond, Syropolous and Winters [12].

anti-trade forces (import competitors)." Levy [28] presents a model in which this need not be the case: a PTA can lessen the incentives for pro-trade forces to lobby against inter-bloc protection and raise the incentive for import competitors to lobby for protection.

A common assumption in dynamic approaches is that multilateral liberalization is feasible as an ultimate goal. However, an argument in favor of PTAs is that multilateral possibilities have been either temporarily or permanently exhausted. This was one explanation for the United States' interest in regionalism in the 1980s, when it was unable to start a successor to the Tokyo Round of GATT negotiations.[9] Srinivasan [41, p. 4] ascribes recent developing country interest in North–South PTAs to pessimism about multilateral prospects: "This revival of interest was understandable when the prospects of concluding the (Uruguay Round) seemed dim and it looked as if the global trading system would collapse into a set of warring trade blocs. It then made sense for a developing country to seek to become part of a trade bloc with some industrialized country to avoid being marginalized."

I will return to this topic in the discussions of Mercosur and the proposed Free Trade Agreement of the Americas below.

9.2.2 South–South agreements

Deeper integration. One frequent argument in favor of PTAs is that the Vinerian calculus described above no longer applies. The lowering of trade barriers may be secondary, the argument goes, to such issues as harmonization of investment standards and the regulation of competition. These considerations are grouped under the rubric "deep integration," whereas in contrast the removal of tariffs and quotas constitute "shallow integration."[10] Lawrence [26, p. 17] makes the case:

> Emerging regional arrangements . . . are motivated by the desire to facilitate international investment and the operations of multinational firms as much as the desire to promote trade. Although liberalization to permit trade requires the removal of border barriers — a relatively shallow form of integration — the development of regional production systems and the promotion of service investment require deeper forms of international integration, for example, the elimination of differences in national production and product standards that make regionally integrated production costly.

[9] Baldwin [3] disputes the importance of this effect for US interest in regionalism. He notes a series of regional overtures that the United States had made over the decades.

[10] Bond, Syropolous and Winters [12] use a different definition of deep integration. For them, deepening means a steady lowering of internal barriers within a customs union. They show how the resulting external policy of the customs union depends on the nature of consumer preferences.

Lawrence argues that there is no particular reason why the nation–state is the optimal level at which to set policy. Nor, for that matter, need the world be the optimal level. As an example, he cites the literature on optimal currency areas.

Even Bhagwati, a staunch opponent of PTAs, allows for the potential desirability of more complete integration among a group of developing nations. He writes [8, p. 70] that discriminatory liberalization may be acceptable if "a smaller group of countries wants to develop a Common Market. In this case, not just trade, but also investment and migration barriers are eventually eliminated just as in a federal state, and the full economic and political advantages of such integration follow."

Park [34] is more skeptical of Lawrence's argument. While he acknowledges the theoretical possibility that the optimal area for setting common standards may be greater than the nation-state, he argues that Lawrence has not met the burden of demonstrating that larger groupings are, in fact, optimal.

I return to this topic in the discussion of leverage and new issues in North–South agreements below.

Bargaining power. This is an argument cited in favor of South–South agreements. There is a history of developing countries functioning as a bloc in GATT negotiations. Srinivasan [41] describes the initial solidarity of developing countries which predominated in the 1960s and 1970s but dissipated during the Uruguay Round discussions of the 1980s.

For PTAs, it is usually suggested that rather than enter into North–South agreements individually with terms most likely to be set by the developed country, the developing countries should first form a bloc of their own and then bargain from a position of strength. This argument raises several important questions. First, will there be a commonality of interests between the participating developing countries? If the goal of each is to make themselves uniquely attractive for foreign investment, they would seem to have different goals.[11] If we assume a commonality of interest, a second question is how the bargaining power of a developing country bloc would compare to the power of a broader group in WTO negotiations. This would likely depend on the constituencies of the blocs. It is worth noting that measures on environmental and labor standards have enjoyed less success on the WTO agenda than on the agenda of some PTAs involving developed countries. A final question raised by the bargaining argument is, what are these countries bargaining over anyway? A naive observer, thinking of shallow integration, might note that under GATT Article XXIV there is supposed to be liberal-

[11] For example, when Mexico got preferential access to the US market through NAFTA, there was a negative reaction from Caribbean nations that became less attractive in relative terms.

ization on "substantially all trade." If we are considering future Free Trade Agreements between developing and developed countries, why can they not just drop barriers on trade between them? Of course, there would be limited phase-ins for sensitive sectors, but is this sufficient to justify South–South PTAs on bargaining grounds? This issue serves to highlight the extent to which PTAs bring preferential rather than free trade. In practice, of course, there are negotiations over rules of origin and the range of deep integration issues discussed above.[12] This topic will be taken up again in the discussion of the Free Trade Agreement of the Americas.

Negotiating resources. A final point that may be particularly relevant to South–South agreements, although it would apply to developing countries more generally, is the question of resources used in negotiation and administration of PTAs. In her discussion of the administrative capacity of developing countries, Krueger [24, p. 65] writes, "in most developing countries, there is a relatively thin layer of highly competent and well-educated administrators at the top of the bureaucracy." Krueger makes the point in the context of the adoption of "deep integration" measures. However, it is relevant to a discussion of the advisability of developing country PTAs, given the tendency of some developing nations to participate in multiple, overlapping PTAs. Not only would there be a high demand for negotiators for each agreement, but the resulting set of multiple rules could be exceedingly complex to administer (imagine, e.g., different standards on government procurement). The simplicity of a common set of rules is sometimes cited as a reason for the pursuit of deep integration, but if a country is trying to administer different versions of these rules simultaneously, the simplicity is lost. If there is a single set of rules and procedures that all countries administer, then we have multilateral integration.

9.2.3 North–South agreements

I turn now to those arguments that are particularly relevant for agreements between developing and developed countries.

Constraint on administered protection. Of course, a major motivation for developing countries to pursue PTAs with developed nations is the hope of getting market access beyond what they can achieve in multilateral negotiations. Given the extent of developed nations' commitments through the Uruguay Round of multilateral liberalization, it is relatively unlikely that developed countries' tariff bindings are a major concern. The WTO restricts

[12] Baldwin [3, p. 868] writes: "Most recent regional schemes . . . involve shallow integration, often little more than tariff and quota liberalisation." He notes NAFTA is different. See Krueger [23] on the protectionist potential of rules of origin.

use of quantitative measures and the Uruguay Round included restrictions on so-called "gray area" measures, such as voluntary export restraints. So what else remains of concern?

One major item is the use of administered protection. This is a category of protection permitted under the WTO by which a country can impose protection either temporarily or in response to a finding of malfeasance on the part of a trading partner.[13] The most prominent such measure in recent years has been "anti-dumping" (AD) protection. If a country finds that one or more trading partners have been selling goods below a "fair value" and that such dumping has caused serious injury to a domestic industry, the country can impose a tariff against the offending partner or partners. In practice, there is sufficient flexibility in the application of these rules that they offer countries a ready excuse for protection.

Perhaps because developing countries retained more of their explicit tariffs, the major practitioners of AD protection have been the United States and Europe. In the recently completed Uruguay Round, the United States refused numerous requests to limit the use of AD policies. If, for example, the motivation of the United States or Europe were to preserve AD measures for use against Japan, developing nations might hope that they could win exemption from such measures through a PTA. I return to this topic in the context of NAFTA.

Commitment. One of the most commonly cited arguments in favor of developing country integration with developed countries is the potential to commit to recent economic reforms. The argument is that developing countries with a history of protectionist policies, nationalizations, and unstable macroeconomic environments will not see an influx of foreign investment upon reform if foreign investors do not see these reforms as credible.[14] If a developing country joins a PTA, however, this supposedly offers an imprimatur of certitude that should instill confidence in foreign investors.

This argument merits further consideration, since it may be the strongest of the reasons for developing country participation in PTAs. As an illustration of the weight given to this argument, consider Rodrik [38].[15] He expresses pages of skepticism about the value of deep integration for developing nations, but then notes that "there is also an upside for LDCs. The 'harmonization

[13] There is a great deal written on this topic. For an overview of the legal situation, see Jackson [22, Ch. 10]. For a brief discussion of Uruguay Round results, see Srinivasan [41, pp. 53–55].

[14] The classic statement of the "time consistency" problem of promising reforms and then reneging is Kydland and Prescott [25]. For a statement in the reform context, see Rodrik [37].

[15] Maggi and Rodríguez-Clare [29] pursue the commitment argument formally. They assume that trade agreements allow commitment and show how this can be valuable in overcoming distortions that emerge from the political process.

in domestic laws and institutions' entailed by deep integration presents an opportunity for reformist governments in developing countries to 'lock in' their reforms and render them irreversible."

He invites the reader to consider what might happen if Bolivia and Ghana were to "overnight harmonize their trade and industrial policymaking machinery with that of the EU." Despite the distortions they would have to adopt, Rodrik [38, p. 111] argues that "economic performance in these two countries would improve considerably . . . because harmonization would probably result in an enhancement of the private sector's expectations about the stability and predictability of the policy regime, as well as an enhancement of the respect for private enterprise, property, enforcement of contracts, and the rule of law."

Thus, the value of commitment would exceed the costs imposed on developing countries by deep integration.

A first question to pose along these lines is what the threat might be against which investors require assurance. It seems there are multiple possibilities and the usefulness of undertaking a PTA might depend on which of them is the issue. If the concern is exchange rate stability, for example, a PTA would not be necessary. Hong Kong and Argentina have successfully used currency boards to commit to an exchange rate peg and endogenize their monetary policies. A second possible concern is that in the face of an adverse shock, the existing government of a developing country would resort to an array of objectionable measures: investment restrictions, capital controls, taxation, protection, or expropriation. A third possible concern is that the present government would remain committed to its reform package but after an election, a new government would take power and implement objectionable policies.

It seems to be the second or third possibility against which PTAs are intended. We can then ask what protection a PTA with a developed country might offer. The presumption seems to be that if a developing country were to renege on commitments made under a PTA then two punishments might befall it: (1) the developed country could become enraged and take some harsh action against the developing country and (2) a clear signal would be sent to investors that the government in the developing country was of a "bad type" so foreign investment would then flee the country.

While either or both of these punishments might ensue, it is not at all clear how they would differ from punishments in the absence of a PTA. The first argument seems to hinge on the assumption that the developed country partners would take umbrage at the LDC withdrawal from an agreement above and beyond their offense at the objectionable policies themselves. For example, the United States has never been in nor contemplated a PTA with India, but it has not hesitated to threaten India with penalties under its "Super 301" trade legislation in response to objectionable Indian policies (see

Haggard [19, pp. 33–36]). If there were to be such additional umbrage, it would presumably be due to the violation of the agreement establishing the PTA. This leads to two questions: (1) Would a developing country have to violate a PTA in order to implement objectionable measures? This is best addressed in the context of NAFTA below. (2) Why do we assume that developed countries would react with fury at the violation of a PTA but would remain calm at the violation of a multilateral accord? This will be addressed below in a discussion of possible multilateral reforms.

An alternative to the theory of developed country's revenge is the signaling argument. One could argue that with the more complete liberalization required by PTAs, any deviation from a developing country's reform program would be more readily apparent (and probably monitored more closely by developed country partners). For this to apply, though, one must have great faith in Article XXIV of the GATT, which requires that customs unions and free trade areas liberalize "substantially all" trade among participants (Jackson [22, pp. 165–66]). In practice, however, not even Europe, the oldest and most developed of PTAs, has completely removed barriers to internal trade (e.g., agriculture under the Common Agricultural Policy). Thus, it seems unlikely that sufficient purity would be achieved to render deviations obvious. If the deviation itself were visibly objectionable — e.g., the nationalization of an industry — the signal would be perfectly clear with or without an agreement. There does not seem to be a strong signaling argument for PTAs.[16]

Thus, the commitment argument for North–South PTAs relies upon the steep penalties that would presumably result from violation of an agreement and on the assumption that any country wanting to implement objectionable policies would have to violate an agreement.[17]

Leverage and new issues. A flip side of the bargaining argument for South–South PTAs is the leverage that developed nations can have in North–South negotiations. To the extent that the developed participants are pushing for economically rational measures, this pressure can enhance developing country

[16] Fernández [16] offers a somewhat different signaling argument which relies on high entry costs rather than high exit costs. She suggests that if there is a significant information asymmetry about a country's policies or economy, and entry costs are significant, then the joining of a PTA will demonstrate the government's type. Examples of high entry costs might be negotiating time and effort or the expenditure of political capital in overcoming opposition. This argument would help with the second type of commitment described above — against the current government reneging — but would not help much with respect to future governments.

[17] McLaren [30] offers an altogether different argument about the role of commitment in trade agreements. He shows how investors in a small country may anticipate the agreement and invest accordingly. If these investments are fixed, this undercuts the country's bargaining power and may cause the country to lose from an agreement, if the "strategic disadvantage" is not overcome by the classical gains from trade.

welfare (Haggard [19]). However, such leverage also can be damaging if used to impose inappropriate rules or standards.

The most prominent recent example has been the push for harmonization of labor and environmental standards. Krueger [24] lays out a strong case for the potential damage that such standards could do to developing countries; for example, the imposition of a high, binding minimum wage in a country with low labor productivity. Srinivasan [40] presents an analytic case for the optimal diversity of such standards across countries. The general argument is that issues such as child labor present countries with trade-offs. In that case, the trade-off may be between the desire that children be free from work and able to pursue an education versus the desire that there be sufficient household income to feed household members. The nature of the trade-offs and the nature of preferences can legitimately differ across countries, so the optimal rules would also differ.

To the extent that North–South PTAs serve as a vehicle for imposing inappropriate harmonization of these standards, they could be very costly.

9.3 Assessing the Theoretical Arguments on the Basis of Recent Experience

Let us now turn to several prominent PTAs involving developing nations as a means of further exploring the arguments presented above.

9.3.1 Asia

Asia can be considered first, because it is perhaps the easiest. There is no free trade agreement involving Asia (see Panagariya [33] for a pessimistic view of Asian integration prior to the more recent APEC developments). There has been ample discussion of such agreements and one in particular, the Asia Pacific Economic Cooperation (APEC) forum, has been described as the principal contender for an Asian free trade agreement. APEC includes the United States, Japan, Canada, Australia and New Zealand, as well as a large number of developing nations including China, South Korea, Indonesia, Malaysia and Thailand. In 1994, at a meeting in Bogor, the participating countries pledged to have free trade within the region by 2010 for developed nations and 2020 for developing nations (Bergsten [4]).

The difficulty, as noted by numerous authors, is that for Asian nations, trade with countries outside the proposed bloc is as important as trade within the bloc (see, e.g., Haggard [19]). Therefore, it is of the utmost importance to these countries to preserve the multilateral trading system. There is thus strong opposition within Asia to creating a *preferential* trading arrangement of the conventional sort. Instead, "open regionalism" has been proposed.[18] Under one interpretation of such a scheme, whatever liberalization is undertaken within the group would be extended on an MFN basis to those outside the group. Bhagwati [8, p. 73] notes the difficulty with this concept:

> Is this free trade for APEC members to be for the members only, or is it to be on an MFN basis? If the former, then the only way for this to happen consistent with WTO, is for APEC to seek to be an FTA and to get Article XXIV exemption from MFN requirements, with all the damage that yet another gigantic FTA would impose on the multilateral trading system. If the latter, then there are distinct problems. It is hard to imagine that APEC countries would make more than token MFN reductions in their trade barriers without reciprocal concessions by non-APEC members.

Using Bhagwati's dichotomy, if APEC is to be an FTA for members only, this would be a PTA but it is very unlikely that many member countries would be willing to participate (Haggard [19]). Some of those countries would prefer a multilateral system and others would be very reluctant to offer free trade to fellow members (e.g., the United States to Japan and South Korea). If the region is to be "open" in the sense that liberalization is extended on an MFN basis, then this is not a PTA at all, but rather unilateral liberalization (not that there's anything wrong with that). It seems even less likely for the United States and Japan to offer unilateral trade liberalization.[19]

This all begs the question of why we see regional or preferential arrangements rather than multilateral liberalization. One answer might be that there is an interest in deep integration, as described above. However, there is still unfinished "shallow integration" in the form of lowering trade barriers that countries could pursue. Presumably, the countries are striking deals with a subset of the trading nations because they feel such deals are not feasible with the complementary set. So, if the United States feels that Europe's offers for reciprocal liberalization are insufficient in multilateral negotiations, it is very hard to see why unilateral liberalization toward Europe through APEC with no reciprocity would be acceptable.

On top of this, despite the impressive Bogor declaration of intentions, participating countries such as Malaysia quickly began to "clarify" and explain that they had not undertaken a firm commitment to liberalize. Thus, APEC is more a perplexing idea than a PTA.

---

[18] Although he is more restrained in his writing, in regular conversation or in seminar T.N. Srinivasan cannot hear this term without exclaiming "Oxymoron!"

[19] Another common interpretation of "open regionalism" is an approach whereby it would be easy for new countries to join by accepting the agreement's provisions. This constitutes a sort of conditional MFN status. See Srinivasan [42, pp. 339–43) for a detailed discussion of "open regionalism" interpretations.

9.3.2 Mercosur

The only major South–South agreement in existence is Mercosur, which grouped Argentina, Brazil, Paraguay and Uruguay beginning in 1991. It has since added Chile as an associate member and has been engaged in expansion negotiations with Andean Pact countries. Mercosur is also the only example of a customs union among the major agreements discussed here. It offers a useful illustration of a number of the theoretical arguments presented above.

One of the more interesting and controversial recent empirical tests for trade diversion and trade creation used Mercosur as its subject. Yeats [48] concludes that Mercosur's member trade patterns within the bloc differ significantly from the trade patterns with non-bloc members. There was a sharp growth in intra-bloc trade in products that were not being successfully exported to the rest of the world. This is taken as evidence of trade diversion — partner countries' products became attractive only because of tariff preferences. Yeats' finding is in contrast with some of the more benign findings for Europe (see Frankel [18, Ch. 5]).

Mercosur is also an interesting example of the potential bargaining power exercised by South–South blocs. Brazil has been explicit about its desire to form a bloc of developing countries prior to engaging in negotiations with the United States for hemispheric liberalization. As another facet of the bargaining question, when there have been delays in progress toward hemispheric liberalization, Mercosur has pursued negotiations with the European Union.[20] However, this is not unambiguous evidence in favor of the positive dynamic aspects of PTAs. Progress in liberalization with the Andean Pact nations has been slow and no agreement could be reached with Mexico (*Economist*, Dec. 20, 1997). Kevin Hall [20] writes: "through Mercosur, Brazil has forcefully imposed a go-slow approach toward the hemispheric trade talks, arguing that it must first deepen and strengthen its trading bloc with (its) neighbors."

Finally, Mercosur serves as evidence that the commitment arguments described above apply primarily to North–South agreements. In the wake of the Asian crisis of late-1997, Mercosur moved to raise its common external tariff by 25 percent on thousands of items (Hall [21]).

9.3.3 NAFTA

NAFTA is the principal example of a major PTA between a developing nation and developed ones.[21] It was the successor to the bilateral Canada–US Free

[20] This may not be simply a bargaining ploy. Panagariya [33] argues that Mercosur should prefer a PTA with the E.U. to one with NAFTA, since the Mercosur nations already enjoy better access into NAFTA.

[21] One could argue that Portugal, Spain and Greece undertook a similar endeavor upon their accession to the European Community (see Krueger [24, pp. 66-72]). They certainly had lower per capita incomes than the earlier members of the EC, but they were still more

Trade Agreement, which came into effect in 1989. Mexico had instituted a program of economic reforms in the mid-1980s which included accession to the GATT. In 1990, the President of Mexico, Carlos Salinas, requested, and the United States agreed to, NAFTA negotiations.[22] NAFTA came into force in 1994 after a sharp debate in the United States.

Why did Mexico want to participate in a PTA with the rest of North America? Mexico already faced relatively low trade barriers from the United States and did not trade much with Canada. It could have liberalized its remaining trade barriers and investment restrictions unilaterally. It certainly was not anxious to undertake "deep integration" in the areas of labor and environmental regulation. According to Tornell and Esquivel [43, p. 3]:

> NAFTA's greatest importance lies in its use as a commitment device. NAFTA is a commitment by the Mexican government to eliminate protection in agriculture and services within the next fifteen years (the agreement also entails a marginal reduction in protection for the already-liberalized manufacturing sector), in exchange for a decrease in U.S. and Canadian protectionist barriers, and a reduction in the uncertainty associated with trade disputes."

We can ask whether Mexico achieved its objectives. I focus on NAFTA's value as a commitment device and the uncertainty associated with trade disputes, drawing on the theoretical discussions above.

On the face of it, NAFTA seems to have lent credibility to Mexico's reforms. Direct foreign investment flows into Mexico almost doubled once NAFTA negotiations began (Lawrence [26, p. 68]). Mexico then suffered an economic crisis in 1994 and 1995 but did not reinstate protection (at least not against NAFTA partners, it did against non-NAFTA countries, see Bhagwati [8]) nor nationalize industries. It is hard to say, though, what would have happened in the absence of NAFTA.[23] As it happened, the government of Ernesto Zedillo that replaced Salinas had a very similar economic philosophy. Even without NAFTA, President Zedillo might have refrained from implementing investor-unfriendly measures for fear of the damage that could be caused to Mexico's reputation.[24]

If we choose to believe, though, that NAFTA really did serve to "lock in" Mexican reforms, we can look for the punishment mechanism that enforces the

developed than the nations of Latin America.

[22] Canada then asked to be included *ex post.*

[23] Panagariya [33] makes a different argument. He contends that tariff bindings under the WTO are clearly superior to those under NAFTA, for example. Therefore, Mexico's earlier reforms should already have been credible.

[24] Note that it is flawed reasoning to argue that previous Mexican administrations had imposed investor-unfriendly measures and therefore NAFTA's restrictions were binding. Previous administrations had different economic philosophies. It is not at all clear what a PRD government would do if faced with a similar situation.

agreement. I return to the question raised above: would a developing country government have to violate an agreement if it wished to reimpose objectionable measures? In the NAFTA case, oddly enough, the answer seems to be 'no.' Article 2205 of the NAFTA [31] text reads: "A Party may withdraw from this Agreement six months after it provides written notice of withdrawal to the other Parties. If a Party withdraws, the Agreement shall remain in force for the remaining Parties."[25]

If Mexico wished to nationalize industry and reinstate protection, it could do so without violating the GATT by giving six months notice. Presumably this could be done either by an existing government with a change of heart or by a newly elected government with a different philosophy. The United States could not take special umbrage at the violation of the agreement, since there would not be any such violation. The United States might well object to specific measures Mexico might undertake. If those objectionable actions violated commitments Mexico had made under the WTO, the United States could pursue those objections under WTO dispute settlement procedures and potentially be granted the right to make a reciprocal withdrawal of concessions. If the objectionable Mexican actions did not violate WTO commitments, the United States would be left to decide whether it wished to retaliate. If so, it could choose between retaliatory measures that were or were not restricted under the WTO.

The essential point is that under this scenario, the US options for retaliation and incentives to retaliate would be identical to those had NAFTA never been negotiated.

In addition to the commitment argument, the other Mexican motivation described above was protection from the uncertainties associated with trade disputes. One interpretation of this is that Mexico wished to restrict US use of administered protection measures, such as the anti-dumping code.[26] If so, it must be disappointed, since the US anti-dumping code remains intact. Moreover, the United States has not even felt constrained by the loose rules on administered protection. In 1996, well after NAFTA came into effect, the United States pressured Mexico into an agreement whereby Mexico would raise the price of winter tomatoes it sold into the United States from $3 per box to $5.17 per box (Postlewaite [36]). The Florida tomato industry had been pressing for protection for decades with little success. The Florida

[25] I would guess that such a clause was requested by the United States. A key issue in the NAFTA and Uruguay Round debates was the effect on US sovereignty. In the WTO context, there was consideration of a US Congressional review panel for dispute settlement findings. If the United States were found to have been treated "unfairly" on three successive occasions, the idea would have mandated US withdrawal from WTO, which also requires six months notice (Srinivasan [41, p. 99]).

[26] This was certainly an objective of Canada's in pursuing a PTA with the United States. See Whalley [44, pp. 17–18]).

growers had pressed a dumping case before the US International Trade Commission, but the ITC ruled that Mexican tomatoes had not caused the 'serious injury' required under anti-dumping rules (Berlau [5]). The Administration was backing congressional legislation that would have likely forced the ITC to reverse its ruling when the settlement with Mexico was reached. The lawyer for the Florida tomato growers was quoted as saying: "One of the cards I had up my sleeve was the commitment the president made in 1993 to congressman Tom Lewis of Florida to protect and preserve Florida's production of vegetables. Finally we got through politically" (quoted in Postlewaite [36]).

Aside from demonstrating Mexico's lack of protection from the capriciousness of US trade policy, the episode also demonstrates the very limited extent to which international agreements bind policy in the face of strong political pressures and the incomplete nature of trade liberalization under PTAs (the presidential commitment to trade protection was allegedly made during the debate over the passage of NAFTA).

One final point to note on NAFTA concerns the passions that were stirred in the US debate over its passage. Prior to NAFTA, the United States and the other major trading nations had lowered trade barriers on an MFN basis through successive rounds of GATT negotiations. Developing nations had received additional preferences through provisions on "Special and differential treatment" (Srinivasan [41]). Mexico had won the right to liberal treatment by the United States with its accession to the GATT in 1986. Therefore, the dangers that the US protectionists were inveighing against should already have occurred.[27] Yet, there was vehement opposition to the idea of trading with a low-wage developing country!

One wonders why these objections were not raised when the true liberalization took place earlier. One possibility is that slow growth in the United States in 1992 and 1993 created a more protectionist climate. However, the same arguments against PTAs with developing countries recurred in the "fast track" debate of 1997, when the United States was experiencing strong growth and record-low unemployment (see FTAA discussion below). A second alternative explanation, more psychological than economic, is that PTAs serve to focus developed country voters on the idea of trading with a developing nation in a way that multilateral liberalization does not, even when developing nations participate in the multilateral liberalization. If so, this would be another argument against North–South PTAs.

In any case, one implication of the debate was that Mexico was compelled to sign side agreements to the NAFTA accord covering labor and environmental regulation. These agreements themselves appeared relatively innocuous,

[27] To be sure, the developed countries had maintained protection in key areas of interest to developing nations, such as textiles and apparel and agriculture, and administered protection remained intact. However, the uproar over NAFTA involved much broader claims that devastation would ensue for all US industry. See Perot and Choate [35].

since they mostly called on the three member nations to enforce their own laws. However, a precedent was established which has carried over into discussions of an FTAA.

9.3.4 FTAA

The current discussion about a Free Trade Agreement of the Americas stems from a proposal by President George Bush for an "Enterprise for the Americas" initiative (Lawrence [26, p. 76]). This was announced almost contemporaneously with the beginning of NAFTA negotiations, partly to allay fears of exclusion on the part of Latin American countries. In a 1994 Miami summit, the leaders of 34 Western Hemisphere countries pledged to create a Free Trade Agreement of the Americas by 2005.

Plans have faltered, however, as President Bill Clinton has failed to obtain "fast-track" negotiating authority from the US Congress.[28] That effectively precludes progress involving the United States, although Canada and Mexico can participate. Among the reasons for the failure was the importance attached by Democrats to the incorporation of environmental and labor standards into any agreement (Clinton refused, so the Democrats withheld their votes). Another reason for Congressional rejection seemed to be the impression that NAFTA had not worked well. This may have been due to the Clinton Administration's 1993 arguments that NAFTA was desirable because the United States ran a bilateral trade surplus with Mexico. After the Mexican economic crisis of 1994–95, that surplus turned into a bilateral deficit.

This highlights a major difficulty with the domino theory of regionalism. First, while the theory's prediction that other nations will wish to join a growing bloc seems to hold, the FTAA serves as a stark reminder that the expansion can be blocked by a single member (albeit an important one).[29] The incident may also provide a new element of consideration for the "dynamic time-path argument" if member nations draw improper inferences from experiments with preferential liberalization.

---

[28] Fast-track authority requires the Congress to vote for or against a negotiated agreement without introducing any amendments. In the absence of such authority, negotiating partners fear that they will reveal their positions in initial negotiations and then be forced to renegotiate as the agreement moves through Congress.

[29] Again, this may not be such a bad thing from the perspective of the excluded countries. Panagariya [33] argues that an FTAA would likely have had a negative static welfare effect on Latin American entrants, since they would be granting preferential access to the United States while receiving little in return, given the relative openness of the US market.

9.4 Policy Implications

9.4.1 Does the developing world's approach matter?

Before proceeding to policy implications, we can consider whether developing countries' stance on preferential liberalization matters. After all, there is a history of inconsequential developing country accords from the 1950s and 1960s which seemed to have little negative impact on the multilateral system. If it is the case that the developed countries are the major impetus behind the recent popularity of regionalism, there might be little that developing countries can do.

Baldwin's [3] arguments suggest the importance that developing countries can play.[30] He describes a history of US and European interest in preferential liberalization that does not seem to have changed much over the post-war era. Japan has been relatively disinterested throughout. Thus, the recent flurry of agreements and initiatives must be due to interest from developing countries. Even if one views the history differently, it is hard to see what agreements developed countries could pursue if developing nations held out for multilateral liberalization. There has been discussion of a Trans-Atlantic Free Trade Agreement between the United States and Europe as well as APEC, which would involve the United States and Japan. However, both of these are relatively implausible, since they would require the participants to overcome the very obstacles that seem to be delaying further multilateral progress.

Srinivasan [41] describes the stake that developing nations have in a healthy multilateral trading system. It is curious to note how recently the developing countries began to participate actively in that system. This leads one to ask whether some of the beneficial attributes of PTAs could be achieved within the multilateral system through judicious reforms.

9.4.2 Suggested reforms

The most obvious reform of the multilateral system would be to graduate developing countries to the status of full participants and to renounce the system of "Special and Differential Treatment." This system was created when it was thought that trade protection was an important development tool. If developing countries now wish to demonstrate their commitment to economic liberalization, they should be able to do so through multilateral liberalization.

[30] It may be worth noting, in the context of this conference, a sharp philosophical difference between Baldwin and Srinivasan in modeling approach. Baldwin [3, p. 886] invokes Sherlock Holmes' quotation: "It is a capital mistake to theorise before one has all the facts. Inevitably one twists the facts to fit the theory, rather than the theory to fit the facts." In contrast, we have Srinivasan's Dictum [39, p. 86]: "What is the fun in modeling, if one can't choose one's own assumptions? It is too bad for the real world if it does not fit my assumptions!"

In a WTO forum, they could also use such willingness to extract concessions from the developed nations.[31]

A second WTO measure is already being undertaken: deeper integration can be pursued multilaterally. The Uruguay Round negotiations included discussions on intellectual property rights, trade-related investment measures, government procurement, and sanitary and phyto-sanitary regulations, among other topics. Given the argument that harmonization makes developing countries more attractive for foreign investment by reducing the costs to multinationals, it would seem that the minimum cost would be achieved through uniformity of regulation. If there is a desire to cover deep integration issues, they should be covered at the multilateral level absent a strong argument for a narrower application. The desirability of this, naturally, is conditional on the desirability of harmonization, which may well be exaggerated.

A more controversial proposal stems from the requirement, under WTO Article III, that the WTO cooperate with the World Bank and the IMF. That, in itself, is not controversial, but one might consider the idea of cross-institutional conditionality — whereby trade liberalization commitments to the WTO might be enforced as a condition of Bank or Fund lending. Srinivasan [41, p. 96] does consider this idea and rejects it, presumably as a violation of developing country sovereignty. However, in the commitment arguments described above, it seems that some developing nations are clamoring for just such a limitation on their sovereignty and seeking a stronger mechanism than the WTO currently offers. If so, this is a remediable failure of the multilateral system, rather than an argument for PTAs.

Finally, there is another policy on which Srinivasan and I disagree. While we both dislike preferential liberalization, he advocates measures that would strengthen WTO restrictions on PTAs to the point where they are no longer preferential. For example, he suggests [42, p. 343] "a precise time limit (say five years) within which . . . any and all preferences (tariff and non-tariff) that are included in any existing or proposed PTAs are required to be extended to all members of the WTO on an MFN basis."

Our disagreement comes over the nature of the WTO. I believe the failure to enforce Article XXIV restrictions on PTAs to date comes from the inability of the organization (be it GATT or WTO) to force unpopular measures on member countries. After all, these organizations are little more than their member countries. Had the GATT tried to block the European Coal and Steel Community in the 1950s, it would have been the United States and Europe trying to block themselves!

Thus, I consider it essential to advocate reforms which will lead countries

[31] Srinivasan [41] makes this point. Note also that my call for graduation only involves developing country trade protection, not necessarily WTO assistance to developing countries in meeting administrative burdens (see Finger and Winters [17, p. 390]).

voluntarily toward multilateral processes and away from the distortions of PTAs.

9.5 Conclusion

In this paper I have tried to cover some of the major arguments for and against preferential trade liberalization from the perspective of developing countries. Taken in sum, the theoretical arguments offer a number of means by which PTAs can do harm, several means by which they might be harmless, and relatively few explicit arguments for positive effects. One must keep in mind that one should not count an effect, such as trade creation, in favor of a PTA since the same effect can be achieved through multilateral liberalization. The choice to pursue a PTA is as much a choice to retain barriers against non-members as it is to lower barriers against members. This can be seen most clearly in the reluctance of some of the APEC participants to proceed with a conventional agreement.

Where preferential arrangements do offer features that the multilateral system cannot, such as in coverage of some new issues and in the opportunity for a developing country to make a credible commitment to an economic reform program, these seem to be arguments for strengthening the WTO rather than for turning to preferential agreements.

References

1 Bagwell, K. and R.W. Staiger, Regionalism and multilateral tariff co-operation, NBER Working Paper No. 5921, 1997.

2 Baldwin, Richard E., A domino theory of regionalism, NBER Working Paper No. 4465, 1993.

3 Baldwin, Richard E., The causes of regionalism, World Economy, 20(7) (1997) 865–88.

4 Bergsten, C. Fred, APEC, The Bogor Declaration and the Path Ahead, Institute for International Economics, Washington, DC, 1994.

5 Berlau, John, 1996 Squishing free trade; stopping imports of Mexican tomatoes to the U.S., Consumers' Research Magazine, 79(11) (1996) 35.

6 Bhagwati, Jagdish, The World Trading System at Risk, Princeton, 1991.

7 Bhagwati, Jagdish, Regionalism and multilateralism: An overview, in: Jaime De Melo and Arvind Panagariya (eds.), New Dimensions in Regional Integration, Cambridge University Press, Cambridge, 1993, Ch. 2.

8 Bhagwati, Jagdish, The world trading system: The new challenges, in: Arvind Panagariya, M.G. Quibria and Narhari Rao (eds.), The Global

Trading System and Developing Asia, Oxford University Press, Oxford, 1997, Ch. 2.

9 Bhagwati, Jagdish, Fast track to nowhere, The Economist, October 18, 1997, p. 21.

10 Bhagwati, Jagdish and Anne O. Krueger, The Dangerous Drift to Preferential Trade Agreements, American Enterprise Institute, Washington, DC, 1995.

11 Bhagwati, Jagdish, Pravin Krishna and Arvind Panagariya (eds.), 1998, Trade Blocs: Alternative Approaches to Analyzing Preferential Trading Agreements, MIT Press, Cambridge, 1998.

12 Bond, Eric, Costas Syropolous and L. Alan Winters, Deepening of regional integration and multilateral trade agreements, mimeo, January 1998.

13 De Melo, Jaime, Arvind Panagariya and Dani Rodrik, 1993, The new regionalism: A country perspective, in: Jaime De Melo and Arvind Panagariya (eds.), New Dimensions in Regional Integration, Cambridge: Cambridge University Press, Cambridge, 1993, Ch. 6.

14 Economist, Mercosur, Back and forth, December 20, 1997, p. 32.

15 Edwards, Sebastian, Comment, in: Stephan Haggard, Developing Nations and the Politics of Global Integration, The Brookings Institution, Washington, DC, 1995.

16 Fernández, Raquel, Returns to regionalism: An evaluation of non-traditional gains from RTAs, NBER Working Paper No. 5970, 1997.

17 Finger, J. Michael and L. Alan Winters, What can the WTO do for developing countries?, in: Anne O. Krueger (ed.), The WTO as an International Organization, University of Chicago Press, Chicago, 1998, Ch. 14.

18 Frankel, Jeffrey A., Regional Trading Blocs in the World Economic System, Institute for International Economics, Washington, DC, 1997.

19 Haggard, Stephan, Developing Nations and the Politics of Global Integration, The Brookings Institution, Washington, DC, 1995.

20 Hall, Kevin G., Free-trade summit begins today in Brazil amid doubts, Journal of Commerce, May 13, 1997, p. 1A.

21 Hall, Kevin G., Traders hit roof over Argentine import rules, Journal of Commerce, December 3 1997, p. 1A.

22 Jackson, John, The World Trading System: Law and Policy of International Economic Relations, 2nd ed., Cambridge: MIT Press, Cambridge, 1997.

23 Krueger, Anne O., Free trade agreements as protectionist devices: Rules of origin, NBER Working Paper No. 4352, 1993.

24 Krueger, Anne O., Trade Policies and Developing Nations, The Brookings Institution, Washington, DC, 1995.

25 Kydland, Finn and Edward Prescott, Rules rather than discretion: The inconsistency of optimal plans, Journal of Political Economy, 85(3) (1977) 437–91.

26 Lawrence, Robert Z., Regionalism, Multilateralism, and Deeper Integration, The Brookings Institution, Washington, DC, 1996.

27 Levy, Philip I., A political-economic analysis of free trade agreements, American Economic Review, 87(4) (1997).

28 Levy, Philip I., Free trade agreements and inter-bloc tariffs, mimeo, Yale University, 1997.

29 Maggi, Giovanni and Andrés Rodríguez-Clare, The value of trade agreements in the presence of political pressures, Journal of Political Economy, 106(3) (1993) June, 574–601.

30 McLaren, John, Size, sunk costs, and judge Bowker's objection to free trade, American Economic Review, 87(3) (1997) 400–20.

31 North American Free Trade Agreement, Text available from the Organization of American States: http://www.sice.oas.org/trade/nafta, 1993.

32 Panagariya, Arvind, East Asia and the new regionalism, World Economy, 17(6) (1994) 817–39.

33 Panagariya, Arvind, The free trade area of the Americas: Good for Latin America?, World Economy, 19(5) (1996) 485–515.

34 Park, Yung Chul, Comment, in: Robert Z. Lawrence, Regionalism, Multilateralism, and Deeper Integration, The Brookings Institution, Washington, DC, 1996.

35 Perot, H. Ross and Pat Choate, Save your Job, Save our Country: Why NAFTA Must be Stopped — Now!, Hyperion, 1993.

36 Postlewaite, Susan, Tomatoes: Mexico vs. Florida, Miami Daily Business Review, November 8, 1996, p. A12.

37 Rodrik, Dani, Promises, promises: Credible policy reform via signaling, Economic Journal, 99 (1989) 756–72.

38 Rodrik, Dani, Comment, in: Anne O. Krueger, Trade Policies and Developing Nations, The Brookings Institution, Washington, DC, 1995, pp. 101–11.

39 Srinivasan, T.N., Comment on Paul Krugman: Regionalism versus multilateralism: Analytical notes, in: Jaime De Melo and Arvind Panagariya (eds.), New Dimensions in Regional Integration, Cambridge University Press, Cambridge, 1993.

40 Srinivasan, T.N., Trade and human rights, Economic Growth Center Discussion Paper No. 765, 1996.

41 Srinivasan, T.N., Developing Countries and the Multilateral Trading System: From the GATT to the Uruguay Round and the Future, Westview Press, Boulder, CO, 1998.

42 Srinivasan, T.N., Regionalism and the WTO: Is nondiscrimination passé?, in: Anne O. Krueger (ed.), The WTO as an International Organization, Chicago: University of Chicago Press, Chicago, 1998, pp. 329–49.

43 Tornell, Aaron and Gerardo Esquivel, The political economy of Mexico's entry to NAFTA, NBER Working Paper No. 5322, 1995.

44 Whalley, John, Why do countries seek regional trade agreements, NBER Working Paper No. 5552, 1996.

45 Winters, L. Alan, Regionalism versus multilateralism, World Bank Policy Research Working Paper Series, No. 1687, 1996.

46 Winters, L. Alan., What can European experience teach developing countries about integration? World Economy, 20(7) (1997) 889–912.

47 WTO, Regionalism and the World Trading System, World Trade Organization, Geneva, 1995.

48 Yeats, Alexander, Does Mercosur's trade performance raise concerns about the effects of regional trade arrangements?, World Bank Policy Research Working Paper No. 1729, 1997.

Trade, Growth, and Development
G. Ranis and L.K. Raut

CHAPTER 10

The Impact of Regionalism on Agricultural Trade: APEC and Japanese Rice Imports[1]

Junichi Goto

Research Institute for Economics and Business, Kobe University,
2-1 Rokkodai-cho, Nada-ku, Kobe 657 Japan

10.1 Introduction

Since the late 1980s, we have observed the emergence of a new regionalism in various parts of the world. Europe has a long history of regionalism. Ever since the European Economic Community (EEC) was established in 1958, it has expanded its membership and deepened the degree of integration. In 1973, the union of the original six members admitted three former EFTA members: Denmark, Ireland, and the United Kingdom. Greece joined the EC in 1981 and Spain and Portugal were admitted in 1986. During this period, the European integration deepened too. In 1968, the EC formed a customs union, and the controversial common agricultural policy (CAP) was initiated. The degree of integration in Europe was further increased in the early 1990s, when the EC countries tried to form a single market by lifting various obstacles to the movement of goods and services within the region. This fairly successful attempt is known as "EC92." Recently, the European Union admitted Austria, Finland, and Sweden as new members, and the Union is also moving toward monetary integration.

In North America, Canada and the United States have a long history of strong economic ties. As early as 1965, the two countries signed the Canada–United States Automotive Products Agreement, which enabled nearly free movement of motor vehicles and parts between the two countries. The Canada–United States Free Trade Agreement (CUFTA) was signed in 1988 and put into effect the following year. Mexico also wanted to have closer economic

[1] The author is grateful to Takamasa Akiyama, John Baffes, Koichi Hamada, Will Martin, Donald Mitchell, Panos Varangis, Alan Winters, and seminar participants at the World Bank and Yale University for helpful comments and suggestions on earlier versions of the paper.

ties with the US, and President Salinas of Mexico and President Bush of the US agreed in 1990 that a free trade agreement between the US and Mexico would substantially benefit the economies of the two countries. Since Canada did not want to be left behind, she tried hard to be included in the new US–Mexican agreement. As a result, the North American Free Trade Agreement (NAFTA) was signed in December 1992. After overcoming some opposition in the US Congress, NAFTA was put into effect in January 1994. Under NAFTA, Canada, Mexico, and the United States agreed to abolish tariff and nontariff barriers in the region by the 2009.

In contrast to the development in Europe and North America, there have been few attempts to form free trade areas in Asia until recently. While Indonesia, Malaysia, the Philippines, Singapore, and Thailand were united into the Association of South East Asian Nations (ASEAN) in 1967, the union started as an anti-communist political and military association rather than as an economic bloc. However, in the 1990s, attempts to form economic unions have become common in Asia too. In 1990, Premier Mahathir of Malaysia advocated that Asian countries, including Japan, form their own economic bloc, such as the East Asian Economic Caucus (EAEC), to counterbalance possible adverse effects of economic integration outside of Asia. This plan did not please the United States or other Asian countries which are heavily dependent upon their exports to the US. Another alternative considered by the Southeast Asian nations was to form a more open regional union consisting of broader membership, such as the Asia Pacific Economic Cooperation (APEC). The United States supports the formation of a broader regional union that would include non-Asian countries such as Australia and New Zealand, and notably, the United States itself. In fact, as manifested in the Bogor Declaration of 1994, APEC members agreed to achieve free and open trade and investment in Asia and the Pacific by 2010 for industrialized countries and by 2020 for developing countries. In November 1996, the eighteen APEC members gathered in Manila and presented individual action plans to achieve this goal.

In view of the increased importance of regionalism in the world economy, the purpose of this paper is to examine the *ex post* and *ex ante* effects of regional integration on international trade flows. The paper will focus on agricultural trade because, as shown below, the impact of regional integration on agricultural trade has often been stronger than on manufacturing trade.

In Section 10.2, the salient features of agricultural trade will be examined in comparison with manufacturing trade. Generally speaking, international flows of agricultural goods are more heavily protected by tariff and nontariff barriers than those of manufacturing goods. In addition, we argue that, contrary to popular belief, agricultural trade is far from the flow of homogeneous products. While the degree of product differentiation of agricultural goods on the whole is probably smaller than that of manufacturing goods, some agricultural goods, such as rice in Japan, are highly differentiated, and therefore, any attempt to measure the degree of impact of regional economic integration under the assumption of homogeneous product seems to be misleading.

In Section 10.3, we develop a simple model for the analysis of the impact of regional economic integration on agricultural trade flows. We develop a Dixit–Stiglitz–Krugman-type product differentiation model with tariff distortions because it captures more realities of agricultural trade than a homogeneous product model for various reasons discussed below, and therefore, it gives deeper insights into the likely effect of regional economic integration on agricultural trade flows. Using the model, we derive two intuitively appealing propositions: (a) *the impact of regional integration is stronger when the degree of pre-integration protection is higher* and (b) *the impact is stronger when the degree of product differentiation is lower.* Taken together, these propositions suggest that regional integration has a stronger impact on agricultural trade than on manufacturing trade because, in general, trade barriers on agricultural products are higher and the degree of product differentiation is lower for agricultural trade, in comparison with manufacturing goods.

In Section 10.4, the validity of the two propositions is tested against the actual data, taking two incidents of the EC expansion as examples — the admission of Greece in 1981 and Spain and Portugal in 1986. We examine whether agricultural trade was more strongly affected than manufacturing trade after these two incidents of progress toward regional integration in Europe. As discussed in detail below, *the intra-regional trade in agricultural products increased sharply after 1981 and 1986, although such a jump cannot be observed for manufacturing trade.* Furthermore, a careful examination reveals that trade flows of some agricultural products were more strongly affected than others, depending on the magnitude of initial trade barriers and the degree of product differentiation.

In Section 10.5, the model developed in Section 10.3 is applied to the examination of a possible *ex ante* impact of regional integration. In view of the fact that APEC countries are actively attempting to realize a free trade regime, we evaluate the likely impact of future liberalization of agricultural trade under the framework of the APEC free trade agreement. One of the most controversial commodities, rice in the Japanese market, will be used in the calibration exercise as an example.

Section 10.6 summarizes the major findings of the paper, and proposes an agenda for future research.

10.2 Salient Features of Agricultural Trade

10.2.1 Heavy protection

As shown in Table 10.1, trade barriers imposed on manufacturing trade, perhaps with the exception of textiles and clothing, have been greatly reduced through a series of tariff negotiations under the GATT, and as of 1989, the average tariffs of advanced countries on manufacturing trade (MFN tariffs) are minimal at 3–6 percent. Tariff rates imposed on agricultural products, however, are much higher. Table 10.2 lists the post-Uruguay Round tariff rates by commodity. Tariff rates on agricultural goods (see "Agriculture, excluding

Table 10.1: Average MFN tariffs on manufactured goods (%)

	1962	1970	1989
European community	11	8	6
United States	12	9	5
Japan	16	12	3

Source: Pohl and Sorsa [9], p. 13.

Table 10.2: MFN tariff rates before and after the Uruguay Round (%)

	Levels and changes weighted by imports from the world excluding FTA[a]		
	Post-UR applied rate	Tariff reduction[b]	Post-UR bound rate
Agriculture, excluding fish: Estimate 1	25.0	32.4	32.4
Agriculture, excluding fish: Estimate 2	7.6	4.5	18.5
Fish and fish products	4.4	4.4	5.2
Petroleum oils	1.7	1.5	3.1
Wood, pulp, paper and furniture	1.2	5.1	2.2
Textiles and clothing	9.8	3.1	12.4
Leather and rubber footwear	6.4	3.1	7.8
Metals	2.9	4.3	4.1
Chemical and photographic supplies	4.8	5.1	7.3
Transport equipment	6.0	3.3	6.9
Nonelectric machinery	3.7	3.6	5.1
Electric machinery	4.6	4.0	5.7
Mineral products, precious stones/metal	1.6	2.8	2.6
Manufactured articles n.e.s.	2.8	4.1	4.1
Industrial goods (lines 5–14)	4.2	3.9	5.7
All merchandise trade (lines 2–14)	4.3	3.9	6.5

a. Average absed on the 40 GATT members.

b. Value of imports from partner countries that do not participate in free trade agreements in the world.

c. Weighted average tariff reduction measured by $dT/(1 + T)$ in percent.

Source: Finger, Ingco and Reincke [3].

fish: Estimate 2" in the table) are 7.6–18.5 percent. In addition to tariff protection, agricultural goods are also heavily protected by various nontariff barriers. For example, until very recently, Japan imposed an almost total ban on rice imports. In the European Community, most of the agricultural products are heavily protected by the infamous Common Agricultural Policy (CAP). As a result, the tariff equivalency of such nontariff barriers on agricultural trade is very high. The numbers for "Agriculture excluding fish: Estimate 1" in Table 10.2 show a combined rate of external barriers (i.e., tariff plus tariff-equivalency of nontariff barriers), of 25.0–32.4 percent. Furthermore, as Table 10.3 shows, external protection rates for certain agricultural products are very high. The tariff equivalency is more than 100 percent in some cases.

Table 10.3: EC agricultural protection and trade patterns, 1989 (%)

SITC code		Tariff-equivalent of protection and subsidies
01	Meat	6–270
02	Dairy products	0–200
03	Fish	0–30
04	Cereals	20–130
05	Vegetables and fruit	0–30
06	Sugar	180
07	Coffee, tea, spices	0–18
08	Animal foodstuffs	0–50

Source: Pohl and Sorsa [9], p. 22.

10.2.2 Some evidence of product differentiation

Contrary to the popular belief that agricultural trade is an international exchange of homogeneous commodities, there is some evidence that agricultural trade is differentiated. Note that under the traditional Heckscher–Ohlin framework, all goods are treated as homogeneous products (e.g., a car is a car). However, after many distinguished economists, including Dixit, Helpman and, most notably Krugman, noticed that international trade can be generated by increasing returns to scale and product differentiation, people came to believe that most manufacturing goods are more or less differentiated (e.g., a Japanese car is differentiated from an American car). But, agricultural products have continued to be regarded as homogeneous, and therefore the assumption has been that consumers only care about their price.

However, there are several reasons to believe that agricultural products are also differentiated, as discussed in detail below. While manufacturing products are generally more strongly differentiated than agricultural products, some agricultural goods (e.g., rice in the Japanese market) are more differentiated than some manufacturing goods (e.g., cotton yarn).

Apples and oranges as the same product. When we analyze trade statistics using available data, or any statistics for that matter, we are observing data which are aggregated at least to some extent. For the analysis of international trade, we often use trade flow data classified according to the SITC. Many studies, including the present paper, rely on the 2-digit or 3-digit SITC data. As discussed in detail below, this kind of aggregation makes it all the more important to analyze the data under the framework of product differentiation assumption rather than under homogeneous product assumption.

Note that the degree of product differentiation depends on how we define the product. If we define the product with sufficient disaggregation, almost any product can be treated as a homogeneous product. To better understand this point, let us take a passenger car, for example, a typical differentiated product. Obviously, a Ford Escort and a Porche are quite differentiated from each other, and consumers do care about the product differentiation. However, if we define a product with sufficient disaggregation as, say, a "new 1998 white 4-door Ford Escort LX with automatic transmission, power steering, air conditioning, etc., which is delivered to New Haven, Connecticut on March 27, 1998," it can be regarded as an almost homogeneous product. In this case, consumers do not care about which particular unit of the product they are buying, and would buy any unit which is cheaper than other units.

Similar things can be said about agricultural products. When we look at trade flow data at SITC 1-digit, 2-digit, or 3-digit level, as most studies do, we are looking, for example, at the product of "food and live animals (SITC 1-digit)," "fruits and vegetables (SITC 2-digit)," or "fresh fruits and fresh or dried nuts (SITC 3-digit)." In other words, in any of the above three classifications, "apples" and "oranges" fall in the same product category. It is only when we define the product at SITC 4-digit level that apples and oranges become different products. Even when we disaggregate the product in such detail, a certain variety of apples can command a much higher price than other apples, because consumers do care which variety of apples they are buying. In fact, even at the SITC 5-digit level, long-grain Indica rice is treated as the same product as short-grain Japonica rice. The same applies to other agricultural products.

Thus the popular belief that manufacturing goods are differentiated but agricultural goods are homogeneous is fairly misleading. Both goods are more or less differentiated. The only difference is the degree of differentiation, which varies from product-to-product and depends on the *degree* of disaggregation when a product is defined.

Real examples: Rice in Japan. Rice in the Japanese market is a typical example of how differentiated certain agricultural products can be. As discussed in detail in Section 10.5 below, the Japanese rice market has been very much closed to foreign rice, except for certain emergency imports. Until the Uruguay Round agreement was put into effect in 1995, Japan imposed an almost total ban on foreign rice, and even after 1995 foreign rice is imported only up to the minimum access level (4 percent of domestic consumption in 1995, which is to be gradually increased to 8 percent by the year 2000). Almost all rice sold in Japan is similar, short-grain, Japonica-type rice grown in Japan, and most foreigners and even many Japanese, will not be able to tell the difference. In spite of such apparent similarity, however, rice is generally highly differentiated in the eyes of Japanese consumers. At the Japanese rice exchange market, prices are quoted according to brand and production sites. Table 10.4 shows a *partial listing* of prices of various brands of rice at the Japanese rice exchange market in 1995. While in the 1960s and 1970s the majority of rice was distributed through government channels, nongovernmental rice distribution has become increasingly popular over time; its share of total rice distribution in 1995 was 70.2 percent. Table 10.4, prepared by the Japanese Ministry of Agriculture and Fishery, shows 62 listings (a partial listing) of brands distributed through nongovernmental channels. While prices for these brands are substantially higher than those for standard government rice (16,392 yen per 60 kg), the price varies from one brand to another. Generally speaking, *Koshihikari* brands command the highest price (around 21,000–25,000 yen per 60 kg). Furthermore, *Koshihikari*-brand rice produced in certain areas commands much higher prices than those produced in other areas. *Koshihikari* rice produced in *Uonuma*, a tiny city in Niigata Prefecture, is considered to be the best of the best. Table 10.5 shows the price difference in such rices in 1996. As shown in the table, *Koshihikari* rice produced in Niigata Prefecture is 55 percent more expensive than standard rice, and the price of *Uonuma–Koshihikari* rice is almost double of that of standard rice. Thus, Japanese consumers perceive a high degree of product differentiation in rice and are willing to pay a huge premium on a certain brand of rice.

The emergency rice import in 1993–94 is another example of how rice is differentiated in the eyes of the Japanese consumers. In the summer of 1993, Japan was unusually cool and the rice harvest that year was only 74 percent of the normal harvest. To cope with a possible food shortage, the Japanese government decided to import 2.59 million tons of rice from Australia, China, Thailand, and the United States. However, Japanese consumers did not like the idea of purchasing foreign rice. Housewives waited patiently for several hours in line to obtain scarce domestic rice, even though foreign rice was readily available at a cheaper price. Faced with the unpopularity of foreign rice, the government sold it as a package with domestic rice. In other words, consumers were forced to buy a certain amount of unwanted foreign rice in order to buy the domestic rice they really wanted. In spite of such a desperate

Table 10.4: Benchmark prices of various brands of rice in Japan, 1995 (yen/60 kg)

Brand/Site	Tokyo	Osaka	Brand/Site	Tokyo	Osaka
Koshihikari			Akitakomachi		
Niigata	25,087	24,863	Akita	21,741	21,859
Toyama	22,885	22,772	Iwate	21,025	20,877
Ishikawa	22,748	22,736	Hanamomai		
Fukui	—	22,640	Yamagata	20,304	20,394
Hyogo	—	22,482	Fukuhikari		
Shimane	—	22,312	Fukui	—	20,501
Fukushima	22,634	—	Yamahikari		
Tottori	—	22,018	Tottori	—	20,558
Nagano	21,905	21,883	Yamaguchi	—	20,345
Fukuoka	—	21,949	Nihonbare		
Kumamoto	—	21,755	Shiga	—	20,109
Shiga	—	21,680	Hyogo	—	20,125
Mie	—	21,718	Asahi		
Ibaragi	21,808	—	Okayama	—	21,222
Okayama	—	21,479	Hatsushimo		
Tochigi	21,705	—	Gifu	—	21,068
Kagawa	—	21,313	Yukinosei		
Chiba	21,640	—	Niigata	20,819	21,031
Gifu	—	21,062	Notohikari		
Aichi	—	21,041	Ishikawa	—	21,007
Sasanishiki			Echigowase		
Miyagi	21,980	22,157	Niigata	20,863	20,804
Akita	21,319	21,363	Tsugaruotome		
Fukushima	21,273	—	Aomori	20,357	20,393
Yamagata (Shonai)	21,208	21,302	Mutsukaori		
Iwate	21,168	21,116	Aomori	20,371	20,413
Yamagata	20,878	20,928	Niigatawase		
Hinohikari			Niigata	20,119	20,150
Kumamoto	—	20,936	Todorokiwase		
Oita	—	20,584	Niigata	19,981	19,946
Saga	—	20,476	Yamahoushi		
Fukuoka	—	20,501	Yamaguchi	—	19,871
Kinuhikari			Akebono		
Ibaragi	21,550	—	Okayama	—	19,901
Shiga	—	20,571	Kirara-397		
Fukuoka	—	20,400	Hokkaido	19,501	19,652
Hitomebore			Mutsuhomare		
Iwate	21,696	21,779	Aomori	19,059	18,803
Miyagi	21,696	21,779	Soraiku-125-Go		
Fukushima	21,696	—	Hokkaido	18,565	18,797
Hatsuhoshi			Yukihikari		
Fukushima	20,765	20,810	Hokkaido	18,588	18,767
Tochigi	20,608	—			
Chiba	20,180	—	Standard rice	16,392	16,392

Source: Japanese Ministry of Agriculture and Fisheries.

Table 10.5: Price of Japanese rice, 1996

Brand name	Price (yen/60 kg)	Index (standard = 100)
Uonuma Koshihikari	31,779	194
Niigata Koshihikari	25,359	155
Standard Domestic	16,392	100

Source: Japanese Ministry of Agriculture and Fisheries.

effort by the government to sell foreign rice, about 1 million tons (or 38%) of the imported rice remained unsold, and the government was obliged either to ship it back to foreign countries as food aid under official development assistance (ODA), or to feed it to animals.

Intraindustry trade in agriculture. If the product in question is purely homogeneous, international trade in the product should theoretically be one-way trade. As described in a traditional textbook of international trade theory, for example, Portugal exports wine to England, and England exports cloth to Portugal in return. However, a brief look at actual trade flow data reveals that agricultural trade, as well as manufacturing trade, is far from one-way trade.

To examine the magnitude of intraindustry trade, we calculated intraindustry trade indices, which were first used by Grubel and Lloyd [7] for agricultural and manufacturing trade in Europe. As in Grubel and Lloyd, the intraindustry trade index of the ith industry product (ITI_i) is defined as

$$ITI_i = \left\{1 - \frac{|X_i - M_i|}{X_i + M_i}\right\} \times 100,$$

where X_i and M_i are the value of exports and the value of imports, respectively, of the ith industry good. A higher ITI value means that the degree of intraindustry trade is higher.

Table 10.6 summarizes intraindustry trade indices for agricultural trade and manufacturing trade in Europe, which were calculated using the above formula. The indices vary from one combination of countries to another, but two things are noteworthy. First, intraindustry trade indices of agricultural trade are far from zero (note that if agricultural goods were completely homogeneous, the indices would all have become zero). For example, in the table, the ITI of agricultural products between Great Britain and Portugal is as high as 97.8, which means exports and imports almost completely overlap. Second, intraindustry trade indices of manufacturing products are generally higher than those of agricultural goods. A simple ITI average for manufacturing trade is 86.1, higher than the 76.3 for agricultural trade. Taken together, Table 10.6 suggests that agricultural trade should also be analyzed under the framework of product differentiation rather than under homogeneous product, although the degree of product differentiation of agricultural products is, on average, lower than that of manufacturing goods.

Table 10.6: Intraindustry trade index, 1995

AGRICULTURE	BLX	DNK	GER	FRA	ITA	IRL	GBR	NLD	GRC	ESP
Belgium-Luxemburg										
Denmark	80.1									
Germany	89.9	60.5								
France	87.2	61.9	79.9							
Italy	69.3	27.2	94.2	53.9						
Ireland	48.4	60.4	11.9	37.7	35.2					
Great Britain	81.4	35.7	68.3	85.1	96.2	68.4				
Netherlands	81.6	72.7	63.6	72.0	29.5	52.2	57.3			
Greece	34.7	9.1	94.0	39.3	58.7	12.9	82.0	18.1		
Spain	93.1	58.6	56.3	88.8	65.9	30.1	96.4	82.3	62.2	
Portugal	89.6	46.0	86.2	68.0	80.9	28.3	97.8	80.5	56.0	56.6
SIMPLE AVERAGE = 76.3										
MANUFACTURING	BLX	DNK	GER	FRA	ITA	IRL	GBR	NLD	GRC	ESP
Belgium-Luxemburg										
Denmark	64.6									
Germany	94.1	80.6								
France	93.0	90.2	94.0							
Italy	81.8	60.3	99.9	92.2						
Ireland	36.0	73.1	72.8	53.1	52.6					
Great Britain	95.2	91.5	77.5	89.1	92.4	90.5				
Netherlands	90.6	66.9	84.0	96.6	77.6	62.3	92.9			
Greece	20.9	46.2	55.6	40.0	21.3	21.6	46.5	19.9		
Spain	76.5	81.4	81.2	89.0	80.4	46.5	90.5	75.7	31.0	
Portugal	71.0	46.7	96.6	92.1	38.8	71.3	93.2	74.8	45.5	65.3
SIMPLE AVERAGE = 86.1										

Source: Author's calculations using United Nations Trade Statistics.

10.3 Regional Integration and Agricultural Trade — A Theory

10.3.1 The general model

In the general model, the situation of a representative country k ($k = 1, 2, 3, ..., M$) is as follows. Consumers possess the individualistic social utility function (U_k) in which

$$U_k = \left[\sum_{i=1}^{N} C_{ik}^{\beta} \right]^{1/\beta}, \quad 0 < \beta < 1, \tag{10.1}$$

where C_{ik} is the amount of consumption of the ith differentiated product in country k, and N is the number of types of differentiated products available to consumers. Some of the differentiated products are domestically produced while others are imported.

Consumers maximize their utility subject to the budget constraint

$$\sum_{i=1}^{N} P_{ik} C_{ik} = Y_k, \tag{10.2}$$

where P_{ik} is the domestic price (i.e., tariff-inclusive price) of the ith differentiated product in country k, and Y_k is the national income of country k.

From the utility maximization, we obtain the inverse demand functions

$$P_{ik} = \frac{C_{ik}^{\beta-1} Y_k}{Z_k}, \text{ where} \tag{10.3}$$

$$Z_k = \sum_{i=1}^{N} C_{ik}^{\beta}. \tag{10.4}$$

From (10.3), the elasticity of demand for the ith differentiated product (ε_{ik}) is

$$\varepsilon_{ik} = \frac{1}{(1-\beta) + \frac{\beta C_{ik}^{\beta}}{Z_k}}. \tag{10.5}$$

If we assume, following Krugman [8] and Dixit and Norman [2], a large number for N and the symmetry of each differentiated product, we can neglect the second term of the denominator on the right-hand side, and (10.5) reduces to

$$\varepsilon = \frac{1}{1-\beta}. \tag{10.5$'$}$$

In equation (10.5′) we omit the subscript i and k for ε, because the demand elasticity turns out to be identical for all products, due to the assumptions of symmetry and the large number for N.

The producer of the ith differentiated product in country k is characterized by the cost function

$$TC_{ik} = W_k F + W_k m \left(\sum_{j=1}^{M} C_{ij} \right), \tag{10.6}$$

where TC_{ik} and W_k are, respectively, total cost of the ith producer and wage rate in country k, and m is the labor input requirement per unit of output, while F is a fixed labor input necessary for any positive amount of production.

Due to the fixed cost $W_k F$, the production technology exhibits increasing returns to scale. The producer maximizes the profit function

$$\pi_{ik} = \sum_{j=1}^{M} \frac{P_{ij}}{1+t_{ij}} C_{ij} - \left[W_k F + W_k m \left(\sum_{j=1}^{M} C_{ij} \right) \right], \tag{10.7}$$

where π_{ik} is the profit of the ith producer in country k, and t_{ij} is the tariff rate imposed by country j on the ith differentiated product. When country j is the home country, the tariff rate is zero. From the profit maximization, we obtain the following pricing rule for the ith producer in country k, facing a demand curve with elasticity $1/(1-\beta)$ as

$$P_{ij} = \frac{W_k m(1+t_{ij})}{\beta}. \tag{10.8}$$

Furthermore, we assume free entry and free exit. Therefore, the profit of each existing firm is forced to zero in equilibrium. Hence, in equilibrium, we have

$$\pi_{ik} = \sum_{j=1}^{M} \frac{P_{ij}}{1+t_{ij}} C_{ij} - \left[W_k F + W_k m \left(\sum_{j=1}^{M} C_{ij} \right) \right] = 0. \tag{10.9}$$

Applying Shepard lemma to equation (10.6), the demand for labor input by the ith producer (ℓ_i) is obtained as

$$\ell_i = F + m \sum_{j=1}^{M} C_{ij}, \tag{10.10}$$

The domestic labor supply (L_k) is assumed to be constant, and in equilibrium we have

$$\sum_{i=1}^{N_k} \ell_i = L_k, \tag{10.11}$$

where N_k is the number of firms in country k.

The tariff revenue is distributed to domestic consumers in a lump-sum fashion. Hence, the national income consists of factor payments and tariff revenues as in

$$W_k L_k + \sum_{i=N_{k+1}}^{N} \frac{t_{ik}}{1+t_{ik}} P_{ik} C_{ik} = Y_k. \tag{10.12}$$

The above model is complete, and the above specification gives equilibrium conditions for a representative country k. We can solve the model which consists of M countries, once the values of the parameters (m, F, β, t_{ik}, L_k, M, and N) are identified. Note that this general model can accommodate not only any number of countries (M) and commodities (N), but also the differences in country sizes (L_k) and tariff rates (t_{ik}).

10.3.2 Implications of the assumption of constant elasticity

So far, we have presented a variant of the Krugman model in which all products are differentiated, although in the following discussions in this section and Section 10.5 we use a simpler version of the product differentiation model, based on Armington [1]. In this model, products produced in the same location — called a "province" — are assumed to be perfect substitutes, while products produced in different locations are differentiated. In this subsection, we will show that the Krugman model with constant elasticity is essentially equivalent to Armington model, and therefore, in this paper, it might be called a Krugman–Armington model.

As mentioned above, we assume the number of the firms to be large enough to justify the constant elasticity of the demand function. Because of this simplification, even though the decreasing cost plays an important role in determining the equilibrium, the resulting scale of production and the equilibrium average cost are not changed by the existence of trade. Suppose that the utility function is expressed as equation (1) above and the production function of the ith producer is characterized by

$$x_i = f(\ell_i). \tag{10.12a}$$

Then, the amount of factor input (ℓ_i) and the amount of output (x_i) are completely determined by the following conditions (see Gros [6] for a proof):

$$\frac{df(\ell_i)}{d\ell_i} \cdot \frac{\ell_i}{f(\ell_i)} = \frac{1}{\beta}. \tag{10.12b}$$

The left-hand side of equation (10.12b) can be interpreted as the *degree of economy of scale* in production, while the right-hand side can be interpreted as the *degree of product differentiation* from the consumers' viewpoint.

Applying this relationship to the model developed above, or substituting equation (10.12b) into (10.10) and (10.11), we know that the amount of production of the ith firm (x_i) and the number of firms in country k, (N_k), are invariant and expressed as

$$x_i = \frac{F\beta}{m(1-\beta)}, \tag{10.12c}$$

and

$$N_k = \frac{1-\beta}{F} L_k. \tag{10.12d}$$

Therefore, the total amount of production of differentiated products in country k becomes

$$x_i N_k = \frac{F\beta}{m(1-\beta)} \frac{1-\beta}{F} L_k = \frac{\beta}{m} L_k. \tag{10.12e}$$

Hence, by the choice of the unit, the total amount of production in country k (and that of every country under symmetry) can be normalized as unity. Thus, in the specification below, each country is assumed to produce *one unit of a type of the differentiated product.* This is essentially equivalent to the general specification above, under the assumption of constant elasticity.

10.3.3 Determinants of the impact of regionalism on trade

Let us examine the impact of regional economic integration on trade flows, using the framework developed above. In this section we will use a little simpler framework to keep the theoretical analysis manageable. In this section, the world is assumed to consist of a large number (N) of identical countries, of which n countries form an economic bloc while the other $(N-n)$ countries are left out. Figure 10.1 shows the basic framework of the analysis in this section. Trade within the bloc is subject to no tariffs, while other trade is subject to a constant tariff t. Since all countries are assumed to be identical, without loss of generality, each country is assumed to produce one unit of a type of differentiated product (for a discussion of the "Krugman–Armington" model, see the previous subsection).

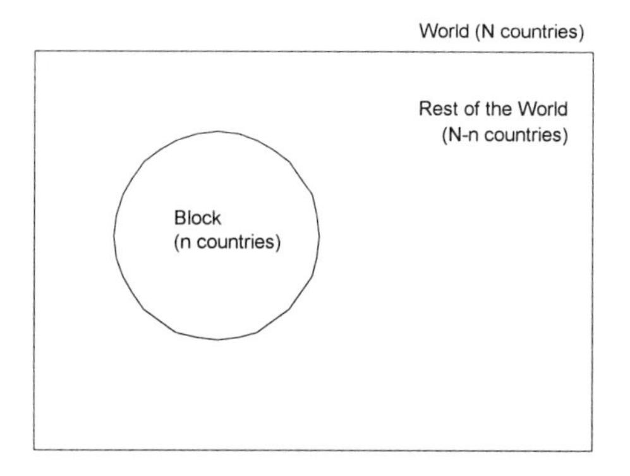

Figure 10.1: Framework of the analysis

In this simple model, the representative country in the economic bloc is characterized by the individualistic social utility function

$$U_B = \left[nC_{BB}^{\beta} + (N-n)_{RB}^{\beta} \right]^{1/\beta}, \tag{10.13}$$

where U_B is the utility of the representative country in the economic bloc, and C_{BB} and C_{RB} are, respectively, the amounts of consumption of each type of bloc good and rest of the world (ROW) good in the representative country in the economic bloc. Consumers maximize their utility subject to the budget constraint

$$PnC_{BB} + (1+t)(N-n)C_{RB} = Y_B, \tag{10.14}$$

where P is the producer price of the goods produced in a country in the economic bloc, This price is the same as the consumer price within the bloc, because no tariffs are imposed by the bloc country on bloc goods. The producer price of the goods produced in the rest of the world is set to unity as a numeraire. Since tariff t is imposed on the ROW good, the consumer price of the ROW good in the bloc country is $(1+t)$. Y_B is the national income of the representative country in the economic bloc.

Maximizing the utility in equation (10.13) subject to budget constraint (10.14), we obtain

$$\left[\frac{C_{BB}}{C_{RB}}\right]^{1/\beta} = \frac{(1+t)}{P}. \tag{10.15}$$

On the other hand, the consumers of the representative country in the rest of the world are characterized by the individualistic social utility function

$$U_R = \left[nC_{BR}^{\beta} + (N-n-1)C_{fR}^{\beta} + C_{hR}^{\beta}\right]^{1/\beta}, \tag{10.16}$$

where U_R is the utility of the representative country in the rest of the world, and C_{BR} is the amount of consumption of each type of bloc good in the representative country in the rest of the world. C_{fR} is the imported amount of each type of ROW good, and C_{hR} is the amount of the home good consumed in the representative country in the ROW. Consumers maximize their utility subject to the budget constraint

$$(1+t)PnC_{BR} + (1+t)(N-n-1)C_{fR} + C_{hR} = Y_R. \tag{10.17}$$

Note that in the representative ROW country, bloc goods as well as imported ROW goods are subject to tariff t.

Solving the utility maximization problem in the above, we obtain

$$\left[\frac{C_{fR}}{C_{BR}}\right]^{1-\beta} = P, \tag{10.18}$$

$$\left[\frac{C_{hR}}{C_{fR}}\right]^{1-\beta} = 1+t, \tag{10.19}$$

and

$$\left[\frac{C_{hR}}{C_{BR}}\right]^{1-\beta} = P(1+t). \tag{10.20}$$

Further, from the world market clearing conditions for bloc goods and ROW goods, we have

$$nC_{BB} + (N-n)C_{BR} = 1, \text{ and} \tag{10.21}$$

$$nC_{RB} + (N-n-1)C_{fR} + C_{hR} = 1. \tag{10.22}$$

Since the trade has to be balanced in equilibrium, we have

$$PC_{BR} = C_{RB}. \tag{10.23}$$

By Walras's law, one of the above eleven equations is redundant. So, ten independent equations determine ten endogenous variables ($U_B, U_R, Y_B, Y_R, C_{BB}, C_{RB}, C_{BR}, C_{fR}, C_{hR}$, and P).

Now, let us suppose that a country, say Greece, is admitted to the economic bloc (e.g., European Community). Since Greece's pre-accession level of export to the EC is C_{RB} and its post-accession export level is C_{BB}, we can evaluate the impact of the Greece accession on its exports to the EC by examining the following index of the change in exports (CE), which is defined as[2]

$$CE = \frac{C_{BB}}{C_{RB}}. \tag{10.24}$$

First, let us examine how CE is affected by different values of t. By repeated substitution using equations (15), (10.18)–(10.23), we can derive the following equation:

$$\begin{aligned} &n(C_{RB}^{-1} - n)^{-1/\beta} C_{RB}(1+t)^{1/(1-\beta)} \Big[(N-n-1) + (1+t)^{1/(1-\beta)}\Big]^{1/\beta} \\ &+ (N-n)(C_{RB}^{-1} - n)^{-(1-\beta)/\beta} C_{RB} \Big[(N-n-1) + (1+t)^{1/(1-\beta)}\Big]^{(1-\beta)/\beta} = 1 \end{aligned} \tag{10.25}$$

Although equation (10.25) looks very complicated, we can notice the following: (i) Left-hand side (LHS) is a monotonically increasing function of t, (ii) LHS is a monotonically increasing function of C_{RB}, and (iii) Right-hand side (RHS) is constant. From (i)–(iii), it is clear that if equation (10.25) is to hold with equality, a larger value of t must be accompanied by a smaller value of C_{RB}. Thus, we have proved the following condition:

$$\frac{\partial C_{RB}}{\partial t} < 0. \tag{10.26}$$

[2] Exactly speaking, this may not be entirely correct, because we are neglecting the impact of the new accession on C_{BB}. By assuming that the new member is small, we are implicitly assuming that C_{BB} is not changed by the admission. While we add this assumption in the theoretical analysis here for algebraic simplification, we use variable C_{BB} in the simulation exercise.

Similarly, by manipulating the equilibrium conditions, we obtain

$$nC_{BB} + (N - n)C_{BB}^{1-\beta}C_{RB}^{\beta}(1 + t)^{-1} = 1. \quad (10.27)$$

By inspecting equation (10.27), we notice the following: (i) LHS is a monotonically decreasing function of t, (ii) LHS is a monotonically increasing function of C_{BB}, (iii) LHS is a monotonically increasing function of C_{RB}, and (iv) RHS is constant. Hence, from (10.26), and (i)–(iv), it is clear that if equation (10.27) is to hold with equality, larger values of t must be accompanied by the larger values of C_{BB}. In other words, we must have the following condition:

$$\frac{\partial C_{BB}}{\partial t} > 0. \quad (10.28)$$

From (10.26) and (10.28), it is clear that we have

$$\frac{\partial CE}{\partial t} > 0. \quad (10.29)$$

Thus, we have proved the following:

Proposition 1. *The degree of increase in exports from a newly admitted member to the old members of trade bloc is larger when the initial trade barrier was larger.*

In other words, Proposition 1 means that, for example, when Greece is admitted to the EC (and when EC tariffs on the goods coming from Greece are lifted), Greek exports of heavily protected products before integration increase more than those of less protected products.

Second, let us examine the relationship between CE and the degree of product differentiation, which can be measured by the elasticity of substitution (σ). Since we have

$$\sigma = \frac{1}{1 - \beta}, \quad (10.30)$$

in order to examine the magnitude of the impact of degree of product differentiation (σ) on CE, all we have to do is determine the sign of $(\partial CE/\partial \beta)$, because σ is a monotonically increasing function of β for the range of $0 < \beta < 1$. Note that a higher value of β means a smaller degree of product differentiation.

First, manipulating the above equilibrium conditions, we can show

$$\frac{\partial P}{\partial \beta} < 0. \quad (10.31)$$

Inequality (10.31) means that the price markup of bloc goods over ROW goods is smaller when the degree of product differentiation is weaker. From

(10.31) and (10.15), it is clear that a larger value of β must be accompanied by a larger value of (C_{BB}/C_{RB}). Hence, we have

$$\frac{\partial CE}{\partial \beta} > 0. \tag{10.32}$$

Thus, we have proved the following.

Proposition 2. *The degree of increase in exports from a newly admitted member to the old members is larger when the degree of product differentiation is smaller.*

In other words, Proposition 2 means that when Greece is admitted to the EC (and when EC tariffs are lifted on goods coming from Greece), exports of less-differentiated products tend to increase more than those of highly-differentiated products.

Taken together, the above two propositions imply that the impact of regional integration on agricultural trade is likely to be stronger than that on manufacturing trade, because agricultural trade is heavily protected and because agricultural products are, generally speaking, less differentiated than manufacturing goods (as discussed in Section 10.2). Also, regional integration would have different magnitudes of impact on different categories of agricultural products, depending on t and σ of each product category.

10.4 The Impact of EC Expansion and Trade Flows — An Ex Post Analysis

10.4.1 Agricultural trade and manufacturing trade

In what follows, we examine whether the above hypothesis is supported by the actual change in trade flows after regional integration, taking two cases of the EC expansion as examples: the accession of Greece to the EC in 1981 and the accession of Spain and Portugal to the EC in 1986.[3] Since the price and quantities of commodity trade fluctuate widely every year, I will use the share figures, rather than raw figures, to eliminate the effect of universal fluctuations. As shown below, the above theoretical predictions are generally supported by the data.

First, let us examine the impact of the accession of Greece to the EC in 1981. Figure 10.2 plots the EC9's imports from Greece as a share of its

[3] We examine the change in trade flows in these two cases, because they are clear cases of the expansion of a trading bloc, and because sufficient time has elapsed for the full impact of the regional integration to be revealed. Note that NAFTA was signed in 1993 but aims to realize free trade by 2009, and that the APEC has just agreed to achieve a free trade regime by 2010 for developed countries and by 2020 for developing countries.

total imports for both agricultural products and manufacturing products. As Figure 10.2 shows, until Greece joined the EC in 1981, the Greek share in EC's agricultural imports stayed consistently around 0.6 percent, with no increasing trend. However, as soon as Greece was admitted to EC membership, this share began to increase dramatically, and it was more than double the pre-accession level by the end of the 1980s. Such a remarkable increase in the share of intraregional trade cannot be observed for manufacturing trade, however. In fact, the Greek share in EC9's manufacturing imports has been declining since the end of the 1970s.

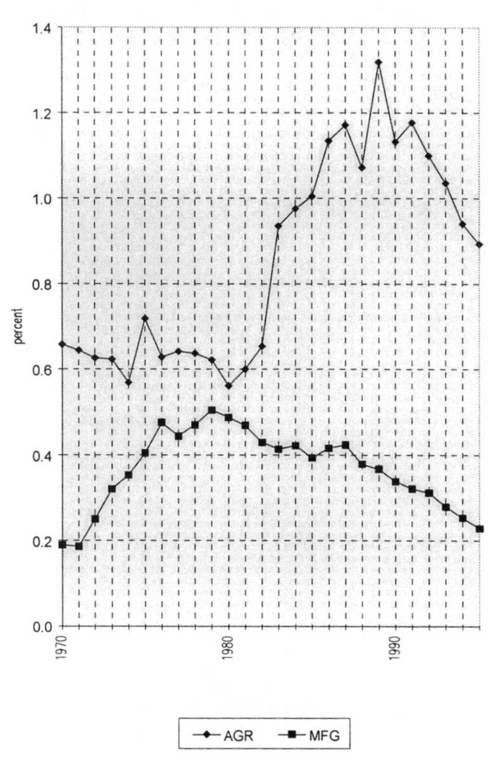

Figure 10.2: EC9, share of import from Greece, 1970–95

The share of imports from EC9 in total Greek imports shows similar trends. As depicted in Figure 10.3, as soon as Greece was admitted to the EC in 1981, the share of EC9's agricultural products in Greece's total agricultural imports jumped from 31 percent in 1980 to 56 percent in 1981. The share continued to increase to more than 70 percent in the 1990s. On the other hand, the EC9's share in total manufacturing imports by Greece has stayed around 60 percent, and there is no sign of increase.

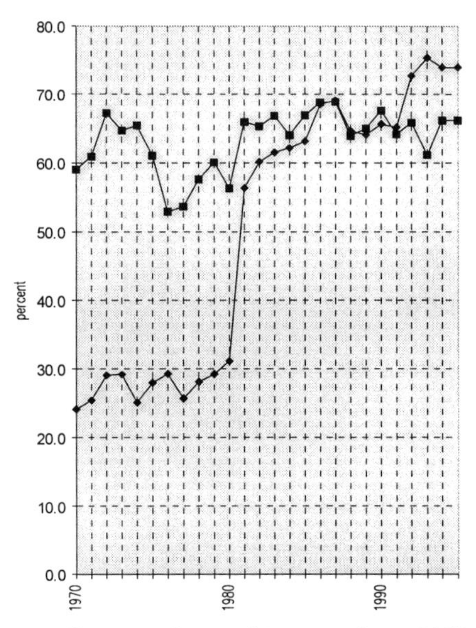

Figure 10.3: Greece, share of import from EC9, 1970–95

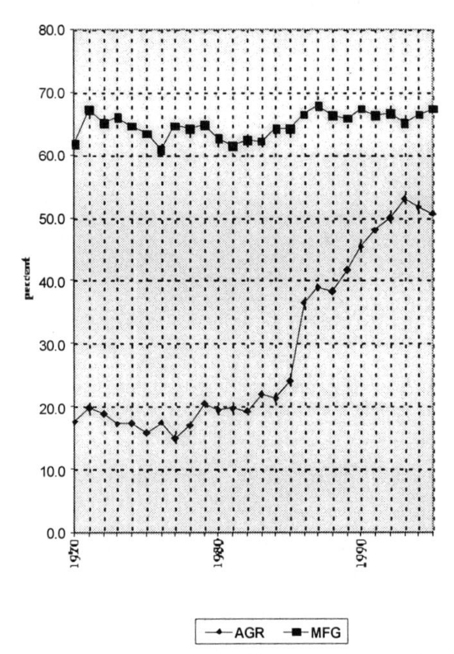

Figure 10.4: Spain, Share of import from EC10, 1970–95

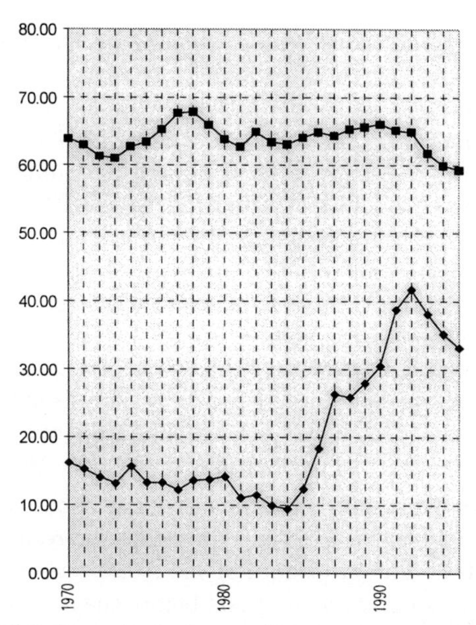

Figure 10.5: Portugal, share of import from EC10, 1970–95

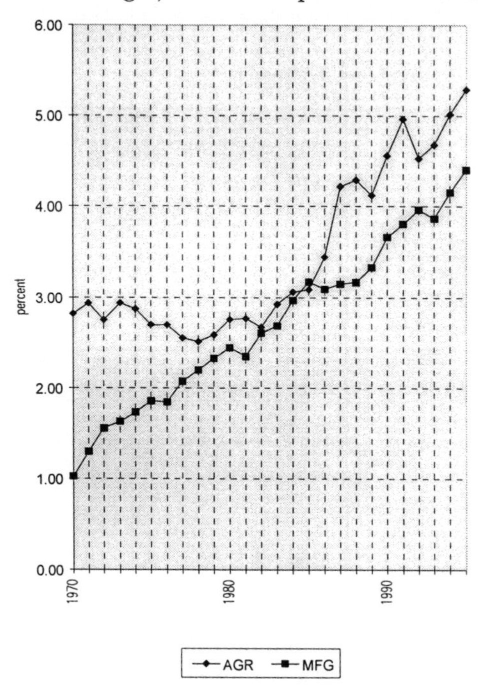

Figure 10.6: EC10, share of import from Spain and Portugal, 1970–95

Second, let us examine the situation when Spain and Portugal joined the EC in 1986. Figures 10.4–10.6 show a similar trend. Figure 10.4 plots the share of EC10's products in total imports by Spain for both agricultural products and manufacturing products. As the figure shows, while the share of EC10's agricultural products in total agricultural imports by Spain was around 20 percent until Spain was admitted to the EC in 1986, the share shows a big jump after the accession. By the end of the 1980s, the share more than doubled to become about 50 percent of Spain's total agricultural imports. On the other hand, there is no increasing trend for manufacturing products. Figure 10.5 shows very similar trends for the share of EC10's products in Portugal's imports. However, the share of imports from Spain and Portugal in total imports by EC10 shows increasing trends in both agricultural and manufacturing products (see Figure 10.6), and we can observe a significant jump in the share of agricultural trade after 1988.

To test somewhat rigorously the above statement based on Figures 10.2–10.6, we performed a t-test for difference in mean, the result of which is summarized on Table 10.7. The procedure of the t-test was as follows. First, we calculated the average share over ten years before the accession and that after the accession for each case. Second, we calculated the difference in means for each case, and ran a t-test for each case to see whether the difference in mean

Table 10.7: Test for difference in means

	Value 1971–80 average	1981–90 average ($mil)	Share 1971–80 average before	1981–90 average after (%)	Difference	t-statistics	
CASE 1: EC9's import from Greece							
Agriculture	439.51	2,240.1	0.62728	1.00018	0.3729	5.13836	A
Manufacturing	874.25	3,458.94	0.39025	0.40596	0.01571	0.43097	
CASE 2: Greece's import from EC9							
Agriculture	251.98	1,486.58	28.0235	63.5014	35.4778	26.0782	A
Manufacturing	2,151.31	4,901.78	59.9696	66.2924	6.32284	3.835	B
	Value 1976–85 Average	1986–95 average ($mil)	Share 1986–95 average before	1986–95 average after (%)	Difference	t-statistics	
CASE 3: EC10's import from Spain and Portugal							
Agriculture	2,691.47	7,863.8	2.72996	3.51329	0.78333	4.33583	A
Manufacturing	8,005.94	32,637.9	1.89638	3.01757	1.12119	4.63569	A
CASE 4: EC10's import from Spain and Portugal							
Agriculture	1,122.49	6,887.01	16.8723	27.5678	10.6955	4.50586	A
Manufacturing	9,108.58	46,442.5	64.2141	64.7192	0.50513	0.7957	

A: Significant at 0.1 percent level.

B: Significant at 1 percent level.

is statistically significant. As Table 10.7 shows, the difference in mean for agricultural trade is significant even at the 0.1 percent level for all four cases. In other words, the intraregional intensity of agricultural trade after integration is higher, with clear statistical significance, than that before integration. On the other hand, as for manufacturing trade, EC9's manufacturing import (Case 1) and manufacturing import by Spain and Portugal (Case 4) are not significant at all, although it is significant for Case 2 and Case 3.

These findings seem to support our general statement that regional economic integration has a greater impact on agricultural trade than on manufacturing trade, probably because agricultural goods are subject to higher trade barriers and are less differentiated.

10.4.2 Varying magnitude of the impact of regionalism on different agricultural products

In the above subsection, we have found that agricultural trade flows are influenced by regional economic integration more than manufacturing trade flows are. The question, then, is whether regional economic integration has a different degree of impact on different agricultural products. Figures 7–10 address this question. In the figures, the shares of trade between old and new member(s) of the EC before and after the new members are integrated are plotted. Although the magnitude of the impact varies from case-to-case, as well as from commodity-to-commodity, it seems that the impact of regional integration is stronger on trade flows of meat, cereal, animal feeding stuff, oil and fat than it is on fish, fruit and vegetables, beverages, tobacco, etc. Although the estimated figures on the degree of product differentiation of these products are not available, commodities with a larger impact seem to be intuitively less differentiated than those with a smaller impact. For example, the degree of product differentiation seems to be smaller for animal feeding stuff, oil and fat, etc., while beverages, which include French wine and German beer, appear to be highly differentiated.

10.5 Regional Integration and Japanese Rice Imports — An Ex Ante Analysis

In this section, we will conduct an *ex ante* analysis of the effect of regional economic integration on agricultural trade flows. Using the framework developed in Section 10.3, we will examine the impact of APEC-wide free trade agreement on rice imports by Japan, one of the most heatedly debated issues in that country in recent years. The Japanese people are very sensitive to imported rice, and due to the almost total control by the government, the price

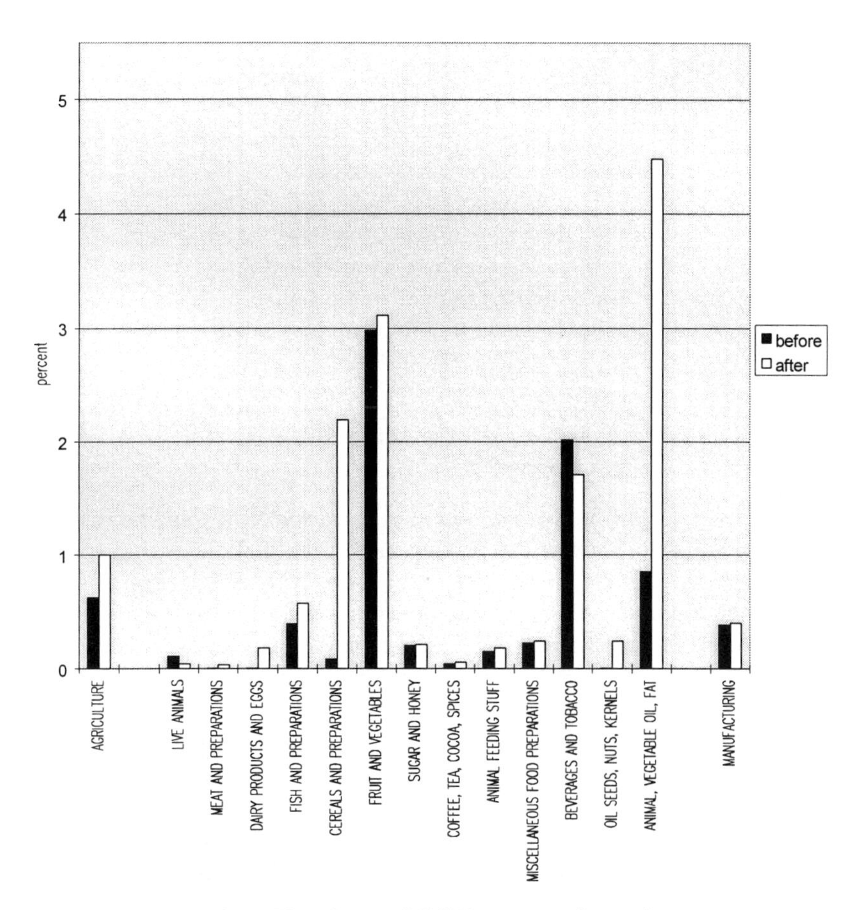

Figure 10.7: The share of EC9's import from Greece by commodity before and after integration

of rice in Japan is much higher than the price on the international market. Because of the huge gap between the price of Japanese rice and the international price, many people in Japan fear that opening up the Japanese rice market would almost wipe out rice production in Japan. The Japanese government adamantly resisted the pressure for liberalization during the Uruguay Round negotiation, and did not accept the tariffication of imports of foreign rice into Japan. Instead, Japan only promised a minimum opening to foreign rice of four percent of domestic consumption (379 thousand tons) in 1995, which will gradually increase to eight percent (758 thousand tons) by the year 2000.

In addition to global liberalization under the Uruguay Round, Japanese rice producers face another challenge. In 1994, leaders of the APEC countries agreed that in order to achieve free trade in the region, developed and developing countries in the APEC area will implement free trade by 2010 and 2020, respectively. This might cause a serious threat to Japanese rice producers because the share of APEC production in total world rice production is more than 50 percent and because APEC includes major rice exporters such as Thailand, the United States, China, and Australia.

Thus, it is feared that if Japan accepts free import (or even preferential import) of rice from APEC countries, the liberalization will have a profound impact on the Japanese rice market. In spite of the importance of the issue, there are few, if any, formal studies that examine the likely impact of an APEC free trade arrangement on the Japanese rice market, and therefore the argument on the issue is often emotional. The following simulation is intended to fill the gap to some extent, even though the estimation is preliminary and depends on various simplifying assumptions.

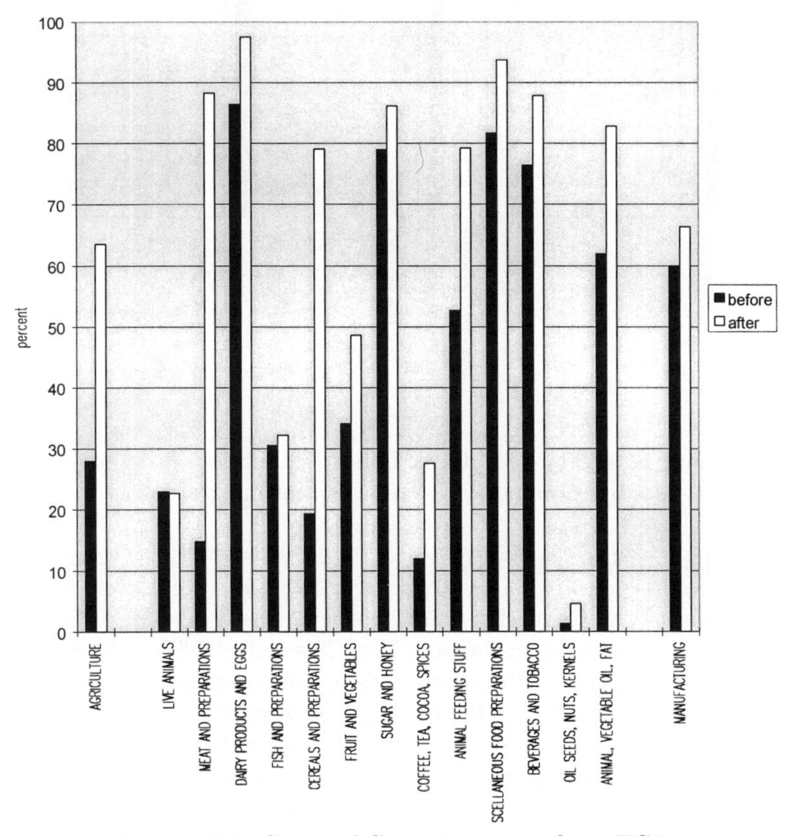

Figure 10.8: Share of Greece's import from EC9 by commodity before and after integration

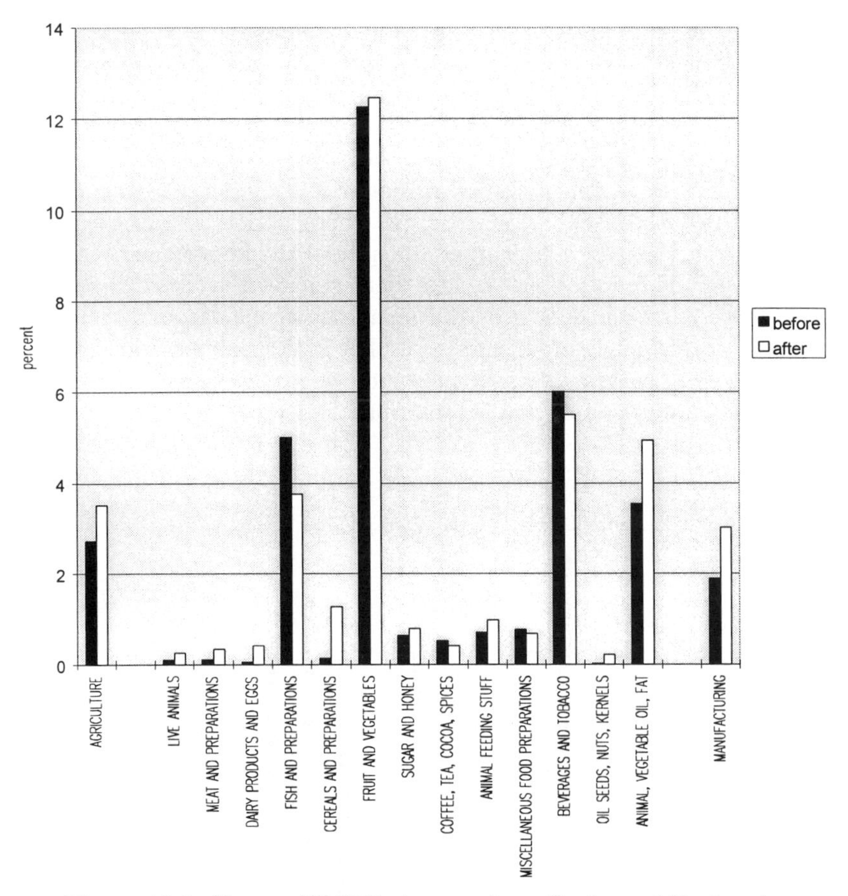

Figure 10.9: Share of EC10's import from Spain and Portugal by commodity before and after integration

10.5.1 Rice market in Japan

Before going to the simulation exercise, let us briefly discuss the salient features of the rice market in Japan. Until November 1995,[4] the production, distribution, and pricing of rice were totally controlled by the government under the Food Control Act of 1942, which had been enacted to cope with the severe food shortage during World War II. Under that law, the government announced the estimated amount of total rice consumption in Japan every year. In order to satisfy the estimated demand, the government decided on

[4] At this time, the new Food Law was put into effect. Although the new law was intended to deregulate the Japanese rice market, imported rice is still almost totally controlled by the government.

the amount of rice to purchase from each local unit, which in turn decided how much to purchase from each farmer. Since farmers had an obligation to sell rice to the government, the planned purchase amount constituted the limit on rice production. The purchasing price from farmers and the selling price to consumers were also decided by the government every year. Until the new law was enacted in 1995, only licenced wholesalers and retailers were allowed to handle the distribution of rice (i.e., it was *illegal* for a supermarket to sell rice).

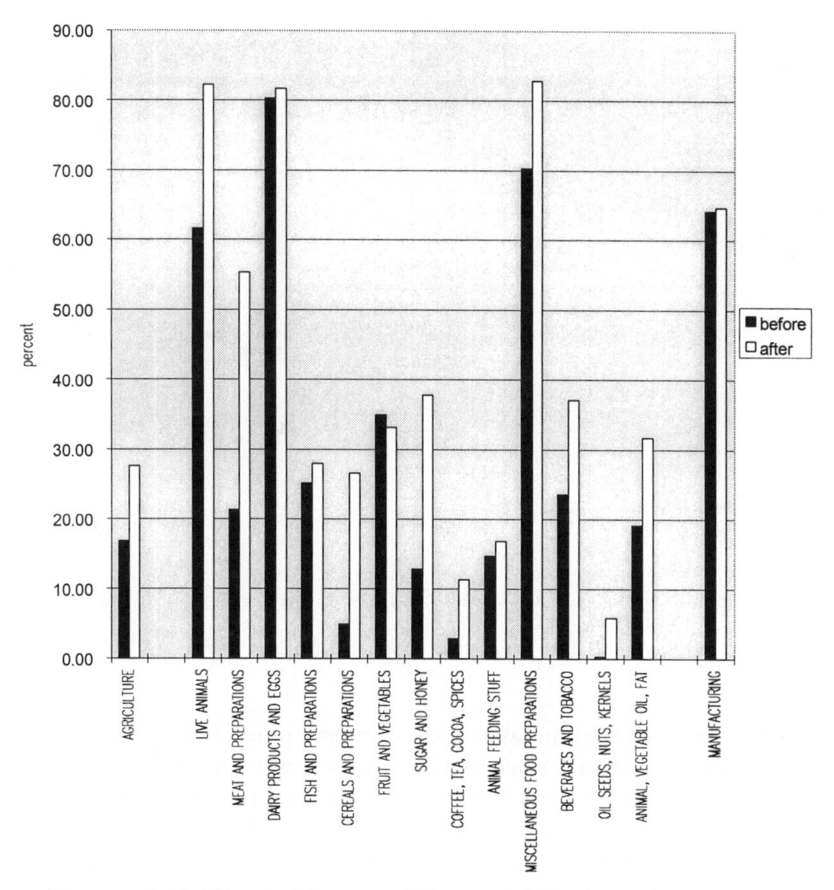

Figure 10.10: Share of import of Spain and Portugal from EC10 by commodity before and after integration

Although the initial purpose of the Food Control Act of 1942 was to protect consumers from severe food shortage during the war, it gradually changed into an income support program for farmers. As urban workers achieved double-digit annual wage increases during the high economic growth of the 1960s, the Basic Agriculture Act was enacted in 1961, with the major purpose of narrowing the income gap between the rural and urban sectors. As a result, rice prices were raised rapidly to provide farmers with an income equivalent to that of urban workers. In early 1960s, the price of the Japanese rice (25 cents per kg) was not very different from the price on the international market (19 cents per kg), but by the end of the 1980s, the former ($1.98) had become more than six times higher than the latter ($0.30).

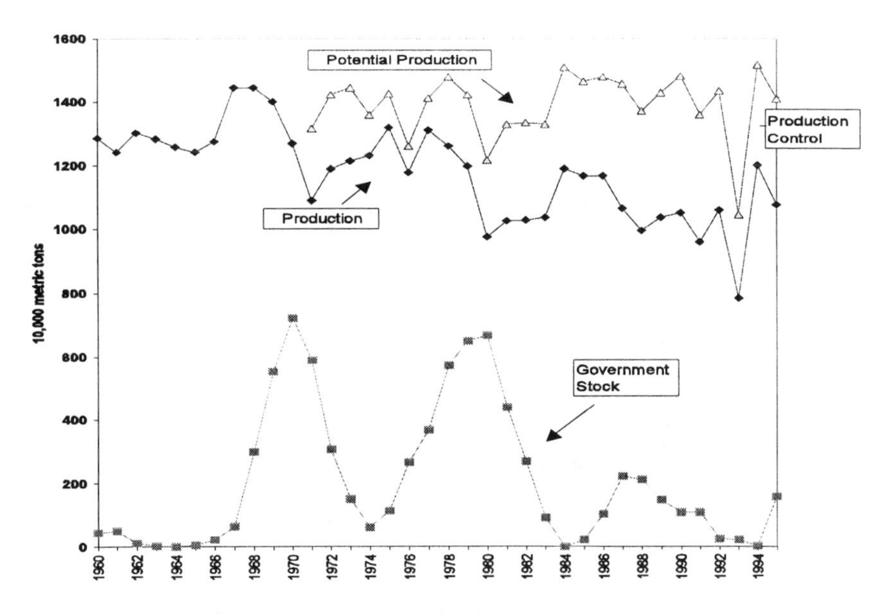

Sources: Compiled from statistics of the Japanese Ministry of Agriculture and Fisheries.

Figure 10.11: Rice production in Japan, 1960–1995

Obviously, the rapid increase in the domestic price of rice encouraged production and discouraged consumption. As a result, the stockpile of rice in Japan grew dramatically in the late 1960s. Figure 10.11 shows the actual and potential production level and the amount of the stockpile of rice in Japan since 1960. As shown in the figure, in 1970 the amount of unsold rice in Japan exceeded 7 million metric tons, which was more than half the annual production. Faced with this huge stockpile, the government began exporting rice (at the much cheaper international price) or gave it away as official development assistance. In addition, in order to cope with the long-term overproduction, the government initiated production controls in 1971. By various measures, such as subsidies and punishment, the planned production control has been strictly enforced. As Figure 10.11 shows, the difference between potential and actual production since 1980 has been around 30 percent of actual production. In short, through strict government control, the high price of the Japanese rice has been maintained, and production controls have been strictly enforced to avoid overproduction.

As mentioned at the beginning of this section, the Japanese rice market has recently been changing. In compliance with the Uruguay Round agreement, the government was obliged to give up the policy of "no single piece of foreign rice in Japan," and in 1995 it imported 379,000 metric tons of foreign rice, an amount equal to four percent of domestic consumption. This amount is to increase to 758,000 metric tons by the year 2000. Although the government successfully avoided tariffication of rice, the opposition by farmers to the abolishment of the "no single piece of foreign rice policy" was enormous. Since such an opening up of the rice market is expected to lead to a lower price for rice (and therefore to a decline in farmers' incomes), the government also in the same Diet session that ratified the Uruguay Round agreement, appropriated more than 6 trillion yen (about 60 billion US dollars) to increase farmers' incomes.[5]

Under the new Food Law, the government gave up total control of rice distribution, and rice distributed outside of government control was legalized. However, the import of rice is still under almost total control of the government. With a minor exception called the SBS (simultaneous buy and sell) formula,[6] only the government can import foreign rice. When the government buys foreign rice, it is allowed to impose a surcharge of up to 292 yen per kilogram of imported rice when the government sells it in the Japanese market. Since the surcharge of 292 yen per kilogram is equivalent to an almost 800 percent surcharge, the government can, in effect, raise the price of foreign rice sold in Japan as much as it wishes. Note that when the full 292 yen surcharge is imposed, the price of foreign rice becomes $2.64 per kilogram, which is even higher than the current price of domestic rice ($1.98 per kg). Thus, even under the Uruguay Round agreement and the new Food Law, the government can set the price of government-distributed rice, both domestic and foreign, at whatever level it wishes.

[5]This 6 trillion yen is not for a direct cash payment to the farmers; it is intended to increase farmers' income indirectly through various measures including improving the infrastructure in rural areas, enhancing agricultural technology, making loans available to farmers, etc.

[6]Under the SBS formula, the importer and wholesaler file a joint application to the government, in which they have to specify the buying price of the importer from the foreign producer and the selling price of the importer to the domestic wholesaler. The government allocates the import quota to the application with the biggest gap between the buying price and selling price, and the difference goes to the government. SBS imports are limited to about ten percent of total rice imports in Japan.

10.5.2 APEC and Japanese rice – An illustrative simulation

Basic simulation strategy. Now, let us examine the impact on the Japanese rice market when the government gives up control of the quantity of rice imports. In this hypothetical case, the government controls the price of foreign rice in the Japanese market only through tariffs, while it can still control the price of domestically produced rice. The simulation is conducted using the product differentiation model developed in Section 10.3, in view of the fact that rice is a fairly differentiated product in Japan (as discussed in Section 10.2). More specifically, the following maximization problem is solved to examine the impact of the APEC free trade agreement on the Japanese market, which is the adapted version of equations (10.1) and (10.2) in Section 10.3. In other words, the objective function of the utility maximization problem of the Japanese consumers is

$$U = \left\{(n_d C_d^\beta + n_a C_a^\beta + n_r C_r^\beta)^{1/\beta}\right\}^\alpha Q^{1-\alpha}, \; 0 < \alpha < 1, \; 0 < \beta < 1, \quad (10.33)$$

where C_d, C_a, and C_r are the consumption of each type of domestically produced rice, rice imported from APEC countries, and rice imported from the rest of the world, respectively. Q is the amount of consumption of numeraire goods. n_d, n_a, and n_r are the number of types of domestically produced rice, rice imported from APEC countries, and rice imported from the rest of the world, respectively. U is the utility of the Japanese consumers.

Japanese consumers maximize their utility subject to the budget constraint

$$P_d n_d C_d + P_a(1 + t_a) n_a C_a + P_r(1 + t_r) n_r C_r + Q = Y, \quad (10.34)$$

where P_d, P_a, and P_r are the producer price of each type of domestically produced rice, rice imported from APEC countries, and rice imported from the rest of the world, respectively. In the equation, t_a and t_r are the tariff rates imposed on the import of rice from APEC and the rest of the world, respectively, and Y is the amount of income. Note that since consumption of rice and consumption of numeraire goods are expressed by a Cobb–Douglas function, the share of income spent on rice purchase is invariant at α; and that, if Y is exogenously given, the amount of money spent on rice purchase is also exogenous (y).

When we identify the parameters n_d, n_a, n_r, P_d, P_a, P_r, Y, t_a and t_r, we can solve the model for the welfare maximizing amount of consumption of each type of rice (C_d, C_a and C_r). Then, to examine the magnitude of the impact of the APEC free trade agreement on the Japanese rice market, all we have to do is obtain, by simulation, the values of C_d, C_a, and C_r for reduced values of t_a, and compare them with corresponding values with no tariff concessions. Needless to say, when the rice imports from APEC countries are totally liberalized, t_a becomes zero.[7]

Identification of parameter values. Using data from Food and Agriculture Organization (FAO), we obtained $Pa = 0.69$ (dollar price per kilogram of rice) and $P_n = 0.69$.[8] For the price of the domestic rice, we adjusted the price of domestic rice in the year 2000 by taking into consideration the recent slight decline in buying price by the government, and we obtained $P_d = 1.79$. Using statistics from the Japanese Ministry of Agriculture and Fishery, we obtained a total buying price for domestic rice $y = 14,928$ (million dollars). While actual data are available for P_d, P_a, P_r, and y, other parameters have to be calculated indirectly. For example, n_d is obtained by dividing the amount of total production of rice in Japan by the number of brands of rice listed in the Annual White Paper of the Ministry of Agriculture and Fishery,[9] i.e., $n_d = 21$, and $n_a(n_r)$ is obtained by dividing the amount of total exports of rice from APEC countries (the rest of the world) by the average amount of production of domestic rice.

In view of the fact that the capacity to export is constrained by the need to feed its own population, we assumed that the amount of rice which could potentially be brought into the Japanese market was limited to the amount already in the international market in 1990–95. Therefore, the calculation of n_a and n_r was based on the amount of exports rather than on the amount of total production, and we obtained $n_a = 19$ and $n_r = 17$.

For the value of β, we used 0.6, which was used in Goto and Hamada [5]; they in turn based their figure on the estimate by Stern, Francis and Schumacher [10]. Since $\beta = 0.6$ is not a decisive number, we conducted sensitivity analysis using different values of β, as reported in the appendix. A brief look at Figure 10.12 and two appendix figures reveals that the thrust of the argument in the next subsection is quite insensitive to variations of β.

We calculated t_a and t_r as the tariff rates which keep the import of rice at the agreed minimal access level (8 percent of domestic consumption) in 2000, and obtained t_a $(t_r) = 7.54$. In other words, to keep the amount of foreign rice at the minimum access level, Japan has to impose a 754 percent tariff on foreign rice.

[7] Since total liberalization of such a sensitive agricultural product as rice in Japan is unlikely, the effects of various levels of tariff reductions, as well as the case of zero tariff, are simulated below.

[8] Note that the price of rice imported to Japan ($0.69 per kg) is substantially higher than the average price of world rice ($0.38 per kg), probably because Japan imported higher quality rice.

[9] The brands of rice below the cut-off point in production (15,000 ha) are omitted from the calculation.

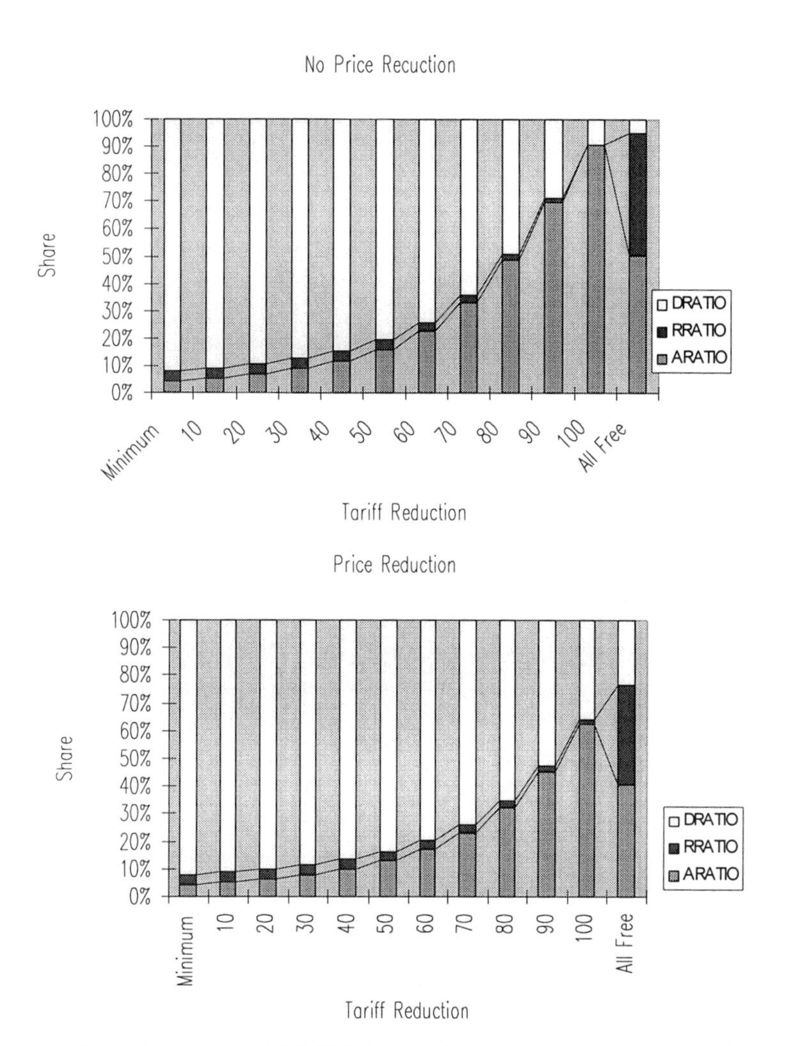

Figure 10.12: Impact of APEC FTA on the Japanese rice market ($\beta = 0.6$)

Result of simulation. Now that we have equations for the optimization problem and all of the parameter values needed, we can simulate how the trade liberalization under the APEC agreement affects the rice market in Japan. Let us look at the upper panel of Figure 10.12, which summarizes the simulation results. The horizontal axis plots the degree of trade liberalization under the APEC free trade agreement. "Minimum" means the tariff rate required to keep foreign rice at the minimum access level. In order to keep foreign rice at this level, the government has to impose a 754 percent tariff on imported rice. "10 percent," "20 percent," . . . means the situations where the initial tariff (754%) imposed on APEC rice is reduced by "10 percent," "20 percent," and so forth. When the initial tariff on APEC rice is reduced by 10 percent (20%), the new tariff rate becomes 679 percent (603%), and so forth. "All Free" means the situation where tariffs imposed on foreign rice, both APEC and non-APEC rice, are totally abolished.

Keeping the above in mind, let us examine the simulation result. Contrary to the popular argument in Japan, the impact of preferential liberalization of APEC rice is not so large, although the completely free import of APEC rice has a big impact. As the figure shows, even if the minimum access tariff level (754%) is halved (377%), the share of imported rice increases by 11.2 percent to become 19.2 percent, although if total liberalization of the import of APEC rice is realized, the share of imported rice becomes as high as 90.4 percent. If the tariff reduction to APEC rice remains at 10–30 percent, the impact on the domestically produced rice is small. For example, when the tariff on APEC rice is reduced by 20 percent, the share of foreign rice in the Japanese market increases only by 2.4 percent to become 10.4 percent. Thus, the simulation result reported here, which is based on the product differentiation framework rather than the homogeneous product assumption, suggests that the impact of *partial* liberalization of rice imports from APEC countries is too small to wipe out the Japanese rice production.

Furthermore, note that the simulation result in the upper panel of Figure 10.12 is based on the assumption that the current high price of domestic rice is maintained without major change. However, since the mid-90s, when the Japanese government gave up the policy of "no single piece of foreign rice," support for the idea of maintaining farmers' income through excessively high prices seems to be fading. If the Japanese government can reduce the support price of domestically produced rice, the impact of a tariff reduction on APEC rice becomes smaller. The lower panel of Figure 10.12 shows the simulation result when the support price is halved from the benchmark price level used for the simulation in the upper panel. If the government succeeds in reducing the price of domestic rice by half, the tariff necessary to keep imported rice to the minimum access level is 327 percent. In this case, the impact of a 10–30 percent reduction of tariffs on APEC rice is minimal. Even when the tariff rate on APEC rice is reduced by 50 percent, the share of imported rice increases by a mere 8.3 percent to 16.3 percent. However, if the tariff on APEC rice is totally abolished, the market share of domestic producers is reduced to 36 percent.

Thus, according to the simulation result reported here, which incorporates product differentiation, the impact on Japan of partial liberalization of the rice market seems to be much smaller than feared.[10]

Furthermore, it should be noted that the welfare level of the Japanese

[10] The appendices suggest that the above result is rather insensitive to variations of β.

consumers consistently increases as the Japanese rice market becomes open to a foreign competition. Figure 10.13 plots the welfare level for each stage of liberalization of the Japanese rice market. The figure shows that, from a consumer's viewpoint, the best policy is the complete liberalization of rice imports.

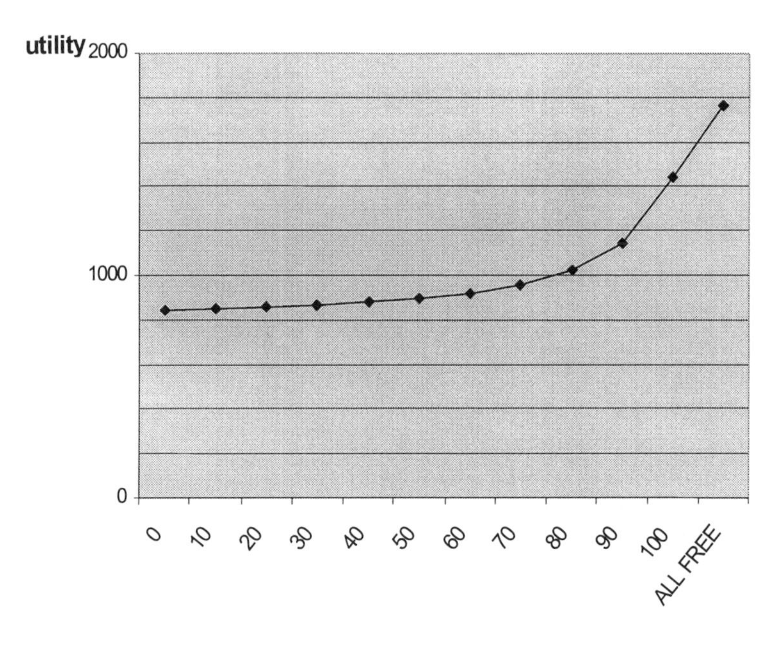

Figure 10.13: APEC FTA and welfare

10.6 Concluding Remarks

Using a simple trade model with product differentiation, we have analyzed the impact of regional economic integration on agricultural trade, comparing it with the impact on manufacturing trade. Using the model developed in the paper, we have found that the degree of impact of the free trade agreement on trade flows depends on two key parameters: *the degree of initial protection* (t) *and the degree of product differentiation* (σ). The impact of the expansion of the FTA on agricultural trade flows tends to be larger than that on manufacturing trade, because the initial protection level on agricultural trade is larger and because, generally speaking, agricultural goods are less differentiated than manufacturing goods.

Examining the data on the change in trade flows for two incidents of EC expansion, we have confirmed the validity of the theoretical prediction. When the EC expanded, agricultural trade between the new member(s) and old members increased dramatically, while no comparable jumps were observed for the manufacturing trade. Furthermore, among agricultural goods, the impact on certain products with less product differentiation, such as meat, animal feeding stuff, and fat, was stronger than the impact on more differentiated products, such as fruits and vegetables and beverages.

After the *ex post* analysis, the model is applied to an *ex ante* analysis of the impact of the APEC Free Trade Agreement (FTA) on one of the most controversial commodities, rice in the Japanese market. Many people in Japan are arguing that if their rice market is opened up to foreign rice, Japanese rice production will be wiped out. In spite of this emotional argument, there have been very few, if any, objective studies on the impact of liberalization on the Japanese rice market. In view of this, one of the purposes of this study is to fill in the gap by presenting an objective simulation result. The simulation result in this paper suggests that the impact will not be as large as many people fear.

It seems that the fear of liberalization emphasizes only one of the two parameters mentioned above; i.e., the degree of initial protection. Of course, the impact of the APEC FTA on the Japanese rice market tends to be large. because the current protection level on rice in Japan is very high (the tariff equivalency of the protection on rice is more than 700%). However, it should be noted that in the Japanese market, rice is a highly differentiated product, and consumers are willing to pay a high premium on certain brands of domestic rice. As examined in the theoretical part of the paper, such a high degree of product differentiation tends to reduce the impact of the FTA on the domestic market. Thus, two conflicting forces determine the impact of the APEC FTA on Japanese rice imports. According to the simulation result, the overall impact of the *partial* liberalization is rather small, although the *complete* FTA has a profound impact on Japanese rice producers. Moreover, it should be noted that the welfare of the Japanese consumers monotonically increases as the Japanese rice market becomes opened to the foreign competition.

As mentioned above, the simulation result is still at a preliminary stage because it depends on various simplifying assumptions, such as the lack of adjustment by the APEC producers. If the APEC producers expand their production for the liberalized Japanese market, the impact of the APEC FTA would become larger than the simulation result suggests. On the other hand, while the simulation result assumes that the current inefficiency in Japanese production continues, external pressure after the Uruguay Round will certainly push Japanese farmers to adopt larger-scale, more efficient production technology. If this happens, the impact of the APEC FTA on rice imports to Japan would be lessened by the strengthened competitiveness of the Japanese rice producers.

Appendix 1

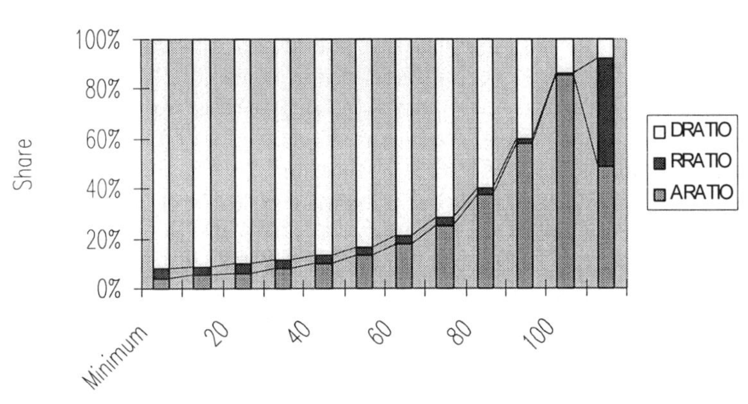

Tariff Reduction

Price Reduction

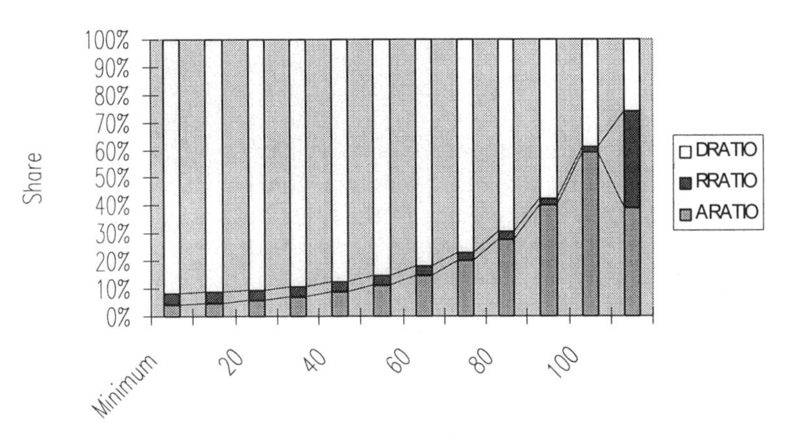

Tariff Reduction

Figure A1: Impact of APEC FTA on the Japanese rice market ($\beta = 0.5$)

Appendix 2

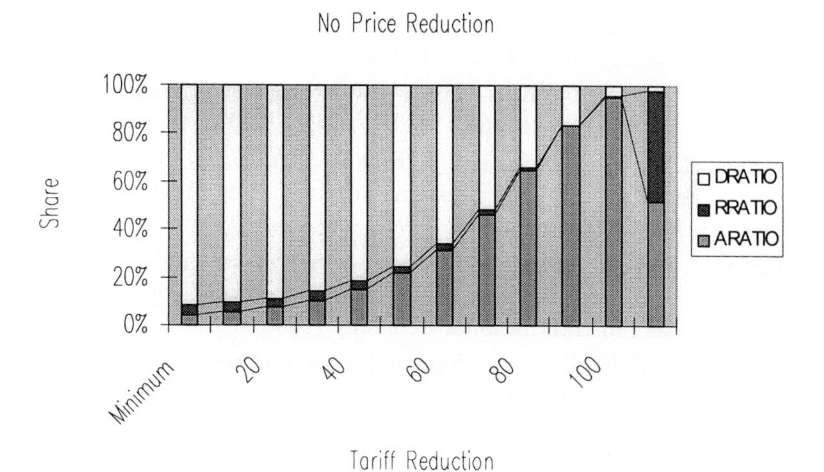

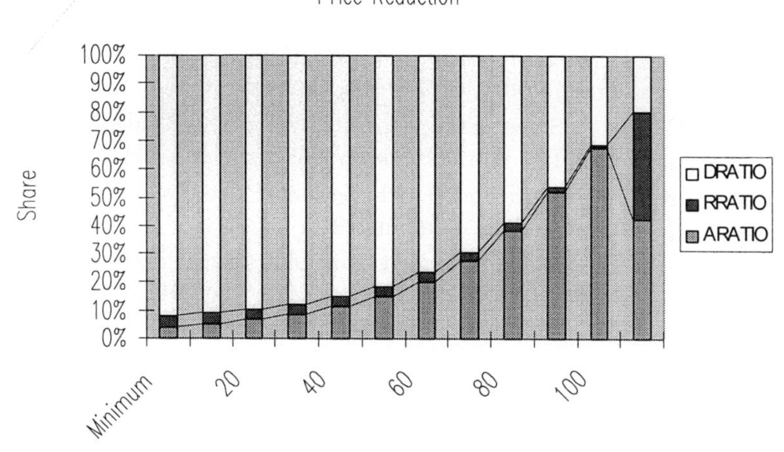

Figure A2: Impact of APEC FTA on the Japanese rice market ($\beta = 0.7$)

References

1 Armington, P.S., A theory of demand for products distinguished by place of production, IMF Staff Papers, 1969.

2 Dixit, A. and V. Norman, Theory of International Trade, Cambridge University Press, Cambridge, 1980.

3 Finger, J.M., M.D. Ingco and U. Reincke, The Uruguay Round: Statistics on tariff concessions given and received, The World Bank, Washington, DC, 1996.

4 Goto, J. and K. Hamada, Economic integration and the welfare of those who are left behind: An incentive theoretic approach, Journal of Japanese and International Economy (1998).

5 Goto, J. and K. Hamada, Regional economic integration and Article XXIV of the GATT, Review of International Economics (1999).

6 Gros, D., A note on the optimal tariff, retaliation and the welfare loss from tariff wars in a framework with intra-industry trade, Journal of International Economics (1987).

7 Grubel, H.G. and P.J. Lloyd, Intra-industry Trade, John Wiley & Sons, New York, 1975.

8 Krugman, P., Is bilateralism bad?, in: Helpman and Razin (eds.), Interantional Trade Policy, MIT Press, Cambridge, MA, 1991.

9 Pohl, G. and P. Sorsa, European Integration and Trade with The Developing World, The World Bank, Washington, DC, 1992.

10 Stern, R.M, J. Francis and B. Schumacher, Price Elasticities in International Trade, Macmillan, London. 1976.

11 Winters, A.L., Regionalism versus multilateralism, Policy Research Working Paper #1687, The World Bank, Washington, DC, 1996.

Part III

General Equilibrium and Development Policy

Trade, Growth, and Development
G. Ranis and L.K. Raut

CHAPTER 11

Virtual Reality in Economy-wide Models Some Reflections on Hope and Despair with Reference to India[1]

N.S.S. Narayana[a] and Kirit S. Parikh[b]

[a]Indian Statistical Institute, Mysore Road, RVCE Post, Bangalore - 560 059, India

[b]Indira Gandhi Institute of Development Research, Gen. A.K. Vaidya Marg, Goregaon (East), Mumbai-400 065, India

> A single new observation may call for a more comprehensive theory.
> —P.C. Mahalanobis

11.1 Introduction

Srinivasan (T.N., as he is usually fondly referred to) frequently comments that in the history, religion rarely ever helped economic development. But do planning models help? Perhaps he would not be as affirmatively negative towards them as he is towards religion, unless he is now totally disenchanted with the process of planning.

T.N. was still in his teens when India achieved her independence in 1947, and all around, there were great hopes expressed for rapid social and economic development. These hopes were greatly echoed in Pandit Jawaharlal Nehru's [52] famous "Tryst with Destiny" speech delivered on the eve of India's independence. He elaborated later,

> A new star rises, the star of freedom in the East, a new hope comes into being, a vision long cherished materializes. May the star never set and that hope never be betrayed! . . . The future beckons to us. Whither do we go and what shall be our endeavour? To bring freedom

[1]Thanks are due to Meenakshi Rajeev and G. Ravindran for their comments on an earlier draft of this paper. H.M. Rajashekara, Bertlyn Reynolds, and Patric Lewis provided secretarial assistance.

> and opportunity to the common man, to the peasants and workers of India, to fight and end poverty and ignorance and disease; to build up a prosperous, democratic and progressive nation, and to create social, economic and political institutions which will ensure justice and fullness of life to every man and woman.

Rising to the occasion were several committed scientists, engineers, industrialists, agriculturists, and economists, among others. T.N. was one of them. Using his formal training in economics both at the Indian Statistical Institute, Calcutta and later at Yale University, New Haven, he contributed enormously to planning exercises for India — initially under the stewardship of Pitambar Pant, who was rated as one of the great visionaries of planning in India. India is one of the earliest, if not the first, countries to have developed a full fledged economy-wide model. The Indian Statistical Institute where T.N. worked at that time largely contributed to this effort. Though he left India for the USA in late 1976, he has never ceased to write on various Indian economic problems, examining issues such as yields of cereal crops, industrial performance, poverty and income distribution, demography, project appraisal, trade liberalization, general equilibrium, economic reforms, privatization, direct foreign investment, and many others.

Of the numerous topics in economics that he wrote on, T.N. contributed significantly to the literature on general equilibrium models — especially applied-computable types. This again expresses his concern and commitment to development. When the authors of this paper were involved in the early 1980s in developing a computable general equilibrium model for India at the International Institute for Applied Systems Analysis (IIASA) in Laxenburg (Austria), T.N. once stopped over in Vienna. He saw what we were trying to do and became deeply interested. Subsequently, he became a member of the working team. This work is reported in Narayana, Parikh and Srinivasan [51].

In celebrating T.N.'s 65th birthday, we share some of his concern for development and reflect on the evolution of planning paradigms and models in India. As our understanding of planning problems improved, so did the sophistication of economy-wide models. The most sophisticated models today are CGE models. We present some of the views expressed on the inadequacies of the equilibrium (based) models. Later we attempt to highlight some difficulties in modeling the government processes which seem to dominate the process of economic development, and thus show why we are required to make the models more realistic.

11.2 If Only Plan Models Were Available!

With a conviction that planning is necessary for systematic economic development, India was foremost among the developing countries to embark upon such exercises. The first five year plan was launched in 1951 covering the period from 1951–52 to 1955–56. The basic notion behind the plan was a conviction that government must undertake a large number of development projects and tell the people of India that a new era has begun. Given the most pressing problem of India, its uncertain agriculture, the plan initiated a number of major multi-purpose schemes to provide irrigation, generate power, and help in flood control. One of these schemes was the Bhakra dam in Punjab, which later on proved to be the cornerstone of Punjab's agricultural development. The First Five Year Plan (1951–56) did not have a formal planning model but stressed investment for capital accumulation in the spirit of the one sector Harrod–Domar Model. It argued that production requires capital and that capital can be accumulated through investment. The faster one accumulates, the higher the growth rate. From this point of view, the plan was considered to be a success, as most of the plan projects were initiated and made considerable progress over the plan period. The investment rate increased over the plan period from 5 percent of GDP to 7 percent of GDP. The target of private investment of Rs. 16,000 million was realized, but the public sector investment target of Rs. 240,000 million was not fully achieved. Investment in large infrastructure and heavy industries, "the temples of modern India" as Nehru called them, required initiative by the public sector. The role of the public sector began to crystallize and got formalized in the industrial policy resolution.

Planning generated lots of excitement and attracted some of the leading intellectuals in the country. They felt the need for a formal plan model. The history of the economy-wide planning model for India began with Professor P.C. Mahalanobis's Plan Frame for the Second Five Year Plan in 1955. The planning techniques in those times were not as developed or at as disaggregated a level as they are today. However, Mahalanobis' [44] innovative ideas led to the distinction of capital goods to produce capital goods, from capital goods to produce consumer goods. This distinction helped people to understand the extent of trade-off that exists between the levels of immediate and future consumption/well-being. This was a clearer formulation of Marx's ideas in *Das Kapital.* Because of its similarity to a model developed by the Russian economist, Feldman, it is now referred to as Feldman–Mahalanobis model.

The Mahalanobis model suggested that if a country wants to rapidly reach a high standard on consumption, the best strategy may be to invest first in building capacity to produce capital goods for some years. After you have accumulated enough of such capacity you use it to augment capacity to produce consumer goods. Thus, the emphasis on heavy industries and capital goods was theoretically justified and formed the basis of the second plan. The Second Plan raised the investment rate to 8 percent of GDP.

The Mahalanobis Model, on which the second five-year plan was based, provided the real intellectual articulation of India's development strategy. In fact, T.N. Srinivasan's Ph.D. thesis at Yale was a more rigorous formulation of Mahalanobis's ideas and led to more insights.[2] Subsequent plan models, in a sense, only further elaborated that philosophy. The basic insights of the Mahalanobis model are the following:

(i) The one-sector Harrod–Domar's model assumes that all savings can be converted into investment. However, to convert savings into investment, one needs investment goods. Thus, one must look at sectoral details.

(ii) An emphasis on basic capital and heavy industries (machines to build machines to build machines . . . to build fertilizer plants) in the early years lead to much higher consumption in the future.

The Mahalanobis model had four sectors. Investment allocation to different sectors was exogenously prescribed. The need for greater sectoral disaggregation and a model for optimal allocation of investments was felt. Economists rose to the challenge and subsequent research led to development of planning models from an aggregate level to more and more disaggregated levels. During the course of such development India witnessed visits by several renowned economists such as Oscar Lange, Ragnar Frisch, Charles Bettleheim, Jan Tinbergen, Nicholas Kaldor, Kenneth Galbraith, Paul Baron, Richard Goodwin, Alan Manne, Jan Sandee, Richard Eckaus, Luis Lefeber, etc.

Goodwin helped construct an inter-industry table; Frisch aided in developing a multi-sectoral programming model as one of the earliest attempts to apply linear programming model for economy-wide planning; Bettleheim helped prepare material balances; and similarly many others also participated in this research.

While research and development, both in India and abroad, was going on, the technical development of planning models progressed gradually from aggregate consistency models to static and dynamic input–output models, linear programming based models, and consistency and dynamic optimization models. The theoretical construct behind the Third Plan, 1960–62 to 1965–66, became more sophisticated compared to the Mahalanobis model of the Second Plan. The perception that planning required sectoral allocations of investment was further refined in subsequent plan models. Multi-sectoral input–output models were used to determine intersectorally consistent investment allocations, beginning with the Third Plan. Academic economists developed even more elaborate intertemporal models for optimal allocations that accounted

[2] For example, the Golden Rule which became famous later was proved first by T.N. [65] in his thesis. True to his character however, T.N. did not go to town heralding it as a great new insight. To T.N., new insights are seldom great: usually they are only minor logical extensions of neoclassical principles.

for gestation lags (see Eckaus and Parikh [23]). The central issues addressed in these models whether optimization type or consistency type, were mobilization and optimum use of resources, taking into account certain production and behavioral constraints, consistencies and balances. Taylor [69] and Rudra [60] contain excellent reviews of the theoretical foundations and technical limitations of these models.

11.3 If Only Theories Were Available!

In spite of the increasing sophistication of planning models, the performance of the Indian economy was very modest and in any case did not match the expectations of the plans. This generated two types of responses: the first blamed the failure of the plans on implementation, while the second sought to improve the models. There has always been a question bugging the minds of the modellers, relating to (in)adequacy of the planning methods and how well the models represent the reality.

One notes this feeling, for example, when Eckaus and Parikh [23] say,

> The published economic plans for many countries often treat many aspects of development and contain a great amount of detail. Nevertheless such plans are, for the most part, collections of separate programs the real interactions of which are not fully worked out and taken into account in the making of policy. The inadequacies of the planning methods practiced are only partly due to the limitations of the development theories available, and most development plans could be improved by a more consistent and intensive application of these theories.

Kanaan [38] argues that the Mahalanobis framework could have been enriched by incorporating Arrow's [4] insights on learning by doing and emphasizing human capital formation, as was done two decades later by Romer [59]. However, these would have merely strengthened the case for planning, for an activist government and for a greater role for the public sector. No fundamental change in the strategy of development pursued by India would have resulted.

Discussing systematically the theoretical bases of economy wide planning Taylor [69] says,

> . . . much of the richness of mathematical planning technique comes from its close relationship with economic systems theory. On the other hand, in so far as currently accepted economic theory is an inaccurate description of what really goes on in developing economies, formal planning is likely to be irrelevant. Technical specialists and non-mathematical planners must understand both sides of the models —

where they are elegant and where shabby, where realistic and where surrealistic — before they can make a rational decision about whether to use them or not.

Ashok Rudra [60] is clear on this elegance and shabbiness, though his views are more with respect to the empirical content of a model. He was a staunch leftist economist and did not like the capitalistic mode of development. (By the way, T.N. was one of the early students of Professor Rudra.) He understood clearly the roots of the Indian economic system and how it works. Certainly, many imperfections exist, for which a proper theory is still missing. However, this aspect did not get in the way of Rudra's work on economy-wide planning models. His work, along with Alan S. Manne and others on a consistency model for India's fourth plan, is well noted. He makes a list of tasks expected of plan models as follows:

(a) to provide a framework for the assessment of the soundness of the targets of a plan that might have been set by some less formal methods, for example, those of an official plan prepared by the planning authority;

(b) to enable the making of quantitative projections for the economy over the plan period;

(c) to provide a framework for the selection or preparation of projects for being integrated into the plan; and

(d) to yield such lessons as to better equip the planning authority to make policy decisions regarding the official plan.

Based on this, he distinguishes a class of plan models which are "constructed not so much to make any projections with their help as to provide helpful information to the policy makers." He calls such plan models "Policy Models." This elaboration helps us to understand that all plan models are not of the same class, and in particular, that they may not or even need not be adequate in every sense of the working of an economy.

Anyone with serious concern about the theoretical weaknesses would have thought the era of plan models was going to be over forever. For example, Ginsburgh and Waelbroeck [29] felt that Taylor [69] "gave the coup de grace." But it was not to be.

Some economists feet that planners cannot absolve themselves by blaming poor implementation for plan failures. Thus Parikh [53] argues that planners should have accounted for the existing inadequacy of implementation capability or the probabilities of events such as droughts. If they did not do it, they must share part of the blame for failures of a plan to achieve its targets.

The main problem with the planning models was the perception of the problem and the premises behind the strategy of development that was adopted. It was assumed that mobilization and efficient allocation of investment were the most critical problems of development: planners can know all they needed to know; markets cannot be trusted; everyone but politicians and bureaucrats are selfish; and government is all-powerful to attain whatever it proposes. With these perceptions, incentives and policies did not seem important.

These assumptions are not realistic for the democratic polity and mixed economy of India. It is true, however, that the plan documents usually stated that appropriate policies would be devised to reach the targets. But individuals, consumers, producers, traders, bureaucrats and politicians all adapt their behavior to protect their own interests in response to government policy. Whether any kind of policy could have led to a realization of the targets was not obvious. For example, plan documents set targets for production of wheat, rice, etc. But food is grown by some 70 million odd farmers in the country. What policies can induce the farmers want and make them able to grow as per the targets, is a question never addressed in Indian plans. Narayana and Parikh [50] argued that the required price policies to induce farmers to fulfill the sixth plan targets were not spelt out in the plan document and were unlikely to be followed. Thus, they argued that the neglect of behavioral reactions in the Indian plan models raised doubts about the implementability of the plans. Even a committed and efficient administration could not have realized the targets of some of India's plans. One felt that we could plan better if only better plan theories were available.

11.4 If Only Theories Were Right!

Meanwhile, Debreu, Arrow, and Hahn, as well as others, arrived with rigorous treatment of the Walrasian-general equilibrium theory. One of the main attractions of the general equilibrium framework is the incorporation of behaviors of economic agents and the endogeneity of prices and facility with which growth and income distribution can be analyzed. This resulted in the formulation of price endogenous economy-wide models for several countries, models now generally referred to as computable-applied general equilibrium models. These models usually identify various sectors of production and various groups of consumers and producers who interact and exchange goods in a market characterized by perfect competition and under at the influence of government policies usually exogenously specified.

Theorists, however, expressed the view that competitive equilibrium analysis cannot answer several down-to-earth issues. Axiomatic basis, treatment of time, uncertainty, presence of an auctioneer, and price-taking behavior, etc., are some of the main issues which made general equilibrium theory controversial.

Nicholas Kaldor [37], exposing the irrelevance of equilibrium economics and its explanation of price determination says, "By the term 'explanation'

Debreu means a set of theorems that are logically deducible from precisely formulated assumptions; and the purpose of the exercise is to find the minimum 'basic assumptions' necessary for establishing the existence of an 'equilibrium' set of prices (and output/input matrixes) that is (a) unique, (b) stable, (c) satisfies the conditions of Pareto optimality. The whole progress of mathematical economics in the last thirty to fifty years lay in clarifying the minimum requirements in terms of 'basic assumptions' more precisely: without any attempt at verifying the realism of those assumptions, and without any investigation of whether the resulting theory of 'equilibrium prices' has any explanatory power or relevance in relation to actual prices."

Harcourt [33] states that the strand of general equilibrium theory "is not concerned with actual description at all and certainly not with general process analysis." He elaborates, "the fatal flaw in general equilibrium theory *for the present purposes* relates to the fiction of the auctioneer (in his role of disseminating freely information) and so to the absence of 'time', together with the related concept of uncertainty about an unknowable future, from general equilibrium models. The assumption of the auctioneer ensures that all prices actually set or setup . . . are market clearing prices. These postulates rule out the treatment of disequilibrium sequences and monetary exchanges, whether or not an actual equilibrium ultimately will be attained following a disturbance."

For further discussion, see several articles on the core and extensions of exchange — oriented theories in Baranzini and Scazzieri [5], and also Dorfman, Samuelson and Solow [21], Weintraub [70] and Loasby [43].

11.5 If Only Models Were Less Mathematical!

Simultaneously, the mathematics of modeling also became quite complicated. Ragnar Frisch [26] reveals his experience:

> Mathematical programming is nothing but systematized common sense.
>
> In the history of economic thought it is curious to note how the *fight* against the application of mathematics to economic phenomena has gone on. The picture is everywhere the same. . . . This applies equally to the countries which are now classified as belonging to the "West" and to those that are classified as belonging to the "East" . . . In my own country Norway, the application which I have made of mathematics has been criticized because some say that it means "introducing planning into economics," and in the "East" the application of mathematics to economics has been criticized on the grounds that it means the introduction of "bourgeois" methods!

> The truth is, of course, that mathematics is a politically completely neutral tool which may be applied wherever it comes in handy.

We shall return to this point later.

While Frisch seems to wonder why mathematics can't be used to understand economic problems, Kaldor (as we already referred to above) seems to think that the mathematics of general equilibrium theory is nothing but to prove existence theorems for a nonexisting economic world (i.e., null set).

Hahn [32], however, views the notion of equilibrium as an important base point in economic theory with reference to which several economic phenomena can be usefully analyzed and understood. He is highly critical of certain confusions prevailing in the current economic thought. "To many economists, Keynesian economics deals with important relevant problems and General Equilibrium theory deals with no relevant problems at all. This view is often the consequence of the ease of learning Keynesian macro-arithmetic compared with reading Debreu. But it also has, alas, an element of truth. This is quite simply that General Equilibrium theorists have been unable to deliver one-half at least of the required story: how does General Equilibrium come to be established? Closely related to this lacuna is the question of what signals are perceived and transmitted in a decentralized economy and how? The importance of Keynesian economics to the General Equilibrium theorist is two fold. It seems to be addressed to just these kinds of questions and it is plainly in need of proper theoretical foundations" (Hahn [31]). Hahn elaborates on some recent theoretical developments including short-period Walrasian equilibrium: "a set of current prices and associated expected prices such that the preferred action of every agent on current markets can be carried out." He asks whether each and every departure from Walras is Keynesian and points out for instance that "economics without the auctioneer requires non-perfect competition economics and not Keynesian economics."

The criticism on the usefulness of models based on general equilibrium theory continues. Substantial theoretical developments have been taking place in the area of disequilibrium economics. Some well-known concepts under this area of research are Dreze equilibrium, Benassy (K-) equilibrium, and Hahn's Rational Conjectural Equilibrium. For more discussion on the strengths and weaknesses of these theoretical developments, and theoretical controversies of disequilibrium and Keynesian economics (in closed and open economies) see Drazen [22], Hey [34], Cuddington et al. [19], Hahn [32], Benassy [10], Grandmont [30], Loasby [43], etc.

While these developments continue, practical planners would be willing to use general equilibrium theory, with all its limitations and advantages, if they could only compute the equilibrium.

11.6 If Only We Could Compute!

Modellers have indeed been conscious of the inadequacies of the theoretical framework provided by Walrasian economic systems. However, it is rather surprising that none of the dissent (only some of which we noted above) on general equilibrium theory and its applicability to real life situations came in the way of developing applied-computable general equilibrium models. Writing in 1975, Taylor [69] defends the models, saying that "any model based on current economic theory has to assume perfect competition, or something very close to it (i.e., simple forms of non-competitive behavior such as monopoly or discrimination monopoly can also be included). . . . If we had a better theory of prices and economic power than the Walrasian one, model builders would clearly use it. . . . If competition is basically the only game in town, you might as well play it with elegance."

Modellers are still looking for theorists to deliver an appropriate theory, but until then, they keep playing the same game again and again in some context or other, one time distribution policies, another time tax policies, and still another time something else.

Thus, the development of CGE models did not wait until everything turned out to be theoretically perfect and commended by one and all. The CGE modeling got a great boost when Scarf published his algorithm to compute the equilibrium. Once assured by Scarf that the equilibrium can indeed be computed, many new algorithms, generally heuristic but much more rapid, have been developed and used by modellers.. The models went ahead by incorporating various quantity constraints and price rigidities whenever it was felt necessary to at least partly reflect reality, essentially still retaining the Walrasian flavor. It seems worth asking why. We will return to this point towards the end.

The precursors of CGE models are Johansen [35, 36], Kelley, Williamson and Cheetham [40], Ginsburgh and Waelbroeck [28], Manne [45], Adelman and Robinson [1], etc. Development of computable general equilibrium models spread almost like wildfire all over the world. In a matter of two decades or so, hundreds of these models were developed for several countries. The wide range of applications of these models included mainly policies related to monetary, trade, energy, environment, income distribution, foreign investment, and asset markets. Bergman, Jorgenson and Zalai [11], Scarf and Shoven [62], Piggott and Whalley [55], Dervis, De Melo and Robinson [20], Starr [68], Srinivasan and Whalley [67], Fischer et al. [25], and Parikh et al. [54] provide surveys of theoretical bases and technical aspects of such models, and present a good number of country studies.

No wonder, Bell and Srinivasan [8] observe, "Find an economist with a keen interest in development planning and almost certainly there will be an economy-wide model in the offing." But each one of these applications has a novelty with regard to technical specification of the models. As Taylor [69] said, "it almost seems that as many ways of closing models have been used as there are model builders." Similar is the art with respect to CGE modeling. This is not surprising as a closure rule defines a particular policy regime, and as there are as many different policy regimes as there are governments.

One question relating to mathematical modeling becomes of interest now. If a set of model equations representing an economy is given without revealing the identity of the country, should it be possible to tell whether the equations correspond to a capitalist, socialist or mixed economy? The answer would come easier with optimization models where the objective function of the model yields clues.

Variables x, y, z, etc., may be neutral. Given a function of x, y and z, it yields the same answers to everyone everywhere. But it is the particular system prevailing in the economy that determines what variables are to be included in the model and whether the variables are to be multiplied or divided, etc. In this sense, applied mathematics is not all that neutral; pure mathematics may be.

Similar is the case with the CGE models. The answer to our question seems to be that simply based on the specified degree of price and quantity rigidities placed and the closure rule used, one can make out which economic system the model represents.

Using this logic, the CGE models to date moved far from Walrasian framework. Some of the interesting applications include those on socialist countries, as well as those on developing economies characterized by several market distortions. The framework of these models is broadened by incorporating price and quantity rigidities. For a discussion on two CGE models applied to Yugoslavia, Hungary and other socialist countries, see Kis, Robinson and Tyson [41]. Also, the CGE models developed by the Food and Agriculture Program of the International Institute for Applied Systems Analysis include those for Poland, Hungary, and China (see Fischer et al. [25] and Parikh et al. [54]). Narayana, Parikh and Srinivasan [51] contains a CGE model applied to India. This model also incorporates many structuralist features in the sense of bounds on taxes, imports and exports, balance of trade, foreign aid, and consumption levels. Thus, the situation does not seem to be as bad as Kaldor puts it. In fact, in view of the recent theoretical developments involving non-Walrasian equilibria, one feels that Kaldor [37] was rather pessimistic when he said, "the ship appears to be much further away from the shore now than it appeared to its originators in the nineteenth century."

Can a CGE model, at least in a theoretical sense, be used as a planning model? The answer is a definite "yes," subject to how the model is specified and solved (see Ginsburgh and Waelbroeck [29] and Karlin [39]). Basically, a planning model consists of a solution that maximizes a welfare function subject to feasibility constraints. A competitive equilibrium model consists of market-clearing prices where consumers maximize utility subject to budget constraints, producers maximize profit subject to technological constraints, and the government maximizes its objective subject to its budget constraint. Thus, planning model can be treated as subsumed within a competitive equilibrium model where the solution not only maximizes government's objective but also identifies policies that will lead to that solution.

Rudra [61] was one of the early economists who questioned the basic philosophies implied by the models. He strongly criticized the premises of economic and political philosophy pursued in India, which he thought were totally wrong. As he felt more and more disappointed with the later day plan models, he says, "official circles in the Government of India have continued to adhere to the make-belief of planning, but it has not deceived even themselves, let alone the public. The academic community of economists who worked up high euphoria over the Second Five Year plan did not find, after the early sixties, any problems of plan modeling worth their attention." Of course, Rudra's concern about "worth attention" is mainly related to philosophical bases of the Indian plan models, and not with respect to technical sophistication. However, he also identifies many technical inadequacies some of which still remain. One of them is in fact modeling government operations which we will take up later.

Thus, we find that a section of the critics demand that economic theory (theories) must explain the reality. Otherwise it cannot be a proper theory. Models, particularly the plan models, must explain reality, and when simulated, must be able to reproduce all the economic data with reasonable accuracy. Chow and Poolo [18] contains a good collection of papers on evaluation of models. Also see Mansur and Whalley [47]. One estimates the equations, calibrates the parameters, validates the model, conducts *ex ante* and *ex post* goodness of fit tests and so on and yet something seems to be amiss. Over the years CGE models have provided many policy insights — yet these are seldom followed. The question arises: why? The same criticism we made earlier about not accounting for implementation competence may apply. Models ought to account for political constraints. Models could be more useful if only we could model government.

11.7 If Only Governments Could Be Ignored!

Systems and schemes of quantity rationing are some of the important concepts developed in the recent literature on disequilibrium economics. In this context, Benassy [9, 10] makes an important assumption called "voluntary exchange," under which no agent is forced to buy more than what he needs and no agent is forced to sell more than what he wants to. He also mentions,

"Most markets in reality meet this condition (except perhaps some labor markets), and we shall henceforth assume that it is always satisfied." Under this situation, according to him, a rationing system is efficient if one never finds rationed demanders and rationed suppliers simultaneously in the market. Should that be the situation, then mutual exchange between individuals would be beneficial.

However, we find that the assumption of voluntary exchange sometimes fails in reality. We give an example from India. With respect to the food grain's market, there exists a rationing scheme, with the Government of India (GOI) being the centralized rationing authority. The GOI procures a part of food grains production from farmers directly or indirectly at prespecified prices, stores and transports them to various places all over India (some are even exported), and distributes to consumers a fixed quantity at ration prices, which are much below the market prices. Usually when harvests are good, the government procurement prices are expected to be higher than the open market prices (thus, farmers' incomes are protected by the GOI), and farmers prefer to sell to GOI. However, GOI sometimes does not buy all the food grains offered by the farmers for the following reasons: (a) it has no adequate storage space, (b) it does not have adequate transport facility for want of railway wagons, and (c) more ration sales at prefixed ration prices (below market prices) mean an additional burden on government budget. This is a situation where supplies are rationed for genuine reasons. But then the farmers are indirectly forced to sell more in the open market than they want, since they don't have their own storage facility for speculation purposes. The story is similar even when harvests are bad. Then, open market prices are expected to be higher than the government procurement price. Now the supply constraints act in the reverse way; i.e., farmers are forced to sell (supply) a part of their production to the GOI though they do not prefer to. This implies that farmers are forced to supply less than they want to in the open market. Thus, they are now taxed. Demand-side the rations are fixed. The assumption of voluntary exchange cannot be assumed here, and the applicability of the Benassy notion of efficiency is not straightforward. One may consider the producers and the government vis-á-vis consumers as representing two different markets. Yet the question arises, should this market be rated as "inefficient"? India's food grain rationing, formally known as the Public Distribution System (PDS), is quite inefficient for a thousand other reasons; but we do not propose to delve into those aspects now.

It seems that several such examples can be cited. India exports superfine rice, called Basmati, which is produced only in some parts of India. Due to exports, domestic supply is constrained, and so is the demand for it, though explicit quotas are not specified. Consequently, the domestic price of this rice variety is far higher than that of other varieties. Voluntary exchange is not even permitted here due to export obligations and the constraint on foreign exchange earnings. All the cases of "production for exports only" fall into this category.

Benassy [9] says, "Clearly a number of trades that satisfy the voluntary exchange assumption are very unsatisfactory. For example, zero trade always satisfies voluntary exchange." Consider the zero trade case of intoxicating drugs which are totally banned. Thus, supply is constrained to nil; but a black market exists, which is an indication of (indirectly) forced supplies and demands in that market. Maybe one day, tobacco gets banned too. Here, the concern for health comes in the way of free supplies and demands.[3]

Thus, we find that whenever a third party (such as government) that can impose certain constraints is involved in the market, or whenever several markets are interrelated when segmented, voluntary exchange itself may become impermissible and the above definition of efficient rationing becomes difficult to apply. Basically, any system running on quantity controls by a government results in suppression of voluntary exchange right away. Why then do we need these controls?

"Economic development does not take place in a vacuum. An economy is not a mechanical system. There are socio-cultural forces which are inextricably intertwined in the functioning of an economic system and its growth. The type of government, the legal system, the standards of education and health, the role of family, the inter-play of caste, creed and religion, the psychological sub-stratum of the masses which reflects itself in their outlook and approach to growth are all highly critical variables involved in the process of growth" (Fernando [24]).

This is almost like asking to produce nonlinearities when one fits a linear model. One cannot explain socio-political equilibria with models of economic equilibria. One may think that the present day models are unable to reflect reality because several socio-political structural features are yet to be incorporated. One such issue is the possibility of representing governmental activities. What are these activities? Government consists of whom? What are their objectives?

Several modellers have developed economy-wide models with a nominal representation of government activity. The reason is mainly that the government is considered as a relatively insignificant part of the total economy. This is to say, the orchestra and the symphony are considered more impor-

[3] It may not be out of place to mention here that India's Finance Ministers, while presenting the annual budgets, often resort to hiking up cigarette prices, saying that the nation's health is to be cared for, as though they just awoke to such a need. This is utterly pretentious. What the government really needs is revenue. Habitual smokers would, in any case, shift to cheap-quality cigarettes which are even more dangerous for health or they would continue to buy the same quality at higher prices, thus forgoing consumption of some other health-promoting nutritious food. This is common sense!

tant than the conductor! But the modellers do not have any pretensions of being able to reproduce reality. Most of the time they are concerned with the directional changes that would result in the economy if a set of policies — either in isolation or as a combination of them — is implemented. Nothing more is promised by modellers.. As we shall see later, in fact, they cannot even do this.

"It is ridiculous to claim that numerical general equilibrium analysis can definitely 'solve' policy problems, especially as the assumptions (let alone the numerical applications) can be and are frequently questioned. However, taken with the appropriate grain of salt at the right time, we believe there is enormous potential for both extending and raising the level of debate on many social issues" (Mansur and Whalley [47]).

However, several modellers have stressed the importance of the governmental operations. Rudra [61], reviewing some of the earlier Five Year Plans of India, mentions some important theoretical limitations and also some assumptions turned out to be ultimately wrong. One of the limitations is the inadequate treatment of government investment and expenditure in certain areas of the economy. He lists them: education, health, research and development, exploration of natural resources, public transport, civil administration, defence, hospitals, office equipment, etc. Further, he draws attention to the relative size of the governmental activity in India in these areas: "Investments in these areas constituted in 1965 as much as a quarter of all investments in the country. These investments and public expenditures (and therefore the inputs required by them) being treated exogenously in the multisector models, the targets for activities in the rest of the economy can hardly be regarded as being endogenously determined in any full and satisfactory sense" (Rudra [61]).

". . . in some cases there are strong theoretical reasons for specifying behavioral relations at the macro level. For example, certain economic actors, such as the government and the central bank, actually base their behavior on aggregate information. Typically, the behavior of policy makers in models is specified exogenously, but there is a growing concern, for example, about the need to 'endogenize' government" (Robinson and Tyson [58]).

Loasby [43] is more elaborate on this aspect: ". . . it is important to recognize that a theory of government intervention is as necessary as a theory of economic development and disequilibrium. To treat the government as the equivalent of the auctioneer, but calling quantities instead of prices, does not aid our understanding" . . .

Loabsy continues, "To postulate the maximization of social welfare as a governmental objective is less plausible than the attribution of profit maximization to firms: the political and administrative processes, the personal and professional interests of ministers, civil servants, advisers and lobbyists cannot be so simply summarized. As well as being less plausible, it is even less operational. How does one discover what actions are implied? Can the structural disequilibrium be identified, and if so, how is it to be dealt with?"

Thus, planners and policy analysts realize the need to model government behavior. To endogenize government behavior into policy models is to characterize it as rational in some sense. We need to understand that rationality.

11.8 If Only We Could Understand the Rationality of Government Behavior!

Literature, especially relating to trade theory, identifies certain economic activities as Directly Unproductive Profit Seeking (DUP). These DUP activities involve profit/income making without the production of goods, directly or indirectly. However, they do end up using real resources, by diverting them from the production. All kinds of rent seeking, political lobbying, tax evasion, etc., fall into this category. For some early literature in this context, see Srinivasan [66], Bhagwati and Srinivasan [16, 17], Krueger [42], Bhagwati [13], Bhagwati, Brecher and Srinivasan [15], and Mayer [49]. Krueger [42] focuses on the effects of rent seeking competition for import licenses under a quantitative restriction of imports and shows that the welfare cost of these restrictions equals that of their tariff equivalents plus the value of rents. Bhagwati, Brecher and Srinivasan [15] examine the implications of DUP activities in the context of welfare analysis. The results on welfare effects of policies are shown to be sensitive to whether DUP activities are built into a model or not. Mayer [49] relates actual tariff policy to the underlying distribution of factor ownership via two links: an economic link based on production structure and a political link through which economic interests are translated into actual tariff policy. Tariff formation is modeled as a function of voting behavior. He then analyses the sensitivity of tariff policies to changing voter eligibility rules and voter participation costs under majority voting.

Becker's [6] theory of political market equilibrium deals with competition among pressure groups for political influence. The basic assumption of his analysis is that taxes, subsidies regulations, and other political instruments are used to raise the welfare of more influential groups. Equilibrium levels of taxes and subsidies are ultimately determined by the competition among the pressure groups (also see Becker [7]).

However, these types of models have not seen the light of the computation world yet. Far from being incorporated into economy-wide models, even sectoral level empirical models are few. Later we take into account some literature that exclusively considers the issue of corruption, based on some empirical data. For the present, let us note that the rationale of government activities may go beyond pure economics.

Government in a country with India's diversity is likely to be a coalition of many interests, either in the form of an explicitly multi-party government, as at present, or in the form of single party which accommodates groups of varied interests, as the Congress party did for many decades. The political power of the constituent groups changes from time to time. Also, ministers in charge of different ministries may have different agendas. Even bureaucrats, who are routinely shifted periodically, may be sympathetic to different ideologies. In any case, short term political concerns may dominate long term economic interests. Thus, the changing behavior of government may reflect changing objectives, and attempts to objectively unravel government from its observed actions may fail.

Formal models where the activities of government sector had been incorporated are very few. Among the Indian models, Narayana, Parikh and Srinivasan [51] postulated some behavioral equations with respect to fixing the procurement quantity and rationing quantity under PDS and desired total investment in the economy. But these were reduced form equations and did not involve optimizing behavior on the part of governments. Such reduced form characterization of government behavior is a far cry from what goes on in the government process — an idea which Rudra [61] has mentioned above. However, Rudra's account is also not complete, as we shall see.

It may seem that no economy-wide model can ever hope to simulate reality, even if theories are as perfect as perfect competition and the government is endogenized. What is missing? The discrepancy is not simply a white noise kind of unexplained random error component in econometrics. It refers to a lot of political systems lying hidden, of which no modeller would ever hope to have complete knowledge; yet they have substantial economic impacts.

A substantial part of economic decision-making is driven by political motives. For example, though in a different context, Amartya Sen viewed that in India, famines did not recur after independence not only because of agricultural achievements, but also because of the existence of a free press. The ruling parties within India's democratic set-up fear a loss of votes and criticism by media and opposition parties. The situation is similar in the affluent countries of Europe and America, as well as in poor countries such as Botswana and Zimbabwe (see Sen [63, 64]).

The real problem behind modeling government behavior is the need to bring in not just economic but also political variables. Even economic variables have to be (dis)aggregated, or classified differently, depending on the purpose or perspective. For example, to start with population, it is no longer adequate to classify people into mere expenditure/income (low, average and high) groups. They need to be classified into power groups, pressure groups (lobbyists), the bureaucracy, powerless workers and so on. Income generation and consumer expenditures have to be modeled accordingly. Unless there is a one to one correspondence between economic classification and socio-political classification (which is hard to assume), the available data would be difficult to use. In addition, even when one is prepared to bring political variables into a model, such data are not available. Who would give, for instance, data on how much each political party received as donations from big industrial groups? Yet such donations could have substantial impact on economic policies decided by the parties and implemented by the bureaucracy.

Thus, while plenty of data on economic variables may be available, they may not be adequate for the purposes of modeling political economic market equilibrium or disequilibrium. Data problems on some political variables are even more serious.

Given a country, what are the issues that the country's theory of political economy must incorporate? Ragnar Frisch [27] suggests ways of constructing a Social Preference Function which is supposed to be maximized by the state. The method involves collecting information from government officials and people's representatives with regard to how much weight they attach to various economic and social issues. He provided an example that he constructed by interviewing officials on issues such as the number of unemployed, the annual growth rate, the inequality of incomes, the change in consumer price index, the trade balance, and government expenditures on health, education, and research. One can be sure that Indian members of parliament and state legislative assemblies can give very impressive answers to such questions, whether they believe in them or not. Also, given the changing objectives of government pointed out above, how stable such a social preference function will be, over time, is unclear.

When politicians' actions reflect the interests of their constituencies, one may be able to detect a pattern; but when their actions are guided by private greed, understanding government behavior becomes more daunting. Corruption, which is basically an unauthorized income transfer, has become widespread in India. Bribes are paid to officers in return for various things: the movement of files; the ability to ignore excise taxes, sales taxes, income taxes, or customs duties; the distortion of policies; or the awarding of favorable contracts or lucrative appointments.

Economists have modeled corruption. Consider an example where an agent has to pay a tax of Rupees X, and he wants to avoid paying it. He could get away with giving a bribe of aX $(0 < a < 1)$ to an officer. What would be the economic consequences? The officer's income goes up by aX, the agent's retained income is $(1 - a)X$, and the government loses revenue by X. The income effects for the agent and the officer would depend on how large is X. If X is not substantial, (aX) of the officer and $(1 - a)X$ of the agent would perhaps result in immediate consumptions. If X is neither too small nor large it may result in some kind of term-deposits. On the other hand, if it is too large an amount, it may result either in the form of capital investment or outflow to foreign countries. Coming to the government side, if all such X's are aggregated over all bribes that occur over a year or so in the entire economy, they add up to a huge amount. This loss results in a reduction of public expenditure for social consumption or investment, an increase in borrowings, externally or internally, or an increase in taxes. Needless to say, price-effects would follow too.

Some studies are now available that analyze economic effects of corruption and bribery. Studies by Besley and McLaren [12], Mauro [48], and Ades and Di Tella [2, 3] are notable in this context. Also see Raj [57]. Mauro's study, based on several countries' data, highlights the negative impact of corruption and bureaucratic inefficiency on investment and economic growth. Besley and McLaren [12] analyze the impacts of alternative wage payment schemes (reservation wage, efficiency wage, capitulation wage) for tax inspectors, in the presence of corruption, on total tax revenues, net of administrative costs. In a recent study, Ades and Di Tella [3], perform regression analyses on the pooled data of 32 countries including India for the years 1989–92, and show that active industrial policies foster corruption. They use in their study two different corruption indexes, one of which is obtained from the World Competitiveness Report of the EMF Foundation in Geneva. Notable conclusions drawn in this study are: (a) The total effect of industrial policies on investment has two components: (i) a direct positive effect and (ii) an indirect negative effect operating through corruption. The negative component is so dominant that the total effect falls down to anywhere between 84 to 56 percent of the direct effect. Thus, corruption leads to substantial loss of benefits of industrial policies, particularly less investment and growth. (b) Whenever corruption has a deleterious effect on investment, optimal subsidies tend to be larger. Ades and Di Tella, therefore, caution that in the context of enthusiastically pursuing interventionist industrial policies, cost–benefit analyses should take into account the presence of corruption.

Given the economic relevance of corruption, let us see how extensive it is in India. We should quickly point out that corruption and bribery are universal phenomena, prevailing both in developed and developing countries. According to a survey by the World Economic forum, India is the fifth most corrupt country in the world.

No recent quantitative estimates of sums involved in corruption are available.[4] Yet one can get an idea from the number of court cases initiated and the amount involved in the various known scams. The number of government

[4]Krueger [42], based on the 1964 Santhanam Committee Report on the Prevention of Corruption and some other studies, estimated that total value of rents covering public investment, imports, controlled commodities, credit rationing, and railways forms about 7.3 percent of India's national income. She further says, ". . . rents must be judged large relative to India's problems in attempting to raise her savings rate."

officers involved in corruption is large. Only when the charges are substantial do they result in a court case. A court case means, an additional burden on government expenditure, in terms of investigations. We understand that the number of court cases against politicians and civil servants, including Indian Administrative Service and Indian Police Service officers, is now running in the thousands. How many more corruption cases, seriously distorting the economy have not come to the people's knowledge? Only heaven — nay, hell knows.

Politicians, particularly of the ruling parties that form the governments in the States and at the Centre, easily outdo officials in terms of the amounts involved. Recently a story was reported in one of the news columns on the cancer of corruption (see Vijaya Pushkarna's report in *The Week*, May 25, 1997). The story reported that a Chief Minister of a state tried to bribe every member of a committee set up by the Central Government, in order to get a favorable decision on an inter-state dispute. A classic case of government bribing government!

Though the charges are yet to be proved (courts take their own time), a large number of financial scams involving politicians have come to light in the last ten or fifteen years. Some of them are Oil, Bofors, Sugar, Telecom, Urea, Fodder, Fodder Machine, Land, Wheat, Cobbler, Securities, Hawala and so on and so forth, to mention those that have occurred in recent times. Many of these scams involve the siphoning of several hundreds of millions of rupees of government money into private coffers.

A typical example relating to Rs. 1330 million, the Urea scam, shows how brazen the politicians and bureaucrats have become.

Prasannan [56] explains:

> . . . the acting chairman of National Fertilisers, had been brought into the panel of candidates by a ministerial decision, through the backdoor, which was later regularised by the appointments committee of the cabinet.
>
> Not only that, the entire urea episode was engineered from the beginning to facilitate a scam which makes one feel that various wings of the government had colluded to create such a urea scarcity that it would necessitate large-scale imports. The whole process began in 1994–95 when the annual plan outlay and budgets were finalised. Curiously, the fertiliser ministry then did not specify production targets. In fact, since April 1993 fertiliser units in the country were working without production targets! This when virtually everyone in the fertiliser sector knew that there would be a major shortfall in urea production.
>
> More mischievous was what was happening in National Fertilisers itself. The company had obtained a sanction to double the capacity of

> its urea ammonia plant at Vajaipur at a cost of Rs. 900 crore. Curiously, the money was delayed. The result was a massive shortage of urea.
>
> Even the 1993–94 import of urea was not exactly above board. After the inconvenient Chinta Mohan, who was handling the fertiliser portfolio as minister of state, was dropped, the ministry projected an import requirement of 22 lakh tonnes, but actually imported 27.44 lakh tonnes!
>
> There was much more loss to the exchequer in that deal. The contracts were signed in early 1993, based on the official exchange rate of Rs. 26 a dollar. On March 1, 1993, the rupee was made partly convertible. The foreign suppliers had to be paid at the new rate of Rs. 31.75 a dollar.
>
> In the latest deal, it was a conveniently headless State Bank that sent out in November 1995 about $37 lakh to its New York branch without permission from the Reserve Bank.

Let us ask: which government wing is not involved? Which crucial ministry is not involved: commerce, trade, industry, agriculture, planning, finance, fertilizers? Banks? And why are the projection targets developed by plan models of the Planning Commission? Why is there a Planning Commission at all, the size of which in terms of number of members, has been increasing over time? The whole story sounds like a planned anti-development exercise.

No one who visits Bangalore and looks at the Vidhana Soudha, the architectural marvel of the Karnataka State Assembly, fails to note the engraved message, "Government work is God's work." Now we know what God's work may involve.

Political activities and government processes of economic decision-making often get intertwined and become almost indistinguishable. Thus, the apparent irrationality of political processes get reflected in economic policies that mystify.

Thus, Indian politicians continue to subsidize electricity for agricultural users in a way that is not even in the interest of the farmers. As a consequence of this subsidy, the state electricity boards are financially bankrupt and unable to expand supply to meet growing demand. As a result, farmers often get electricity for only eight to ten hours a day and that, too, at night. With an uninterrupted power supply the farmer would be able to run his tube-well round the clock, sell surplus water to his neighbor, and make twice as much money as he would pay for electricity at a reasonable price. Yet politicians are reluctant to reform the electricity price for farmers. Many examples of such policies exist.

With political instability, in which different coalitions come to power for short periods of time, the situation becomes worse. Then, different parties compete to offer more bribes, in the name of subsidies to their support groups.

In such a political environment, downward rigidity of subsidies is even stronger than downward rigidity of prices. Thus, government budget has two important roles which the theory hardly recognizes: (a) to support the political bribes thrown to people under the name of subsidies and (b) to support the private coffers of politicians through scams.

The spirit of the individual politicians seems to be more towards "make hay while the sun shines." Make as much as possible while in power, and by any means. Politics seems to have become a system of random economic opportunities. As Raj [57] asks, "how is a country driven by corruption to reform itself when the professing reformers are themselves corrupt?"

All these frustrate the policy analyst and planner. Is it possible to model government behavior? Can we endogenize it in a policy planning model? And if not, is any rational policy analysis possible?

11.9 Hope and Despair

Does this mean that there is no usefulness of computable models? We do not think so. Today's state of the art models are quite powerful and they continue to incorporate new theoretical developments. In particular, equilibrium based multisectoral models have a tremendous advantage of providing insights with regard to the extent of interdependence between various policies. These are not only with respect to aggregate level macro issues such as growth and redistribution, but also with respect to several specific policies concerning rural development, terms of trade, etc. Another advantage is the possibility of studying the impacts of a number of policies in combination.

This advantage actually makes us ask policy questions more precisely and leads policy debates in the right direction. For example, suppose an economy decides to explore the consequences of accelerated irrigation development. If it is "irrigation development" *per se*, one would feel that the consequences on outputs, etc., are obvious. Viewed as a policy to be implemented, the question is not that simple. An equilibrium model requires specification of where funds come from: are they created raising additional taxes, cutting down on public consumption expenditure (if possible!), internal borrowing, or foreign transfers? Similarly, which crops get how much of the additional irrigation created? What would the fertilizer demand be? The solution, containing the implications of resulting prices, differs from specification to specification, including the mode of financing. Thus, analysis with such models helps sharpen the policy questions. Basically, there is no free lunch available to anyone in these models; thus, interdependence gets accounted for in a reasonable fashion. Since the available literature already discusses these aspects, we avoid repetition and further elaboration here.

From the Harrod–Domar model to the modern general equilibrium models which permit a variety of policy instruments and endogenous regime switches, economy-wide models have come a long way. Yet many shortcomings exist, even from the point of view of economists. The models do not satisfactorily incorporate industrial structure and economies of scale. Structural rigidities are often exogenously specified and do not emerge from the models as consequences of optimizing behavior of agents. Truly dynamic behavior is missing. Money is yet to be brought into these models in the true general equilibrium spirit. Parameters are not all estimated econometrically, nor are they "deep" parameters so that the Lucas critique remains relevant. And of course, the general equilibrium paradigm itself is questioned.

Yet, these are not insurmountable difficulties. One can imagine that temporary equilibrium models with all of these features are made operational in future. The work that remains to be done is perhaps not more than the distance economists and modellers have already covered.

However, endogenizing government seems far too complex. But as its importance is realized, the problem will be tackled. There will be many hopeful beginnings, ending in despair as the next mountain range comes in view. Yet, we believe that they will be climbed, simply "because they are there."

The T.N. Srinivasans, with their deep understanding of theoretical requirements, will lead theorists to the next range to be conquered, and with their abiding interests and concern for development, they will assist and encourage empiricists to develop increasingly more useful models.

Why are we hopeful?

Pandit Nehru seems to have been asked immediately after he assumed the office of Prime Minister (the first time) how many problems were there for him to solve — and his reply was, 340 million! That was the population in 1947. We have, to date, more than 950 million. But each one of us, we believe, is quite optimistic about India's future. and that counts a lot.

References

1 Adelman, I. and S. Robinson, Income Distribution Policy in Developing Countries, London, Oxford University Press, 1978.

2 Ades, Alberto and Rafael Di Tella, Competition and corruption; mimeo, Oxford University, 1994.

3 Ades, Alberto and Rafael Di Tella, Rafael, National champions and corruption: Some unpleasant interventionist arithmetic, Economic Journal, (July 1997).

4 Arrow, Kenneth J., The economic implication of learning by doing, Review of Economic Studies, 29 (1962).

5 Baranzini, Mauro and R. Scazzieri (eds.), Foundations of Economics: Structures of Inquiry and Economic Theory, Basil Blackwell, 1986.

6 Becker, Gary S., A theory of competition among pressure groups for political influence, Quarterly Journal of Economics (1983).

7 Becker, Gary S., Public policies, pressure groups, and deadweight costs, Journal of Public Economics (1985).

8 Bell, C. and T.N. Srinivasan, On the uses and abuses of economy-wide models in development policy analysis, in: M. Syrquin, L. Taylor and L.E. Westphal (eds.), Economic Structure and Performance: Essays in Honor of Hollis B. Chenery, Academic Press Inc., 1984.

9 Benassy. Jean-Pascal, The Economics of Market Disequilibrium; Academic Press, 1982.

10 Benassy, Jean-Pascal, Macroeconomics: An introduction to the Non-Walrasian Approach, Academic Press, 1986.

11 Bergman, L., D.W. Jorgenson and E. Zalai, General Equilibrium Modelling and Economic Policy Analysis, Basil Blackwell, 1990.

12 Besley, T. and J. McLaren, Taxes and bribery: The role of wage incentives, Economic Journal (1993).

13 Bhagwati, J.N. (ed.), Illegal Transactions in International Trade, North-Holland/American Elsevier, 1974.

14 Bhagwati, J.N. (ed.), International Trade: Selected Readings; MIT Press, 1987.

15 Bhagwati, J.N., R.A. Brecher and T.N. Srinivasan, DUP Activities and Economic Theory; European Economic Review (1984). Also reproduced in: J.N. Bhagwati (ed.) International Trade: Selected Readings; MIT Press, 1987.

16 Bhagwati, J.N. and T.N. Srinivasan, Smuggling and trade policy, Journal of Public Economics (1973).

17 Bhagwati, J.N. and T.N. Srinivasan, The welfare consequences of directly unproductive profit seeking (DUP) lobbying activities — Price versus quality distortions, Journal of International Economics, (August 1982).

18 Chow, Gregory C. and Corsi Poolo (eds.), Evaluating the Reliability of Macroeconomic Models, John Wiley & Sons, 1982.

19 Cuddington, T. John, Per-Olov Johansson and Karl-Gustaf Lofgren, Disequilibrium Macroeconomics in Open Economies, Basil Blackwell, 1984.

20 Dervis, K., J. De Melo and S. Robinson, General Equilibrium Models for Development Policy, Cambridge University Press, 1982.

21 Dorfman, R., P. Samuelson and R. Solow, Linear Programming and Economic Analysis, McGraw Hill, 1958.

22 Drazen, Atlan, Recent Developments in Macroeconomic Disequilibrium Theory, Econometrica (March 1980).

23 Eckaus, Richard S. and Kirit S. Parikh, Planning for Growth: Multisectoral, Intertemporal Models Applied to India; The MIT Press, 1968.

24 Fernando, A.C., Economic development without the side-effects, Hindu-Open Page 21 (October 1997).

25 Fischer, G., K. Frohberg, M.A. Keyzer and K.S. Parikh, Linked National Models: A Tool for International Food Policy Analysis; IIASA and Kluwer Academic, 1988.

26 Frisch, Ragnar, An implementation system for optimal national economic planning without detailed quantity fixation from a central authority, in: Economic Planning Studies; Collection of Essays by Ragnar Frisch, D. Reidel Publishing Company, 1976, Ch. 4.

27 Frisch, Ragnar A.K., Cooperation between politicians and econometricians on the formalization of political preferences, in: Economic Planning Studies; Collection of Essays by Ragnar Frisch, D. Reidel Publishing Company, 1976.

28 Ginsburgh, V. and J. Waelbroeck, A general equilibrium model of world trade, Part 1: Full format computation of economic equilibria, Cowles Foundation Discussion Paper No. 412, 1975.

29 Ginsburgh, V. and J. Waelbroeck, Planning models and general equilibrium activity analysis, in: Herbert E. Scarf and John B. Shoven (eds.), Applied General Equilibrium Analysis, Cambridge University Press, 1984.

30 Grandmont, J.M. (ed.), Temporary Equilibrium: Selected Readings, Academic Press, 1988.

31 Hahn, F.H., Keynesian economics and general equilibrium theory: Reflections on some current debates, in: G.C. Harcourt (ed.), The Microeconomic Foundations of Macroeconomics, The International Economic Association, Macmillan Press, 1977.

32 Hahn, Frank, Equilibrium and Macroeconomics, Basil Blackwell, Oxford, 1984.

33 Harcourt, G.C. (ed.), The Microeconomic Foundations of Macroeconomics, The International Economic Association; Macmillan Press, 1977.

34 Hey, John D., Economics in Disequilibrium, Martin Robertson, Oxford, 1981.

35 Johansen, Leif, A Multisectoral Study of Economic Growth, North-Holland, 1960.

36 Johansen, Leif, Explorations in long term projections for the Norwegian economy, Economics of Planning, 8 (1968).

37 Kaldor, Nicholas, The irrelevance of equilibrium economics, Economic Journal, 82 (1972).

38 Kanaan, Oussama, (1999): Comment on "Virtual Reality in Economy-wide Models," by N.S.S. Narayana and Kirit S. Parikh in this volume.

39 Karlin, S., Mathematical Methods and Theory in Games, Programming, and Economics; Pergamon Press, 1959.

40 Kelley, A.J., J.G. Williamson and R.J. Cheethan, Economic Dualism in Theory and History, Chicago University Press, 1972.

41 Kis, Peter, S. Robinson and L.D. Tyson, Computable general equilibrium models for socialist economics, in: L. Bergman, D.W. Jorgenson and E. Zalai, General Equilibrium Modelling and Economic Policy Analysis, Basil Blackwell, 1990.

42 Krueger, Anne O., The political economy of the rent-seeking society, American Economic Review (June 1974).

43 Loasby, Brian J., Disequilibrium states and adjustment processes, in: Brian J. Loasby The Mind and Method of the Economist, Edward Elgar Publishing Ltd., 1989.

44 Mahalanobis, P.C., Some observations on the process of growth of national income, Sankhya, 15 (September 1953).

45 Manne, A., General equilibrium with activity analysis, in: C. Hitch (ed.), Modeling Energy — Economy Interactions: Five Approaches, Resources for the Future, Washington, DC, 1977.

46 Manne, Alan S., A. Rudra and others, A consistency model for India's Fourth Plan, Sankhya, 27, Series B (1965).

47 Mansur, A. and J. Whalley, Numerical specification of applied general equilibrium models: Estimation, calibration, and data, in: Herbert E. Scarf and John B. Shoven (eds.), Applied General Equilibrium Analysis, Cambridge University Press, 1984.

48 Mauro, P., Corruption and growth, Quarterly Journal of Economics (1995).

49 Mayer, W., Endogenous tariff formation, American Economic Review (December 1984).

50 Narayana, N.S.S. and K.S. Parikh., Agricultural planning and policy in draft sixth plan: Will farmers fulfill planners' expectations?, Economic and Political Weekly, 14(30–32), Special Number (August 1979).

51 Narayana, N.S.S., Kirit S. Parikh and T.N. Srinivasan, Agriculture, Growth and Redistribution of Income; North Holland/Allied Publishers, 1991.

52 Nehru, Pandit, Tryst and destiny, reproduced in Times of India, August 15, 1997.

53 Parikh, K.S., Planning without policy, in: S. Guhan and M.R. Shroff (eds.), Essays on Economic Progress and Welfare in Honour of I.G. Patel, Oxford University Press, 1985.

54 Parikh, K.S.S., G. Fischer, K. Frohberg and O. Gulbrandsen, Towards Free Trade in Agriculture, IIASA and Martinus Nijhoff Pub, Kluwer Academic, 1988.

55 Piggott, John and John Whalley, New Developments in Applied General Equilibrium Analysis, Cambridge University Press, 1985.

56 Prasannan, R., Son's eclipse: Shadow on father, The Week, June 30, 1996.

57 Raj, K.N., Corruption in public life, The Hindu, November 11 & 12, 1997.

58 Robinson, S. and L.D. Tyson, Modelling structural adjustment: Micro and macro elements in general equilibrium framework, in: Herbert E. Scarf and John B. Shoven (eds.), Applied General Equilibrium Analysis, Cambridge University Press, 1984.

59 Romer, Paul M., Increasing returns and long-run growth, Journal of Political Economy, 94(5) (1986).

60 Rudra, Ashok, Indian Plan Models, Allied Publishing Pvt. Ltd., 1975.

61 Rudra, Ashok, Plan models in India — Evolution and evaluation, in: D.K. Bose (ed.) Review of the Indian Planning Process, Indian Statistical Institute, Calcutta, 1986.

62 Scarf, Herbert E. and John B. Shoven, Applied General Equilibrium Analysis, Cambridge University Press, 1984.

63 Sen, Amartya, Round table discussion on development strategies: The roles of state and the private sector, in: Proceedings of the World Bank Annual Conference on Development Economics, 1990, supplement to the World Bank Economic Review and World Bank Research Observer.

64 Sen, Amartya, Development thinking at the beginning of the 21st century, DERP No. 2, London School of Economics, 1997.

65 Srinivasan, T.N., Investment criteria and choice of techniques of production, Ph.D. Thesis, Yale University, 1962.

66 Srinivasan, T.N., Tax evasion — A model, Journal of Public Economics (November 1973).

67 Srinivasan, T.N. and John Whalley (eds.), General Equilibrium Trade Policy Modelling, MIT Press, 1986.

68 Starr, Ross M. (ed.), General Equilibrium Models of Monetary Economics: Studies in the Static Foundations of Monetary Theory, Academic Press, 1989.

69 Taylor, L., Theoretical foundations and technical implications, in: C.R. Blitzer, P.B. Clark and Lance Taylor (eds.), Economy-wide Models and Development Planning, Published for World Bank by Oxford University Press, 1975.

70 Weintraub, Roy E., On the brittleness of the orange equilibrium, in: A. Klamer, D.N. McCloskey and R.M. Solow (ed.), The consequences of Economic Rhetoric, Cambridge University Press, 1988.

Trade, Growth, and Development
G. Ranis and L.K. Raut

CHAPTER 12

Trade Liberalization, Fiscal Adjustment and Exchange Rate Policy in India[1]

Delfin S. Go[a] and Pradeep Mitra[b]

[a,b]The World Bank, 1818 H Street, NW, Washington, DC 20433

12.1 Introduction

12.1.1 Dedication

Nearly a quarter century ago, T.N. Srinivasan [5] (together with his illustrious colleague Jagdish Bhagwati) pioneered a landmark study of India's foreign trade regime. Since then, he has been indefatigable in urging a thoroughgoing reform of trade and industrial policy in India. Some of these reforms began to be implemented in the 1980s. But it was not until 1991 that India embarked on a more comprehensive program of structural reforms, together with macroeconomic stabilization, at a time when the country was in an economic crisis. It is therefore singularly appropriate to honor T.N. by presenting a paper which undertakes an analysis of trade liberalization, fiscal adjustment, and exchange rate policy in India.

12.1.2 The problem

India was in the throes of a serious foreign exchange crisis in 1991. The profligate fiscal policy of the 1980s had already contributed to a fiscal deficit amounting to nearly 10 percent of GDP and a current account deficit of around 3 percent of GDP in 1987/88 (this refers to the fiscal year April 1, 1987–March 31, 1988). The trade regime was among the most restrictive in the

[1]This paper is based on a larger study of India's trade regime undertaken prior to the reform process which started in 1991. Aspects of the original framework have since been updated in support of more recent studies and used in the results presented in Section 4. We thank Ataman Aksoy, Robert J. Anderson, Francois Ettori, Javad Khalilzadeh-Shirazi, Helena Tang and Roberto Zagha for many helpful discussions, and Michelle Connolly, the discussant at the conference, for several useful comments. Views expressed are those of the authors and do not necessarily reflect those of the World Bank or its affiliated organizations.

nonsocialist world. The average collection rate from import tariffs was around 60 percent and is estimated to have conferred extraordinarily high effective rates of protection on certain sectors of the economy (Aksoy and Ettori [3], for example, report that the effective rate of protection was as high as 585 percent in the capital goods sector). At the same time, import tariffs were contributing some 24 percent of revenue. Quantitative restrictions in the form of import licensing, though extensive, appeared to enjoy premia of the order of some 10 percent.[2]

In 1991, the newly elected government embarked on the urgent task of reducing the underlying fiscal and current account imbalances. However, the economic reform program also recognized that sustaining the resulting macroeconomic gains would require wide-ranging structural reforms in the country's trade and fiscal regimes as well. The foreign exchange crisis was seen as an opportunity to lower tariffs and quantitative restrictions on imports. But the importance of tariffs in public revenue and in imparting a pronounced anti-export bias to the system required that trade liberalization be coordinated both with fiscal adjustment — a combination of trade-neutral tax increases and expenditure reduction — *and* with a policy of exchange rate changes.

This paper asks the following questions: What is the impact of a reduction in the fiscal deficit characteristic of stabilization programs on tax and expenditure levels, and on the real exchange rate and the current account deficit? What is the effect of a significant trade liberalization, without additional external financing, on macroeconomic variables such as the required degree of fiscal adjustment of the real exchange rate and, at a more disaggregated level, on output levels in different export-oriented and import-substituting sectors of the economy? What would the impact of such trade liberalization look like, should substantive external financing become available without the need for domestic fiscal adjustment? These questions are explored using a general equilibrium model of the Indian economy that focuses on the consequences of trade policy reform. Policy makers are, however, also interested in knowing how various import-substituting industries would be adversely affected by trade liberalization as well as how particular export-oriented industries would gain from it. These objectives are reconciled by the innovative expedient of implementing two models on a common data base: (i) a disaggregated 72-sector (price sensitive) input–output version that makes simplified assumptions regarding certain economy wide relationships; and (ii) an aggregated 6-sector version that pays careful attention to those relationships and can suggest what corrections ought therefore to be made to the results of the sectorally disaggregated analysis.

[2] Available evidence, as quoted in Kishor [16], suggests that the premium on import replenishment licenses given to exporters had fallen to around 5 percent in the 1980s, largely due to a shift to a more active exchange rate policy and increased tariffs on imports, thus limiting the revenue gains to which relaxing nontariff import licenses could give rise.

The policy questions posed above were answered on the eve of the 1991 economic reform program launched by India's policy makers. Actual developments in the principal macroeconomic aggregates occurring during the first two years of the liberalization process, were then compared with the outcomes of the model and generally found to correspond quite closely. This finding encouraged an updating of the model to the fiscal year 1992–93 and its deployment to analyze the consequences of a set of further economic reforms for subsequent years.

More generally, the paper, while primarily shedding light on economic reforms in India, also develops an empirical methodology at different levels of aggregation for economies attempting a transition to outward orientation and closer integration into the world economy, in the face of revenue and balance-of-payments constraints.

12.1.3 Relationship to the literature

Empirical work on tax reform in developing countries has broadly followed either of two approaches. On the one hand, exercises of the computable general equilibrium type — Dahl, Devarajan and van Wijnbergen [8] and Mitra [21], Dahl and Mitra [9] — have focused on the macroeconomic consequences of tax design and reform at the expense of sectoral detail. On the other hand, more sectorally disaggregated studies, such as those of Ahmad and Stern [2] and Jha and Srinivasan [15] for India, make strong macroeconomic assumptions, notably that of fixed factor prices. This, together with constant returns to scale and no joint production, implies that producer prices are fixed and therefore that indirect taxes are fully shifted forward into consumer prices. While this approach obviates the need for modeling production and labor markets, it is for the same reason unable to analyze the impact of changes in taxes, tariffs, and quantitative restrictions on factor prices.

This study derives the macroeconomic and sectoral consequences of trade liberalization by combining the two approaches outlined above. The aggregated model is of the computable general equilibrium type whose analytical basis is provided by the absorption reduction-cum-switching model standard in open-economy macroeconomics (see, i.e., Corden [7]). In calculating the economy-wide consequences of particular policy reforms, it provides such information as resulting changes in factor prices, foreign exchange rates, and scarcity premia on imports subject to quantitative restrictions. The values of these variables are treated as parameters of the disaggregated model. The latter, which is implemented on the same database, contains essentially the same equations and is separated, using constant returns to scale in production, into a cost–price module and a fix-price quantity module in order to avoid a full general equilibrium calculation. Given new (i.e., policy-induced) estimates of factor prices and other key parameters from the aggregated model, the cost–

price module calculates new prices for specific industries and, with the new information on production costs, updates the (price-sensitive) coefficients of a detailed input–output matrix. In the next step, the quantity module derives sectoral gross outputs necessary to meet intermediate and final demands. This approach, which is described in Section 12.2.7 below, retains the simplicity of input–output analysis while allowing technical substitution in response to changing cost conditions. Finally, the framework developed in the paper provides some estimates, based on cross-country relationships, of the productivity improvements and growth consequences that could be expected from greater outward orientation.

12.1.4 Plan of the paper

Section 12.2 sets out the model in some detail, including a description of salient features of the Indian economy. Section 12.3 answers the policy questions posed in the introduction to the paper using data pertaining to the pre-reform period. Section 12.4 compares the outcomes generated by the model with actual developments in the major macroeconomic aggregates occurring up to the year 1992–93, and updates the model to that year to explore further rounds of economic reform. Section 12.5 brings together some concluding observations.

12.2 The Framework

This section provides a heuristic description of the model, which is similar in many respects to the six-sector models developed in Mitra [21] to examine the economic performance of oil-importing developing countries in response to external shocks during the seventies. The differences lie mainly in the special features developed for this study and in the addition of a disaggregated version for sectoral analysis. Moreover, using a method well-suited to multi-sectoral analysis, it outlines how total factor productivity (TFP) changes endogenously in the model as a result of increased outward orientation. A complete list of equations and a glossary of terms are included in Appendix A.

The data set is compiled from disparate sources that were made mutually consistent with one another and with the national income figures for 1987–88. The information thus assembled includes detailed revenue data on customs and excise taxes that describe the complex tax and trade protection system in India, an input–output table updated to 1987–88, and household expenditure information from the 38th round of the National Sample Survey (see Appendix B). The broad macroeconomic aggregates are shown in Table 12.1.

Table 12.1: India before recent reforms: GDP and Expenditures in 1987–88

	Amount in Rs billion	Percent share of GDP
GDP at factor prices	2,944.08	88.53
Agriculture	916.55	27.56
Industry	845.73	25.43
Mining	71.13	2.14
Manufacturing	541.60	16.29
Construction	170.08	5.11
Electricity	62.68	1.88
Services	1,181.80	35.54
Indirect taxes	381.45	11.48
GDP at market prices	3,325.53	100.00
Resource gap $(M - X)$	85.90	2.58
Imports $(g + nfs)$	296.19	8.91
Exports $(g + nfs)$	210.28	6.32
Total expenditure	3,411.43	102.58
Consumption	2,650.03	79.69
Private	2,239.69	67.35
General government	410.34	12.34
Investment	761.40	22.90
Fixed investment	674.51	20.28
Private sector	320.47	9.64
Public sector	354.04	10.65
Change in stocks	86.89	2.61
Government net revenue[a]	478.90	14.40
Taxes	569.70	17.13
Government expenditures[b]	764.38	22.99
Deficit	285.43	8.58
Per capita GDP	326.88	
Population (million)	785.00	
Average exchange rate (Rs/USS)	12.99	

a. Government net revenue = Tax revenue less net transfer plus other net income.

b. Government expenditures = Government consumption plus investment

12.2.1 Production

Six productive sectors are identified in the aggregated model. These sectors, with their distribution in value added appearing in parenthesis, are agriculture (31.1%), consumer goods (7.6%), intermediate goods (9.3%), capital goods (3.9%), construction (5.8%), and services (4.2%). They are further divided into 72 subsectors in the disaggregated model in the following manner: four agricultural sectors, five mining sectors, 57 manufacturing sectors, and six service sectors.

In each sector, a fixed value share for inputs at various levels (a nested

Cobb–Douglas structure) is used for domestic production.[3] The corresponding cost functions and input demand equations are shown as equations (12.1)–(12.6) in Appendix A.

Value added and net government and foreign transfers are mapped according to fixed rules into a single rural and a single urban household group [equations (12.7)–(12.8)].[4]

12.2.2 Demand

The components of final demand, with their shares in GDP at market prices appearing in parentheses, are private (67.4%) and public consumption (12.3%), private (9.6%) and public investment (10.6%), and exports (6.3%). Household incomes are divided into savings and private consumption. Private consumption is split into demand for the output of the six broad sectors, following an estimated linear expenditure system that allows subsistence expenditures to be satisfied before allocating the remainder across sectors according to fixed marginal expenditure shares [equation (12.9)–(12.10)]. In the absence of highly disaggregated estimated demand systems, household demand for more specific commodities, at the level of 72 sectors, is defined as fixed expenditure shares of the demands for the six aggregated goods [equation (12.11)–(12.12)]. Total investment is the sum of fixed private investment, fixed public investment, and changes in stocks [equation (12.13)]. Changes in stocks are assumed to be constant, while fixed investment is almost entirely directed in fixed quantity shares at sectors producing capital goods and construction (equation 12.14).

In addition, the demand for domestically produced intermediates, by commodity is given by a fixed-quantity-share breakdown of the total use of domestic inputs across sectors [equation (12.15)]. Service sectors also enjoy an extra source of demand, arising from the imposition of trade and transport margins in all sectors of the economy [see equation (12.16)]. The domestic component of final demand consists of consumption, investment, and the demand for trade and transportation margins [equation (12.17)].

[3] The choice of production structure was conditioned by the absence of reliable estimates of substitution parameters and the simplification required by disaggregation. Krueger [17], for example, argues that a Cobb–Douglas formulation is a reasonable choice. Moreover, the specification does not impose unduly high levels of price responsiveness of demand for inputs, especially the imported kind. In a large country like India, the implied demand elasticities of imported inputs are in fact small, since the cost shares of intermediate imports used as material input are small (see figures in Section 12.2.2).

[4] Thus, intra-rural and intra-urban distributional issues are not emphasized here. Earlier work with a similar framework, Mitra and Tendulkar [23], suggests that these are not significant in tariff reform at a broad level of aggregation. Data limitations preclude distributional matters from being analyzed at the 72-sector level of disaggregation.

12.2.3 Foreign trade

Imports, quotas, and supply of goods. The trade side incorporates price-responsive import relationships and attempts to take into account the various import restrictions prevailing in the economy. Import prices are given, so that the country is small in the relevant market. While the nontariff import licensing regime in India is complex and not susceptible to easy analytical characterization, its essence has been modeled as follows. "Competitive" imports in each sector — i.e., those that are broadly similar to domestic production — are assumed to be subject to quantitative restrictions through a variety of licenses. On the other hand, intermediate imports in each sector — i.e., those that are inputs to domestic production — are assumed to be importable [via Open General License (OGL)] and hence are subject to no nontariff restrictions. In 1987–88, the total c.i.f. value of imports of goods and services equaled Rs. 296.2 billion, or 8.9 percent of GDP at market prices. Their sectoral breakdown is as follows: agriculture (3.2%), consumer goods (10%), intermediate goods (46.2%), capital goods (20.8%), construction (20.8%), and services (19.8%). Of this amount, about 40 percent were competitive imports used in final demand and 60 percent were intermediate imports used in production. The proportion of intermediate imports in the total material input of production in the different sectors is: agricultural goods (1.5%), consumer goods (5.1%), intermediate goods (18.3%), capital goods (13.3%), and services (16%).

The demand for intermediate imports in each sector depends on the level of material input required in production and its import price relative to that of the domestically produced variety. Domestic and imported material inputs, though broadly similar, are not identical; they make up the aggregate material input as part of a fixed value-share production structure [equations (12.18)–(12.20)]. Each material input is a fixed quantity-share bundle of domestically produced and imported intermediates, respectively.

Competitive imports are more substitutable with domestically produced goods than noncompetitive intermediate imports are, but they are subject to quantitative restrictions. Since they are restricted or subject to imports only by parastatals, their levels are taken to be policy-determined. Demand for those goods has to be rationed by some form of quota prices. The latter are modeled using "virtual" prices; i.e., those prices of imports which would induce an unrationed economic agent to demand the observed quantity of rationed imports (see Neary and Roberts [25]). The virtual prices of imports differ from their purchased prices by a wedge created by the presence of quota premia [equations (12.24)–(12.25)]. The presence of quota premia raises the prices of domestic import substitutes, thus providing nontariff protection to producers. If the policy-determined rationed levels are changed, for example during trade liberalization, the premia, and consequently the virtual prices of competitive imports will also adjust to ensure that demands equal the new quotas.

Exports. Exports are negatively related to export prices relative to prices of international competitors, so that the country is assumed to be able, within limits, to vary its export sales by changing its export prices. Export demand also depends positively on incomes in the rest of the world [equation (12.26)]. The f.o.b. value of exports of goods and services totaled Rs. 210.3 billion in 1987–88, or around 6.3 percent of GDP at market prices. The share of the different sectors in exports are agriculture (7.0%), consumer goods (38.7%), intermediate goods (27%), capital goods (4.4%), and services (22.9%).

12.2.4 External debt

In 1987–88, India's external debt stood at $56.4 billion. Total debt service was estimated at $6 billion, of which around 91 percent, or $5.5 billion, was accounted for or guaranteed by the public sector. This represented about 17 percent of consolidated tax revenue, or 24 percent of the fiscal deficit. The need to meet debt service obligations, assumed to be denominated in dollars, would add to the government's fiscal burden in the event of a devaluation undertaken as part of a policy reform package.

12.2.5 Tax-cum-tariff system

The Union or central government raised tax revenue equaling Rs. 376.6 billion in 1987/88, or 11.3 percent of GDP at market prices. The various state governments collected another Rs. 193.1 billion, yielding a consolidated total revenue of Rs. 569.8 billion, or 17.1 percent of GDP at market prices. Indirect taxes accounted for 79 percent of Union revenues and approximately the same proportion of consolidated Union and State revenues. Table 12.2 reports the contribution of the various taxes to the revenue of the Union and that of the Union and States.

The tariff structure is divided into (1) basic and auxiliary customs duties and (2) additional or "countervailing" customs duties (CVD). The former set of duties is protective, while the latter matches the Union excise tax on domestic production. Both Union excise taxes and CVDs are part of the modified value added tax (MODVAT), which applies to the manufacturing sector, excluding petroleum, tobacco, and textile products (for an account of MODVAT as it then operated, see Narayana, Bagchi and Gupta [24]). The MODVAT credits producers in the manufacturing sector for excise taxes and for CVDs paid on inputs of raw materials. Revenues reported under the Union excise tax and CVD are in fact MODVAT revenue net of credits.

Table 12.2: India before recent reforms: Composition of indirect tax revenue, 1987–88*

	Imported Goods	Domestic Goods	Total
A. Union			
Protective import duty	31.23		31.23
Countervailing import duty	4.10		4.10
Union excise tax		43.64	43.64
Total	35.33	43.64	78.97
B. Union and States			
Protective import duty	23.57		23.57
Countervailing import duty	2.74		2.74
Union excise tax		29.11	29.11
State excise tax		4.61	4.61
State sales tax	0.84	18.59	19.43
Total	27.15	52.31	79.46

*Figures are percent of total tax revenue.

The commodity tax rates by broad sector in agriculture and manufacturing are shown in Table 12.3. Those rates are average collections divided by the appropriate tax bases. The protective tariff in 1987/88 was around 60 percent.[5] The tax base for the excise tax-cum-CVD is domestic supply, less untaxed items such as changes in stocks and exports. It also excludes inter-industry purchases in sectors registered under MODVAT which are exempted.[6] These different taxes and the crediting of MODVAT are reflected in various purchasers' prices [equations (12.27)–(12.33)]. The MODVAT does not allow the cost of capital goods to be credited. However, the model is capable of exploring the consequences of reforming the tax so that it does allow such crediting [equation (12.31)], an option which is not explored in the current paper.

12.2.6 Government consumption

Table 12.3 also reports the breakdown of Government consumption by sector. It may be noted that the bulk of it comprises services (83.2%) and construction (6.3%).

[5] The base in this case is the total value of imports reported in the customs statistics (see Government of India [13]), which do not include unclassified items, such as defense-related imports. Inclusion of the latter, such as in the value of total merchandise imports reported in the national income accounts (see Government of India [14]), raises the base by over 25 percent. Their exclusion from the base for the calculation of the protective tariff is justified by the fact that most of the unclassified import items are not subject to import duties.

[6] While not all manufacturers, for administrative and other reasons, avail themselves of the credits, it is assumed that credits are generally taken advantage of and tax rates are calculated accordingly.

Table 12.3: India before recent reforms: Union tax rates and composition of government consumption, 1987–88

	Protective import duty	Excise CVD rate	Government consumption*
Agriculture	0.214		0.0010
Manufacturing	0.526	0.107	0.1040
Consumer Goods	0.498	0.079	0.0168
Intermediate Goods	0.508	0.126	0.0792
Capital Goods	0.581	0.132	0.0080
Construction			0.0629
Services			0.8320

*Figures are sectoral shares in total government consumption.

12.2.7 Market clearing

Equilibrium requires that (1) the demand for goods in each sector equal supply [equation (12.34)], (2) the demand for each type of labor and capital equal their supply (see below), (3) the current account deficit or foreign savings in the balance of payments match foreign exchange outflows with total inflows [equation (12.35)], and (4) government revenue and savings cover public expenditures (see below). It can be shown that the above conditions imply that the savings–investment balance is satisfied [equation (12.36)].

The government budget. Government revenue consists of tax revenues from protective tariffs, CVDs, Union excise taxes, export duties (if any), State sales taxes, and income taxes [equation (12.37)]. Government expenditures include public consumption, public investment, debt service payments by the public sector, and transfers less net income from public enterprises. The difference between government revenue and expenditures equals government savings or deficit [equation (12.38)].

Factor markets. In the labor markets, there are two types of labor in the agriculture sector: own-farm workers and residual farm (landless) workers. In the nonagriculture sectors, there are organized workers and residual nonfarm (informal) workers. Labor supply of all classes except the residual in each sector are responsive to the real wage; i.e., the money wage deflated by the consumer price index [equation (12.39)]. In contrast, members of the residual class may migrate freely into and out of the organized labor class within each region. Since the total numbers of workers in the agricultural and nonagricultural areas are given at any particular time, this formulation [equation (12.40)] implies that each of the residual classes provides a pool of labor which accommodates the demand pressures for other types of labor; i.e., an increase (decrease) in the demand for nonresidual labor decreases (increases) the number of people in the residual classes. There is thus no open unemployment; a contraction in demand pushes people into low-productivity occupations of the kind assumed to be performed by the residual classes. Wages of the residual class are approximately 25 percent lower than those for the nonresidual class.

It is assumed that the nominal wages of nonresidual classes are sticky downwards [equation (12.41)]. This implies that their real wages may be lowered only through upward adjustment in the consumer price index of the kind, for example, that may be brought about through exchange rate devaluation. No stickiness assumption is made for the nominal wages of the residual classes. Equilibrium in the labor market is given by the equality of labor supply with the derived demand for residual and nonresidual workers [equation (12.42)]; this determines the real wages of all classes of labor. This equilibrium is tied to the definition of internal balance, which is presented below.

On the other hand, once installed, capital is fixed in each broad sector and earns a rate of return [equation (12.43)]. In the disaggregated 72-sector version of the model, capital stocks in specific industries are assumed to earn a constant proportion of the rate of return of the broad sector to which they belong.

12.2.8 External and internal balance

External balance is concerned with the attainment of a prescribed value of the current account deficit in the balance-of-payments. The focus of the analysis is to bring about such external balance through a reduction of absorption caused by fiscal adjustment. The first type of adjustment is to raise the average level of trade-neutral taxation (i.e., Union excise taxes and CVD) while keeping government expenditure constant in real terms [equation (12.44a)].[7] Since with fixed expenditures, the government saves all additional income (whereas the private sector saves only part of its additional income), domestic savings is increased by transferring income to the public sector, i.e., by increasing Union excise taxes-cum-CVD. In the second type of domestic adjustment, the government reduces domestic demand by cutting its own noninvestment expenditures [equation (12.44b)].[8] In either case, domestic savings must be raised to meet the difference between investment and the exogenously specified current account deficit of the balance-of-payments, provided the latter is set at a level no higher than that prevailing before the policy change.

Internal balance refers to the maintenance of equilibrium in the labor market. A fiscal contraction (tax increase or expenditure reduction) undertaken in support of a policy reform puts downward pressure on prices. Since nominal

[7] The uniform scaling of excise-cum-CVD could also be accompanied by changes in the sectoral pattern of taxation.

[8] This refers to consolidated government consumption. Transfers from government and abroad are held constant in real terms.

wages of the nonresidual classes are sticky downwards, this raises their real wages to levels incompatible with labor market equilibrium, potentially upsetting internal balance. This situation may be corrected through a devaluation of the exchange rate. Such a policy, by raising domestic prices, depresses the real wage of the nonresidual classes and restores internal balance. There is a transfer of labor into the residual class, where the nominal wage adjusts to clear the labor market. In fact, given a policy such as stabilization or trade liberalization, the model calculates the fiscal and exchange rate adjustments required to bring about external and internal balance.

Implementation of the disaggregated model. The implementation of the disaggregated model is as follows. The aggregated model derives the macroeconomic consequences of a reform and provides key prices and parameters in the economy: factor prices, foreign exchange rate, scarcity premia on imports subject to quantitative restrictions, adjustments of domestic taxes or government consumption, and the demands for the six broad commodities. Given factor prices, no joint production, and the assumption of constant returns to scale, product prices of specific sectors can be derived independently of quantities from the cost-price relationships in the disaggregated version.[9] In the next step, the input–output coefficients are updated by the new vector of prices. Given new product prices, exports and private consumption are estimated, as described in the previous section. The rest of nonintermediate demand, consisting of investment and government consumption, is assumed to be exogenous. Gross outputs of goods are obtained from the familiar Leontief expansion of final demand, based on the inverse matrix involving the price-sensitive input–output coefficients [see equation (12.34)]. While the quantity side retains the simplicity of input–output analysis, the input–output coefficients themselves are dependent on new prices. Given gross outputs and prices, the sectoral quantities such as domestic outputs, competitive imports, domestic and imported intermediates, etc., are estimated from the various input demand functions.[10]

This method of separate solution of the price and quantity modules obviates the need for assembling a complete disaggregated data set consistent with the specifications of a general equilibrium model. This latter task is difficult, given data problems in developing countries. For the application to India, however, the data for the 72 sectors were actually made consistent, thus permitting a disaggregated fully general equilibrium model to be solved simultaneously. By doing so, it was possible to test and obtain some estimates of the savings in computing time and cost when the short-cut in this

[9] The cost–price module contains the following equations: (12.1), (12.4), (12.7), (12.10), and (12.13)–(12.17).

[10] Equations (12.2), (12.3), (12.5), (12.6), (12.8), (12.9), (12.11), and (12.12) for the 72-sector version.

study — the method of separate solution of the price and quantity modules — was implemented. In fact, the two versions gave very similar results. This finding is very encouraging since the separate price-and-quantity calculations, in addition to reducing data work, permit a saving of at least a factor of five in mainframe computer time and cost when compared to a full-fledged disaggregated general equilibrium model. More generally, the use of two models provides an efficient way of examining the macroeconomic and sectoral consequences of policy reform.

12.2.9 Productivity

A large empirical literature points to a strong positive association between outward orientation and the growth of total factor productivity (TFP) in the economy. (See, for example, the surveys by Tybout [29]. Excellent overviews of the new growth literature are also found in Aghion and Howitt [1] and Barro and Sala-i-Martin [4]. A compilation of the key analytical and empirical contributions appears in Grossman [12].) That literature has advanced various possible explanations, such as the impact of R&D and innovation, changing market structure, the exploitation of scale economies, and knowledge spillovers. Much of this analysis is aggregative in nature and does not easily lend itself to the level of disaggregation required for the issues investigated in this paper. The practical approach used here is to make use of the empirical link between the TFP changes and the standard demand-side decomposition of growth into components associated with the expansion of exports and domestic demand (see, i.e., Chenery, Robinson and Syrquin [6]).

The annual growth of TFP in each manufacturing sector is associated with output growth allocated to export expansion and import substitution [see equation (12.45)]. That allocation follows the usual input-output requirements. The change of output between two periods may then be decomposed in terms of components associated with export expansion and import substitution, the latter defined as the growth of domestic output induced by final demand free of import content [equations (12.46)–(12.47)] (the derivation appears in Kubo, Robinson and Syrquin [18]).

It is expected that export expansion would lead to higher TFP growth while import substitution would lower its growth. TFP growth (decline), in turn, will lower (raise) the unit cost of domestic production in manufacturing through an index of productivity [equation (12.1)]. The index of productivity is solved endogenously for the three groups of manufacturing sectors distinguished in the aggregated model. The disaggregated version of the model assumes the productivity index for the specific industries within each group to assume the same value as that for the group as a whole.

12.3 Policy Simulations

All simulations in this section hold investment at its pre-shock level. This is because maintenance of investment is, subject to some reservations, broadly necessary for growth-oriented adjustment. Those reservations have to do with the need to subject investments to rigorous scrutiny with a view to increasing efficiency. Such scrutiny, however, requires a detailed microeconomic analysis which is beyond the scope of this paper. Maintaining investment at a lower level during stabilization and trade liberalization would, *inter alia*, ease the burden of adjustment on public and private consumption–a point that needs to be borne in mind in interpreting the results of the simulations below.

Simulation 1: Reducing the fiscal deficit

In the late 1980s, the Indian economy was characterized by two important macroeconomic imbalances: the current account deficit in the balance-of-payments was nearly 3 percent of GDP, while the Government's fiscal deficit was around 10 percent of GDP. Whether or not deficits of this order are sustainable requires analysis that is beyond the scope of this paper. Instead, we inquire as to the consequences of a prespecified reduction in the fiscal deficit.

With public and private investment fixed and private savings endogenously given as a function of private incomes, an exogenous restriction that the fiscal deficit must be held at some specified ratio requires that the current account deficit in the balance-of-payments be allowed to vary endogenously. This is because the restriction on the fiscal deficit leads to a rise in government savings; financing of the given investment levels thus requires a fall in foreign savings or the current account deficit.

Table 12.4 shows the results of reducing the fiscal deficit from 10 to 7 percent of GDP. Evidently, a one percentage point reduction in the fiscal deficit leads to a roughly one percentage point reduction in the current account deficit. The latter is, therefore, driven to zero, an implication that motivated the choice of a 7 percent fiscal deficit here.[11] This is not to suggest that a zero current account deficit is an appropriate target for India. However, to the extent that a sustainable fiscal deficit lies between the base year ratio (10%) and that which drives the current account deficit to zero (7%), the results of Table 12.4 may be pro-rated to yield results corresponding to different exogenously specified levels for the fiscal deficit.

[11] The assumption that investment is held at its pre-shock level is important here. With private saving being a fraction of private income that does not change very much, the required improvement in the fiscal deficit is brought about through an improvement in public savings which, with unchanged investment, requires a corresponding fall in foreign savings. A change in the savings investment balance of the private sector would modify this result but it is not clear *a priori* in which direction.

Table 12.4: Effects of reducing fiscal deficit to 7% of GDP (base solution: 100)

	Tax adjustment (1)	Public consumption adjustment (2)
Government		
Real government consumption	1.000	0.704
Prices		
Exchange rate (Rs/US$)	1.265	1.264
Producer prices	1.062	1.029
Real exchange rate	1.203	1.235
Consumer price index	1.054	1.014
Real GNP at market prices	0.980	0.980
Output	0.959	0.983
Private consumption	0.928	0.974
Imports	0.864	0.863
Exports	1.407	1.474
Total factor productivity	1.006	1.006
Memo item: Foreign savings/GDP	0.0022	–0.0005

Column (1) of Table 12.4 reports the results of attaining the deficit target through an increase in the average level of the Union excise-tax-cum-CVD rate, while holding investment (public and private), as well as public consumption constant in real terms. That average rate must more than double or, more precisely, increase by 113 percent. Since the fiscal adjustment is contractionary, maintenance of internal balance requires a depreciation of the real exchange rate on the order of 20 percent. Exports rise by 41 percent while imports contract by 14 percent. Alternatively, as reported in column (2), if the fiscal adjustment is accomplished solely through a reduction in public consumption, with investment (public and private) held constant, the required cut averages 30 percent. In interpreting this simulation, it must be noted that our formulation accounts for the costs and hence the budgetary implications of government expenditures on goods and services, but not its benefits. The real exchange rate depreciation in this case is 24 percent. Exports rise by 47 percent while imports contract by 14 percent. Notice that the increase in exports is smaller in the pure tax adjustment case. This is because the increase in taxes raises the cost structure and, *inter alia*, export prices, compared to the pure expenditure reduction case. Inasmuch as a fiscal adjustment would, in practice, involve a combination of tax increases and expenditure reductions, the resulting implications for the macroeconomic aggregates may be worked out by inspecting columns (1) and (2) in Table 12.4. It may also be recalled that the consequences of a smaller reduction in the fiscal deficit may be read off on a roughly pro-rated basis so that, for example, a one percentage point reduction in the fiscal deficit would call for roughly a 7 percent depreciation in the real exchange rate.

Simulation 2: Trade liberalization with fiscal adjustment and no additional external financing

A long term objective of trade reform is to institute a regime without quantitative restrictions and with low and broadly uniform tariffs. (The analytical underpinnings of such a policy regime are explored in Mitra [21].) A first step in that direction, consistent with the need to make India a lower cost economy, would give priority to a reduction of tariffs in capital- and intermediate good-producing sectors, as well as to an elimination of nontariff barriers to imports in those sectors. As reported in Table 12.3, the average protective tariffs for intermediate goods and capital goods are 50.8 percent and 58.1 percent, respectively. To that end, the policy package simulated here reduces protective tariffs — basic-cum-auxiliary customs duties — to a maximum of 40 percent on intermediate goods and a maximum of 25 percent on key machinery subsectors. Those tariffs on intermediates and capital goods (such as fertilizers, coal and lignite, etc.) that are already lower than the recommended rates are kept as they are.[12]

It will be recalled that imports that are broadly competitive with domestic production are restricted through an extensive licensing system. This implies that the scarcity premia on such imports accrue to those with access to licenses. As part of the liberalization effort, quantitative restrictions are relaxed in the intermediate and capital goods sectors. This is interpreted to mean that final imports become free and no premia exist in those sectors after the reform. Estimates of these premia are inherently uncertain and we present the results for a "high" (25%) import premium and a "low" (10%) import premium case.

A reduction in protective tariffs has a negative effect on public revenue and *ceteris paribus* on public savings. However, given that around 60 percent of imports are inputs into the production process, a tariff reduction has a favorable effect on output, private sector income, and hence private savings. But since only a fraction of private sector income finds its way into private savings, the increase in the latter does not completely offset the decline in public savings. With a given current account deficit (foreign savings), total savings in the economy decline, and notwithstanding the fall in the price of investment goods induced by tariff reduction are not sufficient to finance investment expenditures. Restoration of the savings-investment balance requires an increase in domestic savings, which may be brought about either by an increase in the average level of the Union excise-tax-cum-CVD rate or a reduction in government consumption.

Table 12.5 reports results for the high premium case. Column (1) shows

[12] Such a policy can be expected to increase protection for import-substituting final goods, a result that could be offset by intensifying domestic taxation in those sectors. The latter option is not explored in the paper.

the consequences of fiscal adjustment through an increase in the average level of the Union excise-tax-cum-CVD rate. That average rate must increase by 23 percent in order to restore equilibrium. Internal balance is restored through a 13 percent depreciation of the real exchange rate. Exports rise by 29 percent and imports by 7 percent. The greater openness to the external environment induced by the reform increases total factor productivity (TFP) by about 1 percent; its significance may be judged by comparing columns (1) and (2), where the latter takes away the productivity-enhancing effect of outward orientation. Adjustment in this case requires an increase of 29 percent in the average Union excise-cum-CVD rate and a 13 percent real devaluation. While real GNP increases by 1 percent with TFP augmentation [col. (1)] the increase is only 0.3 percent without such augmentation [col. (2)].

Table 12.5: Effects of trade liberalization with fiscal adjustment and devaluation: High (25%) premium case (base solution: 1.00)

	Tax adjustment		Public consumption adjustment	
	with TFP (1)	without (2)	with TFP (3)	without (4)
Government				
Real government consumption	1.000	1.000	0.942	0.927
Excise/CVD tax rate	1.228	1.290	1.000	1.000
Prices				
Exchange rate (Rs/US$)	1.109	1.128	1.105	1.123
Producer prices	0.980	0.991	0.974	0.983
Real exchange rate	1.129	1.129	1.131	1.140
Consumer price index	1.004	1.010	0.996	1.001
Real GNP at market prices	1.009	1.003	1.009	1.003
Output	0.999	0.990	1.002	0.997
Private consumption	0.998	0.989	1.008	1.001
Imports	1.065	1.062	1.068	1.065
Exports	1.292	1.297	1.298	1.304
Total factor productivity	1.007	1.000	1.007	1.000

Column (3) shows the result of fiscal adjustment through a reduction in public consumption with investment (public and private) held constant. The average reduction is nearly 5.8 percent in the presence of a TFP increase and 7.3 percent in its absence. Movements in the exchange rate and in imports and exports are broadly similar to those in the pure tax adjustment case. These results show that if, for example, the fiscal effort were to be divided evenly between tax and expenditure adjustment, it would require an 11.5 percent increase in the average Union excise-tax-cum-CVD rate and a 3 percent reduction in public consumption, together with a real devaluation of 13 percent. Since the average rate of taxation is 13.1 percent, the new higher rate would be 14.6 percent. The magnitude of the fiscal effort therefore appears quite manageable for the trade liberalization in capital and intermediate goods producing sectors.

The case of a high import premium implies that domestic prices are considerably higher than import prices (inclusive of tariffs). Relaxation of quantitative restrictions therefore leads to much higher imports, compared to a low (10%) import premium case presented in Table 12.6. Import increases average 6.5 percent in the high premium case, compared to 5 percent in the low premium case. The above comparisons take into account the productivity-enhancing effects of increased outward orientation; it will be seen that this is higher in the high premium case.

Table 12.6: Effects of trade liberalization with fiscal adjustment and devaluation: Low (10%) premium case (base solution: 1.00)

	Tax adjustment		Public consumption adjustment	
	with TFP growth (1)	without TFP growth (2)	with TFP growth (3)	without TFP growth (4)
Government				
Real government consumption	1.000	1.000	0.940	0.920
Excise/CVD tax rate	1.239	1.289	1.000	1.000
Prices				
Exchange rate (Rs/US$)	1.081	1.097	1.087	1.079
Producer prices	0.982	0.991	0.985	0.980
Real exchange rate	1.099	1.105	1.102	1.110
Consumer price index	1.003	1.008	0.994	0.998
Real GNP at market prices	1.009	1.004	1.008	1.004
Output	1.001	0.995	1.006	1.002
Private consumption	0.999	0.992	1.009	1.001
Imports	1.052	1.049	1.054	1.050
Exports	1.225	1.229	1.231	1.238
Total factor productivity	1.006	1.000	1.006	1.000

The greater reduction in the cost structure of the economy in the high premium case allows increased exports to be generated to finance imports. The increases in exports are 29 and 23 percent in the high and low premium cases, respectively. The more substantial decline in the price of investment goods in the high premium case implies that the pressure to finance investment is less strong, a fact that is reflected in less fiscal adjustment. The magnitudes of fiscal adjustment are broadly comparable. Pure tax and expenditure adjustments are 23 and 6 percent, respectively, in the high premium case, while they are 24 and 6 percent, respectively, in the low premium case. The real devaluation is higher (13%), however, in the high premium case compared to the 10 percent in the low premium case; with the proportionately lower cost structure, this is necessary in order to bring wages relative to other prices in line with the requirements of full employment.

We next examine the sectoral consequences of fiscal adjustment in the presence of increases in total factor productivity. The mapping scheme between the six sectors and the 72 sectors is shown in Table 12.14. By way of background, Table 12.7 provides information regarding the subsectoral structure of the economy. By far the largest share of exports (12.7%) is accounted for by other nonmetallic minerals, which includes gems, ceramics, and glass products. Other important manufactured exports are ready made garments, leather products, miscellaneous food, cotton textiles, and tea and coffee. Crude petroleum is the dominant import item (10.5 percent) followed, among merchandise imports, by other nonelectrical machines, other nonmetallic minerals (which comprise uncut gems), iron and steel foundries, drugs and medicines, and industrial machinery and petroleum products.[13]

Table 12.7 also presents the direct and indirect intermediate and capital goods import content embodied in a unit of output for each sector. Among the most intermediate-and-capital-good-intensive sectors in this sense are ships and boats, office machinery, other nonelectrical machines, petroleum products, machine tools, and iron and steel casting and foundries.

Turning to Table 12.8, it may be seen that the highest excise-cum-CVD tax rates are on nonferrous metals, office machinery, organic and inorganic chemicals, iron and steel foundries, and synthetic fibers and tobacco products, followed by motorcycles, other chemicals, rubber products, and art-silk synthetic textiles.[14] The table also shows that over three quarters of government consumption consists of other services. Thus, restraint in government consumption as part of adjustment would affect this sector very strongly, a consideration that would influence how fiscal adjustment would, in practice, be divided between tax and expenditure adjustment.

Table 12.9 presents the consequences on gross output, by subsector, of fiscal adjustment to tariff reduction in the intermediate and capital goods sector for the cases of tax as well as expenditure adjustment.

The agriculture and consumer goods sectors benefit from the lower input costs arising from tariff reduction and, given no import competition, from the decline in output prices across the sectors of the economy. Outputs of agriculture and consumer goods increase by 0.5 and 3.6 percent, respectively.

[13]It may be noted that "other non-metallic minerals" features prominently on both the export and import sides. This is because even a 72-sector framework does not represent a degree of disaggregation high enough to permit such distinctions.

[14]It will be recollected that the tax base is domestic supply, less untaxed items such as changes in stocks and exports. It also excludes inter-industry purchases in sectors registered under the MODVAT, which are exempted.

Table 12.7: India before recent reforms: Structure of trade and shares of imported intermediates and capital goods in production

Sector	Distribution of exports[a]	Distribution of imports[b]	Direct and indirect intermediate and capital good imports in production[c]
1. Cereal crops	1.83	0.04	2.79
2. Milk	0.00	0.00	0.46
3. Meat and fish	0.70	0.70	0.90
4. Other agriculture	4.45	2.51	1.13
5. Coal and lignite	0.04	0.70	4.32
6. Crude petroleum	0.00	10.50	1.00
7. Iron ore	2.77	0.02	4.44
8. Metallic minerals	0.32	0.09	1.22
9. Nonmetallic minerals	0.80	1.34	2.40
10. Sugar	0.06	0.63	1.70
11. Edible oil	0.95	1.45	2.47
12. Tea and coffee	4.23	0.00	2.17
13. Miscellaneous food	5.13	0.83	2.83
14. Beverages	0.01	0.02	5.60
15. Tobacco products	0.13	0.00	2.14
16. Cotton textiles	4.94	0.09	2.81
17. Woolen textiles	0.07	0.11	4.31
18. Silk textiles	0.63	0.03	1.81
19. Art-silk synthetic textiles	0.53	0.32	7.36
20. Jute, hemp and mesta	1.21	0.02	3.91
21. Carpet weaving	2.50	0.00	2.13
22. Ready-made garments	9.01	0.01	2.95
23. Miscellaneous textiles	1.10	0.18	3.19
24. Wood products	0.09	0.05	2.30
25. Paper and newsprint	0.03	1.71	10.92
26. Printing and publishing	0.13	0.29	9.98
27. Leather products	5.97	0.09	2.35
28. Rubber products	0.62	0.17	7.83
29. Plastic products	0.15	0.20	15.46
30. Petroleum products	3.14	3.41	26.74
31. Coal tar products	0.00	0.10	18.66
32. Inorganic chemicals	0.19	1.39	9.36
33. Organic chemicals	1.31	2.41	13.44
34. Fertilizer	0.01	0.65	16.25
35. Pesticides	0.12	0.17	14.27
36. Paint varnishes	0.24	0.04	14.55
37. Drugs and medicines	1.15	3.78	10.05
38. Soaps and cosmetics	0.48	2.02	5.30
39. Synthetic fibers	0.08	2.12	19.21
40. Other chemicals	0.17	1.89	15.16

Table 12.7 (continued)

Sector	Distribution of		Direct and indirect intermediate and capital good imports in production[c]
	exports[a]	imports[b]	
41. Structural clay	0.02	0.09	4.10
42. Cement	0.00	0.02	4.98
43. Other nonmetal minerals	2.73	7.12	17.44
44. Iron and steel alloys	0.04	1.71	10.66
45. Iron and steel foundries	0.48	4.18	23.07
46. Non-ferrous metals	0.32	2.63	21.45
47. Hand tools hardware	0.52	0.21	14.83
48. Misc. metal products	0.52	0.67	15.06
49. Agricultural implements	0.03	0.14	13.99
50. Industrial machinery	0.55	3.50	37.81
51. Machine tools	0.38	1.18	24.57
52. Office machinery	0.11	0.51	43.44
53. Other nonelectric machines	1.05	7.22	32.36
54. Electrical industrial machinery	0.34	2.40	18.53
55. Electric cables and wires	0.18	0.25	20.24
56. Batteries	0.30	0.09	16.04
57. Electric appliances	0.08	0.61	20.48
58. Communication equipment	0.04	1.05	18.61
59. Electronic equipment	0.62	2.20	22.70
60. Ships and boats	0.01	0.51	46.98
61. Rail machinery	0.04	0.26	15.21
62. Motor vehicle	058	0.69	12.77
63. Motorcycles	0.07	0.22	11.92
64. Bicycles, other transportation	0.47	0.01	10.28
65. Watches and clocks	0.00	0.13	1.79
66. Miscellaneous manufacturing	2.27	2.57	20.10
67. Construction	0.00	0.00	7.78
68. Utilities	0.00	0.00	4.49
69. Rail transportation	1.11	0.00	5.51
70. Other transportation	5.35	7.82	6.16
71. Trade	9.18	0.00	1.02
72. Other services	7.30	11.97	1.10

[a]Percent share in total exports. [b]Percent share in total imports. [c]Percent share in domestic output.

However, outputs of the intermediate and capital goods sectors decrease by 3.1 and 11.9 percent, respectively. Some details at a more disaggregated level follow.

While contraction in the intermediate goods industries is to be expected following trade liberalization, the figures for each subsector should be inter-

Table 12.8: India before recent reforms: Union tax rates and composition of government consumption, 1987–88 (72 sectors)

Sector	Protective import duty	Excise CVD rate	Government consumption (% of total)
1. Cereal crops	0.01		0.0008
2. Milk			
3. Meat and fish	0.20		
4. Other agriculture	0.22		0.0001
5. Coal and lignite	0.04		
6. Crude petroleum	0.57	0.01	
7. Iron ore			
8. Metallic minerals	0.62	0.01	
9. Nonmetallic minerals	0.18	0.04	
10. Sugar	0.46	0.12	
11. Edible oil	0.37	0.01	
12. Tea and coffee		0.09	
13. Miscellaneous food	1.06	.001	0.0004
14. Beverages	1.06	1.10	
15. Tobacco products		1.41	
16. Cotton textiles	0.03	0.02	
17. Woolen textiles	0.07	0.01	
18. Silk textiles	0.15		
19. Art-silk synthetic textiles	0.81	0.36	
20. Jute, hemp and mesta	0.03	0.01	
21. Carpet weaving	0.36	0.03	0.0002
22. Ready-made garments	0.20		0.0002
23. Miscellaneous textiles	0.45		0.0010
24. Wood products	0.44	0.04	0.0009
25. Paper and newsprint	0.17	0.19	0.0013
26. Printing and publishing	0.07		0.0145
27. Leather products	0.12	0.01	0.002
28. Rubber products	0.86	0.39	0.0012
29. Plastic products	0.91	0.30	
30. Petroleum products	0.08	0.30	0.0155
31. Coal tar products	0.43	0.01	
32. Inorganic chemicals	0.32	0.52	
33. Organic chemicals	1.18	0.52	
34. Fertilizer	0.02		0.0002
35. Pesticides	0.30	0.03	
36. Paint varnishes	0.80	0.06	
37. Drugs and medicines	0.39	0.05	
38. Soaps and cosmetics	0.62	0.27	
39. Synthetic fibers	0.71	0.46	
40. Other chemicals	1.30	0.40	
41. Structural clay	0.69	0.07	

Table 12.8 (continued)

Sector	Protective import duty	Excise CVD rate	Government consumption (% of total)
42. Cement	0.25	0.30	
43. Other nonmetallic minerals	0.03	0.09	
44. Iron and steel alloys	0.34	0.12	
45. Iron and steel foundries	0.78	0.xx	0.0001
46. Nonferrous metals	0.95	0.59	
47. Hand tools hardware	0.57	0.11	0.0005
48. Miscellaneous metal products	0.41	0.06	
49. Agricultural Implements	0.82	0.04	0.0003
50. Industrial machinery	0.43	0.11	
51. Machine tools	0.72	0.05	
52. Office machinery	0.30	0.057	0.0013
53. Other nonelectric machines	0.62	0.13	0.0006
54. Electrical industrial machinery	0.85	0.12	0.0007
55. Electric cables and wires	0.45	0.19	
56. Batteries	0.20	0.30	0.0001
57. Electric appliances	0.20	0.15	0.0006
58. Communication equipment	0.61	0.14	0.0006
59. Electronic equipment	0.57	0.15	
60. Ships and boats	0.25	0.18	
61. Rail machinery	0.58	0.06	
62. Motor vehicle	0.64	0.15	0.0036
63. Motorcycles	0.56	0.40	
64. Bicycles, other transportation	0.03	0.03	
65. Watches and clocks	0.94	0.03	0.0001
66. Miscellaneous manufacturing	0.70	0.01	0.0594
67. Construction			0.0629
68. Utilities			0.0262
69. Rail transportation			0.0115
70. Other transportation			0.0096
71. Trade			0.0186
72. Other services			0.7662

preted with some qualifications. This is because of classification problems even at this level of disaggregation. For example, the basic metals industries (sectors 44–46) all show contraction in output. However, among ferrous metal industries (sectors 44–45), mild steel is efficiently produced and internationally competitive, while integrated steel plants, dominated by public enterprises, are more protected and inefficient. The same is true of nonferrous metals (sector 46). While aluminum is produced at close to world prices, this is not the case for copper, for example, where prices are higher than international prices. Metal products (sectors 47–48) constitute a diverse group. The

Table 12.9: Effects of trade liberalization on specified industries by type of fiscal adjustment (base year output = 1.00)

	(1) Tax adjustment	(2) Public consumption adjustment	(3) Average of (1) and (2)
1. Cereal crops	1.006	1.012	1.009
2. Milk	0.997	0.995	0.996
3. Meat and fish	1.012	1.016	1.014
4. Other agriculture	1.001	1.005	1.003
5. Coal and lignite	0.994	1.010	1.002
6. Crude petroleum	1.013	1.054	1.034
7. Iron ore	1.317	1.354	1.336
8. Metallic minerals	0.796	0.810	0.803
9. Nonmetallic minerals	1.063	1.072	1.068
10. Sugar	0.979	0.990	0.985
11. Edible oil	1.001	0.995	0.998
12. Tea and coffee	1.086	1.083	1.085
13. Miscellaneous food	1.020	1.014	1.017
14. Beverages	1.018	1.022	1.020
15. Tobacco products	0.945	1.011	0.978
16. Cotton textiles	1.049	1.068	1.059
17. Woolen textiles	1.033	1.048	1.041
18. Silk textiles	1.061	1.066	1.064
19. Art-silk synthetic textiles	0.996	1.020	1.008
20. Jute, hemp and mesta	1.063	1.078	1.071
21. Carpet weaving	1.265	1.291	1.278
22. Ready-made garments	1.180	1.208	1.194
23. Miscellaneous textiles	1.037	1.041	1.039
24. Wood products	0.983	0.995	0.989
25. Paper and newsprint	0.977	0.987	0.982
26. Printing and publishing	1.011	1.000	1.006
27. Leather products	1.161	1.156	1.159
28. Rubber products	0.992	1.015	1.004
29. Plastic products	0.919	0.931	0.925
30. Petroleum products	0.999	1.017	1.008
31. Coal tar products	0.978	0.986	0.982
32 Inorganic chemicals	0.939	0.960	0.950
33. Organic chemicals	1.089	1.106	1.098
34. Fertilizer	1.020	1.037	1.029
35. Pesticides	1.040	1.053	1.047
36. Paint varnishes	1.009	1.020	1.015
37. Drugs and medicines	1.026	1.019	1.023
38. Soaps and cosmetics	1.021	1.039	1.030
39. Synthetic fibers	0.874	0.897	0.886
40. Other chemicals	0.771	0.777	0.774

Table 12.9 (continued)

Sector	Tax adjustment	Public consumption adjustment	Average of (1) and (2)
41. Structural clay	1.001	1.005	1.003
42. Cement	1.019	1.023	1.021
43. Other non-metal minerals	1.266	1.279	1.273
44. Iron and steel alloys	0.960	0.965	0.963
45. Iron and steel foundries	0.677	0.693	0.685
46. Nonferrous metals	0.708	0.719	0.714
47. Hand tools hardware	0.903	0.922	0.913
48. Miscellaneous metal products	1.006	1.016	1.011
49. Agricultural implements	0.890	0.898	0.894
50. Industrial machinery	0.831	0.839	0.835
51. Machine tools	0.719	0.723	0.721
52. Office machinery	0.888	0.891	0.890
53. Other nonelectric machines	0.767	0.7730	0.770
54. Electrical industrial machinery	0.805	0.808	0.807
55.Electric cables and wires	0.994	0.997	0.996
56. Batteries	1.077	1.107	1.092
57. Electric appliances	1.015	1.029	1.022
58. Communication equipment	0.797	0.802	0.800
59. Electronic equipment	0.863	0.880	0.872
60. Ships and boats	0.937	0.957	0.947
61. Rail machinery	0.955	0.971	0.963
62. Motor vehicle	0.967	0.973	0.970
63. Motorcycles	0.973	0.994	0.984
64. Bicycles and other transportation	1.077	1.085	1.081
65. Watches and clocks	1.009	1.012	1.011
66. Miscellaneous manufacturing	0.895	0.888	0.892
67. Construction	0.999	0.997	0.998
68. Utilities	0.985	0.993	0.989
69. Rail transportation	1.013	1.023	1.018
70. Other transportation	1.010	1.030	1.020
71. Trade	1.018	1.033	1.026
72. Other services	1.019	1.007	1.013
Summary			
Agriculture	1.003	1.007	1.005
Manufacturing	0.976	0.987	0.982
Consumer goods	1.031	1.041	1.036
Intermediate goods	0.962	0.975	0.969
Capital goods	0.877	0.885	0.881
Construction	0.999	0.997	0.998
Services	1.014	1.017	1.016

protection enjoyed by those engaged in casting, forging and foundry is around 80 percent, with many firms using outdated technologies. The petrochemical industries, similarly, exhibit considerable variation. Although well protected, those producing plastic products (sector 29) are generally more competitive than the undersized plants found in aromatics, resins, rubbers, detergents, and synthetic fibers. The other chemical-based industries are also characterized by varying efficiencies. In inorganic chemicals (sector 32), while phosphoric acids and ammonia used in fertilizer production have low tariffs, others have close to 100 percent tariffs. The fertilizer industry (sector 34) operates with very low tariffs but is highly subsidized. Firms in synthetic fibers (sector 39) and other chemicals would generally require major restructuring to survive tariff reform.

Capital goods represent about 10 percent to 13 percent of manufacturing output and value added and constitute a very large group of industries. The 11.9 percent reduction in their output is due to the substantial reduction in average nominal tariff protection (i.e., collection rate) for machinery, which is as high as 67 percent.[15] The high cost of investment has a detrimental impact on other sectors. Thus, Ettori [11] estimates that the cost of capital goods requires compensatory protection of 30 percent to allow industrial projects in India to earn returns comparable to those available under free trade. The impact of the trade liberalization differs by subsectors of capital goods. The highly protected heavy industries, which include a significant number of inefficient public enterprises, are the hardest hit. These include nonelectrical machinery (sectors 49–53), electrical industrial machinery (sector 54), heavy transport equipment such as ships and railways (sectors 60–62), and communications equipment (sector 58).

Simulation 3: Trade liberalization with additional external financing and no fiscal adjustment

A structural reform such as trade liberalization is often accompanied by additional external financing. Availability of the latter diminishes the need for fiscal adjustment to make up any shortfall in domestic savings arising out of tariff reduction. This simulation examines the consequences of adjusting to trade liberalization solely through additional external financing. The results offer insights into a pattern of adjustment polar to that described in Simulation 2. Since policy responses to trade liberalization can, in practice, be expected to include elements of both fiscal adjustment and additional external financing, they are bracketed by the results of this simulation and the previous one.

Columns (1) and (2) of Table 12.10 report the results of undertaking the

[15] The comparable figure in Korea is 9 percent during the late 1970s and early 1980s, 14.5 percent in Pakistan, and 17 and 11 percent for non-electrical machinery in Brazil.

Table 12.10: Effects of trade liberalization without fiscal adjustments and with variable current account (base solution = 1.00)

	With TFP growth (1)	Without TFP growth (2)
Prices		
Exchange rate (Rs/USS)	1.063	1.069
Producer prices	0.969	0.975
Real exchange rate	1.094	0.997
Consumer price index	0.994	0.999
Real GNP at market prices	1.012	1.008
Output	1.004	1.000
Private consumption	1.012	1.008
Imports	1.102	1.108
Exports	1.221	1.210
Total factor productivity	1.006	1.000
Foreign savings/GDP	0.033	0.035

trade liberalization described in Simulation 2 without adjustment in either trade-neutral taxation or in government consumption. Restoration of the savings-investment balance following the drop in public savings brought about as a result of the reduction in protective tariffs therefore requires an increase in foreign savings, which must be made endogenous. Column (1), with endogenous TFP, shows that the ratio of foreign savings increases from 2.8 percent to approximately 3.3 percent of GDP. This is brought about by a 22 percent increase in exports and a nearly 10 percent increase in imports. Since the tariff reduction leads to a fall in producer prices, internal balance requires that real wages be eroded through a devaluation which amounts to 9.4 percent in real terms. TFP increases by 0.6 percent and real GNP by 1.2 percent. In contrast, foreign savings rise to nearly 3.5 percent of GDP in the absence of TFP increases, while GNP rises by 0.8 percent.

Comparison of Table 12.10 with Table 12.5 allows a determination of the range of outcomes bracketed by the fiscal adjustment/fixed current account and the no fiscal adjustment/variable current account cases.

12.4 More Recent Reforms

The framework developed in Section 12.2 and used to simulate trade liberalization and fiscal adjustment in Section 12.3 was calibrated on a database predating the 1991 reforms. Starting that year, India undertook a program of macroeconomic stabilization and pursued an agenda of internal and external liberalization (an assessment of the reforms is provided in World Bank [30]). By 1992–93, the collection rates for protective tariffs and total tariffs on imports had fallen to 33 and 38 percent, respectively. Collection rates for the protective tariffs corresponding to the four broad commodity sectors identified by the model were 16.7 percent in agriculture, 42.5 percent in consumer goods, 29.9 percent in intermediate goods, and 45.8 percent in capital goods. Moreover, many quantitative restrictions applying to imports of intermediates and capital goods had been removed. The stabilization program had reduced the current account deficit to about 2 percent of GDP by 1992–93.

This section examines the ability of the model to reproduce the structure of the economy between 1987/88, the year in which it was calibrated, and 1992–93, the second year of the reform program. Data for those years are used as inputs into the model and its predicted outcomes for the principal macroeconomic aggregates are expressed as a proportion of GDP.[16] Table 12.11 presents the results. The correspondence of consumption, investment, trade, tax revenue and the real exchange rate is seen to be broadly satisfactory, thus enhancing the degree of confidence in the results.

Table 12.11: Structure of the Indian economy, 1992–93, simulated vs actual (in % of GDP at factor prices)

	Model	Actual
Net Tax Revenue	13.83	13.83
Custom Revenue	3.33	3.80
Union Excise/CVD	5.82	5.43
Private Consumption	76.54	75.62
Fixed Investment	24.55	24.64
Public Consumption	11.12	12.65
Exports	11.52	10.78
Imports	12.28	13.12
Real Exchange Rate (87 = 1.00)[a]		

*Defined as the rupee price index of f.o.b. imports divided by the index of domestic inflation.

The database, thus updated to 1992–93, may be used to simulate the effects of a range of policy reforms. The focus of the next simulation is on further trade liberalization. The specific reform implemented comprises the following elements: (1) the elimination of all remaining import quotas in agriculture and consumer goods and the setting of protective tariffs at 30 percent for agricultural goods, 30 percent for consumer goods, 15 percent for intermediate goods (using a nominal rate of 20 percent but assuming that a quarter are exempted because of duty drawback on exports and other policies) and 25 percent for capital goods; and (2) a reduction in the current account deficit of the balance-of-payments to 1 percent of GDP. Investment, as before, is maintained at its pre-shock level.

[16] A more complete procedure of the kind described and implemented in Mitra [22] would have required updating all the parameters and exogenous variables of the model. This was precluded by unavailability of the necessary information.

Column (1) of Table 12.12 reports the results of pursuing this particular trade liberalization agenda through an increase in the average level of the Union excise-tax-cum-CVD rate. The average rate must increase by nearly two-thirds, the actual figure being 62 percent. Maintenance of internal balance requires a depreciation of the real exchange rate by nearly 10 percent. Exports rise by 20 percent, with imports remaining unchanged. Alternatively, as reported in column (2), if the required fiscal adjustment is accomplished solely through a reduction in public consumption, the required cut is of the order of 20 percent. The more favorable domestic cost structure, compared to the pure tax increase case, allows an increase in exports by 22 percent. The required depreciation of the real exchange rate is also somewhat higher.

Table 12.12: Effects of further liberalization with fiscal adjustment and devaluation (base year 1992–93 = 1.00)

	Trade reform		TFP growth, doubled trade elasticities			
					Unchanged current deficit-to-GDP	
	Tax (1)	Public consumption (2)	Tax (3)	Public consumption (4)	Tax (5)	Public consumption (6)
Government						
Real Govt consumption	1.000	0.813	1.000	0.860	1.000	0.947
Excise/CVD rate[a]	1.622	1.000	1.459	1.000	1.171	1.000
Prices						
Exchange Rate (Rs/US$)	1.106	1.103	1.036	1.029	1.012	1.010
Real exchange rate[b]	1.095	1.107	1.051	1.055	1.036	1.038
Producer prices	1.011	0.996	0.985	0.974	0.976	0.972
Consumer prices	1.021	1.000	0.999	0.984	0.988	0.982
Real GNP (market)	0.998	0.999	1.019	1.020	1.022	1.023
Output	0.983	1.002	1.008	1.023	1.019	1.025
Private consumption	0.969	1.000	1.002	1.025	1.021	1.030
Imports	1.000	1.008	1.091	1.101	1.125	1.129
Exports	1.200	1.222	1.269	1.284	1.202	1.207
Current acct/GDP	0.511	0.511	0.511	0.511	1.000	1.000
TFP (manufacturing)	1.000	1.000	1.010	1.010	1.010	1.010

[a]The base year GDP ratios are 0.035 for tariffs, 0.58 for excise/CVC and 0.10 for fiscal deficit.

[b]Nominal exchange rate deflated by product prices.

Columns (3) and (4) of Table 12.12 combine the reform package outlined above with (i) an increase of 1 percent in total factor productivity in the manufacturing sectors, and (ii) a doubling of trade elasticities, as the opening of the economy makes domestically produced goods more substitutable with foreign goods. In this case, the required fiscal adjustment is smaller — 46 percent for pure tax adjustment and 14 percent for the case of reduction in public consumption. A real depreciation of around 5 percent suffices for the restoration of internal balance.

The effect of replacing the assumption of a reduction in the current account deficit with one having no change in relation to GDP, while maintaining the other assumptions of columns (3) and (4), is shown in columns (5) and (6) of Table 12.12 for the cases of tax adjustment and public consumption adjustment, respectively. This is intended to capture the effects of growing foreign investment in response to the reform process. Restoration of the savings-investment balance now calls for a significantly lower fiscal adjustment: an increase of 17 percent for the tax case and a fall of 5 percent in the public consumption case. Internal balance is brought about by a real devaluation amounting to less than 4 percent.

12.5 Conclusion

This paper has developed a framework for examining the consequences of a program of stabilization and trade liberalization in India, where fiscal and current account imbalances needed to be reduced in order to launch the economy on a path of durable growth. In so doing, it gave quantitative expression to tradeoffs between trade liberalization and fiscal adjustment arising from a high degree of dependence of public revenue on tariffs, and to the role of exchange rate policy in restoring internal equilibrium in the face of tariff reduction and fiscal contraction. Moreover, this was accomplished by implementing (i) a modestly aggregated general equilibrium model which captures important economy-wide consequences of labor market adjustment and the price and income effects arising from relaxation of tariffs and quantitative restrictions on imports, and (ii) a highly disaggregated partial equilibrium model which takes the economy-wide effects generated by (i) as inputs and traces the effects of stabilization and liberalization on various subsectors of particular interest to policy makers.

The model was calibrated for the period prior to 1991, the year when Indian economic policy broke from the past. Simulation of some of the actual policies pursued until 1992–93 established a broad correspondence between model results and actual outcomes in the principal macroeconomic aggregates. This encouraging finding led to an updating of the database and an exploration of subsequent rounds of trade liberalization. We conclude by suggesting that the approach developed in the paper has the potential for providing broad indications of the economy-wide and sectoral consequences of pursuing the unfinished agenda of reforms still facing policy makers in India, and indeed more generally, in other developing countries.

Appendix A: List of Equations

Production

Domestic output

$$PQ_i = \left[\frac{\alpha q_i}{\lambda_i}\right] PN_i^{sn_i} PV_i^{sv_i} \tag{12.1}$$

$$PN_i N_i = sn_i PQ_i Q_i \tag{12.2}$$

$$PV_i V_i = sv_i PQ_i Q_i \tag{12.3}$$

Value added

$$PV_i = \alpha v_i r_i^{sk_i} \prod_{\ell} PL_{\ell}^{s\ell_{\ell i}} \tag{12.4}$$

$$PL_{\ell} L_{\ell i} = s\ell_{\ell i} PV_i V_i \tag{12.5}$$

$$r_i K_i = sk_i PV_i V_i \tag{12.6}$$

Income generation

$$\begin{aligned} YM_r pop_r &= \sum_i map1_{i \Rightarrow r} YF_i \\ &+ map2_r(trsCPI + f\ trs\ ER - (1-\theta) irDEBT\ ER \end{aligned} \tag{12.7}$$

$$YF_i = \sum_{\ell} PL_{\ell} L_{i,\ell}(1 - tw_i) + r_i K_i - \delta r_i K_i PK)(1 - tk_i - sf_i) \tag{12.8}$$

Demand

Private consumption

$$U_r = \prod_c (CF_{cr}/pop(r) - \gamma_{cr})^{m_{cr}} \tag{12.9}$$

$$PCD_{rc} CD_{rc} = pop_r[\gamma_{rc} PCD_{rc} + m_{rc}(YM_r(1 - sh_r) - \sum_c \gamma_{rc} PCD_{rc})] \tag{12.10}$$

$$PCD_{rc} = \alpha c_{rc} \prod_{i \in C} [PC_i]^{sc_{i \in (r,c)}} \tag{12.11}$$

$$C_{ri} PC_i = sc_{i \in (r,c)} CD_{rc} PCD_{rc} \tag{12.12}$$

Investment

$$INVEST = \sum_{j=1}^{n} IE_j PK_j + IGPK_n + \sum_i DST_i P_i \tag{12.13}$$

$$ID_i = is_i \left[IG + \sum_j IE_j \right] \tag{12.14}$$

Other demands

$$IND_i = \sum_j ad_{ij} ND_j \tag{12.15}$$

$$\begin{aligned} MARG_s P_s &= \sum_i stm_{si} P_i \left(\sum_r C_{ri} + CG_i + ID_i + E_i \right) \\ &= \sum_i stm_{si} pwm_i ER \left(M_i + \sum_j am_{ij} NM_j \right) \end{aligned} \tag{12.16}$$

$$D_i = \sum_r C_{ri} + G_i + ID_i + DST_i + MARG_s \tag{12.17}$$

Foreign trade

Imported and domestic material input

$$PN_i = on_i PND_i^{sd_i} PNM_i^{sm_i} \tag{12.18}$$

$$ND_i PND_i = sd_i N_i PN_i \tag{12.19}$$

$$NM_i PNM_i = sm_i N_i PN_i \tag{12.20}$$

Competitive imports, quota and supply of goods

$$\tilde{P}_i = \alpha x_i \left[\delta^{\sigma_i} \widetilde{PM}_i^{1-\sigma_i} + (1-\delta_i)^{\sigma_i} PQ_i^{1-\sigma_i} \right]^{-1/(1-\sigma_i)} \tag{12.21}$$

$$\frac{Q_i}{X_i} = \left[\frac{\tilde{P}_i}{PQ_i} (1-\delta_i) \alpha x^{p_i} \right]^{\sigma_i} \tag{12.22}$$

$$\frac{M_i}{X_i} = \left[\frac{\tilde{P}_i}{\widetilde{PM}_i} \delta_i \alpha x^{p_i} \right] \tag{12.23}$$

$$\frac{\widetilde{PM}_i}{PM_i} = 1 + PR_i \tag{12.24}$$

$$P_i = \tilde{P}_i + (PM_i - \widetilde{PM}_i) \frac{M_i}{X_i} \tag{12.25}$$

Exports

$$E_i = \Lambda_i \left[\frac{pwe_i ER}{PE_i} \right]^{\varepsilon_i} \tag{12.26}$$

Taxes and purchasers' prices

$$PM_i = pwm_i^*(1 + tm_i)(1 + trm_i + itm_i) \tag{12.27}$$

$$PND_j = \sum_i ad_{ij} P_i(1 + [td_i]_{\forall j \neq vat} + ts_i) \tag{12.28}$$

$$PNM_j = \sum_i am_{ij} pwm_i^* ER(1 + tm_i)(1 + trm_i + [td_i]_{\forall j \neq vat} + ts_i) \tag{12.29}$$

$$PC_i = P_i(1 + trm_i + td_i + ts_i) \tag{12.30}$$

$$PK_j = \sum_i is_i P_j(1 + trm_i + td_i - [\beta td_i]_{\forall j = vat} + ts_i) \tag{12.31}$$

$$PE_i = P_i(1 + trm_i + te_i) \tag{12.32}$$

$$CPI = \sum_i ch_i P_i(1 + trm_i + td_i + ts_i) \tag{12.33}$$

Market clearing

$$X_i = IND_i + D_i + E_i \tag{12.34}$$

$$\begin{aligned} SAVF &= \sum_i \sum_j pwm_i am_{ij} NM_j + \sum_i pwm_i M_i \\ &\quad - \sum_i E_i PE_i / ER - ftrs - ir\ DEBT \end{aligned} \tag{12.35}$$

$$\begin{aligned} \sum_{j=1}^{n} IE_j PK_j &\equiv SAVG + SAVF\ ER + \sum_r sh_r YM_r pop_r \\ &\quad + \sum_i sf_i(r_i K_i - \delta r_i PK_i K_i) + \sum_i \delta r_i PK_i K_i \end{aligned} \tag{12.36}$$

Government budget

$$\begin{aligned} TAXREV &= \sum_i \sum_j tm_i am_{ij} NM_j pwm_i ER + \sum_I (tm_i + itm_i) M_i pwm_i ER \\ &\quad + \sum_i \sum_j td_i ad_{ij} ND_i P_i + \sum_i te_i E_i P_i \\ &\quad + \sum_i td_i \left(\sum_r C_{ri} + ID_i + CG_i \right) P_i - \sum_i \sum_\ell tw_i L_{\ell i} PL_\ell \\ &\quad + \sum_i tk_i [YK_i - \delta r_i PK_i K_i] \end{aligned} \tag{12.37}$$

$$\begin{aligned} SAVG &= TAXREV - gtrsCPI - PK_nIG \\ &- \sum_i CG_iP_i(1 + trm_i + td_i + ts_i) - \theta ir\ DEBT\ ER \end{aligned} \tag{12.38}$$

Factor markets

$$LS_\ell = LSO_\ell \left[\frac{PL_\ell/CPI}{PLO_\ell/CPIO}\right]^{\zeta\ell},\ \ell = 1, 3 \tag{12.39}$$

$$\sum_\ell LS_\ell = \sum_\ell LSO_t,\ \ell = 1, 2 \text{ and } 1 = 3, 4 \tag{12.40}$$

$$\sum_\ell w_\ell PL_\ell \geq w_\ell PLO_\ell,\ \ell = 1, 3 \tag{12.41}$$

$$\sum_i L_{\ell i} = LS_\ell \tag{12.42}$$

$$K_i = K_i^0 \tag{12.43}$$

Fiscal adjustment

$$\phi td_i = td_i^0 \text{ or} \tag{12.44a}$$

$$\varphi CG_i = CG_i^0 \tag{12.44b}$$

Productivity

$$\lambda_i = 1.0 + \beta_{EE_i} x_{EE_i} = +\beta_{IS_i} x_{IS_i} \tag{12.45}$$

$$x_{EE_i} = \sum_j \upsilon_{ij} \frac{\Delta E_j}{\Delta Q_i} g_i^Q \tag{12.46}$$

$$x_{IS_i} = \sum_j \upsilon_{ij} \frac{\Delta\mu_j^D D_j + \Delta\mu_j^N\,(IND_i + INM_i)}{\Delta Q_i} g_i^Q \tag{12.47}$$

$$\mu_i^D = \frac{D_i}{D_i + M_i} \tag{12.48}$$

$$\mu_i^N = \frac{IND_i}{IND_i + INM_i} \tag{12.49}$$

$$INM_i = \sum_j am_{ij} NM_j \tag{12.50}$$

$$g_i^Q = \frac{\Delta Q_i}{Q_i^O} \tag{12.51}$$

Social welfare function

$$\Omega = \frac{1}{\vartheta} \sum_r pop_r U_r^\vartheta \tag{12.52}$$

Glossary of parameters

ad_{ij} coefficients in the domestic input–output matrix
am_{ij} coefficients in the import flow matrix
αn_i shift parameter in the Cobb–Douglas function for PN_i
αq_i shift parameter in the Cobb–Douglas function for PQ_i
αv_i shift parameter in the Cobb–Douglas function for PV_i
αx_i shift parameter in the CES function for $\tilde{P}_i$
αc_{rc} shift parameter for aggregation of consumer prices
β parameter for extending crediting of MODVAT to capital goods
ch_i weights in the consumer price index
$CPIO$ consumer price index in the base year
δ_i share parameter in the CES function for $\tilde{P}_i$
δr_i depreciation rate
ε_i demand elasticity of exports
$ftrs$ transfers from abroad to households
γ_{cr} committed per capita consumption in the LES demand
$gtrs$ net transfers from government to households
ir_i interest rate on external debt
$is_i m$ allocation of investment expenditure to final demand
Λ_i constant in the export demand function
LSO_ℓ labor supplies in the base year
m_{cr} marginal budget shares in the LES demand
$map1_{i \Rightarrow r}$ allocation of sectoral factor income to households
$map2_r$ allocation of transfers to households
PLO_ℓ wages in the base year
pop_r population by region or household group
pwe_i world prices of exports in US dollars
pwm_i world prices of imports in US dollars
p_i parameter in the CES supply function
σ_i substitution elasticity between Q_i and M_i
sf_i average corporate savings rate by sector
sh_r average savings rate by household
sk_i share parameter for capital input
$s\ell_{\ell i}$ share parameter for labor input of category ℓ
sm_i share parameter for material input
sn_i share parameter for material input
sv_i share parameter for value added
sc_{irc} share parameter for good i in consumption group c by household r
$strm_{si}$ trade or transportation margin for each sector
θ portion of debt servicing accounted by public sector
trs net transfers from government to households
trm_i sum of trade and transportation margins for each sector

tk_i tax rate on capital income
$t\omega_i$ tax rate on wage income
ω_ℓ distribution share of regular workers in each region
ζ_ℓ supply elasticity of each type of regular workers

Glossary of variables

ER foreign exchange rate
CPI consumer price index
C_{ri} consumption of good i by household r
Cd_{cr} consumption of LES composite good c by household r
CG_i government current consumption by commodity
D_i domestic final demand
DST_i changes in stocks by sector of origin
$DEBT$ external debt in US dollars
E_i exports of good i
g_i^Q growth rate of output
ID_i investment demand by sector of origin
IE_j private fixed investment by sector of destination
IG total fixed investment in the public sector
IND_i purchases of domestic intermediates by commodity
INM_i purchases of imported intermediates by commodity
$INVEST$ total investment expenditure
λ_i total factor productivity of domestic output by sector
Ls_ℓ labor supply by category
K_i capital stock of each sector
M_i competitive imports by commodity
$MARG_s$ demand for trade and transport margins
N_i total material input used in sector i
ND_i bundle of domestic intermediates in sector i
NM_i bundle of imported intermediates in sector i
Ω social welfare function
P_i supply price of each good
$\tilde{P}_i$ unconstrained supply price (inclusive of import premia)
PC_i purchase price of good i (including taxes and margins)
PCD_{rc} price of good c consumed by household r
PE_i sales price of exports
PK_j price of capital goods in sector i
PL_ℓ wages by labor category
PM_i domestic price of competitive imports
$\widetilde{PM}_i$ virtual price of competitive imports (including import premia)
PN_i price of material input
PND_i price of domestic material input

PNM_i price of imported material input
PQ_i price of domestic goods
PR_i premium rate of import quota
PV_i price of value added
ϕ scaling variable of domestic taxes
Q_i domestic output by commodity i
r_i rate of return to capital
$SAVF$ foreign savings in US dollars
$SAVG$ government budgetary balance
$TAXREV$ consolidated tax revenue
μ_i^D domestic supply ratio of final demand
μ_i^N domestic supply ratio of intermediate demand
U_r LES utility by household r
V_i value added in sector i
φ scaling variable of government expenditure
X_i supply of commodity i
x_{EE_i} export expansion in the demand decomposition of output growth
x_{IS_i} import substitution in the demand decomposition output growth
YF_i factor income by sector
YM_r per capita income by region

Appendix B

Calibration of the model

The assemblage of data and the calibration of the model follow the usual procedures of what amount to building a detailed Social Accounting Matrix of India and fitting it to the specifications of the model. Using plausible assumptions, disparate sources of information were assembled into a data set consistent with the national income accounts, a recent update of the input–output table, the household expenditure information from the National Sample Survey, the balance of external payments, and, unique to this study, detailed revenue from custom and excise. The base year of the model is 1987/88, the most recent year in which detailed information was available regarding the structure of the Indian economy prior to significant trade liberalization. We briefly describe the key features in what follows.

Trade and tax regime in India

What is unique to this study is the compilation of an enormous amount of information regarding the complicated trade and tax regime in India in the late eighties. From this data set, which is based on a larger study of India's trade regime, we estimated the collection rates of indirect taxes using actual revenue for specific commodities, and in the case of import tariffs, their actual tax bases from custom data.

Commodity tax collections by type and commodity were based on the detailed revenue data for 1987/88 and presented in Table 12.8. The treatment of various taxes in the model is defined in Section 12.2.5. Tariff revenue, classified into protective (basic and auxiliary customs duties) and nonprotective ("countervailing customs duties" or CVD), were collected and compiled carefully from detailed customs data (DGCY&S). Both Union excise taxes and CVDs operate much like a VAT on manufacturing (see MODVAT in Section 12.2.5) and their corresponding tax bases were derived from the input–output table below. The structure of exports and imports were also compiled from the customs data and are presented in Table 12.7. Other taxes, not the focus of this study, received more simple treatment. These included subsidies, states' excise and sales taxes, and a small amount of export taxes. For these taxes, we assumed the distribution reported in the tax table corresponding to the 1978/79 input–output table. In addition, we assumed a single direct tax rate (6.1%) for labor income and another single rate (8.9%) for capital income to generate the reported revenue from NAS. Taxes on labor income are applied only on nonresidual labor income in the nonagricultural sectors.

Input–output table

The input–output transaction matrix corresponding to the 72 sectors was obtained from a 1986/87 update of the 1978/79 115 sector table from CS0-NAS.[17] The 1986/87 update was commissioned by the World Bank in relation to this study and supervised by Saluja [28]. The update was carried out using new estimates of output — value-added and components of final demand — as well as changes in relative prices and in input usage in some sectors since 1978/79. The mapping scheme from the 115 sectors to the 72 sectors is shown in Table 12.13. Given 1987/88 estimates of output, final demand, value added, and taxes (which are described in the next few sections), the 72-sector table was scaled, RASed, and updated further to 1987/88. The 6-sector table is derived from this new matrix. The aggregation scheme is shown in Table 12.14. Of the three manufacturing sectors, the consumer goods industry did not include consumer durables, which was not separable from other capital goods.

In the absence of a more recent table, the import matrix table was derived from the 1978/79 input–output table by applying the simple shares of

[17] The 1978/79 table was the last official matrix constructed at the time of study. While there were some updates to more recent years in the technical notes of various economic plans in India, these updates were often much less disaggregated than the original 115 sectors.

Table 12.13: Mapping scheme for sectoral aggregation from 115 to 72 sectors

Sector	115-sector classification	Sector	115-sector classification
1. Cereal crops	001–006	37. Drugs and medicines	065
2. Milk	018	38. Soaps and cosmetics	066
3. Meat and fish	019–022	39. Synthetic fibers	067
4. Other agriculture	007–018,	40. Other chemicals	068
	022	41. Structural clay	069
5. Coal and lignite	023	42. Cement	070
6. Crude petroleum	024	43. Other nonmetal mins	071
7. Iron ore	025	44. Iron and steel alloys	072
8. Metallic minerals	026–029	45. Iron and steel foundries	073,074
9. Nonmetallic minerals	030–032	46. Nonferrous metals	075
10. Sugar	033,034	47. Hand tools hardware	076
11. Edible oil	035,036	48. Misc. metal products	077
12. Tea and coffee	037	49. Agricultural implements	078
13. Misc. food	038	50. Industrial machinery	079,080
14. Beverages	039	51. Machine tools	081
15. Tobacco products	040	52. Office machinery	082
16. Cotton textiles	041,042	53. Other nonelec.machinery	083
17. Woolen textiles	043	54. Elec.industrial machinery	084,089
18. Silk textiles	044	55. Elec.cables and wires	085
19. Art synthetic textiles	045	56. Batteries	086
20. Jute, hemp and mesta	046	57. Electric appliances	087
21. Carpet weaving	047	58. Communication equip.	088
22. Ready-made garments	048	59. Electronic equipment	090
23. Misc. textiles	049	60. Ships and boats	091
24. Wood products	050, 051	61. Rail machinery	092
25. Paper and newsprint	052	62. Motor vehicles	093
26. Printing and publishing	053	63. Motorcycles	094
27. Leather products	054,055	64. Bicycles and other trans.	095,096
28. Rubber products	056	65. Watches and clocks	097
29. Plastic products	057	66. Misc. manufacturing	098
30. Petroluem products	058	67. Construction	099
31. Coal tar products	059	68. Utilities	100–102
32. Inorganic chemicals	060	69. Rail transportation	103
33. Organic chemicals	061	70. Other transportation	104
34. Fertilizers	062	71. Trade	107
35. Pesticides	063	72. Other services	105,106,
36. Paints and varnishes	064		108–115

intermediate imports to total input use of commodity i by industry j in the new table. These ratios were scaled so that the row sums of the new import flux matrix did not exceed the total amount of imports purchased for each commodity. The split between imports used as raw materials and competitive imports in final demand was broadly reasonable in relation to the distribution between OGL and non-OGL imports.[18]

Table 12.14: Mapping scheme for sectoral aggregation from 72 to 6 sectors

Sector	72-sector classification
1. Agriculture	01 to 04
2. Consumer goods	10–19, 21–23, 26, 27, 37, 38, 64, 65
3. Intermediate goods	05–09, 20, 24, 25, 28–36, 39–48, 66
4. Capital goods	49–63
5. Construction	67
6. Services	68–72

Other items

Household consumption — The structure of private consumption for two households, one urban and one rural, were obtained from the 38th National Sample Survey (NSS). Household expenditures shares in the urban and rural areas were calculated from over 2,000 consumption items, mapped into the 115 input–output sectors, and subsequently aggregated to 72 sectors. The final estimates were adjusted so that the all India figures summed up to the NAS estimates of private consumption for 8 broad items and 39 subitems.

Capital stocks and depreciation rates were obtained from a 1988 study by the Central Statistical Organization (CSO), which provides ratios of capital to net domestic product (NDP) and rates of capital consumption for 18 broad sectors in the Indian economy. Estimates for individual industries were further derived from the survey of industries.

Shares of wage income were derived from 1984/85 CSO-NAS. From the 1984/85 Survey of Industries, we derived the shares of wage income for subsectors in manufacturing and scaled them so that the average share is consistent with the NAS data. In the agriculture sectors, the shares of self-employed/mixed income in net value added were mapped into the income of nonresidual or own-farm workers and the other wage income to residual classes. In the nonagriculture sectors, the informal or residual workers received self-employed income as well as wage income. The distribution of wage income between nonresidual and residual classes was assumed to be the same as the distribution of net value added between organized and nonorganized sectors from CSO-NAS data for 15 broad sectors and 15 subsectors in manufacturing. These estimated shares of labor income for nonresidual and residual classes were then mapped into the 72 sectors. The remaining nonwage income in the net value added of each sector is taken as gross capital income.

[18] In this model, we assume that imports used as raw materials are of the OGL variety and not subject to quantitative restrictions; imports in final demand are competitive with domestic outputs and subject to quota restrictions.

Household incomes and savings — Factor incomes earned in the agriculture sector are mapped to the rural household while those earned in the intermediate and capital sectors go to the urban household. Income earned in the consumer goods, construction, and service sectors is split, with 53.4 percent going to the rural household and the remaining 46.6 percent to the urban household, as in Mitra–Tendulkar [23]. In addition, households receive transfer incomes from the government and abroad based on a fixed allocation scheme. 80 percent of the transfers go to the rural household and 20 percent to the urban household, as in Mitra–Tendukar [23]. The implied saving rate after household expenditures was 9 percent for the urban household and 3 percent for the rural household.

The *distribution of the labor force* is from the 38th round of the National Sample Survey. About 38 percent of the population is in the labor force with 34 percent employed and about 4 percent unemployed. In the agriculture sectors, own-farm workers and regular-farm workers were grouped into nonresidual class workers (115.2 million people) in the model; casual workers were classified as residual workers (53.4 million). In nonagriculture, regular nonfarm workers were grouped into nonresidual class workers (37.9 million); casual workers in public works and nonagriculture, as well as the self-employed (in own nonfarm), were catalogued into workers of the informal type (60.6 million).

A single *wage rate* prevails in each labor market of the model. These wages are obtained by dividing total wage income of each labor class by the amount of labor in that class. In general, the wages in the residual classes are about 25 percent lower than those in the nonresidual classes. Given wage rates and labor income, the distribution of employment by labor type and by sector is derived.

The household demand system is a *Linear Expenditure System* (LES) derived from a study, Models of Complete Expenditure System for India, by Radhakrishna and Murty [27]. The marginal budget shares in study were broadly adjusted for price changes of food and nonfood items relative to the consumer price indices for industrial and agriculture workers since the study was made. The marginal shares are reported in Table 12.15. In the aggregated model, these were further grouped into six sectors (see Table 12.16). Given the marginal budget shares, per capita household income, saving rate, and per capita consumption of goods, the committed quantities in the LES system were derived as residuals.

Certain parameters such as export demand elasticities and elasticities of substitution between competitive imports and domestic goods were not calibrated but postulated to assume values in light of country circumstances. In choosing these parameters, we took note of similar parameters employed in general equilibrium models of countries in the same region — Mitra and

Tendulkar [23] for India, and Dahl and Mitra [9] for Bangladesh; we also conducted numerous sensitivity tests to gauge and understand their impact. Thus, the demand elasticities of exports ε_i were taken to be 1.25 for agriculture products and 2.50 for others; the substitution elasticities between competitive imports and domestic goods σ_i were set at 2.50; the elasticities of labor supply ϕ_ℓ were 0.5; and the price elasticities of investment k_i were 0.8 (estimated by World Bank staff).

Table 12.15: Marginal budget shares in the LES household demand (72-sector model)

	Urban	Rural
Cereals	0.0132	0.1998
Other crops	0.0434	0.1122
Milk	0.0982	0.1324
Meat	0.0262	0.0230
Edible oil	0.0134	0.0286
Other food	0.1217	0.0819
Clothing	0.0967	0.1451
Other consumer goods	0.1113	0.0365
Fuel	0.0299	0.0265
Other manufacturers	0.0585	0.0257
Capital goods	0.0625	0.0258
Services	0.3250	0.1625

Table 12.16: Marginal budget Shares in the LES household demand (6-sector model)

	Urban	Rural
Agriculture	0.1810	0.4674
Consumer goods	0.3430	0.2921
Intermediate goods	0.0715	0.0387
Capital goods	0.0625	0.0258
Construction		
Services	0.3420	0.1760

REFERENCES

1 Aghion, Phillipe and Peter Howitt, Endogenous Growth Theory, MIT Press, Cambridge, MA, 1998.

2 Ahmad, E. and N. Stern, Alternative sources of government revenue: Illustrations from India, 1979–80, in: D. Newbery and N. Stern (eds.), The Theory of Taxation for Developing Countries, Oxford University Press, New York, 1987.

3 Aksoy, Ataman M. and F. Ettori, Protection and industrial structure in India, Policy Research Working Paper No. 990, Country Dept., The World Bank, Washington, DC, 1992.

4 Barro, R.J. and X. Sala-i-Martin, Economic Growth, New York: McGraw-Hill, New York, 1995.

5 Bhagwati, J. and T.N. Srinivasan, Foreign Trade Regimes and Economic Development: India, Columbia University Press, New York and London, 1975.

6 Chenery, H., S. Robinson and M. Syrquin,. Industrialization and Growth: A Comparative Study, Oxford University Press, New York, 1986.

7 Corden, M., Inflation, Exchange Rates and the World Economy: Lectures on International Monetary Economics, Clarendon, Oxford, 1985.

8 Dahl, H.,S. Devarajan and S. van Wijnbergen, Revenue-neutral tariff reform: Theory and an application to Cameroon?, County Policy Department Paper No. 1986–25, The World Bank. Washington, DC, 1986.

9 Dahl, H. and P. Mitra, Does tax and tariff shifting matter for policy?: An application of general equilibrium incidence analysis to Bangladesh, World Bank Research Paper. Washington, DC, mimeo, 1989.

10 Devarajan, S., Fullerton, D. and R. Musgrave, Estimating the distribution of tax burdens, Journal of Public Economics, 13 (1980) 155–82.

11 Ettori, F., The Pervasive effects of high taxation of capital goods in India: Findings and conclusions from a sample of projects, Working Paper Series No. 433, Asia Regional Office, The World Bank, Washington, DC, 1990.

12 Grossman, Gene (ed.), Economic Growth: Theory and Evidence. Volumes 1 and 2, Vermnot: Edward Elgar Publishing Co., Vermont, 1996.

13 India, Government of, Customs Statistics, 1987–88, Directorate General of Commercial Intelligence and Statistics, Calcutta, 1989.

14 India, Government of, National Accounts Statistics, 1989 and Quick Estimate Dated March 10, 1990, Central Statistical Organization, New Delhi, 1990.

15 Jha, S. and P. Srinivasan, Indirect taxes in India: An incidence analysis, Economic and Political Weekly, April 15, 1989.

16 Kishor, Samal C., Foreign exchange market in India: Single currency peg to independent floating — devaluation all the way, Journal of Indian School of Political Economy (India) 6 (1994) 94–155.

17 Krueger, A., Factory Supply and Substitution, National Bureau of Economic Research, Chicago, 1981.

18 Kubo, Y., S. Robinson and M. Syrquin, The methodology of multisectoral comparative analysis, in: H. Chenery, S. Robinson and M. Syrquin (eds.), Industrialization and Growth: A Comparative Study, Oxford University Press, New York, 1986, pp. 121–47.

19 Mitra, P., Protective and revenue raising trade taxes: Theory and an application to India, Country Policy Department Discussion Paper No. 1987–4, World Bank, Washington, DC, 1987.

20 Mitra, P., Tariff design and reform in a revenue-constrained economy: Theory an illustration from India, Journal of Public Economics, 47 (1992) 227–51.

21 Mitra, P., The coordinated reform of tariffs and indirect taxes, The World Bank Research Observer, 7 (1992) 195–218.

22 Mitra, P., Adjustment in Oil-importing Developing Countries: A Comparative Economic Analysis, Cambridge University Press, New York, 1994.

23 Mitra, P. and S. Tendulkar, Coping with internal and external exogenous shocks: India, 1973–74 to 1983–84, Country Policy Department No. 1986–21, World Bank, Washington, DC, 1986.

24 Narayana, A.V.L., Bagchi, A. and R.C. Gupta, The Operation of MODVAT. National Institute of Public Finance and Policy, Vikas Publishing House, New Delhi, India, 1991.

25 Neary, J.P. and K.W.S. Roberts, The theory of household behaviour under rationing, European Economic Review 13 (1980) 25–42.

26 Nishimizu, M. and S. Robinson, Productivity growth in manufacturing, in: H. Chenery, S. Robinson and M. Syrquin (eds.), Industrialization and Growth: A Comparative Study, Oxford University Press, New York, 1986, pp. 283–308

27 Radhakrishna, R. and K.N. Murty, Models of complete expenditure systems for India, Working Paper No. WP–80-98, International Institute for Applied Systems Analysis, Austria, 1990.

28 Saluja, M.R., Update of India's input–output table, mimeo, 1989.

29 Tybout, James R., Linking trade and productivity: New research directions, The World Bank Economic Review, 6(2) (1992) 189–211.

30 World Bank, India: Five Years of Stabilization and Reforms and the Challenges Ahead, World Bank Country Study, Washington, DC, 1996.

Trade, Growth, and Development
G. Ranis and L.K. Raut

CHAPTER 13

Identifying Leading Sectors That Accelerate the Optimal Growth Rate: A Computational Approach[1]

Mukul Majumdar[a] and Ilaria Ossella[b]

[a,b]Department of Economics, Cornell University, Ithaca, NY 14853-7601

> Much of this should have a familiar ring ... But while these great economists had to content themselves with verbal description and deductive reasoning, we can measure and we can compute. Therein lies the real difference between the past and the present state of economics.
>
> — W.W. Leontief [22]

13.1 Introduction and a Summary of Results

In this paper we consider a class of closed multi-sector models of dynamic optimization studied previously by McFadden [24], Atsumi [4], and Dasgupta and Mitra [10] [abbreviated as DM]. Our aim is to throw light on a question of sensitivity analysis: "How is the long-run optimal growth factor affected by productivity improvements?" In a *multi-sector* model it is, of course, difficult to obtain unambiguous qualitative results on comparative statics and dynamics, and our main results are based on explicit numerical computations involving Indian data. We believe, however, that the theoretical framework (despite its well-known limitations) may be of interest in a number of contexts in development economics.

The dynamic optimization problem is described formally by a triplet (A, w, δ) (see Section 13.2 for a precise description of the model). There are n commodities, and $A = (a_{ij})$ is the ($n \times n$, non-negative) input requirement

[1]We wish to acknowledge the comments and encouragement from M. Ali Khan, K. Basu, P. Dutta, V. Kelkar, T. Mitra, P. Mongia, L. Raut, S. Sattar, N. Singh, E. Thorbecke, and H. Wan on earlier stages of our research. Professor Mongia provided us with the Indian data for the calculations. Detailed suggestions of J. Hou, A. Mukherji, and J. Roy were useful in revision, but it is our feeling that we have not done justice to some of the issues they raised.

matrix (see Gale [13] and Dorfman, Samuelson and Solow [12]) specifying the Leontief technology (a_{ij} is the quantity of commodity i required to operate the jth sector, or industry, at a level which produces one unit of the jth commodity); w is the one period felicity function, and δ ($0 < \delta < 1$) is the discount factor. The objective function maximized by "the planner" is of the form $U(c) = \sum_{t=0}^{\infty} \delta^t w(c(t))$, where $w(c(t)) = [f(c(t))]^{1-\alpha}$ (with $0 < \alpha < 1$, and f satisfying some homogeneity, continuity, and concavity properties), and $c = (c(t))$ represents a sequence of feasible consumption programs. Under assumptions listed in DM [10], the optimal program of resource allocation from *any initial* stock of commodities has an interesting "turnpike" property: there is a common *long-run optimal growth factor*, say g, for the optimal outputs of all sectors, and g is equal to $(\delta\lambda)^{1/\alpha}$, where λ is the inverse of the largest eigenvalue, the "Perron–Frobenius eigenvalue," of A (see Debreu and Herstein [11, Thm. 1]). We point out, in Proposition 1, that this formula can be established in an optimization model somewhat different from DM (where A is primitive rather than strictly positive) but better suited for our data, and is at the basis of our sensitivity analysis.

A general result on nonnegative, indecomposable matrices[2] tells us that if any element of A decreases, λ will increase (hence, for fixed δ and α, g will increase). In other words, reducing *any* input requirement leads to an improvement of the long-run optimal growth factor. A natural question is to inquire whether one can go beyond this intuitive monotonicity result and provide a sharper characterization of the variations in λ and g, with respect to changes in A of any specific type. For example, consider $A = [a^1, a^2, \ldots, a^n]$, where a^j is the jth column of A (i.e., a^j specifies the input requirements of industry, or sector, j). Now take $A'_j = [a^1, \ldots, \theta a^j, \ldots, a^n]$, where θ is a fraction (say, $\theta = 0.98$); in other words, consider a reduction in the input requirements of the jth sector alone. One would like to derive at least a bound on the improvement of the long-run optimal growth factor g with respect to variations of this type. Our search for mathematical results on the sensitivity of characteristic roots with respect to changes in the entries of a matrix (see Bhatia [7] and Krause [17]) seems to indicate that such qualitative results of economic interest are elusive. This forces us to experiment with a computational approach.

We use an input requirement matrix A prepared by the Central Statistical Organization of the Government of India for 1989. This matrix consists of sixty sectors representing agriculture, animal husbandry, forestry, fishing, mining, manufacturing, and tertiary activities (construction, utilities, transport, communication, wholesale and retail trade, hotels, restaurants, banking, etc.). It is *not* strictly positive (hence, DM does not apply directly). How-

[2] A is indecomposable if for no permutation matrix Π, $\Pi A \Pi^T = \begin{bmatrix} A_{11} & 0 \\ A_{21} & A_{22} \end{bmatrix}$, where A_{11} and A_{22} are square; in particular, a strictly positive matrix is indecomposable.

ever, the matrix *is* primitive (A raised to the third power is strictly positive), satisfying the requirements of Proposition 1.

One at a time, for each sector j, we consider possible 2 and 5 percent decreases in input requirements and explicitly compute the corresponding ratio of the new to old long-run optimal growth factor. *This ratio*, it should be pointed out, *is independent of δ, the discount factor.* Our computations are performed with the use of the computer program *Mathematica.* It appears that, in terms of improving the turnpike growth factor, *electricity, gas, and water supply; iron and steel; and paper and paper products are the top three leading sectors* (the ratios of the growth factor due to 2 and 5 percent reductions in each sector at a time are given, respectively, in Tables 13.1 and 13.2). Although our computations are made with a numerical specification of α, *we show that the ranking of sectors is independent of the value chosen for α.*

Leading combinations of sectors, arising from simultaneous improvements in efficiency for five sectors, are also identified. We do so by first selecting representative sectors for three categories in the economy (agriculture, infrastructure, and manufacturing). Then, for every subset of five sectors within a category, we reduce the input requirements, by both 2 and 5 percent, and calculate the respective ratio of new to old long-run optimal growth factor. The leading combination for a category is simply the one generating the greatest ratio. Our analysis suggests that *the leading combination of sectors within the manufacturing category yields the greatest acceleration in the long-run optimal growth factor, followed by that of the infrastructure category* (the computed ratios are presented in Table 13.3). In addition, we consider an alternative assignment of sectors to the infrastructure and manufacturing categories. Specifically, we remove the iron and steel sector from manufacturing and place it in infrastructure, since iron and steel are both essential inputs in infrastructure. *The optimal growth factor*, in this case, *is increased the most by improvements in efficiency in the leading combination of sectors in infrastructure. The next largest increase is driven by improvements to sectors within manufacturing* (see Table 13.4 for the ratios of the growth factor). It is also possible to reverse the process and compute the percentage reduction in input requirements, for both single as well as combinations of sectors, needed to achieve a desired long-run optimal growth rate (see Ossella [28]).

The use of Leontief-type technology has been extensive in theoretical and empirical work on optimal growth. The masterly expositions of Dorfman, Samuelson, and Solow [12], Gale [13], and Morishima [25] clearly indicate the scope of the theoretical model. In our context, it is important to stress that the model describes an economy in which "labor" poses no constraint on production and the economy is "closed" (there is no flow of goods to/from the rest of the world). In addition, Leontief [23] and Armstrong and Upton [3] provide a review of some empirical applications.

Note that "reduction in input requirement" may — and typically will — involve costly adjustments in research and development/transfer of technology/restructuring of decision making . . . and there is no reason to expect that such costs will be the same, or even similar, across sectors. Nevertheless, the sensitivity analysis is of interest since it provides a formal framework for discussing some issues that have been raised time and again in the development literature by economists assessing alternative strategies.[3] The early papers by Hirschman [14] and Streeten [32] focused on a strategy of "unbalanced" development led by sectors that have the most effective "backward" and "forward" linkages. (For a detailed presentation on balanced versus unbalanced growth theory, see Basu [5, pp. 17–23], Chakravarty [9, Ch. 11], and Ray [29, pp. 138–42].) Scitovsky [30] pointed out the inadequacy of markets to generate appropriate signals to stimulate growth of sectors that generate externalities for others. In the literature on the East Asian miracle, the possibility of effectively "targeting" specific sectors in order to accelerate the overall development program has been at the heart of a lively debate on the role of the state in leading the development process in these economies (Amsden [1, 2], Lall [19, 20], Leipziger [21], Wade [33], World Bank [35], and Woronoff [36] all address this question). A policy maker in an economy that is "opening up" is likely to be hard pressed to decide which sectors ought to be promoted for direct foreign investment and supported with appropriate policies for ensuring a "level playing field."

In the context of India's development efforts, Bhagwati [6, Ch. 2] looked at explanations of India's "disappointing growth rate." During the planning era, an important feature of India's strategy for development was that it was "inward looking": there was a marked pessimism on export prospects (and performance) and there was also a maze of import controls justified, at the time, on a number of grounds. Hence, over time, the trade sector became increasingly marginal.

Taking an aggregative Harrod–Domar approach, Bhagwati [6, p. 40] maintained:

> In essence, the weak growth performance reflects, not a disappointing savings performance, but rather a disappointing productivity performance. The Indian savings rate more than doubled during this period, from roughly 10 percent to approximately 22 percent during 1950–84 . . . but the growth rates did not step up correspondingly.

Alternative explanations have been offered (in addition to Bhagwati [6], see Jalan [15], Joshi and Little [16], and Chakravarty [8] for a spectrum of

[3] After the completion of the present paper we became aware of the theoretical framework explored by Krishna and Pérez [18], which has no overlap with our approach but faces up to the theoretical issues involving balanced versus unbalanced growth.

views) for India's weak performance (relative to the High Performing Asian Economies), and there seems to be a consensus that an acceleration of the growth rate is a desperately needed necessary condition for achieving other laudable objectives pursued by the Indian planners (like poverty eradication).[4] The inefficiency of specific industries (particularly a large number of public sector industries) has figured prominently in the analysis of India's planning. For example Chakravarty [8, p. 40], despite his sympathetic view of India's planning experiments, wrote:

> We take the case of the power sector. It is often held that the state of infrastructure is a major reason for the slow-down of growth in industrial production. Professor Isher Ahluwalia attaches a lot of importance to the infrastructural bottle-necks as a cause of industrial slow-down in the period stretching from the middle of the sixties onwards. That there is an element of truth in this argument is evident from the available data.

Indeed, the inadequacy of India's infrastructure to generate and sustain growth momentum has become an important theme in India's policy debates in the last five years when attempts are being made to "open up" the economy and attract direct foreign investment. It is tempting to follow a partial equilibrium analysis, or simply to rely on common sense or anecdotal evidence — with or without some "scientific measurement" — to identify "leading sectors" that can accelerate the growth rate. But to us, issues like "linkages" and "externalities" make it essential to carry out a formal multi-sector analysis that can serve as a basis for computation and measurement. The goal of searching for an appropriate theoretical framework is perhaps a natural one in a volume in honor of T.N. Srinivasan [31] who reminded us:

> At the very outset let me state my convictions: first, without measurement any pretense at scientific analysis is impossible; second, even to know what to measure, let alone how to measure, a theoretical framework is necessary; and third, measurement or data collection has to be carefully designed.

---

[4]Chakravarty [8, p. 31] pointed out that Pitambar Pant, a well-known member of the Planning Unit "who declared himself to be in favor of poverty reduction as the central concern of planning," came to the conclusion in 1962 "that an average annual rate of growth of 7% per annum sustained over the decade 1965–75 was needed in order to give the poorest three deciles a nutritionally adequate diet." Unfortunately, for the decades of fifties through seventies, India struggled to maintain a 3 to 4 percent yearly growth rate and perhaps not surprisingly failed to provide meaningful content to the slogans on poverty alleviation.

13.2 A Dynamic Model of Optimal Intertemporal Allocation

Consider a "closed" economy in which there are n producible commodities, each produced by one and only one "activity" (sector, industry). In the well-known simple linear production model, the $n \times n$ non-negative matrix $A = (a_{ij})$ is interpreted as the *input requirement matrix*: the amount of commodity i required to produce one unit of the jth commodity is denoted by a_{ij}. The input requirement for producing an output vector y is given by the vector Ay.

Let $\tilde{y} > 0$ be the exogenously given initial stock vector.[5] A *feasible production program* from $\tilde{y}$ is a sequence $\langle x(y), y(t) \rangle$ such that

$$y(0) = \tilde{y}. \tag{13.1}$$

$$0 \leq x(t) \leq y(t) \text{ for all } t \geq 0, \tag{13.2}$$

$$Ay(t+1) \leq x(t) \text{ for all } t \geq 0. \tag{13.3}$$

We interpret $x(t)$ and $y(t)$ as the *input* and *output* vectors in period t. Note that from (13.3), the input requirements $Ay(t+1)$ of the output vector $y(t+1)$ in the period $t+1$ cannot exceed the planned input vector $x(t)$ of the period t. Associated with a production program $\langle x(t), y(t) \rangle$ is a *consumption program* $\langle c(t) \rangle$ defined by:

$$c(t) = y(t) - x(t) \text{ for all } t \geq 0. \tag{13.4}$$

It is clear from (13.2) that $c(t) \geq 0$ for all $t \geq 0$. We refer to $\langle x(t), y(t), c(t) \rangle$ as a program.

The following assumptions on the input requirement matrix A are maintained in this paper (compare with DM [10, Sec. 2b])

Assumption 1 *A is a nonnegative matrix which is primitive* (that is, there exists a positive integer k for which A^k is a strictly positive matrix).

Assumption 2 *A is productive* (that is, there is a nonnegative vector z such that $Az \ll z$).

Note that Assumption 1 is weaker than the assumption in DM that A is strictly positive (i.e., $a_{ij} > 0$ for all $i, j = 1, \ldots, n$). Assumption 2 is equivalent to the celebrated Hawkins–Simon condition in the literature on Leontief technology (see Nikaido [26, p. 90] for equivalent statements of the productivity concept).

[5] An n-vector $y = (y_i)$ is non-negative (written $y \geq 0$) if $y_i \geq 0$ for all i; $y = (y_i)$ is positive (written $y > 0$) if y is non-negative and $y_i > 0$ for some i; $y = (y_i)$ is strictly positive (written $y \gg 0$) if $y_i > 0$ for all i. $\mathbb{R}^n_+$ is the set of all non-negative n vectors.

The felicity function $w : \mathbb{R}^n_+ \to \mathbb{R}$ is assumed to be of an iso-elastic type (as in Atsumi [4]):

Assumption 3 $w(c) = [f(c)]^{1-\alpha}$ for $c \geq 0$, where $0 < \alpha < 1$ and $f : \mathbb{R}^n_+ \to \mathbb{R}_+$ satisfies the following restrictions:
(a) *f is concave and continuous on $\mathbb{R}^n_+$;*
(b) *f is homogenous of degree one;*
(c) *$f(c') \geq f(c)$ when $c' \geq c$; $f(c') > f(c)$ if $c\prime > c$ and $f(c) > 0$;*
(d) *$f(c) > 0$ if and only if $c \gg 0$.*

Let δ be a discount factor ($0 < \delta < 1$). A program $\langle x^*(t), y^*(t), c^*(t)\rangle$ from $\tilde{y}$ is an *optimal program* if

$$\sum_{t=0}^{\infty} \delta^t w(c^*(t)) \geq \sum_{t=0}^{\infty} \delta^t w(c(t))$$

for all programs $\langle x(t), y(t), c(t)\rangle$ from $\tilde{y}$.

We shall not deal with the subtle problem of existence. To settle the existence question, we impose, as in DM [10, Prop. 1], the following restriction:

Assumption 4 $\delta\lambda^{(1-\alpha)} < 1$,

where λ is the inverse of the largest eigenvalue (Perron–Frobenius eigenvalue) of A (Debreu and Herstein [11]). Observe that the productivity condition implies that $\lambda > 1$ (see Nikaido [26]).

The following "strict concavity" condition is used to prove the "turnpike theorem" (again compare with DM [10]):

Assumption 5 *If c, c' are non-negative, c is not proportional to c', and $f(c) > 0$, $f(c') > 0$, then, for $0 < \theta < 1$, $f(\theta c + (1-\theta)c') > \theta f(c) + (1-\theta) f(c')$.*

Let $g = (\delta\lambda)^{1/\alpha}$. The theoretical basis of our computational approach is given by the following relative stability property of optimal programs (see DM (Sec. 4, Prop. 6ii) for the corresponding result with their set of assumptions):

Proposition 1 *Let $\langle x^*(t), y^*(t), c^*(t)\rangle$ be an optimal program from $\tilde{y}$. Then there is a strictly positive vector z^* such that*

$$\lim_{t\to\infty} [y^*(t)/g^t] = z^*.$$

Proof See Ossella [27]. ■

We shall call g the "turnpike" or long-run optimal growth factor of the model specified by the triplet (A, w, δ). Note that g involves λ, a technological parameter; α, a parameter in the felicity function; and δ, the discount factor.

What can we say about the sensitivity of g to changes in A, if δ and α are held fixed? First, g varies "continuously" with respect to changes in A, and we have (Debreu and Herstein [11]):[6]

Proposition 2 *Under Assumptions* 1 *and* 2, λ *increases if any element of* A *decreases.*

For two different matrices A and A', one can assert (see Krause [17]) that the distance between $1/\lambda$ and $1/\lambda'$ is bounded above by a number, D. This bound can be shown to provide an upper bound of the new long-run optimal growth factor (g') induced by A', if, and only if, $D < 1/\lambda$.[7] However, it is always the case that $(1/\lambda) < 1$, due to Assumption 2, and when n is modestly large, D tends to be greater than one. The upper bound D, thus, does not provide any sharp, qualitative results on the sensitivity of g to changes in A.

13.3 Computational Approach

The variations in A that we consider are of the following type: for any $j = 1, \ldots, n$, let $A'_j = [a^1, \ldots, \theta a^j, \ldots, a^n]$, where $0 < \theta < 1$ and a^j is the jth column of A. Such a change in A represents a reduction in all the input requirements of the jth sector alone by a fixed percentage, $100(1 - \theta)$.

Consider an economy specified by (A'_j, w, δ). The corresponding long-run optimal growth factor is then $g'_j = (\delta\lambda'_j)^{1/\alpha}$, where λ'_j is the inverse of the largest (Perron–Frobenius) eigenvalue of A'_j. A sector ℓ is the *leading sector* of the original economy (represented by the input requirement matrix A) if

$$[g'_\ell/g] = \max_j [g'_j/g].$$

It is by no means claimed that there is a unique leading sector. However, in this paper, when we consider a large economy, we freely use expressions like "top three leading sectors" to mean the sectors that have the three largest (but not necessarily the same) values of $[g'_j/g]$. Formally, j_1, j_2, j_3 are the top three leading sectors[8] if

$$[g'_{j_1}/g] \geq [g'_{j_2}/g] \geq [g'_{j_3}/g] \geq [g'_j/g], \text{ for all } j \neq j_1, j_2, j_3.$$

[6] A primitive matrix is indecomposable by definition.

[7] If $|1/\lambda - 1/\lambda'| \leq D$, then $1/\lambda - D \leq 1/\lambda'$, and so $\lambda' \leq (1/\lambda - D)^{-1}$ if and only if $(1/\lambda - D) > 0$.

[8] A similar caveat applies to the expression "top five leading sectors."

Since the numerical values are all tabled, this abuse of terminology should not create any confusion.

The computational procedure for identifying the leading sector of an economy involves calculating the Perron–Frobenius eigenvalue λ'_j, and the respective value of $[g'_j/g]$, for *each* of the matrices A'_j. Now, $g'_j = [\delta\lambda'_j]^{1/\alpha}$. Hence, the ratio

$$[g'_j/g] = [\lambda'_j/\lambda]^{1/\alpha}$$

is *independent of the discount factor of the economy* (A, w, δ) *that we begin with.*

Note that the definition of a leading sector still involves the value of α, which, of course, needs to be specified numerically for computations. However, for any given j, the ratio $[g'_j/g]$ is a decreasing function of α:

$$\frac{d[g'_j/g]}{d\alpha} = \left(\frac{-1}{\alpha^2}\right)\left[\frac{\lambda'_j}{\lambda}\right]^{1/\alpha} \ln\left[\frac{\lambda'_j}{\lambda}\right] < 0, \text{ for all } 0 < \alpha < 1.$$

Variations in the value of α do not alter the leading sector of an economy. Suppose $\alpha \neq \tilde{\alpha}$, and consider the two economies (A, w, δ) and $(A, \tilde{w}, \delta)$, where $\tilde{w}(c) = [f(c)]^{1-\tilde{\alpha}}$. When the input requirements for only the jth sector of each economy are reduced, the corresponding long-run optimal growth factors are, respectively,

$$g'_j = (\delta\lambda'_j)^{1/\alpha} \text{ and } \tilde{g}'_j = (\delta\lambda'_j)^{1/\tilde{\alpha}}.$$

Denote the leading sector(s) for each economy by the following sets:

$$\begin{aligned} J^* &= \{j \in \{1, ..., n\} : j = \arg\max_j [g'_j/g]\} \text{ and} \\ \tilde{J}^* &= \{j \in \{1, ..., n\} : j = \arg\max_j [\tilde{g}'_j/g]\}. \end{aligned}$$

Proposition 3 *Let* (A, w, δ) *and* $(A, \tilde{w}, \delta)$ *be any two economies where* $\alpha \neq \tilde{\alpha}$. *Then* $J^* = \tilde{J}^*$.

Proof For any $j^* \in J^*$, it is true that

$$[g'_{j^*}/g] \geq [g'_j/g] \text{ for all } j \in \{1, ..., n\},$$

or

$$[\lambda'_{j^*}/\lambda]^{1/\alpha} \geq [\lambda'_j/\lambda]^{1/\alpha} \text{ for all } j \in \{1, ..., n\}.$$

Because $f(x) = x^{(\alpha/\tilde{\alpha})}$ is a strictly increasing function for all $x > 0$, $0 < \alpha < 1$, and $0 < \tilde{\alpha} < 1$, and since $1 < \lambda < \lambda'_j$ for all $j \in \{1, \ldots, n\}$,

$$\left\{[\lambda'_{j^*}/\lambda]^{1/\alpha}\right\}^{(\alpha/\tilde{\alpha})} \geq \left\{[\lambda'_j/\lambda]^{1/\alpha}\right\}^{(\alpha/\tilde{\alpha})} \text{ for all } j \in \{1, ..., n\}.$$

So

$$[\lambda'_{j^*}/\lambda]^{1/\tilde{\alpha}} \geq [\lambda'_j/\lambda]^{1/\tilde{\alpha}} \text{ for all } j \in \{1, ..., n\},$$

which means that

$$j^* \in \tilde{J}^*, \text{ for all } j^* \in J^*.$$

Thus,

$$J^* \subset \tilde{J}^*.$$

By the same reasoning,

$$\tilde{J}^* \subset J^*.$$

Therefore, $J^* = \tilde{J}^*$. ■

The method described thus far need not be limited to the identification of the leading sector of an economy. It can also be extended to determine a *leading combination of k sectors*, sectors for which simultaneous efficiency improvements yield the largest increase in the long-run optimal growth factor.

Let $S = \{1, \ldots, n\}$ and C_k be the set consisting of all possible subsets containing k elements from S. Denote an element of C_k by c_k (e.g., c_k could equal $\{1, 2, \ldots, k\}$). Now consider the following changes in A: for any $c_k \in C_k$, let

$$A'_{c_k} = [a^1_{c_k}, ..., a^j_{c_k}, ..., a^n_{c_k}],$$

such that

$$a^j_{c_k} = \begin{cases} \theta a^j & \text{if } j \in c_k \\ a^j & \text{otherwise} \end{cases}$$

where a^j is the jth column of A and $0 < \theta < 1$. Thus, in the extended case, all the input requirements of k sectors are concurrently reduced by the same percentage, $100(1 - \theta)$.

Such changes in technology will generate the economies (A'_{c_k}, w, δ). If λ'_{c_k} is the inverse of the largest (Perron–Frobenius) eigenvalue of A'_{c_k}, then the long-run optimal growth factor for the economy represented by the matrix A'_{c_k} is $g'_{c_k} = (\delta\lambda'_{c_k})^{1/\alpha}$. A *leading combination of k sectors* for the original economy is defined as the combination of k sectors (π) such that:

$$[g'_\pi/g] = \max_{c_k \in C_k} [g'_{c_k}/g].$$

The methodology for identifying a leading combination of k sectors then consists of calculating all the ratios $[g'_{c_k}/g]$ for every possible combination of k sectors in the economy.

13.4 Identifying India's Leading Sectors

We apply the procedure presented in the previous section to identify leading sectors of the Indian economy. The input requirement matrix that we use was prepared by the Central Statistical Organization of the Government of India for the year 1989. It comprises 60 sectors, of which 19 are in primary production, 34 in manufacturing, and seven in tertiary activities. Primary production incorporates agriculture, animal husbandry, forestry, fishing, and mining. The processing of output from primary production is defined as manufacturing, which includes, among others, food products, iron, chemicals, textiles, and wood products. Tertiary activities encompass construction, utilities, transport, communication, wholesale and retail trade, hotels and restaurants, banking, etc.

Checking that this Indian input requirement matrix (A) is actually primitive is not difficult. In fact, A raised to the third power is strictly positive, so Assumption 1 is satisfied. Furthermore, matrix A is productive [Assumption 2 is satisfied]. This data set, therefore, meets the assumptions of our optimization model.

We first attempt to identify the leading sector of the Indian economy. The value α is set at 0.9 and is held constant throughout all subsequent calculations. Assuming a 2 percent reduction in input requirements for the improvement in efficiency, the electricity, gas, and water supply sector shows the largest increase in the long-run optimal growth factor. Hence, it is *the* leading sector. The iron and steel sector follows, yielding a $[g'/g]$ ratio just 0.0003 lower. Paper and paper products, other chemicals, and other manufacturing, respectively, display the next largest increases in optimal growth.[9] The differences between these three sectors are also small, however, there is a relatively large gap between the growth factor ratios of the top two leading sectors and the remaining three. The above results are summarized in Table 13.1.

Table 13.1: Top five leading sectors in India given a 2% reduction in input requirements

Sector	g'/g
Electricity, Gas, and Water Supply	1.0033
Iron and Steel	1.0030
Paper and Paper Products	1.0019
Other Chemicals	1.0016
Other Manufacturing	1.0013

[9] "Other chemicals" includes inorganic and organic heavy chemicals, paints, medicines, explosives, animal oils and fats, etc.; while "other manufacturing" incorporates clocks, jewelry, scientific and athletic equipment, musical instruments, toys, etc.

Furthermore, we determine the leading sector of India in the context of a 5 percent reduction in input requirements. The resulting top five leading sectors remain the same. As before, there is also a sizable gap between the top two and remaining three sectors. It is important to note that, in both cases, the top leading sector is in the area of infrastructure, while the remaining four belong to manufacturing. The actual values for the ratios of the long-run optimal growth factors are presented in Table 13.2.

Table 13.2: Top five leading sectors in India given a 5% reduction in input requirements

Sector	g'/g
Electricity, Gas, and Water Supply	1.0081
Iron and Steel	1.0074
Paper and Paper Products	1.0044
Other Chemicals	1.0038
Other Manufacturing	1.0031

As mentioned earlier, a leading combination of k sectors (where the productivity of each sector is improved simultaneously) can also be identified. For our analysis on India, we let $k = 5$. However, comparing all possible subsets of five elements out of an original set of 60 is an involved process. To render the task more manageable, we divide the economy into three main categories: agriculture, infrastructure, and manufacturing. We then select ten representative sectors for each category, except for infrastructure, which only has seven. For each of these sets of representative sectors, we consequently determine a leading combination of five sectors. This process is conducted for both a 2 and 5 percent reduction in input requirements.

Table 13.3 lists the representative sectors chosen for each category, with their respective leading combination of sectors printed in bold, as well as the resulting ratios of the growth factor. It is interesting to note that, in this case too, the leading combination of sectors is the same regardless of the level of reduction in input requirements. Among the categories, agriculture exhibits the lowest increase in optimal growth, with improvements in infrastructure having a greater effect. The largest increase, however, is generated by the manufacturing category. One may find this result to be curious, considering the fact that electricity, gas, and water supply (part of infrastructure) is the leading sector for India. Furthermore, it is a consensus among many of India's policy makers that improvements in infrastructure are a crucial part of development efforts. Four of the manufacturing category's leading combination of five sectors, nevertheless, place second through fifth in the single sector analysis. So it is of no surprise that these four sectors have enough of a combined effect to place manufacturing above infrastructure.

Further investigation, however, reveals that under certain circumstances,

Table 13.3: Economic categories and their representative and leading sectors

Category	Sectors	g'/g	
		2%	5%
Manufacturing	Cotton textiles; Electric machinery; **Iron and steel; Non-ferrous metals; Other chemicals; Other manufacturing**; Other non-electrical machinery (drills, cranes, road rollers, etc.); **Paper and paper products**; Petroleum products; Rubber products	1.0090	1.0225
Infrastructure	**Coal and lignite**; Communication; **Construction**; Crude petroleum and natural gas; **Electricity, gas, and water supply; Other transport service** (buses, trucks, shipping transport, etc.); **Railway transport service**	1.0065	1.0160
Agriculture	**Animal husbandry; Cotton; Fertilizers**; Forestry and logging; Jute; **Other crops** (tobacco, coconut, tapioca, etc.); **Paddy**; Pesticides; Tea; Wheat	1.0004	1.0009

Table 13.4: Economic categories (with iron and steel under infrastructure) and their representative and leading sectors

Category	Sectors	g'/g	
		2%	5%
Infrastructure	**Coal and lignite**; Communication; **Construction**; Crude petroleum and natural gas; **Electricity, gas, and water supply; Iron and steel; Other transport service** (buses, trucks, shipping transport, etc.); Railway transport service	1.0089	1.0221
Manufacturing	Cotton textiles; Electric machinery; **Non-ferrous metals; Other chemicals; Other manufacturing; Other non-electrical machinery** (drills, cranes, road rollers, etc.); **Paper and paper products**; Petroleum products; Rubber products	1.0070	1.0173

improvements within infrastructure accelerate the optimal growth rate the most. In particular, an argument can be made for including the iron and steel sector in infrastructure, instead of manufacturing, since iron and steel are key inputs for many infrastructural facilities. Making such a switch (with no other changes to the category assignments) results in reversing the ranking of the manufacturing and infrastructure categories. Now, it is infrastructure, followed by manufacturing, whose leading combination of sectors generates the greatest increase in the long-run optimal growth factor. This is true for both a 2 and 5 percent reduction in input requirements. The respective values for the ratios of the growth factor pertaining to the infrastructure and manufacturing categories are presented in Table 13.4. Agriculture is not listed, since no changes to its sector assignments were made and its ranking, relative to the other categories, remains the same. Looking at the results, both before and after the iron and steel sector is moved, we can see that the ratio of the growth factor for the category not possessing iron and steel stays relatively constant (e.g., 1.0065 vs 1.0070 for a 2 percent reduction in input requirements). This indicates that (without iron and steel) an increase in efficiency in manufacturing yields an improvement in the growth factor which is similar to that of infrastructure. Therefore, the iron and steel sector has enough of an influence that its inclusion in either category raises that category's ranking. In closing, it is important to note that the above results do depend, to a certain extent, on the specific assignment of the representative sectors to their respective categories. This variability underscores the importance of further research to determine the correct construction of such categories.

13.5 Some Conclusions from Earlier Calculations

We carried out similar computations for India based on an input requirement matrix for the year 1983. This matrix was also prepared by the Central Statistical Organization of the Government of India, but consists of only 59 sectors. Moreover, the sector classification scheme adopted for this matrix is somewhat different from the 1989 matrix. In general, the 1983 input requirement matrix comprises 11 sectors in primary production, 33 in manufacturing, and 15 in tertiary activities. Due to differences in the classification of sectors, one has to be extremely careful in making any comparisons, beyond the most general, between the computational results of 1989 and 1983. Nevertheless, some interesting differences do occur, so a quick summary of the 1983 results is provided below.

The top five leading sectors, based on the input requirement matrix for 1983, are as follows (numbers listed in parentheses are the ratios of the new to old growth factor arising from a 2 percent reduction in input requirements): iron and steel industries and foundries (1.0051), electricity (1.0031), metal products except machinery (1.0012), railway transport services (1.0010), and other basic metal industry (1.0009).[10] In contrast to the 1989 results, an improvement in efficiency of the iron and steel sector yields a greater increase in the optimal growth factor than an improvement in the electricity sector. It is also interesting to note that the ratios of the new to old growth factor, given

[10] "Metal products except machinery" comprises hand tools, hardware, metal containers, stoves, metal furniture, etc.; and "other basic metal industry" refers to the melting and refining of non-ferrous basic metals.

an improvement in efficiency in electricity, are extremely close for both years. One must realize, however, that due to structural differences between the two matrices, we cannot derive conclusive implications from these comparisons. For example, the electricity, gas, and water supply sectors are combined in the 1989 input requirement matrix, while electricity stands alone in the 1983 matrix.

Two of the top five leading sectors for the 1983 data belong to infrastructure. In fact, railway transport services, in addition to electricity, emerges as one of the top five leading sectors for 1983, but does not do so for 1989. Furthermore, none of the top five leading sectors for both the 1989 and 1983 matrices pertain to agriculture.

In view of the extensive literature on the role of human capital in accelerating economic development, we note that while education and research does figure as a sector in the 1983 matrix, it does not come out as a top five leading sector. Education and research do not even appear, either as individual sectors or a combined sector, in the classification scheme for the 1989 input requirement matrix. We do recognize, however, that the benefits of education go far beyond what can be captured in an input-requirement matrix.

References

1 Amsden, A., Asia's Next Giant: South Korea and Late Industrialization, Oxford University Press, New York, 1989.

2 Amsden, A.,Why isn't the whole world experimenting with the East Asian model to develop?, Review of The East Asian Miracle, World Development, 22 (1994) 627–34.

3 Armstrong, A.G. and D.C. Upton, A review of input–output applications, Bulletin of the International Statistical Institute, 43 (1969) 113–30.

4 Atsumi, H., The efficient capital program for a maintainable utility level, Review of Economic Studies, 36 (1969) 263-87.

5 Basu, K., Analytical Development Economics, The MIT Press, Cambridge, 1997.

6 Bhagwati, J., India in Transition, Freeing the Economy, Clarendon Press, Oxford, 1993.

7 Bhatia, R., Perturbation Bounds for Matrix Eigenvalues, Pitman Research Notes in Mathematics Series 162, Longman Scientific & Technical, New York, 1987.

8 Chakravarty, S., Development Planning, The Indian Experience, Clarendon Press, Oxford, 1987.

9 Chakravarty, S., Writings on Development, Oxford University Press, Delhi, 1997.

10 Dasgupta, S. and T. Mitra, Intertemporal optimality in a closed linear model of production, Journal of Economic Theory, 45 (1988) 288-315.

11 Debreu, G. and I.N. Herstein, Nonnegative square matrices, Econometrica, 21 (1953) 597-607.

12 Dorfman, R., P.A. Samuelson and R.M. Solow, Linear Programming and Economic Analysis, McGraw-Hill Book Company, Inc., New York, 1958.

13 Gale, D., The Theory of Linear Economic Models, McGraw-Hill Book Company, Inc., New York, 1960.

14 Hirschman, A.O., The Strategy of Economic Development, Yale University Press, New Haven, CT, 1958.

15 Jalan, B., India's Economic Crisis: The Way Ahead, Oxford University Press, Delhi, 1991.

16 Joshi, V. and I.M.D. Little, India: Macroeconomics and Political Economy 1964–1991, World Bank, Washington DC, 1994.

17 Krause, G., Bounds for the variation of matrix eigenvalues and polynomial roots, Linear Algebra and its Applications, 208/209 (1994) 73-82.

18 Krishna, K. and C. Pérez, Unbalanced growth, unpublished manuscript, June, 1997.

19 Lall, S., The East Asian miracle: Does the bell toll for industrial strategy?, World Development, 22 (1994) 645–54.

20 Lall, S., Paradigms of development: The East Asian debate, Oxford Development Studies, 24 (1996) 111–31.

21 Leipziger, D. (ed.), Lessons from East Asia, University of Michigan Press, Ann Arbor, 1997.

22 Leontief, W., The Dynamic inverse, in: A.P. Carter and A. Brody (eds.), Contributions to Input–Output Analysis, North-Holland Publishing, Amsterdam, 1970, pp. 17–46 [reprinted as Ch. 14 of Input–Output Economics, 2nd edition, Oxford University Press, New York, 1986].

23 Leontief, W., Input–Output Economics, 2nd edition, Oxford University Press, New York, 1986.

24 McFadden, D., The evaluation of development programs, Review of Economic Studies, 34 (1967) 25–50.

25 Morishima, M., Equilibrium, Stability, and Growth: A Multisectoral Analysis, Oxford University Press, London, 1964.

26 Nikaido, H., Convex Structures and Economic Theory, Academic Press, New York, 1968.

27 Ossella, I., A turnpike property of optimal programs for a class of simple linear models of production, forthcoming in Economic Theory, 1999.

28 Ossella, I., India's leading sectors: A computational approach, Ph.D. Dissertation, Cornell University, Ithaca, New York, 1999.

29 Ray, D., Development Economics, Princeton University Press, Princeton, 1998.

30 Scitovsky, T., Two concepts of external economies, Journal of Political Economy, 62 (1954) 143–51.

31 Srinivasan, T.N., On studying socio-economic change in rural India, in: P. Bardhan (ed.), Conversations Between Economists and Anthropologists, Oxford University Press, Delhi, 1989, pp. 238–49.

32 Streeten, P.P., Unbalanced growth, Oxford Economic Papers, 11 (1959) 167–90.

33 Wade, R., Governing the Market: Economic Theory and the Role of Government in East Asian Industrialization, Princeton University Press, Princeton, 1990.

34 Wolfram, S., Mathematica, A System for Doing Mathematics by Computer, 2nd edition, Addison-Wesley Publishing Company, New York, 1991.

35 World Bank, The East Asian Miracle, Economic Growth and Public Policy, Oxford University Press, New York, 1993.

36 Woronoff, J., Japanese Targeting: Successes, Failures, Lessons, St. Martin's Press, New York, 1992.

Trade, Growth, and Development
G. Ranis and L.K. Raut

CHAPTER 14

General-Equilibrium Cost-Benefit Analysis of Education and Tax Policies[1]

James J. Heckman,[a] Lance Lochner[b] and Christopher Taber[c]

[a]Department of Economics, University of Chicago, 1126 E 59th St., Chicago, IL 60637, and the American Bar Foundation

[b]Department of Economics, University of Rochester, 232 Harkness Hall, Rochester, NY 14627

[c]Department of Economics and Institute for Policy Research, Northwestern University, 2003 Sheridan Road, Evanston, IL, 60208

14.1 Introduction

In his 1977 Walras–Bowley lecture, presented at the Summer Meetings of the North American Econometric Society in Boulder, Colorado, T.N. Srinivasan presented a magisterial survey of the state of the art in social cost-benefit analysis and project evaluation. He stressed the value of general-equilibrium models in analyzing policies and making specific policy recommendations, and at the same time recognized their limitations.

This paper builds on Srinivasan's lecture and considers four of the many topics he discussed: (1) the importance of accounting for general-equilibrium effects of large-scale programs; (2) the importance of accounting for dynamics; (3) the importance of understanding the causes of a problem in evaluating proposed solutions for it; and (4) the importance of accounting for the impact of a program on the personal distribution of income and welfare. Since we consider policy evaluation in a developed economy, many of T.N.'s other

[1]We are grateful for comments received at the conference honoring T.N. Srinivasan, especially those received from Ricardo Barros, T.N. Srinivasan, and L.K. Raut and comments received from participants in a seminar at the University of Western Ontario, and in particular, those from Michael Parkin. This research was supported by a grant from the Russell Sage Foundation and NSF Grant SBR-93-21-048, NSF Grant SBR-97-09-873 and NIH Grants NIH:R01-HD34958-01 and NIH:R01-HD32058-03.

concerns about the importance of understanding market failure and institutional failure — two topics central to development economics — are less central to this paper and are not discussed here. The emphasis in this paper is on understanding the sources of rising wage inequality in the US economy and evaluating proposals that have been made to combat it. Wage inequality has increased substantially in the American economy since the early 80s. Workers with low skills have experienced large declines in their earnings, both absolutely and relative to more skilled workers. Only recently have economists begun to develop models that explain the rise in wage inequality, focusing on explaining the college/high school wage differential. The primary causes for the recent increase in overall wage inequality are still being debated. Despite the lack of a consensus on the cause, there is no shortage of proposed cures. Numerous tax and tuition policies have been proposed to stimulate investment in high-skilled labor to alleviate rising wage inequality by making some of the unskilled into skilled workers who benefit from the rising skill differential, while at the same time making remaining unskilled workers more scarce. These proposals have not been evaluated within the context of articulated economic models that explain the problems that the policies are designed to solve. Moreover, most of the estimates of the rising "return" to education are based on the regression coefficient of schooling in a series of cross-sectional Mincer earnings functions. During a period of transition driven by skill-biased technical change, estimated "rates of return" are poor guides to the true rates of return that will be experienced by any cohort, and are a poor guide to policy.

Our previous work (Heckman, Lochner and Taber [28, 29, 30]) addresses these issues. We develop an empirically based heterogeneous-agent dynamic general-equilibrium model of labor earnings for the US economy. Unlike the standard approach to applied general-equilibrium models, we minimize the role of calibration in obtaining parameter estimates. We heavily rely on micro data in conjunction with macro time series to justify the parameters we use.

We study both the sources of rising earnings inequality and the effects of various policy proposals aimed at increasing skill formation. We demonstrate the danger in using cross-sectional "rates of return" estimated in a period of transition to guide educational policy. Our previous work assumes that the US economy is closed. One major goal of this paper is to relax that assumption and evaluate tax and tuition policies in an open-economy environment. Many predictions of an open-economy version of our model are similar to those derived from a closed-economy version but there are some important differences. When we compare the open-economy model to the closed-economy model, the closed-economy version produces more plausible predictions about the time paths of economic aggregates in the US economy and the effects of policy. A major conclusion of this paper is that for the class of general-equilibrium models we consider, a closed-economy model provides a more accurate characterization of the US economy.

This paper also emphasizes more forcefully than our previous papers have done the crucial distinction between cross-sectional "rates of return" to schooling and the "rates of return" that are experienced by cohorts of persons living through a transition. We demonstrate that the coefficient on schooling in the Mincer equation is only weakly connected to the true rate of return that should guide human capital investment decisions.

We also emphasize the value of accounting for heterogeneity in ability and heterogeneity in the economic history experienced by different cohorts of persons in evaluating policies and discussing the problems created by rising wage inequality. Distributional considerations affect the likelihood that different policies will be favored. Any assessment of policies should account for this.

We demonstrate the value of a general-equilibrium approach to the evaluation of human capital policies instituted at the national level. Partial-equilibrium — "treatment effect" — approaches are very misleading. This is true both in open-economy and closed-economy versions of our model. Heckman, Lochner and Taber [28] meet Srinivasan's desideratum that policy analysis be based on a model that explains the problem that the policy is designed to solve by producing a model that can explain the changing distribution of wages in the US economy over the last thirty years. We show that accounting for the distinction between skill prices and measured wages is important for analyzing the changing wage structure, because the two often move in different directions. This is also true in the open-economy version of our model reported here. The general patterns are the same in both models. However, there are several features of these models, such as the timing of the transitions and patterns across cohorts, that are different in the open- and closed-economy versions. The closed-economy version of the model does a somewhat better job at explaining the changing wage structure and, hence, is a better framework within which to conduct policy analysis. For both open- and closed-economy versions of our model, the partial-equilibrium effects of tax and tuition policies are substantial. However, the general-equilibrium effects are very small. We find that while the effects of tax reform on capital accumulation are drastically affected by the open-economy assumption, the simulated effects of tax reform on human capital are very similar in the open-economy and closed-economy cases.

The structure of this paper is as follows. In Section 14.1, we first present an intuitive introduction to our model and the policy problems considered here. In Section 14.2, we present a more formal discussion of the model. In Section 14.3, we consider methods for determining the parameters of the model to convert it from a theoretical exercise into a framework for quantitative economic policy evaluation. In this section we emphasize the role of microdata.

In Section 14.4, we demonstrate how well the open-economy version of our model explains rising wage inequality over the past 30 years. We compare the predictions of the open-economy version with the closed-economy version of our model. In Section 14.5, we analyze how overall welfare, and the welfare of different ability and schooling groups, is affected during the transition to a new high-skill economy. In Section 14.6 we present a formal analysis of what a Mincer "rate of return" measures. We compare cross-section and cohort "rates of return" to reveal how misleading the Mincer rate of return is as a guide to formulating and evaluating educational policy, especially in a period of change in the technology. In Section 14.7 we use our model to present a general-equilibrium evaluation of tuition policy and compare it with conventional partial-equilibrium treatment effect estimates of the sort routinely generated by labor economists. In Section 14.8, we compare general-equilibrium and partial-equilibrium approaches to the evaluation of tax reforms designed to stimulate human capital production. In Section 14.9, we summarize the evidence that supports the closed-economy version of our model. Section 14.10 summarizes and concludes the paper.

14.2 A General-Equilibrium Model with Heterogeneous Agents and the Policy Problems We Analyze

Our dynamic general-equilibrium model of human capital and physical capital accumulation has several sources of heterogeneity among its agents: (1) Persons differ in initial ability levels indexed below by θ and this ability affects both earnings levels and personal investment decisions. (2) Skills are heterogeneous. Different schooling levels correspond to different skills. The post-school skills acquired at one schooling level are not perfect substitutes for the post-school skills acquired at another schooling level; however, skills are perfect substitutes across age groups within a given schooling level. (3) The model uses an overlapping generations framework to produce heterogeneity among different cohorts as a result of rational investment behavior. In a period of transition, different skill price paths facing different entry cohorts produce important differences in the levels and rates of growth of earnings across cohorts due to differential human capital investment. All three sources of heterogeneity are important in explaining rising wage inequality over time and over cohorts.

Our model considers human capital choices at both the extensive margin (schooling) and the intensive margin (on-the-job training). Schooling enables people to learn on the job and also directly produces market skills. Our model extends the Roy [53] model of self-selection and earnings to allow for investment and embeds it in a dynamic general-equilibrium model in which the prices of heterogenous skills are endogenously determined. It extends the widely used framework of Ben-Porath [6] by permitting different technologies to govern the production of skill in schools and the production of skill on the job, by recognizing that schooling affects both productivity on the job as well as the ability to learn on the job, and by allowing for multiple skill types. Our model extends the schooling models of Willis and Rosen [60] and Keane and Wolpin [39] by making post-schooling on-the-job training endogenous. It also extends those models and the analysis of Siow [55] by embedding both schooling and job training in a general-equilibrium framework.

We relax the efficiency units assumption for the aggregation of labor services that is widely used in macroeconomics (see, e.g., Kydland [42]). This assumption is not consistent with rising wage inequality across skill groups except in the unlikely case when quantities of skill embodied in each group change over time in a fashion that exactly mimics movements in relative wages. Our model introduces human capital accumulation into the overlapping generations framework of Auerbach and Kotlikoff [2] extending the model of Davies and Whalley [15].[2] We consider multiple skill types, rational expectations, heterogeneity in human capital endowments and production, and distinguish between schooling and on-the-job training as separate means for producing skills. We then use our model to examine the effectiveness of various tax and tuition policies that have been proposed to reduce wage inequality in the US economy.

A commonly used approach for assessing the sources of rising wage inequality begins with an equation postulated as a time-differenced demand relationship which connects the changes in the relative wages of skilled (W_{St}) and unskilled (W_{Ut}) workers at time t to the respective quantities of the two factors, Q_{St} and Q_{Ut}:

$$\Delta \log \left[\frac{W_{St}}{W_{Ut}}\right] = \varphi - \frac{1}{\sigma}\Delta \log \left[\frac{Q_{St}}{Q_{Ut}}\right]. \tag{14.1}$$

In this equation, φ is the trend rate of relative wage growth arising from skill-biased technical change and σ is the elasticity of substitution between the two types of labor. Katz and Murphy [38] estimate this equation for the US economy using the measures of skilled and unskilled labor defined in their paper for the period 1963–87.[3] They report $\sigma = 1.41$ with a standard error

[2] Fullerton and Rogers [18] also estimate a general equilibrium model using microdata, but do not incorporate human capital investment.

[3] They define college equivalents and high school equivalents by using estimated earnings functions to weight persons with less than twelve and greater than sixteen years of schooling to obtain fractions or multiples of pure high school or college types. Persons with some college are allocated 50-50 between college equivalents and high school equivalents using their earnings relative to pure college or high school earnings. They assume perfect substitutes or efficiency units within the two skill groups but less than perfect substitutability across skill groups.

of 0.150, although they also suggest that a range of estimates with σ as low as 0.5 are also consistent with the data (Johnson [36] reports an estimate of $\sigma = 1.50$ for the elasticity of substitution between college and high school later). They estimate φ to be 0.033 (standard error 0.007).

Using Katz and Murphy's definition of skill groups, it is necessary to transform approximately 5.4 million unskilled people to college equivalents to reverse the decade-long (1979–87) erosion of real wages for individuals not attending college. Even using their lower range estimate of $\sigma = 0.5$, two million persons need to be shifted from the unskilled to the skilled category to offset the decade-long trend against unskilled labor. Maintaining the skill gap against the secular bias operating against unskilled labor requires that the percentage of persons acquiring post-secondary skills in each year rise by 55 percent (22 percent in the lower bound case) (these calculations are presented in Heckman [25]).

It is tempting to use the demand function (14.1), in conjunction with microestimates of the supply response of skills to tuition or other subsidies, to evaluate government policies. Conventional microeconomic approaches to policy evaluation assume that skill prices remain constant at their pre-subsidy levels when calculating supply responses. They ignore the feedback of induced price changes (created by the increase in the supply of skill) on the supply decisions of agents. Only when these feedback responses are incorporated into supply decisions can equation (14.1) be used as a valid basis for policy evaluation. Informed policy evaluations allow for skill prices to adjust and for agents to anticipate this adjustment and respond appropriately. Such evaluations reveal that the response to a policy evaluated in a microeconomic setting, which holds prices constant, may be a poor guide to the actual response when prices adjust with changes in quantities of aggregate skill. Thus, the effect of any policy characterized by parameter ψ, $\partial[Q_{St}/Q_{Ut}]/\partial\psi$, is not the same when skill prices are held fixed as it is when skill prices are allowed to vary in response to the policy-induced change.

A more convincing evaluation of any policy designed to promote skill formation and alleviate wage inequality requires a model that explains the rising wage inequality in the US labor market. As stressed by Srinivasan, it is potentially dangerous to "solve" problems whose origins are not well understood. These concerns motivate us to develop a dynamic general-equilibrium model of labor earnings that is consistent with evidence from the US labor market (Heckman, Lochner and Taber [28]). We now present that model.

14.2.1 A dynamic general-equilibrium model of earnings, schooling and on-the-job training

Our model extends Ben-Porath's model [6] of skill formation in several ways. (1) In contrast to his model, we distinguish between schooling capital and job training capital at a given schooling level. In our model, schooling human capital is an input to the production of human capital acquired on the job and is also directly productive in the market. However, the tight link between schooling and on-the-job training investments, which is characteristic of Ben-Porath's model, is broken. In his model, human capital obtained from all forms of schooling and job training are perfect substitutes whereas in our model human capital from schooling is distinct from human capital obtained from on-the-job training and persons of different schooling levels obtain post-school skills that are not perfect substitutes across schooling types. (2) Skills produced by different schooling levels command different prices, and wage inequality among persons is generated by differences in skill levels, differences in investment, and in the prices of different skills. In Ben-Porath's model, wage inequality can be generated only by differences in skill levels and investment behavior, because all skill commands the same price. In our model, persons with different levels of schooling invest in different skills through on-the-job training in the post-schooling period. In the aggregate, skills associated with different schooling groups are not perfect substitutes.[4] Within each schooling group, however, persons with different amounts of skill arising from on-the-job training are perfect substitutes.[5] Unlike the Ben-Porath framework, our model of heterogeneous skills captures comparative advantage, which is an important feature of modern labor markets (see the empirical evidence summarized in Sattinger [54]). (3) In addition to the post-school investment stressed by Ben-Porath, persons in our model choose among schooling levels with associated post-school investment functions. (4) We build in heterogeneity within schooling levels. Among persons of the same schooling level, there is heterogeneity both in initial stocks of human capital and in the ability to produce job-specific human capital. (5) We embed our model of individual human capital production into a general-equilibrium setting so that the relationship between the capital market and the markets for human capital of different skill levels is explicitly developed. We extend the open-economy general-equilibrium sectoral-choice model of Heckman and Sedlacek [33] to allow for investment in sector-specific human capital.

14.2.2 The microeconomic model

We first review the optimal consumption, on-the-job investment, and schooling choices for a given individual of ability type θ who takes skill prices as

[4]This specification is consistent with evidence that the large increase in the supply of educated labor consequent from the baby boom depressed the returns to education (see Freeman [17], Autor, Katz and Krueger [3] and Katz and Murphy [38]).

[5]This specification accords with the empirical evidence summarized in Hamermesh [21, p. 123] that persons of different ages but with the same education levels are highly substitutable for each other.

given. We then aggregate the model to a general-equilibrium setting. Until Section 14.8, we simplify the tax code and assume that income taxes are proportional. Individuals live for $\bar{a}$ years and retire after $a_R \leq \bar{a}$ years. Retirement is mandatory. In the first portion of the life cycle, a prospective student decides whether or not to remain in school. Once he leaves school, he cannot return. Individuals choose the schooling option that gives them the highest level of lifetime utility.

Define K_{at}^S as the stock of physical capital held at time t by a person age a of schooling level S; H_{at}^S is the stock of human capital at time t of type S at age a. The optimal life cycle problem can be solved in two stages: first, condition on schooling and solve for the optimal path of consumption (C_{at}^S) and post-school investment (I_{at}^S) for each type of schooling level. Individuals then select among schooling levels to maximize lifetime welfare. Given S, an individual age a at time t has the value function

$$V_{at}(H_{at}^S, K_{at}^S, S) = \max_{C_{a,t}^S, I_{at}^S} U(C_{at}^S) + \delta V_{a+1,t+1}^S(H_{a+1,t+1}^S, K_{a+1,t+1}^S, S), \quad (14.2)$$

where U is a strictly concave and increasing utility function and δ is a time preference discount factor. This function is maximized subject to the budget constraint

$$K_{a+1,t+1}^S \leq K_{a,t}^S(1 + (1-\tau)r_t) + (1-\tau)R_t^S H_{at}^S(1 - I_{at}^S) - C_{at}^S \quad (14.3)$$

where τ is the proportional tax rate on capital and labor earnings, R_t^S is the rental rate on human capital of schooling type S, and r_t is the net return on physical capital at time t. In this paper, we abstract from labor supply. Estimates of intertemporal substitution in labor supply estimated on annual data are small, so ignoring labor supply decisions will not greatly affect our analysis (see Browning et al. [9] or the survey in Heckman [24]).

In the empirical analysis in this paper, we use the conventional power utility specification of preferences

$$U(C_{at}^S) = \frac{(C_{at}^S)^\gamma - 1}{\gamma}. \quad (14.4)$$

On-the-job human capital for a person of schooling level S (years of schooling) accumulates through the human capital production function

$$H_{a+1,t+1}^S = A^S(\theta)(I_{at}^S)^{\alpha_S}(H_{at}^S)^{\beta_S} + (1-\sigma^S)H_{at}^S \quad (14.5)$$

where the conditions $0 < \alpha_S < 1$ and $0 \leq \beta_S \leq 1$ guarantee that the problem is concave in the control variable, and σ^S is the rate of depreciation of job-

S specific human capital. This functional form is widely used in both the empirical literature and the literature on human capital accumulation.[6]

For simplicity, we ignore the input of goods into the production of human capital on the job. We explicitly allow for tuition costs of college, which we denote by D_{at}^S. Ours is a model with one final output. The same good that is used to produce physical capital and final output is used to produce schooling human capital.[7] After completion of schooling, time is allocated to two activities: on-the-job investment, I_{at}^S, and work, $(1 - I_{at}^S)$, both of which must be non-negative. The agent solves a life-cycle optimization problem given initial stocks of human and physical capital, $H^S(\theta)$ and K_0^S, as well as his ability to produce human capital on the job, $A^S(\theta)$.

$H^S(\theta)$ and $A^S(\theta)$ represent ability to "earn" and ability to "learn," respectively, measured after completing school. They embody the contribution of schooling to subsequent learning and earning in the schooling-level S-specific skills as well as any initial endowments. Notably absent from our model are short-run credit constraints that are often featured in the literature on schooling and human capital accumulation. Our model is consistent with the evidence presented in Cameron and Heckman [10, 11] that long-run family background and long-run family income produce child ability θ. Ability θ operating through $A^S(\theta)$ and $H^S(\theta)$ affects schooling. Short-term credit constraints operating on children at the age when they are deciding to enter college are not empirically important. It is the long run factors that account for the empirically well known correlation between schooling attainment and parental income. The mechanism generating the income-schooling relationship is through family-acquired human capital and not credit rationing. The α and β are also permitted to be S-specific, which emphasizes that schooling affects the process of learning on the job in a variety of different ways.

Assuming interior solutions conditional on the choice of schooling, we obtain the following first-order conditions for consumption and investment:

$$U_{C_{at}}^S = \delta \frac{\partial V_{a+1,t+1}^S}{\partial K_{a+1,t+1}^S} \tag{14.6}$$

$$\frac{\partial V_{a+1,t+1}^S}{\partial K_{a+1,t+1}^S} = \frac{\partial V_{a+1,t+1}^S}{\partial H_{a+1,t+1}^S} \left[\frac{A\alpha_S (I_{a,t}^S)^{\alpha_S - 1} (H_{a,t}^S)^{\beta_S}}{R_t^S H_{a,t}^S (1-\tau)} \right]; \tag{14.7}$$

and envelope conditions for physical and human capital:

$$\frac{\partial V_{a,t}^S}{\partial K_{a,t}^S} = \delta \frac{\partial V_{a+1,t+1}^S}{\partial K_{a+1,t+1}^S} (1 + r_t(1-\tau)) \tag{14.8}$$

[6] Uzawa [58] assumes $(\beta_S = 1)$; Ben-Porath [6], Lucas [45] and Ortigueira and Santos [49] assume that $(\alpha_S = \beta_S)$; and Rosen [52] assumes $\alpha_S = 1/2$ and $\beta_S = 1$.

[7] More general specifications which allow the price of schooling inputs to track the price of skilled labor produce almost identical results.

$$\frac{\partial V_{a,t}^S}{\partial H_{a,t}^S} = \delta \frac{\partial V_{a+1,t+1}^S}{\partial K_{a+1,t+1}^S} R_t^S (1 - I_{a,t}^S)(1-\tau) + \delta \frac{\partial V_{a+1,t+1}^S}{\partial H_{a+1,t+1}^S} (A\beta_S (I_{a,t}^S)^{\alpha_S} (H_{a,t}^S)^{\beta_S - 1} + (1 - \sigma^S)). \tag{14.9}$$

At the end of working life, the final term, which is the contribution of human capital to earnings, has zero marginal value. We assume mandatory retirement at age a_R , leaving $\bar{a} - a_R$ as the retirement period during which there are no labor earnings. All of these calculations are conditional on S. We next consider the choice of S to complete the model at the individual level.

At the beginning of life, agents choose the value of S that maximizes lifetime utility:

$$\hat{S} = \operatorname*{Arg\,max}_S \, [PV_{0t}^S(\theta) - D_{0t}^S - \varepsilon^S] \tag{14.10}$$

where $PV_{0t}^S(\theta)$ is the value at age 0 of schooling at level S, D_{0t}^S is the discounted direct cost of schooling and ε^S represents non-pecuniary benefits expressed in value terms or else a tuition shock. Discounting of V_{0t}^S and D_{0t}^S is back to the beginning of life to account for different ages of completing school. Because of the separation between consumption and investment, we can characterize the life-cycle schooling choice in terms of utility or present value comparisons. For convenience, we use the latter. Tuition costs are permitted to change over time so that different cohorts face different environments for schooling costs. Given optimal investment in physical capital, schooling, investment in job-specific human capital, and consumption, we calculate the path of savings. For a given return on capital and rental rates on human capital, the solution to the S-specific optimization problem is unique, given concavity of the production function of (14.5) in terms of I_{at}^S, $(0 < \alpha_S < 1)$; the restriction that human capital be self-productive, but not too strongly $(0 \leq \beta_S \leq 1)$; that investment is in the unit interval $\left(0 \leq I_{at}^S \leq 1\right)$; and concavity of U in terms of C $(\gamma < 1)$.

The choice of S is unique almost surely if ε^S is a continuous random variable, as we assume in our empirical analysis. The dynamic problem is of split-endpoint form.[8] We know the initial condition for human and physical capital, and optimality implies that investment is zero at the end of life. In this paper, we numerically solve this problem using the method of "shooting" (see Lipton et al. [43]). For any terminal value of H^S and K^S, we solve backward to the initial period and obtain the implied initial conditions. We iterate until the simulated initial condition equals the pre-specified value.

[8] The split endpoint problem arises from the need to simultaneously solve for an optimum that satisfies the initial condition on human capital stock (endowment) and a terminal value condition that the marginal value of human capital at the end of working life must be zero.

14.2.3 Aggregating the model

The prices of skills and capital are the derivatives of an aggregate production function. In order to compute rental prices for capital and the different types of human capital, it is necessary to construct aggregates of each of the skills. Given the solution to the individual's problem for each value of θ and each path of prices, we use the distribution of θ, $G(\theta)$, to construct aggregates of human and physical capital. We embed our human capital model into an overlapping generations framework in which the population at any given time is composed of $\bar{a}$ overlapping generations, each with an identical ex-ante distribution of heterogeneity, $G(\theta)$.

Human capital of type S is a perfect substitute for any other human capital of the same schooling type, whatever the age or experience level of the agent, but it is not perfectly substitutable with human capital from other schooling levels. In our model, cohorts differ from each other only because they face different price paths and policy environments within their lifetimes. We assume perfect foresight (as used in Auerbach and Kotlikoff [2]) and not myopic expectations. Let c index cohorts, and denote the calendar date at which cohort c is born by t_c. Their first period of life is at time $t_c + 1$. Let P_{t_c} be the vector of paths for rental prices of physical and human capital confronting cohort c over its lifetime from time $t_c + 1$ to $t_c + \bar{a}$. The rental rate on physical capital at time t is r_t. The rental rate on human capital of type S is R_t^S. The choices made by individuals depend on the prices they face, P_{t_c}; their type, θ, and hence their endowment; and their non-pecuniary costs of schooling, S. Let $H_{at}^S(\theta, P_{t_c})$ and $K_{at}^S(\theta, P_{t_c})$ be the amount of human and physical capital possessed, respectively, and let $I_{at}^S(\theta, P_{t_c})$ be the time devoted to investment by an individual with schooling level S, at age a, of type θ, in cohort c.

By definition, the age at time t of a person born at calendar time t_c is $a = t - t_c$. Let $N^S(\theta, t_c)$ be the number of persons of type θ, in cohort c, of schooling level S. In this notation, the aggregate stock of employed human capital of type S at time t is cumulated over the non-retired cohorts in the economy at time t:

$$\bar{H}_t^S = \sum_{t_c = t - a_R}^{t-1} \int H_{t-t_c,t}^S(\theta, P_{t_c})(1 - I_{t-t_c,t}^S(\theta, P_{t_c}))N^S(\theta, t_c)dG(\theta)$$

where $a = t - t_c$, and $S = 1, \ldots, \bar{S}$, where $\bar{S}$, is the maximum number of years of schooling. The aggregate *potential* stock of human capital of type S is obtained by setting $I_{a_t}^S(\theta, P_{t_c}) = 0$ in the preceding expression:

$$\bar{H}_t^S(\text{potential}) = \sum_{t_c = t - a_R}^{t-1} \int H_{t-t_c,t}^S(\theta, P_{t_c})N^S(\theta, t_c)dG(\theta).$$

The aggregate capital stock is the capital held by persons of all ages:

$$\bar{K}_t = \sum_{t_c=t-\bar{a}}^{t-1} \sum_{s=1}^{\bar{S}} \int K^S_{t-t_c,t}(\theta, P_{t_c}) N^S(\theta, P_{t_c}) dG(\theta).$$

14.2.4 Equilibrium conditions under perfect foresight

To close the model, it is necessary to specify the aggregate production function $F(\bar{H}^1_t, ..., \bar{H}^S_t, \bar{K}_t)$, which is assumed to exhibit constant returns to scale. To simplify the exposition we assume in this section that there is no technical change. The equilibrium conditions require that marginal products equal pre-tax prices $R^S_t = F_{\bar{H}^S_t}(\bar{H}^1_t, ..., \bar{H}^S_t, \bar{K}_t)$, $S = 1, \ldots, \bar{S}$, and $r_t = F_{\bar{K}_t}(\bar{H}^1_t, ..., \bar{H}^S_t, \bar{K}_t)$. In the two-skill economy estimated below, we specialize the production function to

$$\begin{aligned} & F(\bar{H}^1_t, \bar{H}^2_t, \bar{K}_t) \\ &= a_3(a_2(a_1(\bar{H}^1_t)^{\rho_1} + (1-a_1)(\bar{H}^2_t)^{\rho_1})^{\rho_2/\rho_1} + (1-a_2)\bar{K}^{\rho_2}_t)^{1/\rho_2}. \quad (14.11) \end{aligned}$$

When $\rho_1 = \rho_2 = 0$, the technology is Cobb–Douglas.[9] When $\rho_2 = 0$, we obtain a model consistent with the constancy of capital's share irrespective of the value of ρ_1.

14.2.5 Linking the earnings function to prices and market aggregates

The earnings at time t for a person of type θ and age a from cohort c and schooling level S are

$$W^S_{a,t}(c, \theta) = R^S_t H^S_{a,t}(\theta, P_{t_c})(1 - I^S_{a,t}(\theta, P_{t_c})). \quad (14.12)$$

They are determined by aggregate rental rates (R^S_t), individual endowments, $H^S_{a,t}(\theta, P_{t_c})$, and individual investment decisions, $I^S_{a,t}(\theta, P_{t_c})$. The last two components depend on agent expectations of future prices. Different cohorts facing different price paths will invest differently and have different human capital stocks. An essential idea in this paper, which is absent from currently used specifications of earnings equations in labor economics, is that *utilized* skills, and not potential skills, determine earnings. The utilization rate is an object of choice linked to personal investment decisions and is affected both by individual endowments, interest rates, and aggregate skill prices. As the quantity of aggregate skill is changed, so are aggregate skill prices. This affects investment decisions, measured wages, and savings decisions.

[9] Auerbach and Kotlikoff [2] assume efficiency units so different labor skills are perfect substitutes ($\rho_1 = 1$). In addition, they assume a Cobb–Douglas aggregate technology relating human capital and physical capital ($\rho_2 = 0$).

14.3 Determining the Parameters of the Model

This section justifies our choice of parameters for the model. In contrast to the conventional practice in the applied general-equilibrium literature, we use microdata in conjunction with macrodata to obtain the estimates used in our simulation analysis. In estimating skill-specific human capital production functions, we account for individual heterogeneity in technology and endowments. In Heckman, Lochner, and Taber [28], we present a new test of the consistency of the estimates of model parameters used to generate the general equilibrium simulations. We require that the econometric procedure used to produce the micro-based parameters employed in our model (including the implicit assumptions made about the economic environment in implementing any particular econometric procedure) recover the parameters estimated from synthetic microdata sets generated by the model used to simulate the economy. We further require that our assumptions about agent expectations produce the behavior observed in our sample period.

The ideal data set for our purposes would combine microdata on employment by skill type and capital stock of firms, data on the earnings of workers and their investment in training, their life-cycle consumption, and their wealth holdings, and macrodata on prices and aggregates. With such data, we could estimate all the parameters of our model and the distribution of wages, wealth, and earnings. Using microdata joined with aggregate prices, we could estimate the parameters of the micromodel. Using the estimated microfunctions, we can construct aggregates of human capital that can be used to determine the output technology. The estimated aggregates should match measured empirical aggregates and, when inserted in aggregate technology, should also reproduce the market prices used in estimation. This self-consistency property is an important aspect of a general-equilibrium model.

Three obstacles prevent us from implementing this approach: (1) we lack information on individual consumption linked to labor earnings; and (2) we lack direct observations on investment in on-the-job training. (3) As a consequence of (2), the data on market wages do *not* reveal skill prices, as is evident from the distinction between R_t^S and $W_{a,t}^S(c,\theta)$ in equation (14.12). Since prices cannot be directly equated with wages, it would seem impossible to estimate aggregate stocks of human capital to use in determining aggregate technology.

To circumvent the first limitation, we follow practices widely used in the literature on empirical general-equilibrium models by choosing discount and inter-temporal substitution parameters in consumption that are consistent with those reported in the empirical literature and that enable us to reproduce key features of the macrodata like the capital–output ratio. In Heckman, Lochner and Taber [28], we explore the sensitivity of our simulations to alternative choices of these parameters. To circumvent the second problem, we develop methods to estimate investment time. These methods are described more fully in Heckman, Lochner and Taber [28].

To circumvent the third problem, we develop a new method for using wages to infer prices and to estimate skill-specific human capital aggregates. Since calibration methods and sensitivity analysis are widely used in applied general equilibrium analysis, we turn to our more original empirical contributions. The method is described more fully in Heckman, Lochner and Taber [28].

14.3.1 Simple methods for estimating skill prices and aggregate production technology with heterogeneous skills

We first present a method for identifying the aggregate technology and estimating skill-specific human capital stocks by combining micro- and macrodata. It exploits the insight that at older ages, changes in wages are due solely to changes in skill prices and to depreciation. Suppose that for two consecutive ages, a and $a+1$, $I^S_{a,t} = I^S_{a+1,t+1} = 0$. More ages of zero investment only help to identify skill prices, so we present a worst-case analysis. At late ages in the life cycle, $I^S_{a,t} \cong 0$ is an implication of optimality. This condition enables us to identify rental rates up to scale. Note that from the definitions (dropping the θ and c subscripts for simplicity) and from the identifying assumption at ages a and $a+1$, for older cohorts it follows that

$$W^S_{a+1,t+1} \equiv R^S_{t+1} H^S_{a+1,t+1} = R^S_{t+1} H^S_{a,t}(1-\sigma^S)$$

where σ^S is the rate of depreciation for skill S. It is assumed that deflated real wages are used.[10] Then, it follows that

$$\frac{W^S_{a+1,t+1}}{W^S_{a,t}} = \frac{R^S_{t+1}(1-\sigma^S)}{R^S_t}.$$

Normalize $R^S_0 = 1$. In the absence of measurement error in wage ratios, we can identify $R^S_0, ..., R^S_T$ from a time series of cross-sections of individuals ages a and $a+1$ up to scale $(1-\sigma^S)^t$, where t is the time period; i.e., we can identify $R^S_0, R^S_1(1-\sigma^S), ..., R^S_T(1-\sigma^S)^T$. If the ratios have mean zero measurement error, we can estimate the ratios of skill prices without bias. With these rental rates in hand, we can recover utilized human capital stocks up to scale.

Denote WB^S_t as the total wage bill in the economy at time t for schooling level S. This is available from the aggregate information in a time series of cross-sections on wages. Then, it is possible to estimate the aggregate utilized human capital stock of type S up to scale from the equation

$$\frac{WB^S_t}{(1-\sigma^S)^t R^S_t} = \frac{\sum_{a=1}^{A} H^S_{a,t}(1-I^S_{a,t})}{(1-\sigma^S)^t} = \frac{\bar{H}^S_t}{(1-\sigma^S)^t}$$

[10] We use the GNP deflator.

so that we can generate human capital stocks at time t up to scale $(1-\sigma^S)^t$. If there is no depreciation ($\sigma^S = 0$), then we can identify the entire time series of skill prices and stocks with only the initial normalization $R_0^S = 1$. Since this assumption cannot be rejected by the data (Browning, et al. [9]) and it greatly simplifies the analysis, we proceed under this assumption and show how to recover the aggregate technology (see Heckman, Lochner and Taber [28] for the more general case where $\sigma^S \neq 0$).

14.3.2 Identifying the aggregate technology

From aggregate production technology (14.11), and the assumption of market clearing in competitive markets, we obtain the first-order conditions that generate skill prices. To simplify the derivation, we first define

$$Q_t = \left[a_1(\bar{H}_t^1)^{\rho_1} + (1-a_1)(\bar{H}_t^2)^{\rho_1}\right]^{1/\rho_1},$$

so the aggregate technology can be written in terms of this composite:

$$F(\bar{H}_t^1, \bar{H}_t^2, \bar{K}_t) = \left[(1-a_2)Q_t^{\rho_2} + a_2\bar{K}^{\rho_2}\right]^{1/\rho_2}.$$

Let $M_t = [(1-a_2)Q_t^{\rho_2} + a_2\bar{K}_t^{\rho_2}]^{(1-\rho_2)/\rho_2}$. Then in this notation

$$r_t = M_t a_2 \bar{K}_t^{\rho_2 - 1}$$

$$R_t^1 = M_t(1-a_2)a_1 Q_t^{\rho_2-\rho_1}(\bar{H}_t^1)^{\rho_1-1}$$

$$R_t^2 = M_t(1-a_2)(1-a_1)Q_t^{\rho_2-\rho_1}(\bar{H}_t^2)^{\rho_1-1}.$$

The log ratio of the last two optimality conditions is

$$\log\left(\frac{R_t^2}{R_t^1}\right) = \log\left(\frac{1-a_{1t}}{a_{1t}}\right) + (\rho_1 - 1)\log\left(\frac{\bar{H}_t^2}{\bar{H}_t^1}\right) \tag{14.13}$$

where we now permit the a_1 to depend on time. We allow for linear trends in $\log[(1-a_{1t})/a_{1t}]$ so $\log[(1-a_{1t})/a_{1t}] = \log[(1-a_{10})/a_{10}] + \varphi_1 t$ where $t=0$ is the baseline period. Ordinary least squares applied to

$$\log\left(\frac{R_t^2}{R_t^1}\right) = \beta_0 + \beta_1 \log\left(\frac{\bar{H}_t^2}{\bar{H}_t^1}\right) + \varphi_1 t + \varepsilon_t \tag{14.14}$$

consistently estimates $\beta_0 = \log[(1-a_{10})/a_{10}]$, $\beta_1 = \rho_1 - 1$, and φ_1 if the measured shifters are exogenous with respect to ε_t.

To recover the other parameters, observe that from CES algebra the price of the bundle Q_t is

$$R_t^Q = \left((R_t^1)^{\rho_1/(\rho_1-1)}(a_1)^{1/(1-\rho_1)} + (R_t^2)^{\rho_1/(\rho_1-1)}(1-a_1)^{1/(1-\rho_1)}\right)^{(\rho_1-1)/\rho_1}$$

Then, we may write the log ratio of the first order conditions for Q_t and $\bar{K}_t$ as

$$\log\left(\frac{R_t^Q}{r_t}\right) = \log\left(\frac{a_2}{1-a_2}\right) + (\rho_2 - 1)\log\left(\frac{Q_t}{\bar{K}_t}\right) \tag{14.15}$$

Substituting for a_2 we may write the estimating equation for (14.15) as

$$\log\left(\frac{R_t^Q}{r_t}\right) = \log\left(\frac{a_{20}}{1-a_{20}}\right) + (\rho_2 - 1)\log\left(\frac{Q_t}{\bar{K}_t}\right) + \varphi_2 t + v_t. \tag{14.16}$$

Instruments are required for estimation of equations (14.14) and (14.16) if there are demand shocks.[11] A test of $\rho_2 = \rho_1 = 0$ is a test of the Cobb–Douglas specification for aggregate technology using the constructed skill prices and aggregates obtained from the first stage estimation procedure.

14.3.3 Estimating human capital production functions

We estimate the human capital production parameters in (14.5) using NLSY data on white male earnings for the period 1979–93. We follow the literature (Heckman, [22, 23]) and assume that interest rates and the after-tax rental rates on human capital are fixed at constant but empirically concordant values.[12] This ignores the price variation induced by technological change. A remarkable finding of our previous research (Heckman, Lochner, and Taber [28, Appendix B]) is that this misspecification of the economic environment has only slight consequences for the estimation of the curvature parameters of the human capital technology, at least within the range of skill price variation generated by our model of the US economy. Misspecification in share parameters is compensated for by calibration.

We take the real after-tax interest rate r as given and fix it at 0.05, which is in the range of estimates reported by Poterba [51] for our sample period. To produce these estimates, we treat R_t^S as a constant[13] (normalized to 1) for all skill services, following a tradition in the literature. We set $\sigma^S = 0$, an estimate consistent with what is reported in the literature (see Browning, et al. [9]). This estimate is also consistent with the lack of any peak in life

[11] Observe that in forming R_t^Q and Q_t and using them in subsequent estimation, one should correct for parameter estimation error in subsequent steps in order to produce correct standard errors.

[12] One free normalization for a steady-state economy is that $R_t^S = k > 0$ for all t. Below we use $k = 2$.

[13] Recall that for within-sample age ranges of cohorts we know R_t^S using the estimates from the procedure described in subsection 14.3.1. However, for out of sample age ranges these rental rates are not known. As reported in Heckman, Lochner and Taber [28] misspecification of the rental rates does not affect the estimates of the curvature parameters.

cycle wage–age profiles (Meghir and Whitehouse [47]). We use a tax rate of 15 percent which is consistent with the effective rate over our sample period reported by Pechman [50]. For each ability-schooling (θ, S) type, the relevant parameters are $(\alpha^S, \beta^S, A^S(\theta), H^S(\theta))$. We assume that, conditional on measured ability, there is no dependence on unobservables across the schooling and wage equations.[14]

Solving the human capital model backward is easier computationally than simultaneously solving it forward and backward. (We know the terminal value for $\partial V^S_{a_R+1}/\partial H^S_{a_R+1}$ and the initial value for $H^S_0(\theta)$.) Rather than parameterizing the model in terms of initial human capital, we parameterize it in terms of terminal human capital, denoted $H^S_{a_R}(\theta)$. Since there is a one-to-one relationship between initial human capital and terminal human capital, this parameterization is innocuous.

For any particular set of parameters $(\alpha^S, \beta^S, A^S(\theta), H^S_{a_R}(\theta))$, we can simulate the model and form log wage profiles as functions of these parameters. Let the measured wages be $W^*_{i,a} = W_{i,a} + \eta_{i,a}$ where $\eta_{i,a}$ is a mean zero measurement error assumed independent of the true value. For each ability type (θ), we estimate the model by nonlinear least squares, minimizing over individuals (denoted i):

$$\sum_i \sum_a (W^*_{i,a} - W_a(\alpha^S, \beta^S, A^S(\theta), H^S_{a_R}(\theta)))^2$$

and constraining $0 < \alpha^S < 1$, $0 \leq \beta^S \leq 1$ and $A^S(j) > 0$ for two schooling groups. $S = 2$ if a person has completed one year of college; $S = 1$ otherwise. Estimates from this model are presented in the top panel of Table 14.1. Level $\theta = 1$ is the lowest quartile of AFQT ability while $\theta = 4$ is the highest quartile. The estimates of α^S and β^S are quite similar for the two schooling groups. The value of the productivity parameters $A^S(\theta)$ usually increase with AFQT, suggesting that more able people are more efficient in producing human capital. The terminal levels of human capital are higher for college-educated individuals than for persons attending high school.

In Heckman, Lochner and Taber [28], we present a sensitivity analysis of our estimates of the model to misspecification of heterogeneity and the economic environment. Except for the case where we estimate the model under one interest rate and simulate the model under another, the model is surprisingly robust to misspecifications of the economic environment.

In the top panel of Table 14.2, we present the initial levels of human capital for each schooling type by solving the model backward, given the terminal condition estimates reported in Table 14.1. For all four ability types, the job

[14] We used the AFQT score in the NLSY as our measure of ability, θ.

Table 14.1: Estimated parameters for human capital production function and schooling decision (standard errors in parentheses)

	Human capital production	
	$H^S_{a+1} = A^S(\theta)\left(I^S_a\right)^{\alpha_S}\left(H^S_a\right)^{\beta_S} + (1-\sigma^S)H^S_a, \quad S = 1,2$	
	High school ($S = 1$)	College ($S = 2$)
α	0.945 (0.017)	0.939 (0.026)
β	0.832 (0.253)	0.871 (0.343)
$A(1)$	0.081 (0.045)	0.081 (0.072)
$H_{a_R}(2)$	9.530 (0.309)	13.622 (0.977)
$A(2)$	0.085 (0.053)	0.082 (0.074)
$H_{a_R}(2)$	12.074 (0.403)	14.759 (0.931)
$A(3)$	0.087 (0.056)	0.082 (0.077)
$H_{a_R}(2)$	13.525 (0.477)	15.614 (0.909)
$A(4)$	0.086 (0.054)	0.084 (0.083)
$H_{a_R}(4)$	12.650 (0.534)	18.429 (1.095)
	College choice equation	
	$P(d = 1) = \Phi(-\lambda D + v(\theta))$	
	Probit parameters	Average derivatives
λ	0.166 (0.062)	-0.0655 (0.025)
$v(1)$	-1.058 (0.097)	—
$v(2)$	-0.423 (0.087)	0.249 (0.037)
$v(3)$	0.282 (0.089)	0.490 (0.029)
$v(4)$	1.272 (0.101)	0.715 (0.018)
Sample size: Persons	869	1,069
Sample size: Person years	7,996	11,626

(1) D is the discounted tuition cost of attending college.
(2) $v(\theta)$ is the nonparametric estimate of $(1-\tau)[PV^2(\theta) - PV^1(\theta)]$, the monetary value of the gross discounted returns to attending college, plus $\mu(\theta)$ defined in Section 14.3.4.
(3) $d = 1$ if attend college; $d = 0$ otherwise. Φ is the unit normal cdf.

market entry level of human capital increases with college. Except for one case, the initial level of human capital increases with AFQT across schooling groups. The one anomaly in the table is that the initial level of human capital is larger for the high school group of ability type 3 than for the high school group of ability type 4. (This is consistent, however, with the earnings profiles of young men in the NLSY displayed in Figures 14.1a and 14.1b.) The present value of earnings increases with ability for most groups.

The estimates of the human capital production function reported in Tables 14.1 and 14.2 are consistent with the Ben-Porath model ($\alpha^S = \beta^S$) that was widely used in the early literature on estimating human capital technology. The point estimates are remarkably similar to those reported by Heckman [23] for his income maximizing models ($\alpha^S = \beta^S = 0.812$) for males and are consistent with the range of estimates reported by Brown [8] for his sample of young males.[15] The estimated models fit the earnings data rather well for different schooling and ability levels (see Figures 14.1a and 14.1b).

Table 14.2: Derived parameters for human capital production function and schooling decision (in thousands of dollars)

	Human capital production	
	High school ($S = 1$)	College ($S = 2$)
$H^S(1)$	8.042 (0.094)	11.117 (0.424)
$H^S(2)$	10.0634 (0.118)	12.271 (0.325)
$H^3(3)$	11.1273 (0.155)	12.960 (0.272)
$H^S(4)$	10.361 (0.234)	15.095 (0.323)
Present value earnings 1	260.304 (3.939)	289.618 (12.539)
Present value earnings 2	325.966 (5.075)	319.302 (10.510)
Present value earnings 3	360.717 (6.352)	337.260 (9.510)
Present value earnings 4	335.977 (8.453)	393.138 (11.442)
College decision: Attend college if $(1-\tau)PV^2(\theta) - D^2 + \varepsilon_\theta \geq (1-\tau)PV^1(\theta)$, $\varepsilon_\theta \sim N(\mu_\theta, \sigma_\varepsilon)$		
σ_ε (std.deviation of ε)	22.407 (8.425)	
Nonpecuniary costs by ability level		
μ_1 (lowest ability quartile)	-53.0190 (16.770)	
μ_2 (second ability quartile)	-2.8172 (12.760)	
μ_3 (third ability quartile)	29.7712 (11.540)	
μ_4 (highest ability quartile)	-28.6494 (16.966)	

(1) $PV^i(\theta)$ is the monetary value of going to school level i for a person of AFQT quartile θ. $i = 1$ for high school; $i = 2$ for college. We assume $\tau_r = \tau_h = \tau$.
(2) ε_θ is the nonpecuniary benefit of attending college for a person of ability quartile θ.
(3) D^2 is the discounted tuition cost of attenting college.

Using our model and the assumption of no depreciation in skills, we can estimate the contribution of schooling and on-the-job training to lifetime human capital. Using an accounting framework that equates marginal and average rates of return, Mincer [48] estimates that half of all human capital formation is on the job. Using our optimizing framework which distinguishes between marginal and average rates of return, we find that the contribution of OJT to the total human capital stock is much less — on the order of 23 percent — over all ability groups.

[15] Our estimates for women are quite similar to those for males [27]. For them "β^S" and "α^S" are high and we cannot reject the Ben-Porath model as a description of their human capital production function, or that women and men have the same human capital production function. Lochner [44] reports similar estimates for a much more extensive analysis of the human capital production function with controls for heterogeneity.

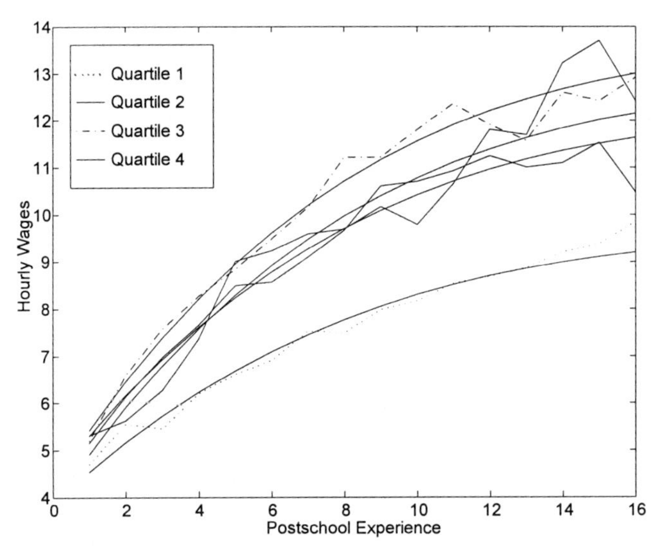

Figure 14.1a: Predicted vs. actual hourly wages (in 1992 dollars) by AFQT quartile (college category)

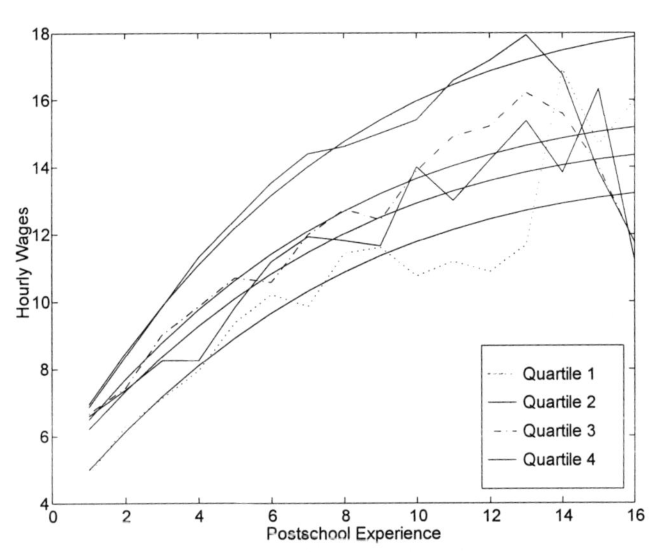

Figure 14.1b: Predicted vs. actual hourly wages (in 1992 dollars) by AFQT quartile (high school category)

14.3.4 Estimating the probability of attending college

Let D_i be the college tuition faced by individual i. The difference in utility between going to college and not going to college for individual i who is of ability type θ can be written as

$$PV_i^*(\theta) = (1-\tau)[PV_i^2(\theta) - PV_i^1(\theta)] - D_i + \varepsilon_i(\theta)$$

where τ is the tax rate, and $\varepsilon_i(\theta) \sim N(\mu_\theta, \sigma_\varepsilon^2)$. For notational simplicity we delete the S superscript. D refers explicitly to college tuition. We estimate μ_θ and σ_ε^2 by probit analysis using a two-step procedure. First, we run a probit regression of college attendance on tuition and dummy variables for each schooling-ability group for each of the seven birth cohorts in the NLSY. (The dummies estimate the difference in the after-tax valuation of schooling for each group.) The tuition variable is measured in units of thousands of dollars. We cannot reject the null hypothesis of no change in the value of attending school over the time period 1975–82 when our sample makes its schooling decisions. We report results (in the bottom panel of Table 14.1) for the case when the value of attending school is constrained to be the same across time.

The third column in the bottom panel of Table 14.1 presents average derivatives. The interpretation of the average derivative is that an increase in tuition of \$1,000 decreases the probability of attending college by about 0.07 on average. This estimate is on the high end of the range of estimates reported in the literature (see e.g., Cameron and Heckman [10] or Kane [37]). In the second stage, we transform the parameters D_i presented in Table 14.1 into the structural parameters of our model. We use the relationships $\lambda = 1/\sigma_\varepsilon$, and $v(\theta) = \{(1-\tau)[PV^2(\theta) - PV^1(\theta)] + \mu(\theta)\}/\sigma_\varepsilon$ and the estimates reported in Table 14.1 to form $[PV^2(\theta) - PV^1(\theta)](1-\tau)$. Then, we obtain

$$\mu(\theta) = v(\theta)\sigma_\varepsilon - (1-\tau)[PV^2(\theta) - PV^1(\theta)]$$

as the mean non-pecuniary return to college for a person of ability level θ. The estimates are reported in the bottom panel of Table 14.2. The only surprise in this table is the negative mean psychic cost of attending college for persons of the highest ability.

14.3.5 Estimating the technology and aggregate stocks of human capital by skill

Using the CPS data for the period 1963–93, we employ the methodology presented in Sections 14.3.1 and 14.3.2 to estimate skill prices and human capital aggregates. The data sources for the macroaggregates are presented in Appendix A. From the constructed aggregates, we estimate aggregate technology (14.11) and test for various specifications of the aggregate technology. To correct for endogeneity of inputs, we use the standard instrumental variables often used in macroeconomics: military expenditures and cohort size. OLS and IV estimates of the technology are reported in Table 14.3 for all possible combinations of the instruments.

Table 14.3: Estimates of aggregate production function estimated from factor demand equations (14.14) and (14.16), 1965–1990, allowing for technical progress through a linear trend (std errors in parentheses)

	(ρ_1)	Implied elasticity of substitution (σ_1)	Time trend (φ_1)	(ρ_2)	Implied elasticity of substitution (σ_2)	Time trend (φ_2)
OLS (base model)	0.306	1.441	0.036	–0.034	0.967	–0.004
	(0.089)	(0.185)	(0.004)	(0.200)	(0.187)	(0.007)
Percent working population < 30 and defense percent of GNP	0.209	1.264	0.039	–0.036	0.965	–0.004
	(0.134)	(0.215)	(0.005)	(0.200)	(0.187)	(0.007)
Defense percent of GNP	0.157	1.186	0.041	–0.171	0.854	–0.008
	(0.125)	(0.175)	(0.004)	(0.815)	(0.594)	(0.024)
Percent working population < 30	0.326	1.484	0.036	0.364	1.572	0.007
	(0.182)	(0.400)	(0.006)	(1.150)	(2.842)	(0.034)

The estimated elasticity of substitution between capital and the labor aggregate Q is $\sigma_2 = 1/(1-\rho_2)$ and is not statistically significantly different from one (see cols. 4 and 5 of Table 14.3.) Our estimates justify excluding capital (or the interest rate) from equation (14.1). Our model produces rising wage inequality without assuming a special complementary relationship between capital and skilled human capital — the centerpiece of the Krussell et al. [41] analysis of wage inequality and other discussions of rising wage inequality. Our estimates are consistent with the near-constancy of the capital share in the US economy but the declining share of unskilled labor (see Figure 14.2). The estimated elasticity of substitution between high-skill and low-skill labor (1.441) is remarkably close to the point estimates reported by Katz and Murphy (1.41) and Johnson (1.50). The instrumental variable estimators do not change this estimate very much. Assuming no depreciation in human capital, we estimate the skill-bias parameter $\varphi_1 = 0.036$, very close to the corresponding estimate of 0.033 reported by Katz and Murphy.

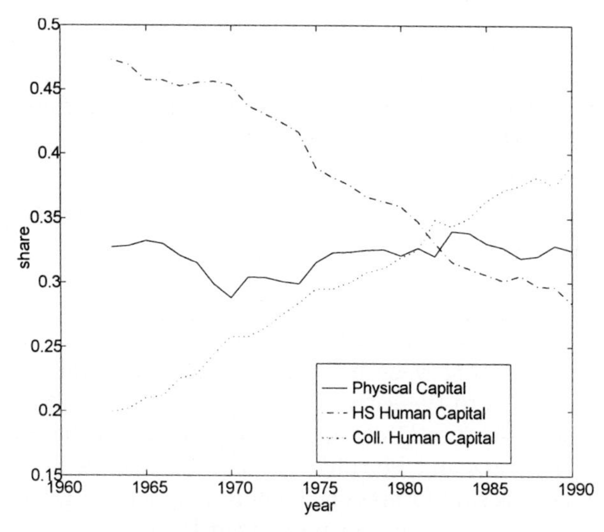

Figure 14.2: Labor and capital shares over time

14.3.6 Calibrating the model

Given our estimates of the human capital production function, we choose an initial steady state which is consistent with the assumptions used to obtain those estimates in the NLSY data. Given a tax rate τ ($= 0.15$) that is suggested by Pechman [50] as an accurate approximation to the true rate over our sample period once itemizations, deductions, and income-contingent benefits are factored in, we calibrate the aggregate production parameters to yield a steady state after-tax interest rate of 0.05 and pre-tax rental rates on human capital of 2. These values are consistent with those used in estimation of the human capital production parameters. Since human capital is measured in terms of hourly wages, earnings from our simulations are annual income measured in thousands of dollars if agents work 2,000 hours per year. The calibration yields shares that are consistent with the NLSY white male sample we use.

It is necessary to make some assumptions about time preference and the inter-temporal elasticity of substitution for consumption in order to determine savings rates and aggregate capital. Given the levels of human capital investment implied by the estimates of our model and levels of initial assets for each individual, we obtain consumption and savings under the assumption that $\delta = 0.96$ and $\gamma = 0.1$. In order to determine initial levels of assets, we partially redistribute physical capital from retiring workers (cohort a_R) to the cohort just entering the labor market so that the capital–output ratio in the economy is 4.[16] This calibration procedure yields an initial steady state which

[16] For each year, transfer X is taken from all workers at retirement age a_R, and the total

emulates the NLSY data and central features of the macroeconomy. In Appendix C of Heckman, Lochner and Taber [28], we test the sensitivity of our simulations to the choice of these parameter values, and find quantitatively similar results across a variety of specifications.

14.4 A Dynamic General-Equilibrium Model of Rising Wage Inequality in the US Economy

In our previous papers, we use our estimated technology and human capital accumulation equations and determine whether or not we can explain the central features of rising wage inequality in the US labor market over the period of the late 70s to the early 90s using a model of skill-biased technical change. The novel feature of our model, in contrast to the model of Bound and Johnson [7], Katz and Murphy [38] and Krussell et al. [41] is that we allow for endogenous skill formation. Ours is a model of a gradual shift in the skill bias of technology to a higher permanent level.

Since we use a human capital production function fitted on young white males from the NLSY, our empirical model may not capture all features of the US labor market. Since our estimates of the human capital production function for females are virtually identical to those of males, we do not think that this is a major source for concern (see the estimates in Heckman, Lochner and Taber [28]). Of potentially greater concern is our use of earnings data fit on the early years of the life cycle for a recent cohort of workers. Since most human capital investment takes place early in the life cycle, we capture the main portion of such investment. One final concern is the possibility of cohort effects in terms of endowments, ability, and human capital investment functions which we ignore in this paper, although our model produces an endogenous cohort effect in a period of transition to a new level of technology.

We consider a permanent shift in technology toward skilled labor using the estimated trend parameter $\varphi_1 = 0.036$ reported in Table 14.3 as our base case. We start from an initial steady state and suppose that the technology reported in Table 14.3 begins to manifest a skill bias in the mid 70s. Greenwood and Yorukoglu [19] claim that 1974 is a watershed year for modern technology. Following their suggestion about the timing of the onset of technical change, $\log[a_1/(1-a_1)]$ is assumed to decline linearly at 3.6 percent per year (as estimated above) starting in the mid-70s and continuing for 30 years. Shifts of longer and shorter duration produce qualitatively similar simulation results within the time period we analyze (see Heckman, Lochner and Taber [28, Appendix C]).

In this paper, we explore wage dynamics in an economy with open capital

amount is equally distributed to all individuals (irrespective of ability) of age 1 in that period. For the simulations reported in this paper, $X \approx \$30,000$.

markets. We hold the after tax interest rate constant at 0.05 for all time periods, so that changes in the supply of skill and "local" savings rates do not affect the price of capital. Capital is assumed to flow in and out of the economy to maintain the constant world interest rate. However, the labor market in this economy is closed, so that the prices of skills are determined nationally by the derivative of the aggregate production technology given the national supply of skill.[17] Factor price equalization theorems do not apply since we have one good and three factors.

To compute the general equilibrium of our model and the implied transition paths, we use the methodology of Auerbach and Kotlikoff [2, p. 213]. Starting from an initial steady state calibrated using the parameters of preferences and technology specified or estimated in Section 14.3, we examine the transitional dynamics to the new steady state, imposing the requirement that convergence occurs in 200 periods or less.[18] Agents make their schooling and skill investment decisions under full rational expectations about future price paths, but they are surprised by the change in technology.

In the period immediately after the introduction of technological change, the price of skilled human capital increases while that of unskilled human capital decreases (see Figure 14.3). This occurs because short-term demand curves for skills shift more quickly than short-term supply curves which take time to shift because the stock of college graduates is slow to grow. This produces a rising college–high school wage differential (see Figure 14.4). The "rate of return" in a Mincer regression increases 40–60 percent over the 10–15 year period after the technology shock begins (see Figure 14.5).[19]

While not shown in the 30-year graphs we present,[20] the skill price paths eventually converge again (though not completely) as aggregate technology and demand for skills stop shifting, while short-term supply curves for skill continue to shift outward until skill supplies settle into their new steady state values. Consider the price path for college skill. The price rises initially as the demand for skill rises quickly, while the number of college graduates rises more slowly. Therefore, short-term demand curves shift out quickly, while supply curves move slowly. Once technology stops changing, demand curves for skill remain fixed, while supply curves continue to shift right as more college graduates enter the market. This causes the college skill price to fall until it reaches its new steady state level, inducing what looks like a cobweb

[17] National prices and quantities refer to the economy of interest — the US — while world prices and quantities are determined outside the model. The distinctive feature of an open economy model is that the interest rate is set in world markets, not that it is constant. For simplicity, we fix the interest rate.

[18] Our models always converge in less than 200 periods so increasing the number of periods would not affect the transition path.

[19] The "rate of return" is the coefficient of schooling in a regression of log earnings on schooling, experience, and experience squared.

[20] Graphs of the full transition are available on request from the authors.

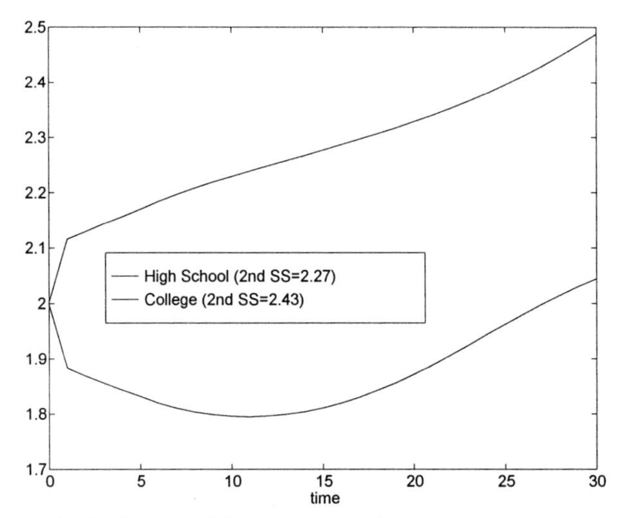

Figure 14.3: Prices of human capital for the open economy

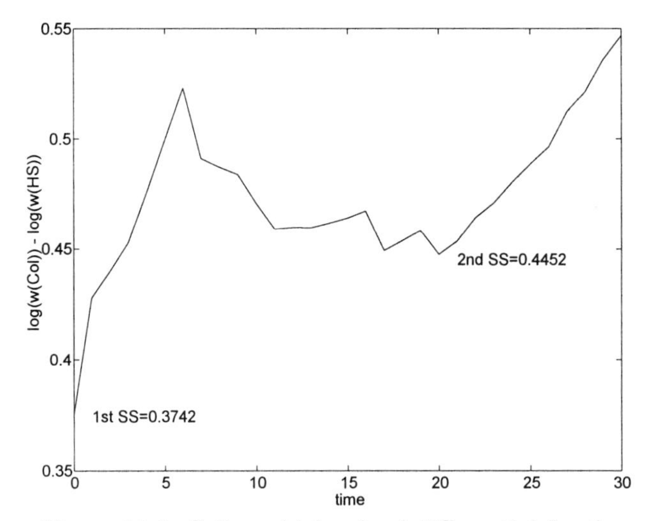

Figure 14.4: College–high school differential for the open-economy model

in a model with perfect foresight and privately optimal skill supplies. The "oversupply" of human capital in the process of adjustment is privately optimal because long-lived agents harvest both the returns when skill is scarce and when it is in glut and make a sufficient return in the period of scarcity to offset the period of glut. Figure 14.6 graphs the trend in the aggregate stock of potential and utilized skills produced from our model. Investment in human capital creates the wedge between the two measures of human capital stock.

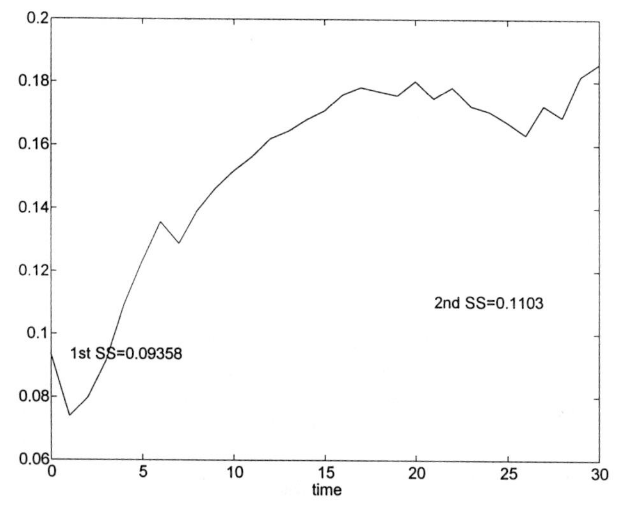

Figure 14.5: Cross-section Mincer rates of return for the open-economy model

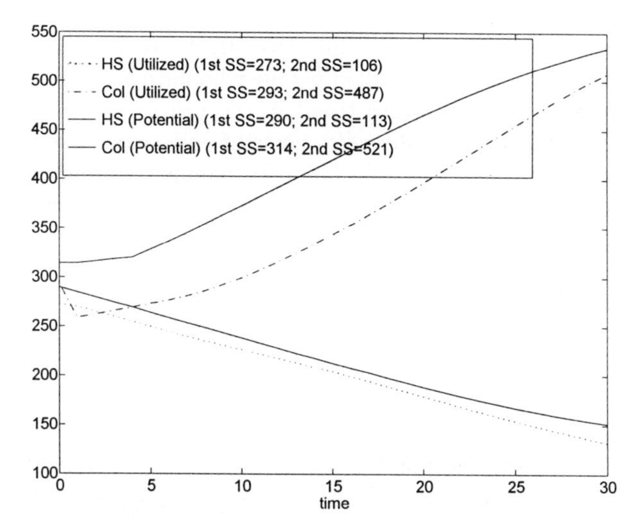

Figure 14.6: Aggregate human capital and human capital utilized for the open economy model

An important feature of our description of the economy of the late 70s and early 80s is that the movement in wage differentials differs from the movement in price differentials, especially for younger workers (see Figure 14.7 where wage differentials at different ages are compared to price differentials). This phenomenon is a consequence of the economics embodied in equation

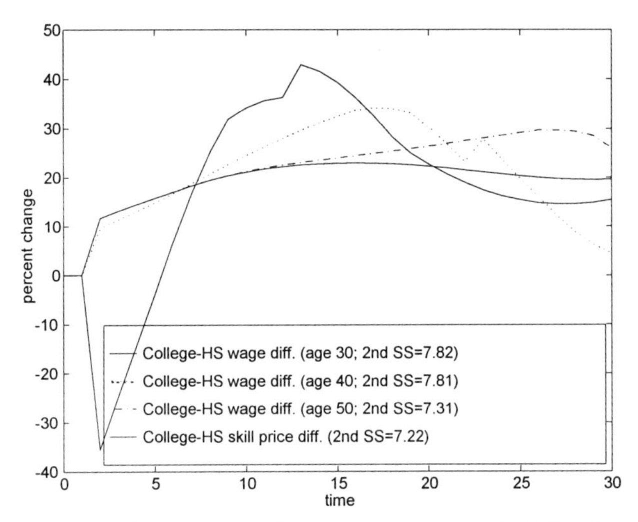

Figure 14.7: Percentage changes in wages and skill prices in the open-economy model

(14.12). In response to the rise in skill prices in the years following the onset of technological change, highly skilled people who have left college invest more on-the-job (foreseeing the upward trend in the price of their skill) and then curtail their investment as opportunity costs of investment rise relative to the payoff. These effects are especially large for younger workers who are more active investors. Cohorts of young skilled workers entering the market after the onset of technical change invest substantially less than earlier cohorts (see Figure 14.8). The story is quite different for low-skill workers. For these workers, the price of skill initially declines and then rises, so new cohorts of young workers invest slightly less in the early years of the transition. Late cohorts invest more as the price of low-skill human capital turns upward (Figure 14.9). As the price of their skill begins to converge to its new steady-state level, still later cohorts reduce their investment.

This differential response in investment by skill groups over time explains the evidence for the 80s presented in Katz and Murphy [38, Table I] that the measured skill differential by education increases more for young persons than it does for older persons (this assumes that the onset of the technology shift is in the mid-70s, as claimed by Greenwood and Yorukoglu [19]). Additionally, Figure 14.7 shows the wage differential declining for young workers but rising for older workers during the first periods of the transition. In the first phase of the transition, differential investment by skill groups narrows the college–high school wage gap for young workers. For older workers, however, the rise in the skill premium is enough to offset changes in investment, so the wage gap rises for them. On average, the wage gap between skilled and unskilled labor

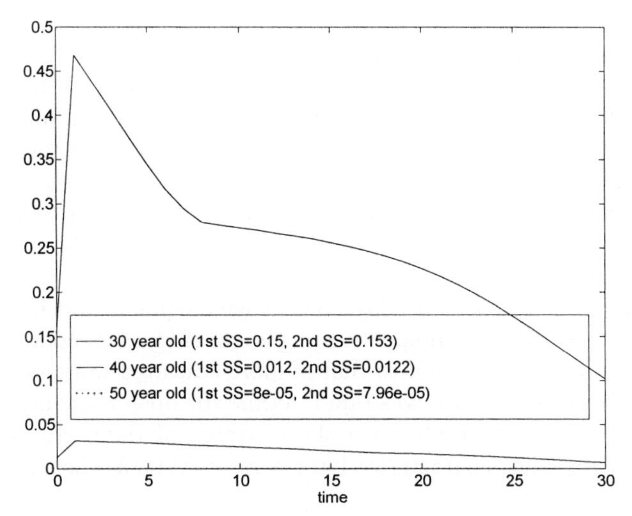

Figure 14.8: Proportion of time spent investing on the job for college educated workers in the open-economy model

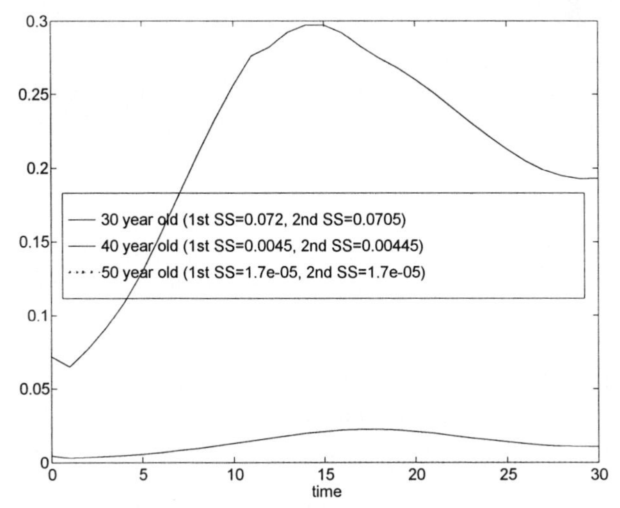

Figure 14.9: Proportion of time spent investing on the job for high school educated workers in the open-economy model

increases over the first six years of the transition as skill prices diverge and the investment differential narrows. As investment by low-skill workers continues to rise, the college–high school wage gap begins to decline for about 15 years, at which time it starts to rise again. Both changes in investment behavior and skill prices are responsible for observed patterns in the skill differential.

These results differ from the closed-economy version of our model, where only two trends in the wage gap are present after the onset of skill-biased technical change: a short decline arising from increased investment by college workers, which more than offsets the rise in their skill price, and then a sustained rise (as the investment differential narrows and skill prices diverge). This difference between the models occurs because in the closed-economy version, interest rates rise and disproportionately choke off the rise in investment for high school workers. The closed-economy version of the model is more consistent with the evidence reported by Bartel and Sicherman [4]. They find that over the period of the mid-80s, when wage inequality was increasing substantially, investment in company training increased for less-educated workers, both absolutely and compared to that of more-educated workers in industries where technological progress was rapid. While the open-economy version presented here yields this result for a short stretch of the early period of the transition, it does not characterize later periods whereas the closed-economy model does.

The model produces a large jump in college enrollments with the onset of technical change. This is an artifact of our perfect foresight assumption. A model in which information about skill bias disseminates more slowly would be more concordant with the data.[21]

While both investment in on the job and schooling increase during early years of the transition, the amount of human capital per worker declines for each skill type in the long run. In the closed-economy version, there is a phase where more people attend college, but both college and high school workers invest less on the job. Thus, movements in skill investment at the extensive margin may differ from those made at the intensive margin.

It is interesting to examine the self-correcting properties of this equilibrium. In response to the new technology regime, the standard deviation of log wages initially rises 75–100 percent but then converges to its original steady state value. The model also explains rising wage inequality at different percentiles of the wage distribution. However, this phenomenon is transient (see Figure 14.10). Wage inequality initially rises quickly for college graduates, but it then declines to approximately its initial steady state value. The story is quite different for high school graduates. For this group, inequality is stable for almost 30 years, at which point within-group inequality begins to rise steeply. On this point, our model is at odds with the stylized facts about wage inequality.[22]

[21] An alternative way to generate a more gradual response in educational enrollment is to endogenize tuition, recognizing that college is skilled-labor intensive so that as skill prices increase, the cost of schooling rises.

[22] Our model is consistent with the more recent evidence. Krueger [personal communication, 1997] suggests that rising wage inequality within narrowly-defined skill groups is no longer increasing for all skill groups. When we alter the model to account for migration

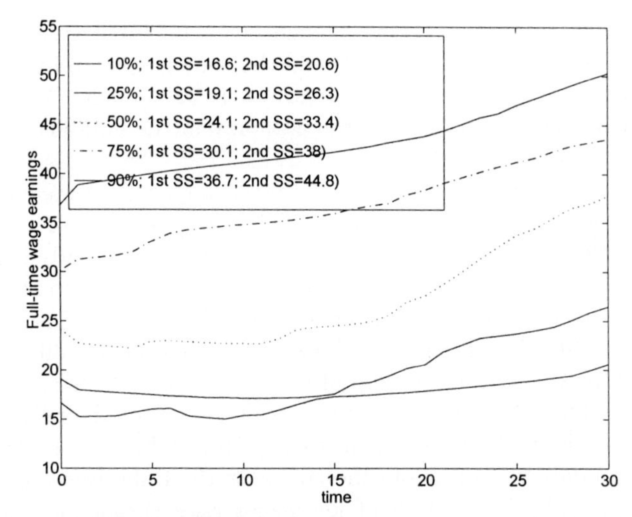

Figure 14.10: Percentiles of cross-section wage distribution: Open-economy case for 30-year trend

The models of Krussell, et al. [41] and Greenwood and Yorukoglu [19] abstract from heterogeneity within skill groups and cohorts and cannot explain the within-skill-class rise in inequality. Their models cannot explain the flattening of wage–experience profiles for high-skill groups or the steepening of wage–experience profiles for low-skill groups. The analyses of Caselli [13] and Violante [59] cannot explain this phenomenon either. For Caselli, individuals work only one period, so his model offers no prediction about experience–wage profiles. Violante's model predicts that when a new technology arrives, the returns to skills learned on previous technologies fall, leading to a decline in the slope of wage–experience profiles for low-skill workers. This prediction is grossly at odds with the data. However, both Caselli's and Violante's models explain the rise in wage inequality within narrowly-defined skill cells. For Violante, this is a consequence of the matching of workers to vintages, accelerated technical change, and induced labor turnover. For Caselli, this is due to increased sorting of high-skill labor with capital, where skill is endogenously determined.

14.4.1 Accounting for the baby boom

As a test of our model, we ask how well we match the past 35 years of US wage history using the aggregate technology and human capital accumulation equations estimated in this paper. We consider an economy in which an

---

of low-ability workers in the economy, we can produce widening wage inequality within all skill groups. However, the required increase in migration is implausibly large.

episode of skilled-biased technical change ($\varphi = 0.036$) begins in 1960 and continues for 30–40 years. To this, we add the demography of the Baby Boom in which cohort sizes increased by approximately 32 percent. We assume that Baby Boom cohorts begin to enter the economy in the mid-60s and continue for a period of 15 years.

Figure 14.11 presents the simulated college–high school wage differential. It captures the essential features of recent US wage history. The differential increases in the 60s, decreases in the 70s and rises again the 80s and 90s. The simulated model predicts that college enrollment rates jump up in the 60s and decline in the 70s and 80s. Predicted real wages of high school graduates fall in the 60s and 70s but rise in the 80s and 90s. The real wages of college graduates rise, fall, and then rise in the period of the 60s, 70s and 80s respectively. The college–high school wage differential at age 30 rises over the period 1963–74, decreases until the mid-80s and then rises again (see Figure 14.12). However, the closed-economy version of our model tracks the Baby Boom more closely. Overall, the standard deviation of log wages rises throughout most of the 30-year period.

Table 14.4 summarizes the properties of our model in a format comparable to Table I of Katz and Murphy [38]. The college–high school wage differential rises in the 60s, falls in the 70s and rises in the 80s in a fashion that mirrors the evidence reported by Katz and Murphy (Table I). Income inequality measured by the standard deviation in log wages increases over all decades, although the largest change is in the 70s. These basic trends are consistent with those of our closed-economy version of the model.

Table 14.4: Simulated changes in wages and wage inequality from 1960–90, includes the estimated trend in technology and entrance of baby boom cohorts from 1965–80 (multiplied by 100)

	College–HS log wage difference	Mean HS log wage		Mean college log wage		Standard deviation of log wages		
		Age 25	Age 50	Age 25	Age 50	HS	College	All
1960–70	7.22	–36.39	–8.43	1.80	0.13	3.16	2.54	4.88
1970–80	–4.38	6.17	–0.74	–14.00	–3.61	20.85	3.12	13.28
1980–90	7.90	7.83	1.66	17.98	–1.04	13.54	–11.48	7.84
1960–90	10.74	–22.39	–7.51	5.78	–4.51	37.55	–5.82	25.99

With the same basic ingredients of human capital investment and aggregate technology, we explain 35 years of US wage history, assuming that skill-biased technology starts around 1960 and continues for 30–40 years and the Baby Boom cohorts enter the work and schooling economy in the mid-60s. The expansion in college enrollment in the 60s can be explained by basic economic forces and not as a consequence of generous tuition policies. In fact, as we show in Section 14.6, tuition policies are likely to have small effects on enrollment and wage inequality once skill prices are allowed to adjust and rational agents respond to these adjustments in making their schooling decisions.

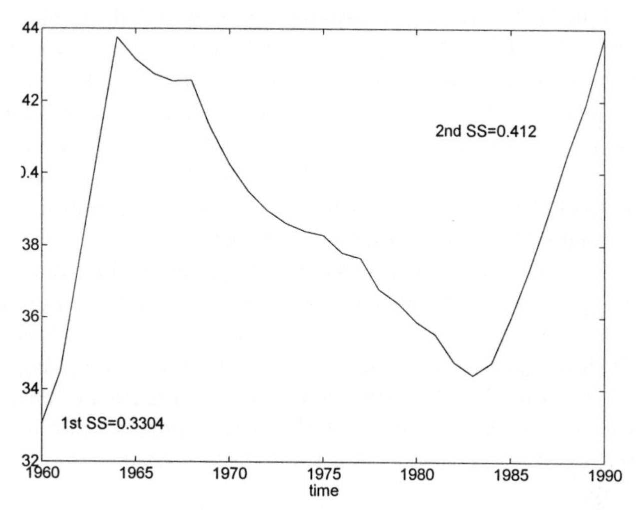

Figure 14.11: Baby boom case: College–high school wage differential — Open-economy model
Baby boom (expansion of cohort size by 32%)
between years 1965–80 and technology bias starting in the late 50s

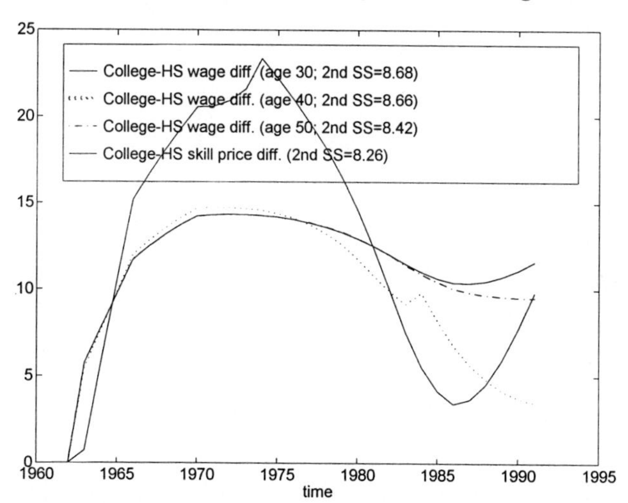

Figure 14.12: Baby boom case: Percentage changes in wage rates and skill prices from initial base state for an open economy
Baby boom (expansion of cohort size by 32%)
between years 1965–80 and technology bias starting in the late 50s

14.5 Comparisons Between Cohorts and Cross Sections in Utility Levels and in Mincer "Rates of Return"

One benefit of an overlapping generations model is that it enables us to compare cross sections with cohort paths, which are economically more interpretable for conducting welfare analysis and for making policy recommendations about human capital investment decisions for new cohorts entering the labor market. We use our model to perform systematic generational accounting in the fashion pioneered by Auerbach and Kotlikoff [2]. Figures 14.13a–d show how overall lifetime utility, lifetime utility by ability, and lifetime utility by ability and schooling type change over cohorts.

The widely used Benthamite measure of aggregate lifetime utility (obtained by summing over the utilities of all persons in the economy at a point in time) rises for cohorts entering the labor market less than 30 years prior to the onset of technological change and jumps up even more for cohorts entering the labor market immediately after the shock (see Figure 14.13a). The date of onset of the technology change is at the time cohort zero enters the labor market. Later cohorts gain even more than those entering immediately after the shock.

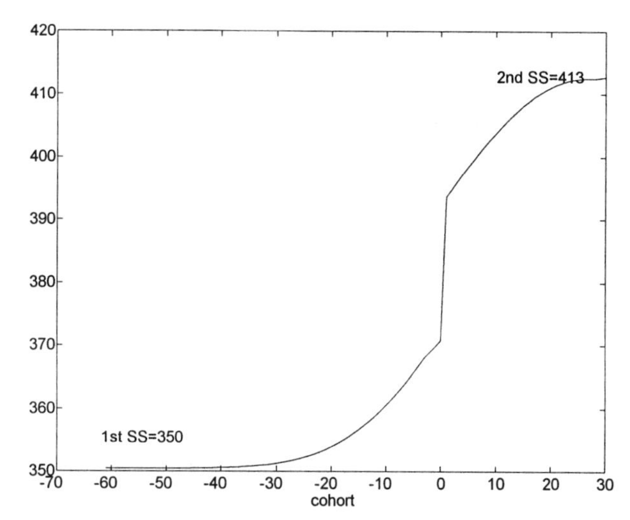

Figure 14.13a: Weighted average of lifetime utility by cohorts in the open-economy case

Disaggregating by ability groups (Figure 14.13b), higher ability workers gain from the new economy both in the long run and in the short run. High-ability persons who enter the economy before the technology shock do somewhat better than their predecessors. Low-ability persons entering the economy before the technology shock do slightly worse. Low-ability cohorts born before the onset of the technology change are hurt for substantial periods of time after the shock occurs. In the long run, however, cohorts of workers of all ability levels are better off.

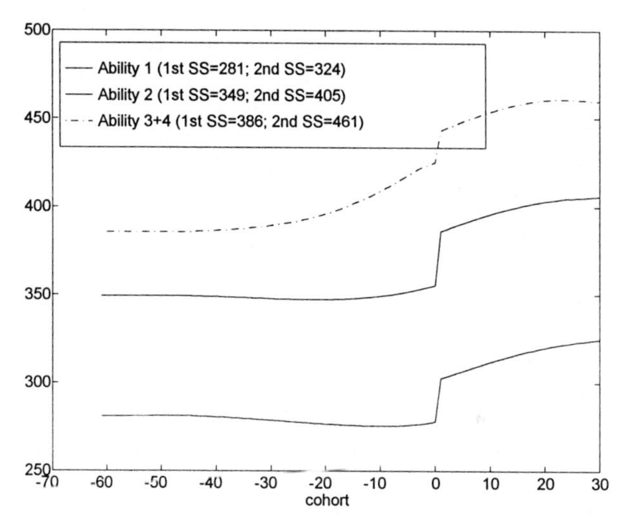

Figure 14.13b: Weighted average of lifetime utility of cohorts by ability group in the open-economy case

Figures 14.13c and 14.13d report results disaggregated by ability and education. For cohorts born after the shock, the utility path for high school-educated workers declines but recovers until it reaches a new higher level (see Figure 14.13c). Cohorts entering the market before the shock have lower utility than the predecessor and successor cohorts. In the new steady state, cohort utility levels are higher. For college-educated persons, the story is different and is not entirely the mirror image of the case for high school graduates (see Figure 14.13d). High-skill persons educated before the advent of technology change capture a large rent due to the unanticipated rise in skill prices. Successor cohorts do not fare as well, though they are still better off than if the economy had remained in the initial steady state. Their lower mean utility can be attributed to the strong distaste for college of the new college entrants. They now attend college, because the decline in their earnings in the unskilled sector is even greater than the tuition and psychic costs of attending college.[23]

In the closed-economy version of our model, both high school and college graduates entering the labor market within 10–15 years after the onset of technological change fare worse than their predecessors and successors. In the

[23] The discontinuity in the utility paths for college graduates arises from the discontinuity in college attendance induced by the onset of technology change and by the greater psychic and tuition costs of attending school by the marginal college entrants.

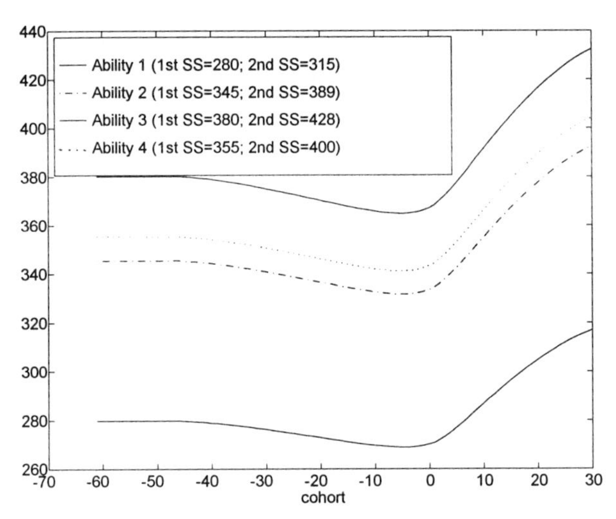

Figure 14.13c: Weighted average of lifetime utility by cohorts by ability and education: High school workers in the open-economy case

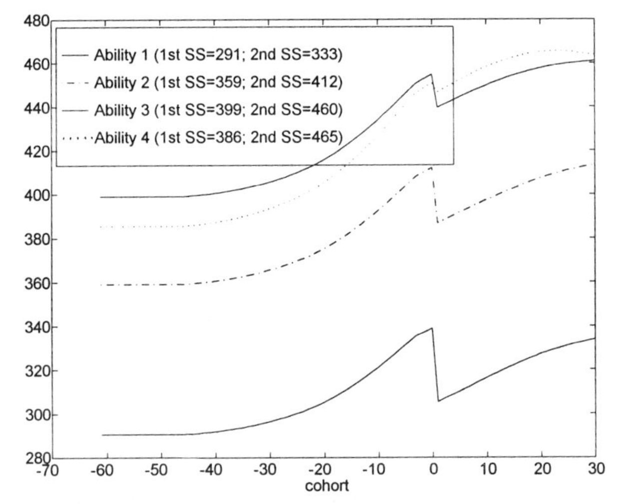

Figure 14.13d: Weighted average of lifetime utility by cohorts by ability and education: College workers in the open-economy case

open-economy version of our model, all cohorts fare much better throughout the transition, and those born after the transition begins fare better than their predecessors. That is because in the open-economy version of our model capital inflows keep the interest rate from rising and choking off investment in skills. The lower interest rate particularly favors high school graduates born early in the transition since they want to borrow and invest heavily as they see the price of their skill rising after a few years into the transition.

Evidence reported by MaCurdy and Mroz [46] and Beaudry and Green [5] that cohorts entering the labor market immediately after the start of technological change do worse than predecessor cohorts is consistent with our closed-economy version but is at odds with the open-capital market version presented here. This is another strike against the open-economy version. Figure 14.14 shows the lifetime earnings of various cohorts of college and high school graduates. Each successive cohort of college graduates earns more than its predecessor until almost thirty years after the change in technology began. High school cohorts born near the beginning of the transition fare the worst in terms of lifetime earnings.

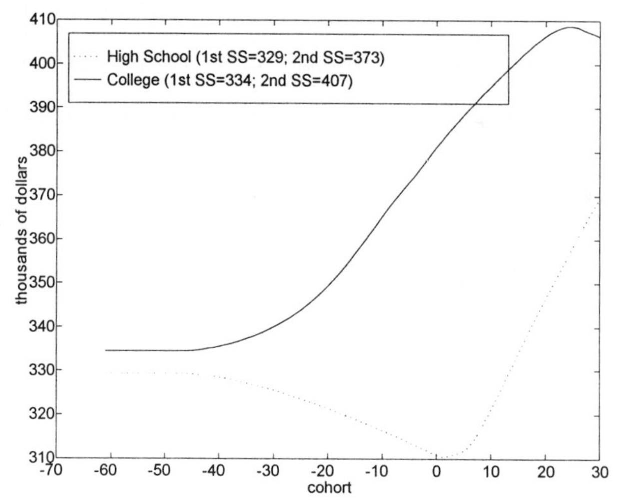

Figure 14.14: Cohort discounted lifetime earnings for the open-economy model

The important role of non-pecuniary costs in explaining college attendance accounts, in part, for the gap between the after-tax opportunity cost of capital (5%) and the conventional cross-section Mincer "return" which ranges from 7.5–18 percent (see Figure 14.5). Non-pecuniary components are 15 percent of the total cost to college attendance for the most able. Also, observe the gap between the cohort "rate of return" (Figure 14.15) and the cross-section rate (Figure 14.5). During early years of the transition, the estimated cross-section "rate of return" is substantially lower than the "rate of return" experienced by any entry cohort. However, in later periods of the transition, the opposite is true. Cross-sectional "rates of return" are not appropriate guides to educational investments for entering cohorts, although they are often used that

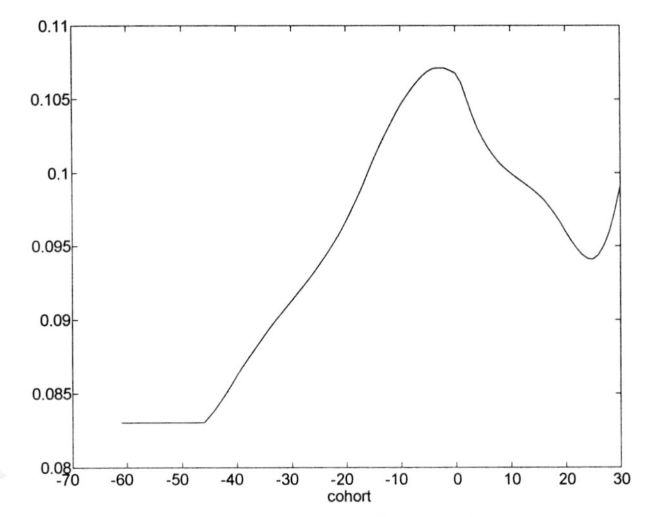

Figure 14.15: Cohort Mincer coefficients for open-economy case 30-year trend

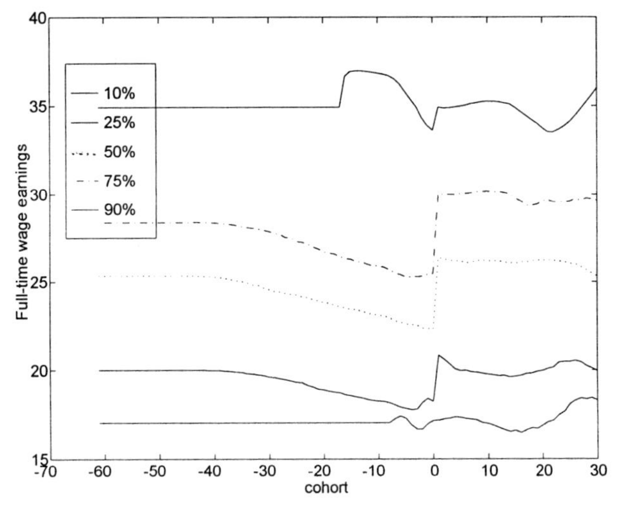

Figure 14.16: Percentiles of cohort wage distributions: Open-economy case for 30-year trend

way. This points to an important weakness in the conventional method of evaluating tuition and other policies, in addition to the more familiar problems that (a) monetary rates of return are not true rates of return (because of psychic benefits), (b) steady-state general-equilibrium adjustments resulting from policies are typically ignored, and (c) taxes are required to raise revenue. During periods of transition to a new skill regime, cross-section rates of return to education present an inaccurate account of the "rate of return" any single cohort can earn. For example, the Mincer coefficient in year 15 is around 16 percent (Figure 14.5) while the Mincer return for the cohort entering in year 15 is 6.5 percent (see Figure 14.15). We discuss this discrepancy further in the next section.

Observe, finally, that within-cohort wage inequality, presented in Figure 14.16, often moves in the opposite direction from aggregate wage inequality. In years immediately after the technology shock, entry level cohorts experience declining wage inequality even though aggregate wage inequality is rising (compare Figures 14.10 and 14.16).

14.6 Understanding Mincer Rates of Return

It is instructive to compare the rate of return in our model with that reported in the standard literature in labor economics: the coefficient of schooling in a log wage regression. The standard approach in labor economics is based on the following model which implicitly assumes that skill prices are stationary. Let $w(s,x)$ = wage income at experience x for schooling level s; $T(s)$ = last age of earnings; v is private tuition costs minus non-pecuniary returns, τ is the proportional tax rate; and r is the before-tax interest rate.

Agents are assumed to maximize the present value of earnings. The criterion maximized at the individual level is

$$\int_0^{[T(s)-s]} (1-\tau)e^{-(1-\tau)r(x+s)}w(s,x)dx - \int_0^s ve^{-(1-\tau)rj}dj. \tag{14.17}$$

This expression embodies an institutional feature of the US economy where income from all sources is taxed but one cannot write off costs of tuition. However, we assume that agents can write off interest on their loans. This is consistent with the institutional feature that persons can deduct mortgage interest, that 70 percent of American families own their own homes, and that mortgage loans can be used to finance college education.

The first order conditions for maximizing present value are

$$\begin{aligned}(1-\tau)[T'(s)-1]&e^{-r(T(s))(1-\tau)}w(s,T(s)-s)\\ &-(1-\tau)r\int_0^{T(s)-s}(1-\tau)e^{-(1-\tau)r(x+s)}w(s,x)dx\\ &+\int_0^{T(s)-s}(1-\tau)e^{-(1-\tau)r(x+s)}\frac{\partial w(s,x)}{\partial s}dx = ve^{-(1-\tau)rs}. \end{aligned}\tag{14.18}$$

Rearranging, we obtain

$$(1-\tau)[T'(s)-1]e^{-(1-\tau)r[T(s)-s]}w(s,T(s)-s)$$

$$- (1-\tau)r \int_0^{T(s)-s} (1-\tau)e^{-(1-\tau)rx} w(s,x)dx$$

$$+ \int_0^{T(s)-s} (1-\tau)e^{-(1-\tau)r(x+s)} \frac{\partial w(s,x)}{\partial s} dx = v. \tag{14.19}$$

Thus,

$$\frac{[T'(s)-1]e^{-(1-\tau)r[T(s)-s]}w(s,T(s)-s)}{\int_0^{T(s)-s} e^{-(1-\tau)rx} w(s,x)dx}$$

$$+ \left[\frac{\int_0^{T(s)-s} e^{-(1-\tau)rx} \left[\frac{1}{w(s,x)} \frac{\partial w(s,x)}{\partial s}\right] w(s,x)dx}{\int_0^{T(s)-s} e^{-(1-\tau)rx} w(s,x)dx} \right]$$

$$- \frac{v}{(1-\tau)\int_0^{T(s)-s} e^{-(1-\tau)rs} w(s,x)dx} = (1-\tau)r. \tag{14.20}$$

The special case assumed in the Mincer model writes $v = 0$, $T'(s) = 1$. (No private tuition costs and no loss of work life from schooling). This simplifies the first order condition to

$$(1-\tau)\int_0^{T(s)-s} e^{-(1-\tau)rx} w(s,x)dx \tag{14.21}$$

$$= \int_0^{T(s)-s} e^{-(1-\tau)rx} \left[\frac{1}{w(s,x)} \frac{\partial w(s,x)}{\partial s}\right] w(s,x)dx.$$

Mincer further assumes multiplicative separability:

$$w(s,x) = \mu(s)\varphi(x) \tag{14.22}$$

so in logs,

$$\log w(s,x) = \log \mu(s) + \log \varphi(x).$$

Thus

$$(1-\tau)r\ \mu(s) \int_0^{T(s)-s} e^{-(1-\tau)rx} \varphi(x)dx$$

$$= \mu'(s) \int_0^{T(s)-s} e^{-(1-\tau)rx} \varphi(x)dx$$

and

$$(1-\tau)r = \frac{\mu'(s)}{\mu(s)}. \tag{14.23}$$

Then the coefficient on schooling in a Mincer equation estimates the rate of return to schooling which is the same as the after tax interest.

In the general case, we obtain

$$\begin{aligned}
&\frac{[T'(s)-1]e^{-(1-\tau)r[T(s)-s]}w(s,T(s)-s)}{\int_0^{T(s)-s} e^{-(1-\tau)rx}w(s,x)dx} \quad \text{(Term 1)} \\
&+ \frac{\int_0^{T(s)-s} e^{-(1-\tau)rx}\left[\frac{\partial \ln w(s,x)}{\partial s}\right] w(s,x)dx}{\int_0^{T(s)-s} e^{-(1-\tau)rx}w(s,x)dx} \quad \text{(Term 2)} \\
&- \frac{v}{(1-\tau)\int_0^{T(s)-s} e^{-(1-\tau)rx}w(s,x)dx} \quad \text{(Term 3)} \\
&= (1-\tau)r \,.
\end{aligned}$$

Term 1 is the earnings-life effect: the value of the change in the present value of earnings due to the change in working-life due to schooling as a fraction of the present value of schooling measured at age s. Term 2 is the weighted (by discounted income over the lifetime) effect of schooling on earnings by age. Term 3 is the cost of tuition expressed as a fraction of lifetime income measured at age s.

Observe that if there is uncertainty about future earnings, then we may replace $w(s,x)$ and $[\partial \ln w(s,x)]/\partial s$ by their expected values. In the special case of the Mincer model, if the shocks to earnings equations are multiplicatively separable, the basic relationship (14.23) is unaffected. Thus, if $w(s,x) = \mu(s)\varphi(x)\varepsilon(s,x)$, the Mincer model is not affected at any age as long as $\varepsilon(s,x) \neq 0$.

The evidence presented in this paper and Heckman, Lochner and Taber [29] argues strongly against assumption (14.22). Log wage profiles are not parallel in work experience across schooling groups (see also the evidence in Heckman, Lochner and Todd [34]). In addition, private tuition costs (v) are not zero and are not offset by work in school for most college students. In addition, we estimate substantial psychic costs-benefits of schooling (see the estimates reported in Table 14.2). The assumption of equal-working lives for people of different schooling levels is at odds with the data presented by Mincer. These factors account for the disparity between the after-tax interest rate of 0.05 ($= (1-\tau)r$) and the steady-state Mincer coefficient obtained from regressing simulated steady-state log earnings on schooling, experience and experience squared. (This is the intercept in Figure 14.5 and is 0.092.) Finally, the least squares estimate does not control for variation in the ability of persons attending college so that classical ability bias may raise the return.[24]

[24] The evidence on the importance of ability bias is mixed. See, e.g., Griliches [20] and Card [12]. The evidence reported in Cawley et al. [14] demonstrates that fundamental identification problems plague studies of the effect of ability on earnings.

More serious, however, is the implicit steady-state assumption built into the Mincer model. In a period of changing demand for different skill groups, cross-sectional and cohort Mincer coefficients disagree, as is evident from comparing Figures 14.5 and 14.15, and as discussed in Section 14.5. High-skilled individuals who acquire their schooling before the onset of technological change earn a rent just as low-skilled individuals who entered the work force before the onset of technical change suffer a capital loss. Cross-section earnings functions capture these rents which are the returns that fresh entrants into the labor force will enjoy. Moreover, persons who become skilled just after the onset of technical change earn super-normal returns that will not be experienced by subsequent cohorts as supplies adjust. Accordingly, widely used cross-sectional "rates of return" to schooling are a poor guide to policy. Fifteen years after the onset of technological change, cross section Mincer rates of return are over 16 percent while cohort rates are under 7 percent. The cross-section rate of return is wildly optimistic and should not be used as a guide to educational policy.

14.7 General-Equilibrium Treatment Effects: A Study of Tuition Policy

This section of the paper uses our model to consider the effects of changes in tuition on schooling and earnings, accounting for general equilibrium effects on skill prices. The typical evaluation estimates the response of college enrollment to tuition variation using geographically dispersed cross-sections of individuals facing different tuition rates. These estimates are then used to determine how subsidies to tuition will raise enrollment. The impact of tuition policies on earnings is evaluated using a schooling-earnings relationship fit on pre-intervention data and does not account for the enrollment effects of the taxes raised to finance the tuition subsidy. Kane [37] exemplifies this approach.

The danger in this widely used practice is that what is true for policies affecting a small number of individuals need not be true for policies that affect the economy at large. A national tuition-reduction policy that stimulates substantial college enrollment will likely compress skill prices, as advocates of the policy claim. However, agents who account for these changes will not enroll in school at the levels calculated from conventional procedures which ignore the impact of the induced enrollment on earnings. As a result, standard policy evaluation practices are likely to be misleading about the effects of tuition policy on schooling attainment and wage inequality. The empirical question is how misleading? We show that these practices lead to estimates of enrollment responses that are ten times larger than the long-run general equilibrium effects. We also improve on current practice in the treatment effects literature by considering both the gross benefits of the program and the tax costs of financing the treatment as borne by different groups.

Evaluating the general-equilibrium effects of a national tuition policy requires more information than the tuition-enrollment parameter that is the centerpiece of partial-equilibrium policy analysis. Most policy proposals extrapolate well outside the range of known experience and ignore the effects of induced changes in skill quantities on skill prices. We apply our open capital market general-equilibrium framework to evaluate tuition policies that attempt to increase college enrollment.

14.7.1 Conventional models of treatment effects

The standard framework for microeconometric program evaluation is partial-equilibrium in character (see Heckman and Robb [32]). For a given individual i, $Y_{1,i}$ is defined to be the outcome the individual receives if he participates in the program, and $Y_{0,i}$ is the outcome he receives if he does not participate. The treatment effect for person i is $\Delta_i = Y_{1,i} - Y_{0,i}$. When interventions have general equilibrium consequences, these effects depend on who else is treated and the market interaction between the treated and the untreated.

To see the problems that arise in the standard framework, consider instituting a national tuition policy. In this case, $Y_{0,i}$ is person i's wage if he does not attend college, and $Y_{1,i}$ is his wage if he does attend. The "parameter" Δ_i then represents the impact of college, and it can be used to estimate the impact of tuition policies on wages. It is a constant, or policy-invariant, parameter only if wages $(Y_{0,i}, Y_{1,i})$ are invariant to the number of college and high school graduates in the economy.

In a general-equilibrium setting, an increase in tuition subsidy increases the number of individuals who attend college, which in turn decreases the relative wages of college attendees, $Y_{1,i}/Y_{0,i}$. In this case, the program not only impacts the wages of individuals who are induced to move by the program, but it also has an impact on the wages of those who do not. For two reasons, then, the "treatment effect" framework is inadequate. First, the parameters of interest depend on who in the economy is "treated" and who is not. Second, these parameters do not measure the full impact of the program. For example, increasing tuition subsidies may increase the earnings of uneducated individuals who do not take advantage of the subsidy. To pay for the subsidy, the highly educated would be taxed and this may affect their investment behavior. In addition, more educated workers enter the market as a result of the policy, depressing the earnings of other college graduates. Conventional methods ignore the effect of the policy on non-participants. In order to account for this effect, it is necessary to conduct a general-equilibrium analysis.

14.7.2 Exploring increases in tuition subsidies in a general-equilibrium model

We first simulate the effects of a revenue-neutral $500 increase in tuition subsidy (financed by a proportional tax) on enrollment in college and wage inequality starting from our baseline economy. The partial-equilibrium increase in college attendance is 5.3 percent in the new steady state. This is in the range of effects reported by Kane [37] and Cameron and Heckman [11]). This analysis holds skill prices, and therefore college and high school wage rates, fixed — a typical assumption in microeconomic "treatment effect" analyses of tuition policies.

When the policy is evaluated in a general-equilibrium setting, the estimated effect falls to 0.49 percent. Because the college–high school wage ratio falls as more individuals attend college, the returns to college are less than when the wage ratio is held fixed. Rational agents understand this effect of the tuition policy on skill prices and adjust their college-going behavior accordingly. Policy analysis of the type offered in the "treatment effect" literature ignores the responses of rational agents to the policies being evaluated. There is substantial attenuation of the effects of tuition policy on capital and the stocks of the different skills in our model. Simulating the effects of this policy under a number of additional alternative assumptions about the parameters of the economic model, including analysis of a case where tuition costs rise with enrollment, reproduces the basic result of substantial partial-equilibrium effects and much weaker general-equilibrium effects.

Our steady state results are long-run effects. When we simulate the model with rational expectations, the short-run enrollment effects are also very small, as agents anticipate the effects of the policy on skill prices and calculate that there is little gain from attending college at higher rates. If we simulate using myopic expectations, the short-run enrollment effects are much closer to the estimated partial-equilibrium effects. All of these results are qualitatively robust to the choice of different tax schedules. Progressive tax schedules choke off skill investment and lead to even lower enrollment responses in general equilibrium.

We next consider the impact of a policy change on discounted earnings and utility. We decompose the total effects into benefits and costs, including tax costs for each group. Table 14.5 compares outcomes in two steady states: (a) the benchmark steady state and (b) the steady state associated with the new tuition policy. Given that the estimated schooling response to a $500 subsidy is small, we instead use an extremely high $5,000 subsidy for the purpose of exploring general-equilibrium effects. The rows High School–High School report the changes in a variety of outcome measures for those persons who would be in high school under the benchmark or new policy regime; the High School–College rows report the changes in the same measures for high school students in the benchmark who are induced to attend college only by the new policy; College–High School outcomes refer to those persons in college in the benchmark economy who only attend high school after the new policy is put in place; and so forth.

Table 14.5: Simulated effects of $5,000 tuition subsidy on different groups, steady state changes in present value of lifetime wealth (thousands of 1995 dollars)

Group (proportion)[a]	After-tax earnings using base tax[b] (1)	After-tax earnings[b] (2)	After-tax earnings net of tuition[b] (3)	Utility[b] (4)
High school–High school (0.5210)	17.520	6.849	6.849	6.849
High school–College (0.023)	9.757	-0.372	14.669	8.263
College–High school (0.0003)	-37.874	-49.528	-45.408	7.287
College–College (0.447)	1.574	-10.233	8.412	8.412
Ability quartile 1				
High school–High school (0.844)	14.696	5.673	5.673	5.673
High school–College (0.045)	30.587	21.043	36.179	8.001
College–High school (0.000)	0.000	0.000	0.000	0.000
College–College (0.111)	1.273	-8.271	10.353	10.353
Ability quartile 2				
High school–High school (0.689)	18.571	7.269	7.269	7.269
High school–College (0.033)	-5.356	-15.874	-0.841	8.440
College–High school (0.000)	0.000	0.000	0.000	0.000
College–College (0.277)	1.308	-9.210	9.428	9.428
Ability quartile 3				
High school–High school (0.446)	20.691	8.181	8.181	8.181
High school–College (0.014)	-22.046	-33.156	-18.409	8.694
College–High school (0.000)	0.000	0.000	0.000	0.000
College–College (0.0541)	1.4010	-9.691	8.946	8.946
Ability quartile 4				
High school–High school (0.139)	19.286	7.633	7.633	7.633
High school–College (0.000)	0.000	0.000	0.000	0.000
College–High school (0.001)	-37.874	-49.528	-45.408	7.287
College–College (0.859)	1.802	-11.152	7.498	7.498

a. The groups denote counterfactual groups. For example, the High school–High school group consists of individuals who would not attend college in either steady state, and the High school–College group would not attend college in the first steady state, but would in the second, etc.

b. Column (1) reports the after-tax present value of earnings in thousands of dollars discounted using the after-tax interest rate where the tax rate used for the second steady state is the base tax rate. Column (1) reports just the effect on earnings, column (2) adds the effect of taxes, column (3) adds the effect of tuition subsidies, and column (4) includes the nonpecuniary costs of college expressed in dollars.

By the measure of the present value of earnings (col. 3), some of those induced to change are worse off. Contrary to the monotonicity assumption built into the LATE parameter of Imbens and Angrist [35], defined in this context as the effect of tuition change on the earnings of those induced to go to college, we find that the tuition policy produces a two-way flow. Some people who would have attended college in the benchmark regime no longer do so. The rest of society also is affected by the policy–again, contrary to the implicit assumption built into LATE that only those who change status are affected by the policy. People who would have gone to college without the policy and continue to do so after the policy are financially worse off for two reasons: (a) the price of their skill is depressed and (b) they must pay higher taxes to finance the policy. However, they now receive a tuition subsidy and for this reason, on net, they are better off both financially and in terms of our monetary value of utility (inclusive of taste and tuition shocks). Those who would abstain from attending college in both steady states are also better off in the second steady state. They pay higher taxes, but their skill becomes more scarce and their wages rise. Those induced to attend college by the policy are better off in terms of utility but are not always better off in terms of income. For example, individuals from ability quartiles 2 and 3 have lower net incomes as a result of the tuition policy; however, their utility rises due to a strong taste for college education (low cost of education). While most groups gain about the same in terms of utility, there is substantial variation in the effects on lifetime earnings. Note that neither category of non-changers is a natural benchmark for a "difference in differences" estimator. The movement in their wages before and after the policy is due to the policy and cannot be attributed to a benchmark "trend" that is independent of the policy. The implicit assumptions that justify the widely-used difference in differences estimator do not apply here. The tax system and the market make the "nontreated" affected by the policy (see the discussion in Heckman, LaLonde and Smith [31]). These conclusions are robust as to whether a closed-economy or open-economy model is used.

14.8 Tax Policy and Human Capital Formation

Missing from recent discussions of tax reform is any systematic analysis of the effects of various tax proposals on skill formation (see the papers in the collection edited by Aaron and Gale [1]; Davies and Whalley [15] and Dupor et al. [16] are notable exceptions). This gap in the literature in empirical public finance is due to the absence of any empirically based general-equilibrium models with both human capital formation and physical capital formation that are consistent with observations on modern labor markets. We use our model to study the impacts on skill formation of proposals to switch from progressive taxes to flat income and consumption taxes, focusing our attention on steady states.

14.8.1 Tax Effects on human capital accumulation

In the absence of labor supply and direct pecuniary or non-pecuniary costs of human capital investment, there is no effect of a proportional wage tax on human capital accumulation. Both marginal returns and costs are scaled down in the same proportion. When untaxed costs or returns to college are added to the model (i.e., non-pecuniary costs/benefits), proportional taxation is no longer neutral. An increase in the tax rate decreases college attendance if the net financial benefit before taxes is positive ($PV^2 - D - PV^1 > 0$). Progressivity reinforces this effect. A progressive wage tax reduces the incentive to accumulate skills, since human capital promotes earnings growth and moves persons to higher tax brackets. As a result, marginal returns on future earnings are reduced more than marginal costs of schooling.

Heckman [23] notes that in a partial-equilibrium model, proportional taxation of interest income with full deductibility of all borrowing costs reduces the after-tax interest rate and hence, promotes human capital accumulation. In a time-separable, representative agent general-equilibrium model, the after-tax interest rate is unaffected by the tax policy in steady state as agents shift to human capital from physical capital (see Trostel [57]). In that framework, flat taxes with full deductibility have no effect on human capital investment. In a dynamic overlapping generations model with heterogeneous agents, endogenous skill formation, and progressive rates, taxes have ambiguous effects on human capital and both their quantitative and qualitative effects can only be resolved by empirical research. We use our empirically grounded model to study alternative proposals for tax reform in this framework.

14.8.2 Analyzing two tax reforms

Following Kotlikoff, Smetters and Walliser [40], we assume that the US income tax can be captured by a progressive tax on labor income and a flat tax on capital income. This assumption greatly simplifies the computations. Similar results are obtained from a more realistic tax schedule. We assume that each earner has 1.22 children and is single. For each additional dollar beyond \$9,660, we assume that there is an increase in itemized deductions of 7.55 cents. Therefore, an individual with labor income Y has taxable income $(Y - \$9,660)(1 - .0755)$. Using the 1995 tax schedule, we compute the taxes paid by income and approximate this schedule by a second order polynomial. We assume a 0.15 flat tax rate on physical capital.

We consider two revenue-neutral tax reforms from this benchmark progressive schedule. The first reform (which we call "Flat Tax") is a revenue-

Table 14.6: Open-economy effects of alternative tax proposals, general-equilibrium (steady state) and partial-equilibrium effects,[a] percentage difference from progressive case[b]

	Flat tax[c]		Flat cons.tax[c]	
	PE	GE	PE	GE
After-tax interest rate	0.00	0.00	17.65	17.65
Skill price college HC	0.00	0.00	0.00	22.87
Skill price HS HC	0.00	−4.13	0.00	21.77
Stock of physical capital	−15.07	−16.21	86.50	124.42
Stock of College HC	22.41	4.49	−15.77	−5.16
Stock of HS HC	-9.94	2.19	1.88	−5.05
Stock of college HC per college graduate	3.04	3.88	−4.08	−3.30
Stock of HS HC per HS graduate	1.84	2.56	−5.23	0.16
Aggregate output	-0.09	−5.14	15.76	42.11
Aggregate consumption	-0.08	−5.81	7.60	34.40
Mean wage college	3.39	−1.09	0.12	24.00
Mean wage HS	2.44	−1.13	0.25	22.31
Standard deviation log wage	4.09	3.45	−1.94	−1.22
College/HS wage premium at 10 yrs exp[d]	1.92	−1.90	3.10	7.19
Fraction attending college	18.79	0.59	−12.18	−1.92
Type 1: Fraction attending college	50.29	−6.85	−42.57	11.47
Type 2: Fraction attending college	28.50	−1.97	−15.60	−17.95
Type 3: Fraction attending college	14.13	−1.82	−5.20	-21.56
Type 4: Fraction attending college	15.27	3.47	−11.77	11.93
Type 1: College HC gain first 10 years[e]	5.81	5.81	−7.53	−7.53
Type 2: College HC gain first 10 years[e]	5.33	5.33	−6.84	−6.84
Type 3: College HC gain first 10 years[e]	5.60	5.60	−6.70	−6.70
Type 4: College HC gain first 10 years[e]	6.85	6.85	−6.41	−6.41
Type 1: HS HC gain first 10 years[e]	3.42	3.42	−7.79	−7.79
Type 2: HS HC gain first 10 years[e]	4.49	4.49	−7.60	−7.60
Type 3: HS HC gain first 10 years[e]	5.36	5.36	−7.62	−7.62
Type 4: HS HC gain first 10 years[e]	5.29	5.29	−7.95	−7.95

a. General-equilibrium (GE) effects allow skill prices to change, while partial-equilibrium (PE) effects hold prices constant.

b. In the progressive case we allow for a progressive tax on labor earnings, but assume a flat tax on capital at 15%

c. In the flat tax regime we hold the tax on capital fixed to the same level as the progressive tax, but the tax on labor income is flat and is calculated to balance the budget in the new GE steady state. This yields a tax rate on labor income of 7.7%. In the consumption regime, we tax only consumption at a 10.0% rate, again balancing the budget in steady states.

d. The college-high school wage premium measures the difference in log mean earnings between college graduates and high school graduates with ten years of experience.

e. These rows present changes in the ratio of human capital at ten years of experience versus human capital upon entering the labor force.

neutral flattening of the tax on labor earnings holding the initial flat tax on capital income constant. The second reform ("Flat Consumption Tax") is a uniform flat tax on consumption. In both flat tax schemes, tuition is not treated as deductible. For each tax, we consider two models: (1) a partial-equilibrium model in which skill prices and interest rates are fixed and (2) an open-economy general-equilibrium model in which skill prices adjust while pretax interest rates remain constant.

Table 14.6 presents both partial-equilibrium and general-equilibrium simulation results measured relative to a benchmark economy with the KSW tax schedule. Table 14.7 presents a corresponding closed-economy version where the interest rate is not fixed in world capital markets. We first discuss the open-economy partial-equilibrium effects of a move to a "Flat Tax," which eliminates progressivity in wages and stimulates skill formation. College attendance rises dramatically as the higher earnings associated with college graduation are no longer taxed away at higher rates. The amount of post-school on-the-job training (OJT) also increases for each skill group (as measured by the stocks of human capital per worker of each skill). While the percentage increases in college enrollment are substantially larger among low ability individuals (reflecting in part their low initial enrollment rates), the amount of skill per worker acquired during the first 10 years of work increases more for the most able who benefit more from the flattening of the wage tax. The aggregate stock of high school and college human capital rises, while the stock of physical capital used in production declines. This reflects the fact that the flatter tax schedule favors human capital formation over the benchmark economy. The college–high school wage differential increases by roughly 2 percent, and the standard deviation of log wages increases by about 4 percent. The partial-equilibrium effects of reform on aggregate consumption and output are inconsequential.

The open-economy general-equilibrium effects of tax reform are noticeably different from the partial-equilibrium effects. The skill prices for both types of human capital now decline, although the price of college human capital drops much more. As a result, the effects on college attendance are negative for low-ability workers, although the overall attendance rate still rises marginally. Since interest rates are the same in both the partial-equilibrium and general-equilibrium flat tax environment, the gains in human capital over the early part of the lifecycle are identical to those of the partial-equilibrium model. The on-the-job investment profiles will be identical in the closed-economy and the open-economy model as long as after-tax interest rates are identical. Flat taxes on labor earnings mark down costs and returns in the same proportions. The greater increases in human capital per worker in the general-equilibrium state are, therefore, due solely to the fact that only the more able are drawn into college compared to the partial-equilibrium state. Aggregate stocks of college and high school human capital both rise in general

Table 14.7: Closed-economy effects of alternative tax proposals, general-equilibrium (steady state) and partial-equilibrium effects,[a] percentage difference from progressive case[b]

	Flat tax[c]		Flat cons.tax[c]	
	PE	GE	PE	GE
After-tax interest rate	0.00	1.96	17.65	3.31
Skill price college HC	0.00	−1.31	0.00	3.38
Skill price HS HC	0.00	−0.01	0.00	4.65
Stock of physical capital	−15.07	−0.79	86.50	19.55
Stock of College HC	22.41	2.82	−15.77	1.85
Stock of HS HC	−9.94	0.90	1.88	0.08
Stock of college HC per college graduate	3.04	2.55	−4.08	1.72
Stock of HS HC per HS graduate	1.84	1.07	−5.23	0.16
Aggregate output	−0.09	1.15	15.76	4.98
Aggregate consumption	−0.08	0.16	7.60	3.66
Mean wage college	3.39	2.60	0.12	6.96
Mean wage HS	2.44	2.44	0.25	6.82
Standard deviation log wage	4.09	1.56	−1.94	0.69
College/HS wage premium at 10 yrs exp[d]	1.92	−0.45	3.10	0.18
Fraction attending college	18.79	0.26	−12.18	−1.92
Type 1: Fraction attending college	50.29	−1.25	−42.57	2.14
Type 2: Fraction attending college	28.50	−5.89	−15.60	−7.88
Type 3: Fraction attending college	14.13	−6.93	−5.20	−9.56
Type 4: Fraction attending college	15.27	6.13	−11.77	7.50
Type 1: College HC gain first 10 years[e]	5.81	3.12	−7.53	1.51
Type 2: College HC gain first 10 years[e]	5.33	2.86	−6.84	1.38
Type 3: College HC gain first 10 years[e]	5.60	3.10	−6.70	1.61
Type 4: College HC gain first 10 years[e]	6.85	4.17	−6.41	2.56
Type 1: HS HC gain first 10 years[e]	3.42	1.06	−7.79	−0.34
Type 2: HS HC gain first 10 years[e]	4.49	1.97	−7.60	0.46
Type 3: HS HC gain first 10 years[e]	5.36	2.67	−7.62	1.06
Type 4: HS HC gain first 10 years[e]	5.29	2.55	−7.95	0.92

See notes to Table 14.6.

equilibrium. Now that fewer workers respond to the tax change by attending college, the aggregate stock of high school human capital rises rather than falls since each high school graduate responds to the flat tax by investing more. In general equilibrium, the physical capital stock still drops substantially, and the declines in output and consumption are much larger than in the partial-equilibrium model, due to the large drop in skill prices. By these measures of welfare, the flat tax is worse than the progressive wage tax even though skill levels are higher per worker. In regard to wage inequality, our two measures show different effects. The standard deviation in log wages rises as it did in the partial-equilibrium case; however, the college–high school wage premium declines by about 2 percent.

The greatest differences in the general-equilibrium effects of reform from our previous closed-economy version of the model deal with physical capital (see the simulation estimates in Table 14.7). In a closed economy, interest rates rise in response to the policy change. This substantially reduces the decline in physical capital that occurs in the open-economy case. Furthermore, the closed-economy version shows little effect of the tax reform on aggregate consumption and output, while our open-economy version shows declines in excess of 5 percent for each, an effect largely due to the huge outflow of capital. The decline in the college–high school wage gap is less in the closed-economy version and the rise in wage inequality, as measured by the standard deviation of log wages, is less in the closed-economy model. College attendance becomes more stratified in the closed-economy version. Because of the rise in the after-tax interest rate, the increase in post-school human capital investment is smaller in the closed-economy version of the model than in the open-economy version.

Next, consider a revenue neutral move to a "flat consumption tax." This reform is more pro-capital and is less favorable to human capital than the flat income tax because it increases the tax on human capital to offset the cut in the tax on physical capital. It raises output, the capital stock and consumption substantially, while it reduces the aggregate stock of college human capital. In general equilibrium, it also reduces the stock of high school human capital. The amount of human capital per worker declines for college graduates, but is nearly constant for high school graduates once general equilibrium-effects are accounted for. Investments on the job decline by about 7–8 percent for all individuals, due to the large increase in the after-tax interest rate (this is the effect stressed by Heckman [23]). The fraction attending college declines substantially for all ability types in partial equilibrium. In our open-economy general equilibrium model, the policy reduces the attendance of middle ability workers and raises attendance rates of individuals at the top and bottom of the ability distribution.[25] The reform raises wage inequality as measured by the college–high school wage premium, but lowers it as measured by the standard deviation of log wages. The general-equilibrium comparison shows more equality (by both measures) than the partial equilibrium. Skill prices and mean wage rates rise substantially as capital flows into the economy, in contrast to the "Flat Tax" economy discussed above. Since capital is a direct complement with both forms of human capital, the increase in capital raises skill prices about equally for both skill groups.

The increase in the physical capital stock (124%) in general equilibrium in the open-economy case raises output (42%) and consumption (34%). The

[25] The rise in college-going at the bottom of the ability distribution is due to the greater tax on the earnings of the less able and the flattening of the tax schedule. They now pay greater taxes and going to college offsets their enhanced tax burden.

scale of most responses seem inflated compared to the closed-economy case when the after-tax interest rate increases after the reform. For that version of our model, the simulated general-equilibrium responses look more credible (see Table 14.7). Capital stock increases by 19.5 percent, consumption by 3.6 percent, and output by 5 percent. The increase in the college–high school wage differential is much less than is predicted in the open-economy case. The decrease in the fraction attending college is the same in both cases. However, in the closed-economy case, movement to a flat consumption tax promotes post-school investment whereas it substantially reduces post-school investment in the open-economy case.

When we introduce deductibility of tuition in both reforms, and preserve revenue neutrality, there is virtually no change in the effects of skill formation (or anything else) in general equilibrium for either the closed economy or the open economy. This is consistent with the simulations reported in the previous section, in which we show that general-equilibrium effects of tuition subsidies are small. The lessons from partial-equilibrium analyses are substantially misleading guides for analyzing the effects of tax and tuition policy on skill formation. The benefits from changing to proportional wage taxation are small and must be weighed against the costs. While our open-economy simulations show that it would increase skill formation, it would also reduce consumption, output, and wages. A change to a flat consumption tax has larger (and positive) effects on output, consumption, and real wages, but it also slightly raises the college–high school wage premium and reduces aggregate skill levels. However these effects are all much weaker in the closed-economy case. The effects of tax reform on physical capital accumulation seem implausibly large in the open-economy case.

14.9 Comparing Open- and Closed-Economy Versions of the Model

Comparing the open-economy model with the closed-economy model, we find that the closed-economy version produces more plausible investment responses for both physical and human capital. For example, the open-economy version predicts unbelievably large OJT responses during the period of transition to the new technology compared to the closed-economy version. The physical capital inflows implied by the open-economy version are also implausibly large. The investment responses to tax reform also seem implausibly large in the open-economy version compared to the closed-economy version. Investment is very sensitive to interest rates in our model. The movement in the interest rate induced by technical change or tax reform in the closed-economy version of our model chokes off these implausibly large investment responses.

Additional support for the closed-economy model comes from the pattern of skill investment by the less skilled during the transition. Bartel and Sicherman [4] report that over the period of the mid-80s, when wage inequality was increasing substantially, investment in company training increased for less educated workers, both absolutely and compared to that of more educated workers in industries where technological progress was rapid. The open-economy version of the model produces this phenomenon only for a brief stretch of time. The closed-economy version of the model produces this phenomenon for the longer periods of time observed in the Bartel–Sicherman study.

Assuming that skill-biased technical change commences in the early 70s, the path of the real interest rate produced by the closed-economy version of the model is in rough agreement with the data. See Figure 14.17, which reveals that in the early stages of the onset of technical change, the real interest rate is predicted to rise as indeed it did in the late 70s and early 80s in the US economy. This is one more bit of evidence in support of the closed-economy version of our model.

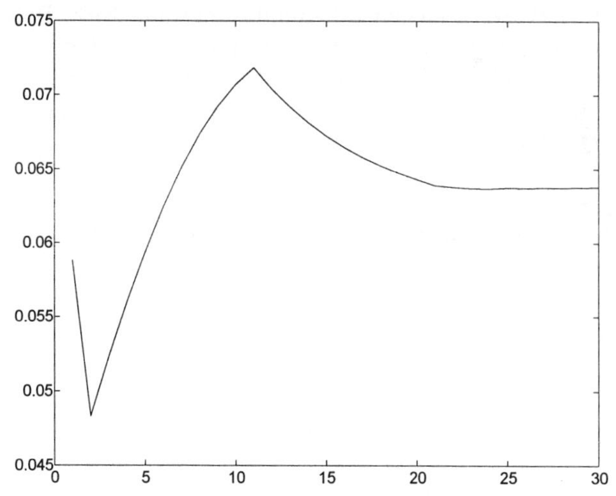

Figure 14.17: Closed economy interest rates for a technology shock originating in the early 70s

14.10 Summary and Conclusions

This paper summarizes the overlapping generations general-equilibrium model of heterogeneous skills developed in Heckman, Lochner and Taber [28] and applies it in an open-capital market environment in order to evaluate proposals designed to foster skill formation. The model has important self-correcting features. Through supply adjustments, substantial wage inequality induced by skill-biased technical change is all but eliminated in the long run. Skill-biased technical change hurts cohorts of unskilled and low-ability workers entering the market at the time of the onset of technical change. In the long run steady state, persons of all ability groups are better off than they are in the steady state prior to the onset of technological change. The major policy problem is intergenerational in character. Certain cohorts are badly hurt by the onset of technical change. Temporary policies that compensate the losers are appropriate — not long term skill subsidies of the sort proposed in recent policy discussions. In results not reported in this paper, tuition subsidies financed by taxes have only a slight effect on improving intergenerational equity.

This paper also defines and estimates general-equilibrium treatment effects. Focusing on the impact of tuition policy, we find that general-equilibrium impacts of tuition on college enrollment are an order of magnitude smaller than those reported in the literature on microeconometric treatment effects. The assumptions used to justify the LATE parameter in a microeconomic setting do not carry over to a general-equilibrium framework. Policy changes, in general, induce two-way flows and violate the monotonicity — or one-way flow — assumption of LATE. We extend the LATE concept to allow for the two-way flows induced by the policies. The effects of the tuition policy on both "treated" and "untreated" persons after accounting for taxes and adjustments in skill prices renders conventional differences in differences estimators invalid. We present a more comprehensive approach to program evaluation by considering both the tax and benefit consequences of the program being evaluated and placing the analysis in a market setting. We have also examined the impact of two proposed tax reforms on skill formation and wage inequality. A shift to a flat consumption tax slightly discourages skill formation but increases the real wages of all skill groups and barely affects the two commonly used measures of wage inequality. The closed-economy version of our model produces more plausible responses. Widely used partial-equilibrium analyses are misleading and reveal the value of our general-equilibrium approach.

We have examined the commonly utilized Mincer "rate of return" to schooling and show that even in a stationary environment it does not reflect the true economic rate of return. In a period of transition induced by skill-biased technical change, the conventionally computed cross-section Mincer coefficient captures rents that accrue to specific cohorts of workers that will not be captured by later generations. Cross-section Mincer rates of return to education do not measure cohort rates of return to new entrants and are a poor guide to policy.

In comparing the open-economy version of our model to the closed-economy version, the closed-economy version seems more plausible. The movement in the interest rate in the closed-economy version chokes off implausibly large investment responses that appear in the open-economy version. The movement in the real interest rate induced by skill-biased technical change originating in the early 70s that is predicted by the closed-economy version of our model is in rough agreement with the actual movement of the real interest rate observed in that period.

Appendix: Data

The data used to estimate the human capital production function is from the National Longitudinal Survey of Youth (NLSY) — a nationally representative sample of individuals that began in 1979 and interviewed youth ages 14 to 22. These same individuals are re-interviewed annually until 1993 — the last year of data that we use. We use a sub-sample of white male civilians from the NLSY and exclude the over-sampling of poor whites. Individuals are included in the sample if they work more than 500 hours in a particular year. Persons with hourly wages above $100.00 and less than $1.00 in 1992 dollars are deleted.

We partition the data into four groups on the basis of their Armed Force Qualifying Test (AFQT) score. These identify the types as used in this paper. In 1980, 94 percent of the sample was given the Armed Service Vocational Aptitude Test which consists of ten standardized tests that are used to assess a variety of skills. Four of these tests are combined to form the AFQT which is used as an admission criterion into the armed forces. We normalize the test by subtracting the mean score for each individual's birth year. We then divide the sub-sample of white males into four equal-sized groups ranked on the basis of their AFQT score.

In estimating the price elasticity of the decision to attend college we use the state average tuition levels for public two-year institutions (from the calendar year in which the sample member turned 18) as our measure of the local price of college. The macroaggregates come from the National Income and Product accounts as presented in the Citibase data and in FRB data sets on capital stock. We define labor's share in the following way:

labor's share = compensation/(GNP – indirect business taxes – proprietors' income),

capital's share = 1 – labor's share.

Note that indirect business taxes and proprietors' income are, equivalently, being excluded from the calculation and are assumed to break down the same way that the rest of GNP does between labor and capital. Indirect business taxes are largely sales taxes, which are "skimmed" off before businesses can allocate the income to capital or labor, and proprietors' income includes income of the self-employed and non-incorporated partnerships (family businesses, law firms, etc.). We do not know how to break down proprietor's income, because it is compensation to both labor and capital.

REFERENCES

1 Aaron, H. and W. Gale, Economic Effects of Fundamental Tax Reform, Brookings Institution Press, Washington, DC, 1996.

2 Auerbach, A. and L. Kotlikoff, Dynamic Fiscal Policy, Cambridge University Press, Cambridge, 1987.

3 Autor, D., L. Katz and A. Krueger, Computing inequality: Have computers changed the labor market, NBER Working Paper No. 5956, 1997.

4 Bartel, A. and N. Sicherman, Technological change and the skill acquisition of young workers, National Bureau of Economic Research, Working Paper No. 5107, 1998; forthcoming, Journal of Labor Economics, October 1998.

5 Beaudry, P. and H. Green, Cohort patterns in Canadian earnings: assessing the Role of skill premia inequality trends, NBER Working Paper No. 6132, 1997.

6 Ben-Porath, Y., The production of human capital and the life cycle of earnings, Journal of Political Economy, 75(4) (1967) 352–65.

7 Bound, J. and G. Johnson, Changes in the structure of wages in the 1980s: An evaluation of alternative explanations, American Economic Review, 82 (1992) 371–92.

8 Brown, C., A model of optimal human-capital accumulation and the wages of young high school graduates, Journal of Political Economy, 84(2) (1976) 299–316.

9 Browning, M., L. Hansen and J. Heckman, Micro data and general equilibrium models, in: J. Taylor and M. Woodford (eds.), Handbook of Macroeconomics, North Holland, Amsterdam, 1999.

10 Cameron, S. and J. Heckman, Life-cycle schooling and educational selectivity: Models and choice, Journal of Political Economy, 108(2) (1998).

11 Cameron, S. and J. Heckman, Can Tuition Policy Combat Rising Wage Inequality?, in M. Kosters, ed., Financing College Tuition: Government Policies and Social Priorities, Washington, D.C.: AEI Press, (1999).

12 Card, D., Schooling, earnings and ability revisited, in: S. Polachek (ed.), Research in Labor Economics, Vol. 14, JAI Press, Greenwich, CT, 1996.

13 Caselli, F., Technological revolutions, University of Chicago Business School, mimeo, 1997.

14 Cawley, J., J. Heckman, L. Lochner and E. Vytlacil, Understanding the role of cognitive ability in accounting for the recent rise in the economic return to education, in: K. Arrow and S. Durlauf (eds.), Meritocracy and Inequality, Princeton University Press, Princeton, 1999.

15 Davies, J. and J. Whalley, Taxes and capital formation: How important is human capital?, in: B. Bernheim and J. Shoven (eds.), National Saving and Economic Performance, University of Chicago Press, Chicago, 1991.

16 Dupor, W., L. Lochner, C. Taber and M.B. Wittekind, Some effects of taxes on schooling and training, American Economic Review, 96(2) (1996) 293–97.

17 Freeman, R., The Overeducated American, Basic Books, New York, 1976.

18 Fullerton, D. and D. Rogers, Who Bears the Lifetime Tax Burden?, The Brookings Institution, Washington, DC, 1993.

19 Greenwood, J. and M. Yorukoglu, Carnegie-Rochester Series on Public Policy, 1997.

20 Griliches, Z., Sibling models and data in economics: Beginnings of a survey, Journal of Political Economy, 87(5) (1979) S37–S64.

21 Hamermesh, D., Labor Demand, Princeton University Press, Princeton, 1993.

22 Heckman, J., Estimates of a human capital production function embedded in a life-cycle model of labor supply, in: N. Terleckyj (ed.), Household Production and Consumption, Columbia University Press, New York, 1975, pp. 227–64.

23 Heckman, J., A life-cycle model of earnings, learning, and consumption, Journal of Political Economy, 84(4) (1976) S11–S44.

24 Heckman, J., What has been learned about labor supply in the past twenty years? American Economic Review Papers and Proceedings, 83(2) (1993) 116–21.

25 Heckman, J., Should the U.S. have a human capital policy and if so, what should it be?, Gilbert Lecture, University of Rochester, April, 1996.

26 Heckman, J. and P. Klenow, Is there underinvestment in human capital?, unpublished manuscript, University of Chicago, 1997.

27 Heckman, J., L. Lochner and C. Taber, Formulating and estimating dynamic general equilibrium models to evaluate policies that promote skill formation, Marschak Lecture, Econometric Society Meetings, Hong Kong, July 25, 1997.

28 Heckman, J., L. Lochner and C. Taber, Explaining rising wage inequality: Explorations with a dynamic general equilibrium model of labor earnings with heterogeneous agents, Review of Economic Dynamics, 1(1) (1998), 1-58.

29 Heckman, J., L. Lochner and C. Taber, General equilibrium treatment effects: A study of tuition policy, American Economic Review, 88(2) (1998) 381–86.

30 Heckman, J., L. Lochner and C. Taber, Tax policy and human capital formation, American Economic Review, 88(2) (1998) 293–97.

31 Heckman, J., R. LaLonde and J. Smith, The economics and econometrics of active labor market programs, in O. Ashenfelter and D. Card (eds.), Handbook of Labor Economics, North Holland, Amsterdam, 1999.

32 Heckman, J. and R. Robb, Alternative methods for estimating the impact of interventions, in: J. Heckman and Singer (eds.), Longitudinal Analysis of Labor Market Data, Cambridge University Press, Cambridge, 1985, pp. 156–245.

33 Heckman, J. and G. Sedlacek, Heterogeneity, aggregation and market wage functions: An empirical model of self-selection in the labor market, Journal of Political Economy, 98(6) (1985) 1077–25.

34 Heckman, J., L. Lochner and P. Todd, 50 Years of Mincer Earnings Functions, unpublished manuscript, University of Chicago, Department of Economics, (1999).

35 Imbens, G. and J. Angrist, The local average treatment effect, Econometrica, 62(2) (1994) 467–75.

36 Johnson, G., The demand for labor by educational category, Southern Economic Journal, 37 (1970) 190–204.

37 Kane, T., College entry by blacks since 1970: The role of college costs, family background and the returns to education, Journal of Political Economy, 102 (1994) 878–911.

38 Katz, L. and K. Murphy, Changes in relative wages, 1963–1987: Supply and demand factors, Quarterly Journal of Economics; 107(1) (1992) 35–78.

39 Keane, M. and K. Wolpin, The career decisions of young men, Journal of Political Economy, 105(3) (1997) 473–522.

40 Kotlikoff, L., K. Smetters and J. Walliser, The economic impact of privatizing social security, unpublished manuscript, Boston University, 1997.

41 Krussell, P., L. Ohanian, J.V. Rios-Rull and G. Violante, Capital skill complementarity and inequality: A macroeconomic analysis, unpublished manuscript, University of Pennsylvania,1996.

42 Kydland, F., Business cycles and aggregates labor market fluctuations, in: T. Cooley (ed.), Frontiers in Business Cycle Research, Princeton University Press, Princeton, 1995.

43 Lipton, P., J. Sacks, J. Poterba and L. Summers, Multiple shooting in rational expectations models, Econometrica, 50 (1982) 1329–34.

44 Lochner, L., A life-cycle model of human capital and crime: Estimating deterrent effects of wage and education subsidies, Ph.D. Dissertation, University of Chicago, 1998.

45 Lucas, R., On the mechanics of economic development, Journal of Monetary Economics, 22 (1988) 3–42.

46 MaCurdy, T. and T. Mroz, Measuring microeconomic shifts in wages from cohort specification, unpublished manuscript, Stanford University, 1995.

47 Meghir, C. and E. Whitehouse, The evolution of wages in the United Kingdom: Evidence from micro data, Journal of Labor Economics, 14(1) (1996) 1–25.

48 Mincer, J., 1962, On the job training: Costs, returns and some implications, Journal of Political Economy, 70(2) (1962) S50–S79.

49 Ortigueira, S. and M. Santos, On convergence in endogenous growth models, Discussion Paper 9409, ITAM, 1994.

50 Pechman, J., Federal Tax Policy, Fifth Edition, Washington: Brookings Institution, 1987.

51 Poterba, J., The rate of return to corporate capital and factor shares: New estimates using data revised national income and capital stock, unpublished manuscript, MIT, October 1997.

52 Rosen, S., A theory of life earnings, Journal of Political Economy, 84(4) (1976) S45–S57.

53 Roy, A.D., Some thoughts on the distribution of earnings, Oxford Economic Papers, 3 (1951) 135–46.

54 Sattinger, M., Assignment models of the distribution of earnings, Journal of Economic Literature, 31(2) (1993) 831–80.

55 Siow, A. Occupational choice under uncertainty, Econometrica, 52(3), (1984), 631-645.

56 Srinivisan, T.N., General equilibrium theory, project evaluation and economic development, in: Gersovitz, Diaz-Alejandro, Ranis and Rosenzweig (eds.), The Theory and Experience of Economic Development, George Allen & Unwin, London, 1982.

57 Trostel, P., The effect of taxation on human capital. Journal of Political Economy, 101(2) (1993) 327–50.

58 Uzawa, H., Optimum technical change in an aggregative model of economic growth, International Economic Review, 6 (1965) 18–31.

59 Violante, G., Equipment investment and skill dynamics: A solution to the wage dispersion puzzle?, unpublished manuscript, UC London, 1996.

60 Willis, R. and S. Rosen, Education and self-selection, Journal of Political Economy, 87(5) (1979) S7–S36.

Part IV

Market Failures and Economic Structure

Trade, Growth, and Development
G. Ranis and L.K. Raut

CHAPTER 15

International Treaties on Global Pollution: A Dynamic Time-Path Analysis

Parkash Chander

Indian Statistical Institute, 7 SJS Sansanwal Marg, New Delhi-110 016 India

15.1 Introduction

In recent years there has been a trend towards the formation of preferential trading areas (PTAs) such as the European Union (EU) and the North American Free Trade Agreement (NAFTA). An important implication of the formation of such coalitions is that they create externalities on nonmembers. For example, the abolition of tariffs on trade among member countries and the readjustment of external tariffs may worsen nonmember countries' terms of trade with member countries (see Srinivasan [11] for a comparative analysis of the alternative approaches to determining the impact on nonmember countries' welfare).

Two opposing views have persisted regarding the formation of PTAs. On the one hand, it is claimed that the ever-expanding PTAs may lead progressively to the state of nondiscriminatory free trade for all and therefore as policy-makers we must encourage their formation. This has been debated, however, and it has been argued that free trade for all may not come about as the PTAs, once formed, may lead to a stagnant state in which the world remains divided into various trading blocks. This argument has given the theory of PTAs what Bhagwati [2] has called, a dynamic time-path dimension.

In this paper we analyze a conceptually similar dynamic time-path dimension in an international economic-ecological setting. In particular, we examine the question whether expanding coalitions can lead progressively to the formation of the grand coalition and therefore to a worldwide treaty on global pollution (typically climate change).[1]

[1] This is a different kind of dynamics than the one considered in Chander and Tulkens [5] and Chander [4]. There the time-path is defined in the space of feasible allocations, while here it is in the space of coalitions.

In a recent paper, Chander and Tulkens [7] examine the likelihood of a worldwide treaty on global pollution. They present the contents of a feasible treaty which not only satisfies optimality but also group rationality, which they call the γ-core property. The negotiations for the conclusion of such a treaty, as in the case of international trade, might proceed via the emergence of coalitions of subsets of countries with common views and aims coming together who may on their own reduce their pollution levels. Such coalitions might be easier to form, be more efficient and certain, and might seem to be a quick, even if a half-way solution to the global pollution problem. It has been also argued that the formation of such coalitions can act as a catalyst and lead eventually to the conclusion of a worldwide treaty (see, for example, Heal [10]).

In this paper, we show that the formation of coalitions by subsets of countries might diminish the likelihood of a successful worldwide treaty. Nonmember countries may be less willing to sign a worldwide treaty than they would be in the absence of such coalitions. In fact, the coalition formation may raise the reservation utility of nonmembers above the worldwide treaty level and thus take away their incentives to sign it. The paper proceeds as follows: in Section 15.2 we state the economic-ecological model. In Section 15.3 we consider the idea of coalition formation and show that though formation of a coalition may be Pareto-improving, it might lead to a situation in which the nonmembers have no incentives to sign the worldwide treaty. We also relate our results to the literature on stable coalitions.

15.2 The Economic-Ecological Model

We adopt the model most commonly used for the economic analysis of global pollution problems. A set $N = \{1, 2, \ldots, n\}$ of countries share a common environmental resource. Each country i's preferences over the consumption of some private good ($x_i \geq 0$) and of some environmental good ($z \leq 0$) are described by the utility function $u_i(x_i, z)$. Define $\pi_i = (\partial u_i/\partial z)/(\partial u_i/\partial x_i) \geq 0$ as country i's willingness to pay (in terms of commodity x) for the environmental good z. We assume the utility function u_i to be of the quasi-linear form, i.e., $u_i(x_i, z) = x_i + v_i(z)$ with v_i concave and increasing. Furthermore, let $y_i = g_i(p_i)$ denote country i's production function, linking its output $y_i \geq 0$ of the private good with its emissions $p_i \geq 0$ of pollutant in the environment. We assume that g_i is strictly concave and $\gamma_i(p_i) = dg_i(p_i)/dp_i > 0$. The derivative γ_i, when taken to the left, is country i's marginal cost (in terms of y_i) of abatement.

The *transfer function* $z = -\Sigma p_i$ specifies how the pollutant emissions of all countries are diffused and transformed by ecological processes into the ambient quantity z.

A *feasible state* of the economy is a vector $(x, p, x) = (x_1, ..., x_n; p_1, ..., p_n; z)$ such that

$$\sum_{i \in N} x_i = \sum_{i \in N} g_i(p_i)$$

and

$$z = -\sum_{i \in N} p_i.$$

A *Pareto-efficient state* of the economy is a feasible state $(x^*, p^*, z^*) = (x_1^*, ..., x_n^*; p_1^*, ..., p_n^*; z^*)$ that maximizes

$$\sum_{i \in N} [x_i + v_i(z)].$$

Given the quasi-linearity of preferences, it is easily seen that in all Pareto-efficient states the emission levels are the same. Only the private good consumption levels might be different across Pareto-efficient states.

A Pareto-efficient state $(x_1^*, ..., x_n^*; p_1^*, ..., p_n^*; z^*)$ is characterized by the following first-order conditions for maximization:

$$\sum_{j \in N} \pi_j(z^*) = \gamma_i(p_i^*), \ i = 1, 2, \ldots, n. \tag{15.1}$$

A *disagreement equilibrium* is a feasible state $(\bar{x}_1, ..., \bar{x}_n; \bar{p}_1, ..., \bar{p}_n; \bar{z})$ such that for each i, $(\bar{x}_i, \bar{p}_i)$ maximizes $x_i + v_i(z)$ subject to $x_i = g_i(p_i)$ and $p_i + z = -\Sigma_{j \neq i} \bar{p}_j$. A disagreement equilibrium describes the state of autarky in which there is no agreement. Each country acts on its own and decides its pollution level without taking into account the welfare impact of its pollution on other countries. A disagreement equilibrium $(\bar{x}_1, ..., \bar{x}_n; \bar{p}_1, ..., \bar{p}_n; \bar{z})$ is characterized by the following first-order conditions for maximization:

$$\pi_i(\bar{z}) = \gamma_i(\bar{p}_i), \ i = 1, 2, \ldots, n. \tag{15.2}$$

It is easily seen from these conditions that a disagreement equilibrium is not a Pareto-efficient state.[2] The environmental efficiency at the world level can be therefore achieved only through some form of cooperation among the countries.

Chander and Tulkens [7] demonstrate the theoretical possibility of such cooperation. For this purpose, they introduce the new concept of γ-core as distinct from the α- and β-cores. That concept is described as follows: if a

[2] The existence and uniqueness of a disagreement equilibrium follow from the standard arguments for the existence and uniqueness of a Nash equilibrium.

coalition $S \subset N$ forms, the highest payoff it can achieve for its members is given by the function:

$$w^{\gamma}(S) = \sum_{i \in S} [\hat{x}_i + v_i(\hat{z})],$$

where $(\hat{x}_1, ..., \hat{x}_n; \hat{p}_1, ..., \hat{p}_n; \hat{z})$ is the solution to the problem:

$$\max_{(x_i, p_i)_{i \in S}} \sum_{i \in S} [x_i + v_i(z)] \text{ subject to}$$

$$\sum_{i \in S} x_i = \sum_{i \in S} g_i(p_i) \text{ and}$$

$$\sum_{i \in S} p_i + z = - \sum_{j \in N \backslash S} \hat{p}_i,$$

where for each $j \in N \backslash S$

$$(\hat{x}_j, \hat{p}_j) \text{ maximizes } x_j + v_j(z) \text{ subject to}$$

$$x_j = g_j(p_j) \text{ and}$$

$$p_j + z = - \sum_{i \neq j} \hat{p}_i.$$

We shall refer the solution $(\hat{x}_1, ..., \hat{x}_n; \hat{p}_1, ..., \hat{p}_n; \hat{z})$ as the *partial agreement equilibrium* with respect to S. Thus, $w^{\gamma}(S) = \sum_{i \in S} [\hat{x}_i + v_i(\hat{z})]$ is the payoff of S corresponding to the partial agreement equilibrium with respect to S. The partial agreement equilibrium $(\hat{x}_1, ..., \hat{x}_n; \hat{p}_1, ..., \hat{p}_n; \hat{z})$ is characterized by the following first-order conditions:

$$\sum_{i \in S} \pi_i(\hat{z}) = \gamma_i(\hat{p}_i), \; i \in S,$$

and

$$\pi_j(\hat{z}) = \gamma_j(\hat{p}_j), \; j \in N \backslash S. \tag{15.3}$$

For the grand coalition N, the highest payoff is given by $w^{\gamma}(N) = \sum_{i \in N} [x_i^* + v_i(z^*)]$ where (x^*, p^*, z^*) is a Pareto optimum. Since the emission levels p_i^*'s are the same across Pareto-efficient states, the aggregate private good output, $\sum_{i \in N} x_i^*$ $(= \sum_{i \in N} g_i(p_i^*))$ is also the same. The stability of an environmentally efficient worldwide treaty therefore depends on how the aggregate world output $\sum_{i \in N} x_i^*$ is distributed among the countries. Chander and Tulkens [7] show that the following distribution rule is sufficient:

$$x_i^* = g_i(\bar{p}_i) - \frac{\pi_i(z^*)}{\sum_{j \in N} \pi_j(z^*)} \left(\sum_{j \in N} g_j(\bar{p}_j) - \sum_{j \in N} g_j(p_j^*) \right), \; i \in N. \tag{15.4}$$

Given this distribution rule, N is stable in the sense that for each $S \subset N$, $w^{\gamma}(S) \le \sum_{i\in S}[x_i^* + v_i(z^*)]$ where x_i^*'s are as defined in (15.4).[3]

As seen above, the definition of $w^{\gamma}(S)$ involves the assumption that there are no non-singleton coalitions, except possibly S.

In order to consider the dynamic time-path question, we may define the payoff of S in the presence of another (possibly non-singleton) coalition T. Given coalitions S and T, a *coalitional equilibrium* with respect to S and T is the state $(\tilde{x}_1, ..., \tilde{x}_n; \tilde{p}_1, ..., \tilde{p}_n; \tilde{z})$ such that:

(i) $(\tilde{x}_i, \tilde{p}_i)_{i\in S}$ maximizes $\sum_{i\in S}[x_i + v_i(z)]$ subject to
$$\sum_{i\in S} x_i = \sum_{i\in S} g_i(p_i), \quad \sum_{i\in S} p_i + z = -\sum_{j\in N\setminus S} \tilde{p}_j;$$

(ii) $(\tilde{x}_j, \tilde{p}_j)_{j\in T}$ maximizes $\sum_{j\in T}[x_j + v_j(z)]$ subject to
$$\sum_{j\in T} x_j = \sum_{j\in T} g_j(p_j), \quad \sum_{j\in T} p_j + z = -\sum_{j\in N\setminus T} \tilde{p}_j; \text{ and}$$

(iii) for each $k \in N\setminus S\setminus T$, $(\tilde{x}_k, \tilde{p}_k)$ maximizes $x_k + v_k(z)$ subject to
$$x_k = g_k(p_k), \quad p_k + z = -\sum_{j\neq k} \tilde{p}_j.$$

A coalitional equilibrium with respect to S and T defines the payoff of S when there is a pre-existing coalition T. Let $w^{\gamma}(S|T)$ denote this payoff:

$$w^{\gamma}(S|T) = \sum_{i\in S}[\tilde{x}_i + v_i(\tilde{z})],$$

where $(\tilde{x}_1, ..., \tilde{x}_n; \tilde{p}_1, ..., \tilde{p}_n; \tilde{z})$ is the coalitional equilibrium with respect to S and T. The coalitional equilibrium with respect to S and T is characterized by the following first-order conditions:[4]

$$\begin{aligned} &\sum_{i\in S} \pi_i(\tilde{z}) = \gamma_i(\tilde{p}_i) && \text{for all } i \in S, \\ &\sum_{j\in T} \pi_j(\tilde{z}) = \gamma_i(\tilde{p}_i) && \text{for all } i \in T, \\ &\pi_j(\tilde{z}) = \gamma_j(\tilde{p}_j) && \text{for all } j \in N\setminus S\setminus T. \end{aligned} \tag{15.5}$$

Theorem *For each coalition $S \subset N$, the payoff $w^{\gamma}(S|T)$ corresponding to the coalitional equilibrium with respect to S and T is at least as large as the*

[3] The allocation (x^*, p^*, z^*) can be given a similar interpretation as the Lindahl equilibrium in public good economies or the Walrasian equilibrium in exchange economies.

[4] The existence and uniqueness of a coalitional equilibrium follow from similar arguments as in the case of partial agreement equilibrium.

payoff $w^\gamma(S)$ corresponding to the partial agreement equilibrium with respect to S.

The theorem implies that coalition formation T creates positive externalities for S. By raising the reservation utility of S, it may reduce the incentives of S to sign the worldwide treaty.

Proof of the Theorem We first show that $\tilde{z} \geq \hat{z}$, where $\tilde{z} = -\sum_{i \in N} \tilde{p}_i$, $\hat{z} = -\sum_{i \in N} \hat{p}_i$, and $(\tilde{p}_1, ..., \tilde{p}_n)$ and $(\hat{p}_1, ..., \hat{p}_n)$ are the emission levels corresponding to the coalitional equilibrium and the partial agreement equilibrium.

Suppose, contrary to the assertion, that $\tilde{z} < \hat{z}$. We must then have $\pi_i(\tilde{z}) \geq \pi_i(\hat{z})$ for each $i \in N$. From the characterization of coalitional equilibrium (see (15.5) above), partial agreement equilibrium with respect to S (see (15.3) above), and disagreement equilibrium (see (15.2) above), it follows that:

$$\begin{aligned} \gamma_i(\tilde{p}_i) &= \sum_{j \in S} \pi_j(\tilde{z}) \geq \sum_{j \in S} \pi_j(\hat{z}) = \gamma_i(\hat{p}_i) \text{ for all } i \in S, \\ \gamma_i(\tilde{p}_i) &= \sum_{j \in T} \pi_j(\tilde{z}) \geq \sum_{j \in T} \pi_i(\hat{z}) = \gamma_i(\hat{p}_i) \text{ for all } i \in T, \end{aligned}$$

and

$$\gamma_i(\tilde{p}_i) = \pi_i(\tilde{z}) \geq \pi_i(\hat{z}) = \gamma_i(\hat{p}_i) \text{ for all } i \in N \backslash S \backslash T.$$

From these inequalities and the strict concavity of each g_i, it follows that $\tilde{p}_i \leq \hat{p}_i$ for each $i \in N$. But this contradicts our supposition that $\tilde{z} < \hat{z}$. Hence we must have $\tilde{z} \geq \hat{z}$.

Now, since $\tilde{z} \geq \hat{z}$,

$$\gamma_i(\tilde{p}_i) = \sum_{j \in S} \pi_j(\tilde{z}) \leq \sum_{j \in S} \pi_j(\hat{z}) = \gamma_i(\hat{p}_i) \text{ for all } i \in S,$$

which, from concavity of each g_i, means $\tilde{p}_i \geq \hat{p}_i$ for all $i \in S$. Therefore,

$$\sum_{i \in S} \tilde{x}_i = \sum_{i \in S} g_i(\tilde{p}_i) \geq \sum_{i \in S} g_i(\hat{p}_i) = \sum_{i \in S} \hat{x}_i.$$

Since $\tilde{z} \geq \hat{z}$ and $\sum_{i \in S} \tilde{x}_i \geq \sum_{i \in S} \hat{x}_i$,

$$w^\gamma(S|T) = \sum_{i \in S} [\tilde{x}_i + v_i(\tilde{z})] \geq \sum_{i \in S} [\hat{x}_i + v_i(\hat{z})] = w^\gamma(S).$$

This completes the proof. ■

We have shown that for any S and T, $w^\gamma(S|T) \geq w^\gamma(S)$ in general. From this it does not follow that $w^\gamma(S|T) > \sum_{i \in S} [x_i^* + v_i(z^*)]$, where $(x_1^*, \ldots, x_n^*;$

$p_1^*, \ldots, p_n^*; z^*)$ is the worldwide treaty as defined in (15.4), which is clearly a stronger requirement since $\sum_{i \in S}[x_i^* + v_i(z^*)] \geq w^\gamma(S)$. Though it need not be satisfied in general, we construct an illustrative example to show that this might indeed be the case in certain circumstances.

There are three identical countries, e.g., $N = \{1, 2, 3\}$ with $u_i(x_i, z) = x_i + z$ and $g_i(p_i) = p_i^{1/2}$. For this economy, the disagreement equilibrium is given by $\bar{x}_i = 1/2$, $\bar{p}_i = 1/4$, $\bar{z} = -3/4$, and $\bar{u}_i = u_i(\bar{x}_i, \bar{z}) = -1/4$. The coalition equilibrium with respect to $S = \{3\}$ and $T = \{1, 2\}$ implies

$$\tilde{x}_1 + \tilde{x}_2 = \tfrac{1}{2},\ \tilde{p}_1 = \tilde{p}_2 = \tfrac{1}{16},$$
$$\tilde{x}_3 = \tfrac{1}{2},\ \tilde{p}_3 = \tfrac{1}{4},\ \tilde{z} = -\tfrac{1}{8} - \tfrac{1}{4} = -\tfrac{3}{8}, \text{ and}$$
$$\tilde{u}_3 = u_3(\tilde{x}_3, \tilde{z}) = \tfrac{1}{2} - \tfrac{3}{8} = \tfrac{1}{8}.$$

Thus, $w^\gamma(\{3\}|\{1, 2\}) = 1/8$. The Pareto-efficient emission levels are $p_i^* = 1/36$, $z^* = -1/12$. The payoff of $\{3\}$ corresponding to the worldwide treaty (see (15.4)) is

$$u_3^* = u_3(x_3^*, z^*) = \tfrac{1}{6} - \tfrac{1}{12} = \tfrac{1}{12}.$$

Since $w^\gamma(\{3\}|\{1, 2\}) = 1/8 > 1/12$, country 3 would not be willing to sign the treaty if countries 1 and 2 were to form a coalition. Country 3 would, however, be willing to sign the treaty if countries 1 and 2 did not form a coalition, since $w^\gamma(\{3\}) = -1/4 < 1/12 = u_3^*$. Note that

$$w^\gamma(\{1, 2\}) = -\tfrac{1}{4} \geq w^\gamma(\{1\}) + w^\gamma(\{2\}) = -\tfrac{1}{2}, \text{ and}$$
$$w^\gamma(\{3\}|\{1, 2\}) = \tfrac{1}{8} \geq w^\gamma(\{3\}) = -\tfrac{1}{4}.$$

This means that formation of coalition $T = \{1, 2\}$ is Pareto-improving compared to the disagreement equilibrium. Yet it is bad for T in the dynamic time-path sense, since country 3 will then be no longer willing to sign the worldwide treaty under which the payoff of T $(= 1/6)$ is still higher. We may summarize the result in terms of the following diagram, which is a suitable adaptation from Bhagwati [2] in the context of PTAs.

The figure illustrates metaphorically the dynamic time-path considered in the example. The formation of the coalition $T = \{1, 2\}$ may Pareto-improve world welfare immediately from $\bar{u}$ $(= \Sigma \bar{u}_i)$ to $\tilde{u}$ $(= \Sigma \tilde{u}_i)$, but not thereafter, and it may never reach u^* — the maximum world welfare — as country 3 will then have no incentive sign the worldwide treaty. Since $u_1^* + u_2^* = 1/6 > w^\gamma(\{1, 2\}) = -1/2$ and $u_3^* = 1/12 > -1/4$, if no such coalition forms, the dynamic time-paths would eventually lead to u^* since each group of countries, including $S = \{3\}$ and $T = \{1, 2\}$, would find it advantageous to sign the worldwide treaty.

Carraro and Siniscalco [3] and Barrett [1] also analyze the possibility of cooperation among countries on global pollution. Their approach is based on

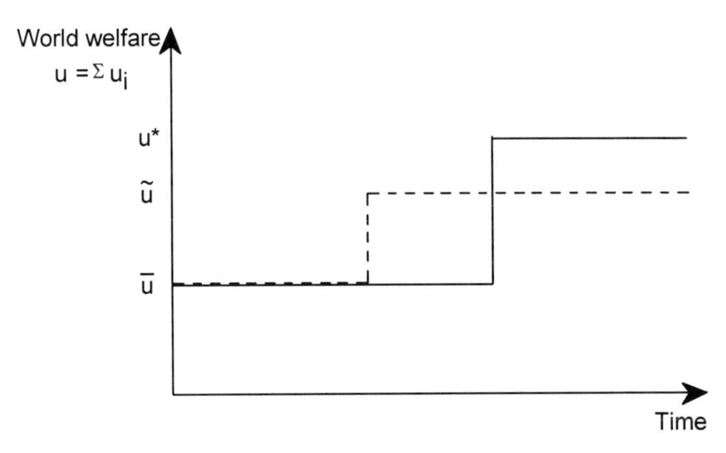

Figure 15.1

the concept of coalition stability (see d'Aspremont and Gabszewicz [8]), borrowed from the industrial organization literature on cartels. They assert that only small subset of countries can ever emerge as coalitions and sign a treaty among themselves (see Tulkens [12] for a candid criticism of this view vis-á-vis the one in Chander and Tulkens [7]). More specifically, as can be shown, in terms of their approach $T = \{1, 2\}$ in the example considered above is a stable coalition. Our analysis above clarifies that this might be so if coalition T is assumed to be concerned only with its immediate (static) payoff and not with the payoff it can achieve in the dynamic time-path sense.[5]

It should however be clarified that all coalition formations are not bad. For example, in the three-countries case considered above, if $u_3(x_3, z) = x_3$ (that is, country 3 does not care about the environmental quality), then formation of $T = \{1, 2\}$ is not bad even in the dynamic time-path sense.

15.3 Conclusion

In this paper we have shown that in an international economic-ecological setting, the formation of a coalition by a subset of countries can imperil the emergence of a worldwide treaty. Such a coalition formation can upset the balance of gains and losses from a worldwide treaty and make it unacceptable to some countries.

Finally, it may be noted that the sign of externalities created by a coalition formation on nonmembers is positive in our context, which is the opposite of

[5] It is interesting to note in this context that an international treaty which has been recently under negotiations, namely the so-called Comprehensive Test Ban Treaty or CTBT, can come into force *only* if *all* the current and potential nuclear powers sign it by a specific date. This is perhaps a case of practice running ahead of our theory.

that in the case of PTAs in the trade context. Accordingly, coalition formation in our case reduces the incentives of nonmembers to join the coalition, whereas in the case of PTAs, it must reduce the incentives of member countries to let new members join the coalition.

References

1 Barrett, S., Self-enforcing international environmental agreements, Oxford Economic Papers, 46 (1994), 878–94.

2 Bhagwati, J., Regionalism and multilateralism: An overview, in: Jaime de Melo and Arvind Panagariya (eds.), New Dimensions in Regional Integration, Cambridge University Press, Cambridge, 1993.

3 Carraro, C. and D. Siniscalco, Strategies for the international protection of the environment, Journal of Public Economics, 52 (1993), 309–28.

4 Chander, P., Dynamic procedures and incentives in public good economies, Econometrica, 61 (1993) 1341–54.

5 Chander, P. and H. Tulkens, Theoretical foundations of negotiations and cost sharing in transfrontier pollution problems, European Economic Review, 36(2/3) (1992), 288–99.

6 Chander, P. and H. Tulkens, A core-theoretic solution for the design of cooperative agreements on transfrontier pollution, International Tax and Public Finance, 2(2) (1993) 279–94.

7 Chander, P. and H. Tulkens, H. The core of an economy with multilateral environmental externalities, International Journal of Game Theory (1997).

8 d'Aspremont, C. and J. Gabszewicz, On the stability of collusion, in: J.E. Stiglitz and G.F. Mathewson (eds.), New Developments in the Analysis of Market Structure, M.I.T. Press, Cambridge, MA, 1986, pp. 243–64.

9 Hart, S. and M. Kurz, Endogenous formation of coalitions, Econometrica, 51(4) (1983) 1047–65.

10 Heal, G., International negotiations on emission control, mimeo, Columbia University, 1993.

11 Srinivasan, T.N., (1995) Common external tariffs of a customs union: The core of identical Cobb–Douglas tastes", mimeo, Yale University, 1995.

12 Tulkens, H., Cooperation vs. free riding in international environmental affairs: Two approaches, in: N. Hanley and H. Folmes (eds), Game Theory and the Environment, Edward Elgar, London, 1997.

Trade, Growth, and Development
G. Ranis and L.K. Raut

CHAPTER 16

Efficiency and Market Structure: Testing for Profit Maximization in African Agriculture[1]

Christopher Udry

Economic Growth Center, 27 Hillhouse Avenue, New Haven, CT 06511

16.1 Introduction

A central question in development economics is the extent to which the rural environment is characterized by competitive markets. The answer has direct implications for the efficiency of the allocation of resources, for the design of economic policy, and for the choice of appropriate analytic methods. Hence, an important task of empirical development economics is to provide a characterization of the market environment in rural areas of poor countries. Much of the relevant literature has been concerned with the existence of well-behaved labor and land markets (e.g., Benjamin [7], Collier [16], Rosenzweig [35], Pitt and Rosenzweig [34], Binswanger and Rosenzweig [10]), but there has also been extensive interest in intertemporal and insurance markets (e.g., Morduch [32], Chaudhuri [14], Deaton [17], Carter [12], Townsend [43]). For historical reasons, the task of an empirical characterization of rural market structure is furthest advanced in Asia and Latin America. Relatively little is known of the extent or characteristics of markets in rural Africa. The goal of this paper is to improve this base of knowledge by using agronomic information to provide evidence on the basic characteristics of rural markets in a variety of African settings.

An implication of the availability of a complete set of competitive markets that greatly simplifies empirical work is *separation*: the household maximizes profit in its production decisions without regard to its preferences (Krishna [30]). The null hypothesis of complete markets may seem incredible and thus

[1] Marcel Fafchamps, Elaina Rose, John Strauss, Ann Velenchik and Fred Zimmerman provided valuable comments and advice on an earlier draft. I thank the International Crop Research Institute for the Semi-Arid Tropics and the International Food Policy Research Institute for making data available. Ellen Payongayong provided vital assistance with the IFPRI Kenya data. Financial assistance from the NSF is gratefully acknowledged.

unworthy of testing. It is true, of course, that nowhere in rural Africa (or anywhere else) does there exist an economy characterized by a complete set of competitive markets. However, a broad variety of spot markets do exist, and in some contexts appear to operate competitively (e.g., Hays [23], Jones [27], Pingali, Bigot and Binswanger [33]). Informal mechanisms exist which might fill the same functions as other competitive markets (Udry [44], Aryeteey [1], Guyer [22]). Moreover, the separation result is robust to the nonexistence of some markets (for example, it is retained if there is no market for one of the factors of production). The most practical reason for testing the separation hypothesis, however, is that separation is commonly invoked in empirical studies of agricultural production in poor countries (Singh, Squire and Strauss [40]).[2]

The separation theorem, therefore, serves as a simple and useful benchmark for further analysis. In addition, much can be learned from rejections of the exclusion restrictions that are implied by separation. The *pattern* of rejections has implications for the structure of rural markets. This pattern, and its variation over space, can provide useful information about the geography of markets.

The empirical work of this paper consists of a series of tests of separation in two African contexts. The African setting is of interest because the quantitative literature on rural factor markets in Africa remains quite limited, and that which does exist contains arresting suggestions of empirical regularities which look quite different from other parts of the world. The most important example concerns a "stylized fact" of traditional agriculture (Bardhan [4]): the inverse relationship between farm size and yield. This pattern has been established in a variety of data sets from Asia and Latin America. The limited evidence from Africa, however, looks different. Hill in West Africa and Kevane [29] in Sudan both show (using small data sets) the opposite relationship — larger farms have higher yields than smaller farms. In contrast, Gavian and Fafchamps [21], and Collier [16] find that in Madagascar, Niger and Kenya, respectively, there is evidence of the conventional inverse relationship.

Section 16.2 describes the separation result, and discusses the exclusion restrictions that it implies. It also raises the most important econometric issue which arises in this context — the possibility that apparent violations of separation are the result of unobserved variation in farm characteristics. Section 16.3 presents the data from a wide variety of sources which are used to implement the tests. The results are presented, country by country, in Section 16.4.

[2] Separation is also invoked, usually without comment or testing, in studies of production in rich countries.

16.2 Separation and Profit Maximization

16.2.1 A review of the separation result and its relation to profit maximization

Consider a household with a conventional utility function over vectors of goods (c_{st}) and leisures (ℓ_{st}) in state $s \in S$ of period $t \in T$ (so c_{st} is the vector of goods consumed if state s occurs in period t). For simplicity, assume that both S and T contain finite numbers of elements. This household faces a complete set of competitive markets. Let p_{st} and w_{st} be the price vectors of the state-contingent goods and labor. Let E_{st} be the household's endowment of time in state s and period t, q_{st} be the price of farm output, L_{st} be the labor used on the household's farm, A_{st} be the vector of land inputs, and r_{st} be the vector of prices of these land inputs (to simplify notation, I have assumed that there are no inputs other than land and labor — these can be accommodated with no substantive change in the analysis). $F_{st}(L_{st}, A_{st})$ is a set of (state-contingent) production functions. Then the household's problem is to

$$\max_{c_{st},\ell_{st},L_{st}} \quad U(c_{st}, \ell_{st}) \qquad \text{subject to} \tag{16.1}$$

$$\sum_{t\in T, s\in S} \{w_{st}E_{st} + \Pi_{st} - p_{st}c_{st} - w_{st}\ell_{st}\} \geq 0 \tag{16.2}$$

$$\Pi_{st} = q_{st}F_{st}(L_{st}, A_{st}) - w_{st}L_{st} - r_{st}A_{st}, \ L_{st}, A_{st} \geq 0 \text{ and} \tag{16.3}$$

$$c_{st}\ell_{st} \geq 0, \ \ell_{st} \leq E_{st}. \tag{16.4}$$

The problem is recursive. With weak conditions on $U(\cdot)$, (16.2) is binding at the solution and the maximized value of $U(\cdot)$ is increasing in Π_{st}. L_{st} and A_{st} appear only in (16.3). Hence (16.1)–(16.4) can be solved by first maximizing profit Π_{st} with respect to L_{st} and A_{st} and then maximizing utility. That is, the problem (16.1)–(16.4) is equivalent to:

$$\max_{c_{st},\ell_{st}} U(c_{st}, \ell_{st}) \qquad \text{subject to} \tag{16.5}$$

$$\sum_{t\in T, s\in S} \{w_{st}E_{st} + \Pi^*_{st}(q_{st}, w_{st}, r_{st}) - p_{st}c_{st} - w_{st}\ell_{st}\} \geq 0, \text{ and} \tag{16.6}$$

$$c_{st}, \ell_{st} \geq 0, \ \ell_{st} \leq E_{st} \tag{16.7}$$

where

$$\Pi^*_{st}(q_{st}, w_{st}, r_{st}) = \max_{L_{st},A_{st}} q_{st}F_{st}(L_{st}, A_{st}) - w_{st}L_{st} - r_{st}A_{st}, L_{st}, A_{st} \geq 0. \tag{16.8}$$

Households maximize profits in each state and on each date and, therefore, production decisions on any plot depend *only* on prices and on the characteristics of that plot.[3] The simplification is extraordinary: from a problem (16.1)–(16.4) of dealing with a (risk averse) household's dynamic behavior in a risky environment, we arrive at a very simple static profit-maximization problem.[4]

Equation (16.8) provides the basis of the empirical strategy of the paper. Within any group of plots subject to the same prices, input choices (and outputs) are identical on identical plots. This paper present two series of empirical tests based on this restriction. The first examines the distribution of yields and inputs on similar plots. Since separation implies that output and inputs are identical on identical plots, we examine the dispersion of yields and input intensities on apparently identical plots. Is this dispersion so large that the null hypothesis of separation must be rejected?

The second series of tests is based on the strong exclusion restrictions on input demand and output supply functions which are implied by (16.8): input demand and output supply functions depend on prices and on plot characteristics but on nothing else. Controlling for prices and plot characteristics, is there a strong enough correlation between input use (or output) and other household characteristics that we must reject the null hypothesis of complete markets?

Now consider production on a particular plot i. From (16.8), yield (output per unit area) on that plot (in state s of period t, but I drop those subscripts to simplify notation) depends only on prices and the characteristics of the plot itself:

$$Q^i = G(A^i, w, r, q). \tag{16.9}$$

[3]Note that the choice of the household as the unit of analysis has no effect on this particular result. If markets are complete, separation holds for all standard models of the household – the cooperative bargaining models, the collective model, and even for most non-cooperative models of the household.

[4]It should be noted that (16.8) does not describe expected profit maximization. Profits are maximized separately in each state of nature. It is unrealistic, of course, to presume that households can adjust labor and land inputs in each state of nature, because some inputs must be committed before the state is realized. To examine the consequences of this fact of life, suppose that in each period t, labor (L_t) and land (A_t) are chosen before the state of nature for that period is realized. Factor prices, therefore, are no longer state-contingent. The budget constraints (16.2) and (16.3) now become:

$$\sum_t [w_t E_t + \Pi_t - w_t \ell_t - \sum_s p_{st} c_{st}] \geq 0 \tag{2'}$$

$$\Pi_t = \sum_s q_{st} F_{st}(L_t, A_t) - w_t L_t - r_t A_t \tag{3'}$$

The problem remains recursive, and (16.8) becomes

$$\Pi_t^* = \max_{L_t, A_t} \sum_s q_{st} F_{st}(L_t, A_t) - w_t L_t - r_t A_t. \tag{8'}$$

Production decisions remain a function only of prices and the characteristics of the plot.

Consider a set of plots K, *each of which is subject to the same prices* (w, r, and q). Then a first order Taylor expansion of (16.9) across plots $i \in K$ implies

$$Q^i - \bar{Q} \approx \partial G(A)/\partial A \cdot [A^i - \bar{A}] \ \ \forall i \in K, \tag{16.10}$$

where $\bar{A}$ is the mean area of the plots planted with the same crop and subject to the same prices. Within groups of plots subject to the same prices (in this paper, I will assume that plots in a particular village face the same prices), the deviation of yield on a plot from the group average yield is a function only of the deviation of the plot's area from the average area of plots in the group. With a flexible specification for the function $\partial G(A)/\partial A$, (16.10) is an approximation to an arbitrary concave production function. If one assumes that the technology is CES, so that $A^i \cdot Q^i = [\delta(A^i)^{-\rho} + (1-\delta)(L^i)^{-\rho}]^{-\nu/\rho}$, then (16.10) becomes

$$\ln(Q^i) - \ln(\bar{Q}) = \frac{\rho \cdot (v-1)}{\rho + v} \cdot [\ln(A^i) - \ln(\bar{A})]. \tag{16.10$'$}$$

The first evidence that I examine is the extent to which agricultural data from Africa matches the theoretical prediction that yield is identical on identical plots. This evidence is summarized by estimates of the distribution of deviations of actual yield from expected yield. The second type of evidence is drawn from tests of the exclusion restrictions implied by (16.10): if markets are complete, then yield depends only on prices and plot characteristics, not on other characteristics of the household which cultivates the plot.[5]

16.2.2 The empirical strategy

Much of the empirical work of this paper is based on a series of tests of the exclusion restrictions implied by (16.10) and (16.10′). Each of the tests has power against different violations of the assumptions which imply separation. The pattern of violations, therefore, can provide information about the structure of rural markets. It is to be expected that the pattern of violations will vary across areas.

Some issues remain before (16.10) or (16.10′) can be estimated. Contrary to (16.10) and (16.10′), of course, A^i is not scalar-valued. Plot i has characteristics in many dimensions — its area, its soil quality, its topography, the particular amount of rain it receives in state s. Moving to a notation more amenable to a discussion of econometric specification, let A^i be composed of two components: X_{vhtci}, which is a vector summarizing the information

[5] Crop choice, as well as input choice conditional on crop, depends only on prices and plot characteristics if markets are complete. Tests based on this fact would provide another avenue through which agronomic information could be used to test the separation hypothesis.

about plot i (planted to crop c in year t by household h in village v) available to the researcher and ϕ_{vhtci}, a set of unobserved characteristics of the plot (including plot level random shocks as well as more permanent unobserved characteristics), so $G(A) \equiv G(X, \phi)$. Prices are assumed to be the same for all plots in the same village v in a particular year t. Equation (16.10) becomes:

$$Q_{vhci} = \bar{Q}_{vtc} + \frac{\partial G(X,\phi)}{\partial X} \cdot [X_{vhtci} - \bar{X}_{vtc}] + \frac{\partial G(X,\phi)}{\partial \phi} \cdot [\phi_{vhtic} - \bar{\phi}_{vtc}]. \quad (16.11)$$

ϕ is unobserved, so the final part of the expression is subsumed in an error term (about which more will be said) and I approximate $\partial G(\ )/\partial X$ with a linear function:

$$\begin{aligned} Q_{vhtchi} &= \bar{Q}_{vtc} + [X_{vhtci} - \bar{X}_{vtc}]\beta + \varepsilon_{vhtci} - \bar{\varepsilon}_{vtc} \\ &= X_{vhtci}\beta + \lambda_{vtc} + \varepsilon_{vhtci}. \end{aligned} \quad (16.12)$$

(16.12) is the basic empirical framework of the paper.[6]

It is now possible to define the two series of empirical tests of the paper. First, is output identical on identical plots? This is a question about the characteristics of ε. If the null hypothesis of complete markets is correct, then the distribution of ε is determined by the distribution of ϕ. Is there a sense in which we can say that in a particular sample, the distribution of estimated residuals is not compatible with the null hypothesis? Such tests can be performed if relatively strong assumptions about ϕ are valid. Second, (16.8) implies a series of very strong exclusion restrictions in (16.12). In particular, input demand and output supply depend on prices and plot characteristics and on nothing else. Other household characteristics such as farm size, household composition, or wealth have no role in (16.12).

The first series of tests is based on the maintained assumption that the allocation of factors of production across the various plots controlled by an individual is efficient, so that the separation hypothesis is true across these plots. Consider the following pair of regressions:

$$Q_{vhtci} = X_{vhtci}\beta + \lambda_{vtc} + \varepsilon^{v}_{vhtci} \quad (16.13a)$$

$$Q_{vhtci} = X_{vhtci}\delta + \lambda_{vhtc} + \varepsilon^{v}_{vhtci}. \quad (16.13b)$$

(16.13a) is identical to (16.12), while (16.13b) replaces the village-year-crop fixed effect with a household-year-crop fixed effect. Attention is restricted

[6] Model specification choices must be made to characterize $\partial G(\ )/\partial X$. The econometric work is cast in per-hectare terms, with log yield and log input intensity as dependant variables. I estimate $\partial G(\ )/\partial X \times X$ as $X\beta$, where X includes log plot area and dummy variables characterizing the plot's topography, location and soil type. In this specification, the coefficient on, say, a dummy variable representing a soil type can be interpreted as the percentage increase in the yield of a plot of that soil type over the base soil type. There is concern about heteroskedasticity and the standard errors are appropriately adjusted.

to one farmer in each household, so that the distribution of ε^h reflects the variation in yields across apparently identical plots cultivated by a single individual. The distribution of ε^h, therefore, is a function of the distribution of ϕ across plots controlled by an individual. Now consider ε^v, the error term defined in (16.13a). If we maintain the hypothesis that the distribution of ϕ across plots within a household-year-crop group is the same across plots within a village-year-crop group, then if separation is valid, the distribution of ε^v should be the same as that of ε^h. The first series of tests, therefore, is based on a comparison of the distributions of ε^v and ε^h. The second line of questions is based on the exclusion restrictions implied by (16.8). Equation (16.8) implies that $\gamma = 0$ in the regression

$$Q_{vhtci} = X_{vhtci}\beta + E_{vhtci}\gamma + \lambda_{vtc} + \varepsilon_{vhtci}, \tag{16.14}$$

where E_{vhtci} is the exclusion restriction under consideration. I consider tests based on three sorts of exclusion restrictions: farm size, household demographics, and wealth or cash flow.

The first exclusion restriction is that the total area cultivated by the household should have no effect on output on a particular plot, if the null hypothesis is correct. This test is closely related to the large literature which finds an inverse relationship between farm size and yield, or farm size and labor demand. (A recent paper in this tradition is Barrett [6]. If agricultural production is governed by a CES technology with constant returns to scale, then the log of farm area should have a coefficient of zero in a regression of the log of yield on the log of farm area (see equation (16.10′), $v = 1$ is the case of constant returns to scale). In fact, it is often found that the coefficient of farm area is negative. The earliest and most popular explanation for this regularity is market failure of some sort — small and large farms face different opportunity costs and hence optimally choose different mixes of inputs on their farms. This type of explanation is sensible and probably correct in many instances, but the connection between the result and the conclusion is not direct. Bhalla [9] and Benjamin [8] argue that unobserved variations in land quality could underlie the oft-observed inverse relationship between farm size and yield — larger farms are less fertile, and thus are optimally farmed less intensively. The analysis in this paper is conducted at the plot level, as specified in (16.12), with E_{vhtci} defined as the area cultivated by the household on plots other than plot i. I show below that this mitigates the problem of omitted variables bias.

The second series of tests is based on household demographic measures. This is a standard test of the assumptions which imply separation. Do larger households (controlling for land area) farm more intensively? Benjamin [8], Kevane [29] and Pitt and Rosenzweig [34] are key recent papers. The third set of tests is concerned with correlations between input demand and nonfarm wealth, income, and cash flow (Swamy [42], Chaudhuri [14], Morduch [32], Rosenzweig and Binswanger [37]).

Unobserved plot quality variation and type I error. Each of these tests, however, is subject to a similar important econometric caveat. Under separation, cropping decisions depend only on prices and plot characteristics. I can account for prices through village fixed effects provided that the model is linear. Plot characteristics, however, are problematic. There is undoubtedly unobserved variation in land quality. Therefore, there is a classic omitted variables problem and the bias induced by this problem could lead us to reject the null hypothesis of separation when separation is in fact true.

Consider the relationship between farm size and output. Suppose that in fact $F(\cdot)$ exhibits constant returns to scale and that separation is true. Suppose, however, that there are unobserved (to the analyst) plot characteristics (soil type, weather shocks, etc.), so that $Q_{vhi} = F(L_{vhi}, T_{vhi})$, where $T_{vhi} = A_{vhi}(1 + \theta_{vhi})$, A_{vhi} is observed plot area, and θ_{vhi} is an unobserved land-augmenting plot characteristic (dropping tc subscripts to shorten the notation). A regression of output on T would yield an estimated coefficient of one, given the validity of the separation hypothesis. However, the regression of output on plot area is subject to omitted variables bias because θ_{vhi} is not included in the regression. The sign and size of that bias depends on the covariance of A and θ. If the unobserved variation in land quality is uncorrelated with the area of the plot then the regression is subject to attenuation bias and we will find a less than proportional relationship between output and plot area.

Worse, there is reason to expect that unobserved land quality is worse on larger plots than on smaller plots. Suppose, for example, that all markets exist and operate smoothly except the labor market. If each household has access to one type of land (θ_{vhi}) and cultivates only one plot, then in each state of each period, each household's production problem is to solve:

$$\max_{A_{vhi} \geq 0} qF(L_{vh}, A_{vhi}(1 + \theta_{vhi})) - r(\theta_{vhi})A_{vhi}, \qquad (16.8'')$$

where L_{vh} is the amount of labor time the household chooses to spend working. Choose units so that for some plot k, $\theta_k = 0$. Then define $r = r(\theta_k)$. Competition in the land market ensures that $r_{vhi} = r(1 + \theta_{vhi})$. By the implicit function theorem, $\delta A/\delta\theta = -[A_{vhi}/(1 + \theta_{vhi})] < 0$. Hence plot quality will be negatively correlated with plot size in the cross-section. A regression of yield on T would produce an estimated coefficient of 0. The omission of land quality from the regression of yield on land area, however, will bias the estimated coefficient on land area down from zero because of the negative correlation between unobserved land quality and land area.

It is also true, of course, that if $F(\ )$ has DRTS [$v < 1$ in (16.10')] then there will be a direct technological explanation for the inverse relationship. In order to eliminate the technological possibility and to attenuate the problem of unobserved land quality, I propose to estimate (16.14) with E_{vhtci} equal to the area of *other* household plots. (Blarel et al. [11] conduct a similar exercise, but don't examine the implications for separation.) Are plots of a given size cultivated differently, depending upon the total area on other plots that are cultivated by the same household?

It is clear that this specification eliminates the potential technological explanation for the inverse relationship. It also mitigates, but does *not* eliminate, the problem of omitted variables bias due to unobserved variation in plot quality. Under the null hypothesis of separation, we expect γ in (16.14) to be zero. Any attenuation bias due to unobserved variation in land quality that is uncorrelated with farm area, therefore, cannot cause a false rejection of the null. However, there is still reason to expect an inverse correlation between the quality of a given plot and the area cultivated by the household on other plots. To see why, consider the household's problem when it cultivates many plots of different qualities:

$$\max_{A_{vhi}, L_{vhi} \geq 0} \sum_{i \in P} [F(_{vhi}, A_{vhi}(1 + \theta_{vhi})) - r \cdot (1 + \theta_{vhi}) A_{vhi}] \qquad (16.8''')$$

$$\text{s.t.} \quad \sum_{i \in P} L_{vhi} \geq E_{vh}$$

If $F(\,)$ is constant returns to scale, (16.8''') implies that $\sum_{i \in P}(A_{vhi} \cdot \theta_{vhi})/E_{vh}$ is a constant which depends only on r. Hence, if for a plot i, θ_{vhi} is particularly low, then the household optimally farms that land with relatively low intensity. At the same time (given E and r), the household optimally acquires more land, either on plot i or on other plots. If some of the land acquired is on other plots, there will be a negative correlation between the quality of plot i and the area cultivated by the household on other plots. We would then estimate $\gamma < 0$ even if separation is valid.

The tests of exclusion restrictions with respect to demographic variables are also subject to misinterpretation as a consequence of unobserved variation in plot quality. Suppose again that labor markets do not work well, but that land rental markets remain perfect, so that separation still holds. It remains the case that under constant returns to scale the ratio of quality adjusted land holdings to household labor availability is a constant. If households can choose the quality of the land that they cultivate (choosing which plots to cultivate), then household size will be positively correlated with unobserved land quality. Unobserved land quality is also a potential cause of misinterpretation of tests of exclusion restrictions with respect to cash flow or wealth. Unobserved good quality land can be associated with high income, wealth and cash flow, hence potentially causing γ to appear positive even if separation is true.

There are a variety of avenues for exploring the potential importance of unobserved variation in plot quality beyond the strategy of attenuating the bias by using plot level information.

(a) The natural solution, of course, is to use IV estimates. Unfortunately, while there are plenty of variables that are correlated with farm size, plot size, household demographics, and nonfarm income, it is very difficult to make the case that any of these are uncorrelated with unobserved land quality. I have not, therefore, been able to identify likely instruments in any of the available datasets.

(b) Unobserved land quality variation is less of an issue in one of the data sets used in this paper than in most data from developing countries. The ICRISAT Burkina Faso data contains rich information on plot characteristics. For the other data, the use of village-crop-year fixed effects, as well as some plot-specific information, permits a finer degree of control for land quality than has been possible in most previous investigations of related issues (Benjamin [7, 8]). In addition, some diagnostic lessions may be drawn from the exercise of conducting tests of the exclusion restrictions with and without a subset of the plot characteristics data. If unobserved land quality is positively correlated with observed land quality, and if inclusion of observed measures of land quality reduces the effect of the test variables, this heightens our concern that unobserved land quality may be causing false rejections of the null of separation.

(c) It is possible (though difficult) to trace plots over time in some of these data. This is implemented in the Kenya data set discussed below. Plot fixed effects, then, can control for unobserved land quality. Many observations are lost due to the difficulty of matching plots, and there is heightened concern regarding measurement error in the variables tested for exclusion (Ashenfelter, Deaton and Solon [2]), but the additional confidence gained with respect to land quality is valuable.

What causes rejections of separation? Suppose that it is found that output on one plot controlled by a household decreases with the amount of land on other plots cultivated by the household, and increases with the size of the household, and that it is concluded that these correlations are not attributable to specification error. It would seem that the implication is that both labor and land markets in the economy under investigation are imperfect, thus preventing households from realizing the gains which are potentially available from trading land or labor to equalize marginal products across plots controlled by different households. It is true, in fact, that (16.14) can detect violations of the assumption of perfect labor and/or land markets.[7] How-

[7] Benjamin [7] provides a thorough discussion of the power of a similar test to detect specific violations of the assumption of perfect labor markets when E contains demographic

ever, Srinivasan [41] (and more recently Feder [20], Eswaran and Kotwal [18], Carter and Wiebe [13], Banerjee and Newman [3], Barrett [6]) show that violations of the assumption of complete insurance or intertemporal markets can also cause rejections of the exclusion restrictions with respect to household demographics or the area of other plots.

To examine the import of this ambiguity, consider the simplest model in which it can arise. There are two states of nature (the probability of state 1 is π) with multiplicative production risk (θ_i). A household with fixed land and labor endowments makes production and labor-leisure decisions before the resolution of output uncertainty. With complete labor and insurance markets, (but no land market) the household's problem is to:

$$\begin{aligned} &\max_{c_1, c_2, \ell, L^f} \pi U(c_1, \ell) + (1-\pi) U(c_2, \ell) \\ &\text{s.t. } p_1 c_1 + p_2 c_2 + w\ell \leq (p_1, \theta_1 + p_2\theta_2) f(L^f) - wL^f + wE \end{aligned} \tag{16.15}$$

where L^f is the amount of labor used on the farm, and E is the household's labor endowment. Separation holds, and the household maximizes profit on the farm. In contrast, if there are complete insurance markets, but no labor market, then the household solves:

$$\begin{aligned} &\max_{c_1, c_2, \ell} \pi U(c_1, \ell) + (1-\pi) U(c_2, \ell) \\ &\text{s.t. } p_1 c_1 + p_2 c_2 \leq (p_1\theta_1 + p_2\theta_2) f(E-\ell). \end{aligned} \tag{16.16}$$

This is analogous to Sen's [39] formulation of an autarkic farm household under certainty and separation is violated. Farm output increases in the household's labor endowment.[8]

Finally, if the labor market is complete, but there is no insurance, the household solves:

$$\begin{aligned} &\max_{c_1, c_2, \ell, L^f} \pi U(c_1, \ell) + (1-\pi) U(c_2, \ell) \\ &\text{s.t. } p_1 c_1 + w\ell \leq p_1\theta_1 f(L^f) - wL^f + wE \\ &\qquad p_2 c_2 + w\ell \leq p_2\theta_2 f(L^f) - wL^f + wE. \end{aligned} \tag{16.17}$$

The first-order conditions of this problem include

$$0 = \lambda_1(\theta_1 p_1 f'(L^f) - w) + \lambda_2(\theta_2 p_2 f'(L^f) - w) \tag{16.18}$$

variables. He examines the specific alternative hypotheses that there is a binding constraint on obtaining off-farm employment, a binding constraint on hiring labor, and the more general possibility that the returns to on- and off-farm labor differ. Bardhan [4] offers a similar treatment of the power of the test when E contains farm size.

[8] To be more precise, if derivatives are denoted by subscripts, λ is the Lagrange multiplier, and $U_{c\ell} = 0$, then $dL^f/dL = [\pi U^1_{\ell\ell} + (1-\pi) U^2_{\ell\ell}]/[\pi U^1_{\ell\ell} + (1-\pi) U^2_{\ell\ell} + \lambda f_{LL}$ which is positive and less than one.

where λ_i is the marginal utility of income in state i. Separation does not hold because input decisions depend on the ratio of the marginal utility of income in the two states. An increase in the household's endowment of labor will affect this ratio (increasing λ_1 relative to λ_2 if $\theta_1 > \theta_2$ and the household has diminishing absolute risk aversion) and will thus change input decisions.

The violation of separation along a particular dimension — in this instance the correlation between household demographics and production — is not conclusive evidence of a particular market failure. However, we can examine the separation hypothesis along various dimensions (in this paper, effects of household demographics, farm size, and nonfarm income). Particular configurations of results that are revealed by this series of tests *can* provide evidence of particular market failures. A complete taxonomy is possible, but it would be lengthy. Instead, I provide this analysis in the discussion of the results of the case studies in Kenya and Burkina Faso (Section 16.4).

16.3 Data

16.3.1 Burkina Faso

The Burkina Faso data used for this study are drawn from a farm household survey conducted by the International Crops Research Institute for the Semi-Arid Tropics (ICRISAT) (see Matlon [31] for documentation of the survey). The survey was a four-year panel study (1981–85) of 150 households in six villages in three different agro-climatic zones of Burkina Faso. This study uses data from the first three agricultural seasons of the survey (1981–83), during which the most detailed agronomic information was collected. During these three seasons, enumerators visited the sample households approximately every 10 days to collect information on farm operations, inputs, and outputs on each of the household's plots since the previous visit. These three seasons of data collection result in 432 household-years of data on agricultural activities, with usable data on a total of 2,576 plots cultivated by household heads.[9] An important advantage of these data for this study is the fact that they contain rich descriptive information concerning the area (measured by the enumerators), topography, location, and soil characteristics of the plots cultivated by the households.

All of the farmers in the survey are poor, with an average income per capita of less than $100 (Fafchamps [19]). The farming system is characteristic of rainfed agriculture in semi-arid Africa: each household simultaneously

[9] Sample households cultivated 4,787 plots over the three years; 132 of these plots did not have their area measured and so were dropped from the sample; while 2,079 plots were excluded because they were cultivated by people other than household heads, thus raising complex issues of intrahousehold resource allocation.

cultivates multiple plots (10 is the median number of plots per household in any year) and many different crops (a median of six different primary crops on the plots farmed by a household in a given year).

16.3.2 Kenya

The Kenya data are drawn from a 1985–87 survey of 617 households conducted by IFPRI in South Nyanza, a sugar-growing area of Kenya (see Cogill, [15] for a description of the survey). Data from two rounds of this survey, covering successive cropping seasons, is used. There is usable data on output on 3,194 cultivated plots over these two seasons. These data do not contain as rich information on plot characteristics as do the data from Burkina Faso, but they do have an important advantage. Many of the plots can be traced over the two seasons, permitting the use of fixed effect estimators to mitigate the problem of unobserved plot characteristics. As in Burkina Faso, the Kenyan households are poor, with an income per capita of about $120 (Kennedy [28]). Households farm multiple plots (an average of six plots per household) and many crops (about 8 different crops), though maize is the dominant crop, accounting for between 15 and 30 percent of all farm area, depending on the season (Cogill [15]). Summary data from both data sets is provided in Table 16.1.

Table 16.1: Descriptive statistics

	Burkina Faso	Kenya
Mean output per ha.	36.13 (1000 CFA)	0.38 (1000 KSh)
Mean labor per ha.	2,127 hous	15 days
Mean manure per ha.	3.28 (1000 Kg)	n.a.
Mean household size	10.17	9.59
Mean cultivated area (ha)	4.79	2.20
Mean nonfarm resource inflows	93.10 (1000 CFA)	2.33 (1000 KSh)
Per capita land (ha/person)		
33rd percentile	0.31	0.09
66th percentile	0.53	0.58

16.4 Testing for Profit Maximization

16.4.1 Burkina Faso

We begin by examining the characteristics of ε^v and ε^h in the two regressions (16.13a) and (16.13b). X_{vhtci} includes the rich set of plot descriptors discussed in Section 16.3.1, and λ_{vtc} and λ_{vhtc} are village-year-crop and household-year-crop fixed effects. The sample is limited to plots cultivated by household heads (rather than all plots within households) because of the evidence presented in Udry [45] that separation is violated across plots controlled by different individuals within households. I maintain the hypothesis that the household head is free to rearrange resources across the plots he controls as an individual, so separation is a maintained hypothesis in (16.13b). ε^h, therefore, is zero in the absence of measurement error in Q, unobserved variation in plot quality, and plot-specific risk (which can be considered a form of unobserved variation in plot quality). In fact, ε^h is not zero: Figure 16.1 reports a kernel estimate of its distribution. Figure 16.2 reports the analogous distribution for plot-specific labor demand.

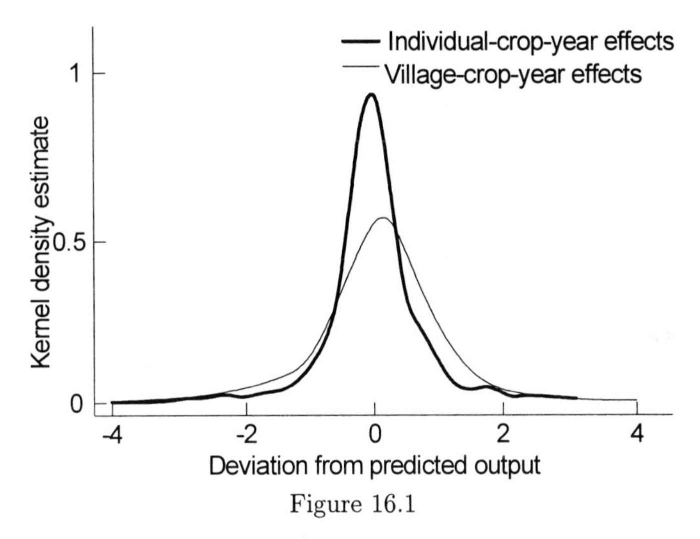

Figure 16.1

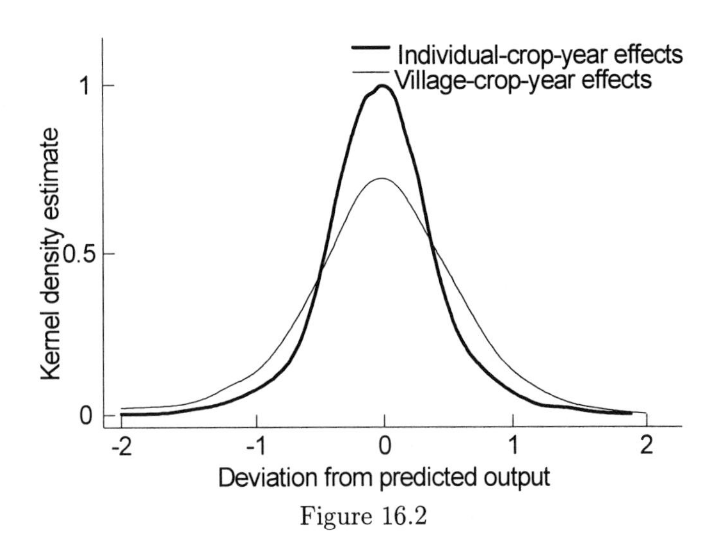

Figure 16.2

Consider the joint null hypothesis that: (1) separation holds within the village and (2) distribution of measurement error in Q and unobserved variation in plot quality is the same across households as it is across plots controlled by an individual household head. Under this null, the distribution of ε^v is the same as that of ε^h; however, the estimates indicate that ε^v has a much more diffuse distribution.[10] There is much more variation in both yield and labor demand per hectare across similar plots controlled by different households within a village than there is across plots controlled by a single individual. Some of this additional variation may be caused by greater unobserved plot quality variation across plots controlled by different households than across plots within a household. The force of this plausible explanation is reduced, however, by recognition of the fact that plots controlled by individual household heads are dispersed throughout the village land, probably as one element in an effort to diversify risk. The much larger dispersion of ε^v than ε^h raises the possibility that the separation hypothesis is violated.

Further evidence that separation is violated is provided by the correlations between plot yield or input intensity and a series of variables which should be orthogonal to yield and input intensity (conditional, of course on plot characteristics and prices). Table 16.2, column 1 reports the results of estimating (16.14) with E_{vhtci} equal to the log of household size. In apparent violation of profit maximization, plot yield is increasing in household size, conditional on prices and all observed characteristics of the plot. The elasticity is about 0.2 and is significantly different from zero at any conventional significance level. It is necessary to interpret these results cautiously because, as noted in Section 16.2.2, they could be an artifact of a correlation between unobserved plot quality and household size. Turn now to an examination of the relationship between plot yield and the area cultivated by the household on other plots. Again in apparent violation of the separation hypothesis, column 2 of Table 16.2 reports a highly significant correlation between plot yield and the area cultivated by the household on other plots. In this instance, however, plot yield increases with the area cultivated on other plots. This is a striking result, because the implication of conventional land and labor market imperfections would have been a negative correlation between farm size and plot yield; similarly (as shown on page 000) unobserved variation in plot quality would be expected to lead to a negative correlation between farm size and plot yield. This result provides further evidence of violations of the separation hypothesis in Burkina Faso, but broadens the range of potential market failures which should be considered: households with larger farms cultivate their land more intensively than other households, raising the possibility of imperfections in credit or insurance markets.

---

[10] Equivalently, the $F(1684, 706)$ test statistic of the hypothesis that $\lambda_{vhtc} = \lambda_{vtc}$ has a value of 1.429 ($p = 0.00$).

Table 16.2: Estimates of the determinants of log plot output per hectare, ICRISAT Burkina Faso data

	Village-year-crop fixed effects			
	Estimate	t	Estimate	t
	(1)	(2)	(3)	(4)
Log plot area	−0.240	−10.85	−0.203	−10.00
Log household size	0.176	5.98		
Log area other plots			0.068	3.74
Log nonfarm income and gifts				
Toposequence				
Top of slope	−0.445	−3.37	−0.418	−3.36
Near top	−0.258	−2.25	−0.249	−2.37
Mid slope	−0.261	−2.31	−0.241	−2.34
Near bottom	−0.225	−2.00	−0.204	−1.99
Soil types				
soil7	−0.163	−1.42	−0.169	−1.46
soil21	0.135	1.00	0.120	1.01
soil31	−0.045	−0.43	−0.065	−0.71
soil32	−0.128	−1.17	−0.131	−1.33
soil33	0.013	0.08	0.058	0.46
soil37	0.001	0.01	−0.004	−0.03
soil35	−0.080	−0.45	0.054	0.41
soil45	−0.081	−0.76	−0.089	−0.89
soil51	0.099	1.22	0.089	1.09
soil1	0.423	3.12	0.412	3.07
soil3	−0.164	−1.78	−0.127	−1.36
soil11	−0.475	−2.19	−0.523	−2.47
soil12	0.082	0.48	0.014	0.09
soil13	0.169	0.71	0.019	0.08
soil34	−0.184	−0.84	−0.226	−1.46
soil46	0.041	0.20	0.003	0.02
soil53	0.265	1.66	0.260	1.60
Compound plot	0.152	2.55	0.193	3.59
Village plot	0.122	2.65	0.125	2.92

Notes: The dependent variable is the log of the value of all plot output per hectare. These are OLS estimates with village-year-crop fixed effects (in the first four columns) or with village-year-crop and household fixed effects (in the final column). The t-ratios are calculated from heteroskedasticity-consistent standard errors.

Village-year-crop fixed effects				Village-year-crop and household fixed effects	
Estimate	t	Estimate	t	Estimate	t
(5)	(6)	(7)	(8)	(9)	(10)
−0.217	−8.71	−0.241	−9.47	−0.303	−10.64
		0.089	2.04	−0.010	−0.07
		0.004	0.11	−0.144	−2.59
0.070	4.10	0.052	2.54	0.109	2.27
−0.477	−3.15	−0.503	−3.28	−0.415	−2.77
−0.369	−2.88	−0.362	−2.76	−0.296	−2.35
−0.308	−2.43	−0.296	−2.27	−0.202	−1.63
−0.371	−2.90	−0.372	−2.81	−0.251	−1.99
−0.100	−0.82	−0.115	−0.93	−0.117	−0.83
−0.132	−0.63	−0.130	−0.52	−0.249	−1.06
−0.159	−1.33	−0.141	−1.18	−0.101	−0.81
−0.199	−1.61	−0.201	−1.63	−0.156	−1.21
0.016	0.10	−0.002	−0.01	0.208	1.05
0.001	0.01	0.014	0.10	0.125	0.87
−0.019	−0.10	−0.043	−0.23	0.010	0.05
−0.238	−1.87	−0.233	−1.83	−0.100	−0.76
0.173	1.78	0.166	1.71	0.101	0.85
0.359	2.49	0.367	2.51	0.380	1.87
−0.147	−1.47	−0.164	−1.65	−0.052	−0.43
−0.792	−2.04	−0.863	−2.17	−1.185	−3.09
0.013	0.05	0.125	0.45	0.349	1.17
0.097	0.27	0.496	1.20	0.224	0.70
−0.181	−0.82	−0.181	−0.81	−0.108	−0.51
−0.046	−0.19	−0.048	−0.21	−0.105	−0.43
0.189	1.01	0.180	0.98	0.285	1.57
0.174	2.52	0.149	2.13	0.098	1.30
0.133	2.51	0.144	2.73	0.125	2.21

Following Chaudhuri [14], I explore the possibility of credit constraints by calculating the correlation between yield (or input intensity) and short-run measures of cash flow. It is possible to use ICRISAT data to calculate income from nonfarm activities and gifts for the four months preceding clearing activities for the single annual crop season. Estimates of (16.14) with E_{vhtci} equal to the log of these resource inflows are reported in Table 16.2 (col. 3) —the results are very similar when nonfarm income and gifts are entered separately. Similar plots within village-crop-year groups of plots are cultivated more intensively by households which receive higher short-term resource inflows.

There is a positive correlation between plot yield and household size, area cultivated by the household on other plots, and short term resource inflows conditional on all observable plot characteristics. The results presented in Table 16.2 (col. 4), however, indicate that conditional on short-term resource inflows and household size (and plot characteristics), plot yield does not vary with the area cultivated on other plots. On the other hand, plot yield is higher for large households with large short-term inflows of nonfarm resources, conditional on other household characteristics.

Despite the rich information available in this data set regarding plot characteristics, omitted variables bias induced by unobserved variation in plot quality remains a concern. One possible interpretation of the results revealed in columns 1–4 of Table 16.2 is that there are certain wealthy households which cultivate large amounts of land, receive a great deal of nonfarm income, and have many members, and that these households also control land of particularly high (unobserved) quality. The panel nature of the data permits us to mitigate this concern by controlling for the average (over time) quality of each household's land. Column 5 of Table 16.2, therefore, reports the results of estimating (16.12) with household fixed effects as well as the village-year-crop fixed effect. The results change in a manner consistent with the interpretation above. Conditional on the household effect, increases in the area cultivated on other plots are now associated with declines in yield on a given plot. The elasticity is large and statistically significant. Increases in inflows of nonfarm resources remain associated with increases in plot yield — the elasticity is 0.1 and statistically significant. Changes in household size have no perceptible relationship to plot yield, which may be a consequence of the relative stability of household size over the three years.

Interpretation: This pattern of results: a negative correlation between yield and the area cultivated on other plots and a positive correlation between yield and short-term inflows of nonfarm resources and (possibly) between yield and household size, has strong implications for the structure of rural markets. First, simple labor and land market imperfections would not lead to this pattern of results. If the only market failure is that labor and land cannot be traded, than an increase in nonfarm resources would be associated with an increase in leisure and hence a reduction in yield.[11] However, a combination of financial market and land market failures leads directly to the pattern of correlations observed in the data. Suppose that the labor market operates freely, but that there is a binding cash-in-advance constraint on production in the first period. Increases in nonfarm resource inflows or household size relax this constraint and permit yields to rise, while an increase in area cultivated on other plots dilutes available inputs and lowers yields.[12] The observed pattern of empirical results, therefore, is consistent with imperfections in the financial and land markets, rather than with labor market imperfections.

Turning to data on farm inputs, Table 16.3 reports estimated demand functions for the two most important farm inputs. Columns 1 and 2 estimate the determinants of log plot labor demand per hectare. In column 1, household effects are omitted and a strong positive correlation is found between plot labor demand and household size, while plot labor demand declines as the area cultivated on other plots increases. When household effects are added to the model, the correlation between household size and plot labor use disappears, but plot labor demand remains strongly decreasing in the area cultivated by the household on other plots. Column 3 reports estimates of the determinants of plot manure demand per hectare. Manure use on a particular plot declines with increases in the area cultivated by the household on other plots. There is no observable effect of inflows of nonfarm resources on either labor or manure demand, despite the strong correlation between such inflows and plot yield. A plausible interpretation of this contrast is that inflows of nonfarm resources affect the timing rather than the level of input application.

---

[11] Consider the simplest model of a labor market imperfection with smoothly operating credit markets. Production takes time, so labor and land are used in period 1 to produce output in period 2. Consumption of goods occurs in both periods, and there is an inflow of nonfarm resources (G) in period 1. The price vector of the consumption bundle in period t is p_t and the output price is q_2. The household has E units of labor, and T of land. The household solves (suppressing the standard non-negativity constraints):

$$\max_{c_1,\ell_1,c_2} U(c_1,\ell_1,c_2)$$
$$\text{s.t. } p_1c_1 + p_2c_2 \leq q_2F(E-\ell_1,T)+G,$$

which has a first order condition of $U_\ell = \lambda q_2 F_1$, where λ is the Lagrange multiplier and $\delta\lambda/\delta G < 0$. Hence, $\delta\ell_1/\delta G > 0$ and yield falls with nonfarm resources.

[12] The utility function is the same as in the previous footnote. The budget constraints now must be met period-by-period:

$$p_1c_1 + w_1(L^f - (E-\ell_1)) \leq G$$
$$p_2c_2 \leq q_2F(L^f,T)$$

where L^f is labor demand on the household's farm. The first order conditions include $\lambda_1/\lambda_2 \cdot w_1 = q_2 \cdot \partial F/\partial L^f$, where λ_t is the Lagrange multiplier on the period t budget constraint and $\delta(\lambda_1/\lambda_2)\delta G < 0$. Hence we have $\delta L^f/\delta G = w_1 \cdot \delta L^f/\delta L > 0$. With $F(\ )$ CRTS, after some algebra one can show $0 < \delta L^f/\delta T < 1$, hence yield declines with larger farm size.

Table 16.3: Estimates of the determinants of log plot demand and plot manure demand per hectare, ICRISAT Burkina Faso data

	Log plot labor demand per hectare				Plot manure demand per hectare (1000 kg)	
	(1)		(2)		(3)	
	Village-year-crop fixed effects		Village-year-crop and household fixed effects		Village-year-crop and household fixed effects	
	Estimate	t	Estimate	t	Estimate	t
Log plot area	–0.301	–11.84	–0.317	–11.91	–1.980	–5.19
Log household size	0.158	4.29	0.097	0.94	–0.452	–0.35
Log area other plots	–0.059	–2.49	–0.164	–4.19	–1.375	–2.65
Log nonfarm income and gifts	0.012	0.73	–0.036	–0.90	–0.342	–0.63
Toposequence						
Top of slope	–0.197	–1.95	–0.179	–1.63	–1.579	–0.86
Near top	–0.274	–3.12	–0.240	–2.48	–2.545	–1.44
Mid slope	–0.286	–3.36	–0.269	–2.87	–2.589	–1.52
Near bottom	–0.228	–2.62	–0.165	–1.72	–2.874	–1.61
Soil types						
soil7	0.067	0.79	–0.087	–0.76	–0.290	–0.17
soil21	–0.063	–0.39	–0.067	–0.41	0.005	0.00
soil31	0.113	1.13	0.089	0.85	–0.651	–0.50
soil32	0.097	0.91	0.084	0.76	–0.808	–0.56
soil33	0.277	1.94	0.412	2.41	0.146	0.07
soil37	0.329	2.76	0.357	2.88	–1.711	–1.08
soil35	0.058	0.35	0.108	0.58	–0.813	–0.52
soil45	0.095	0.88	0.019	0.17	–0.581	–0.44
soil51	0.201	2.68	0.038	0.40	–1.038	–0.71
soil1	–0.075	–0.27	0.146	0.67	–8.494	–1.86
soil3	–0.041	–0.44	0.063	0.57	–2.407	–1.11
soil11	0.111	0.59	0.117	0.67	2.179	1.32
soil12	–0.163	–0.87	0.217	1.16	1.111	0.64
soil13	0.314	1.41	0.145	0.73	–0.820	–0.67
soil34	–0.098	–0.51	–0.129	–0.66	1.135	0.69
soil46	–0.073	–0.43	–0.027	–0.14	1.032	0.30
soil53	0.333	2.36	0.083	0.53	3.050	0.98
Compound plot	–0.085	–1.46	–0.119	–1.88	–0.068	–0.08
Village plot	–0.035	–0.80	–0.003	–0.06	0.590	0.89

Notes: The dependent variable is the log of the sum of male, female, and non-household labor per hectare used on the plot. These are OLS estimates with village-year-crop and household fixed effects. The t-ratios are calculated from heteroskedasticity-consistent standard errors.

There is strong evidence of violations of profit maximization in Burkina Faso, associated in particular with changes in the area cultivated by a household on other plots, and with inflows of nonfarming resources. Increases in the inflow of resources not directly tied to cultivation are associated with more intensive cultivation, and increases in total farm size are associated with declines in the intensity of cultivation on particular plots. The finding with respect to nonfarm resources complements the work of Savadogo, Reardon and Pietola [38] and Reardon, Crawford and Kelley [35], who find in Burkina Faso that households with higher nonfarm income (averaged over a period of years) are more likely to use intensive methods of cultivation (e.g., animal traction). The pattern is strikingly reminiscent of Polly Hill's [26] argument that poor farming households in the savannah of northern Nigeria are "too poor to farm." She argues that a "householder who has an empty granary at the beginning of the farming season may well have to earn his living every day by means of sundry paid occupations, including farm-laboring, so that he has no time to cultivate on his own account. . . . [T]here are many other circumstances in which it ['too poor to farm'] applies, such as when manure is lacking" (pp. 69–70).

These results provide some evidence concerning the source of the market failure. Standard models of labor market imperfections obviously could underlie the inverse correlation between farm size and the intensity of cultivation of plots. The same imperfections, however, would induce a negative correlation between inflows of nonfarm resources and cultivation intensity rather than the positive correlation which is observed in the data. An increased inflow of nonfarm resources would relax the household's budget constraint, inducing the household to consume more leisure and thus (with imperfect labor markets) use less labor on their plots.

In contrast, capital market imperfections could cause both of the correlations observed in the data. With liquidity constraints, an increase in the availability of nonfarm resources permits the household to cultivate more intensively, and an increase in the size of the farm cultivated by the household further dilutes available resources and reduces the intensity of cultivation of a given plot. These results suggest, therefore, that further research on market imperfections in Burkina Faso begin with an examination of financial markets.

16.4.2 Kenya

The analysis begins with an examination of the residuals in (16.13a) and (16.13b). In this instance, X_{vhtci} includes a much less rich set of plot characteristics. I maintain the hypothesis that the household is free to allocate factors of production efficiently across its various plots, hence ε^h would be zero in the absence of measurement error in plot quality or yield. Figure 16.3 reports the distribution of ε^h and the corresponding distribution of ε^v.

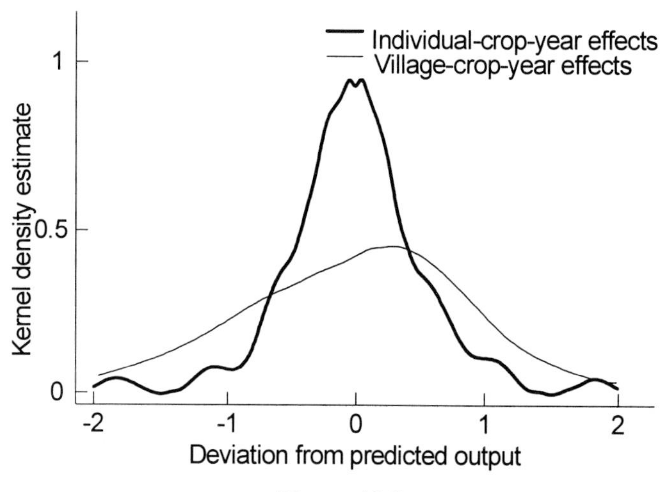

Figure 16.3

Yield is much more dispersed across similar plots cultivated by different households than it is across similar plots cultivated by a single household (the $F(1549, 128)$ test statistic of the hypothesis that $\lambda_{htc} = \lambda_{tc}$ has a value of 1.836 ($p = 0.00$)). As was the case in Burkina Faso, the much larger dispersion of ε^v raises the possibility that the separation hypothesis is violated at the village level. However, given the limited information available about plot characteristics, there is particular reason to be concerned that these results are contaminated by omitted variables bias. The estimates of β from (16.13a) and (16.13b) are presented in columns 1 and 2 of Table 16.4. If there were no unobserved variations in plot quality, and the production function were CES with CRTS, than the coefficient of the log of plot area would be zero. The fact that this coefficient is so negative is striking, and it is difficult to believe that agricultural production is characterized by such strongly decreasing returns to scale. An inverse correlation between plot size and unobserved quality could account for this result. More importantly, the large decline in the coefficient on plot size when moving from the household-crop-season fixed effects to the crop-season fixed effects specification can be interpreted as evidence that the dispersion of these unobserved plot characteristics is larger across plots in the region than it is across plots controlled by individual households. This casts doubt on the interpretation of the larger spread of ε^v than ε^h as an indicator of non-separation.

It is reasonable, therefore, to treat the tests of exclusion restrictions presented in Table 16.5 with caution. Column 1 presents the results of the regression testing the exclusion restriction with respect to household size. Plot output is strongly increasing in household size and the coefficient is statistically significant at any conventional level. One interpretation of this result is

Table 16.4: Estimates of the determinants of log plot output per hectare, IFPRI Kenya data

	(1) Crop-season effects		(2) Household-crop-season effects	
	Estimate	t	Estimate	t
Log plot size	–0.530	–13.60	–0.334	–3.91
Soil type				
Poor	–0.371	-2.98	0.535	0.88
Average	–0.067	1.28	0.108	1.00
Toposequence				
topo1	0.023	0.13	0.117	2.69
topo2	0.069	0.39	0.088	0.40
topo3	0.055	0.29	–0.560	–1.73
topo4	0.018	0.07	0.970	1.51

Notes: These are OLS estimates. The t-ratios are calculated from heteroskadasticity consistent standard errors.

that with imperfect labor markets, larger households farm their plots more intensively, but other market imperfections and omitted land quality correlated with household size should also be considered. Column 2 of Table 16.5 shows that plot output is also increasing with the amount of land cultivated by the household on other plots. As in Burkina Faso, this correlation is opposite of that which would be expected in the presence of conventional land and labor market imperfections or unobserved variation in land quality. Households with larger farms seem to cultivate their land more intensively than other households.

Columns 3 and 4 of Table 16.5 present exclusion restrictions with respect to short-run measures of cash flow. The log of the sum of nonfarm income and gifts and remittances received in the preceding two months is not significantly correlated with plot output, though if attention is restricted to nonfarm income the correlation is significant at the 2 percent level (col. 4). There is weak evidence, therefore, that short-run resource inflows are positively related to the intensity of cultivation, in violation of separation. Finally, the results in column 5 indicate that plot output increases in both household size and the area cultivated by the household, conditional on plot and other household characteristics. As was the case in Burkina Faso, a case could be made that this set of results provides evidence in support of the vision of African rural economies in which poor households are small, have relatively little land, and face resource constraints which prevent them from farming efficiently.

These conclusions, however, have to be drawn with extreme care in this instance. The relatively meager amount of information which is available about plot quality, coupled with the dramatically negative relationship between plot size and plot yield in the data raise serious issues of potential omitted variables bias. It may be that households with large families and large

Table 16.5: Estimates of the determinants of log plot output per hectare crop-season fixed effects, IFPRI Kenya data

	Specification 1		Specification 2		Specification 3		Specification 4		Specification 5	
	Estimate	t	Estimate	t	Estimate	t	Estimate	t	Estimate	t
Log plot size	−0.547	−14.25	−0.563	−14.05	−0.618	−11.67	−0.623	−10.04	−0.673	−10.57
Log household size	0.292								0.292	3.77
Log area other plots			0.022	2.57					0.026	2.00
Log nonfarm income and gifts					0.029	1.41	0.064	2.50	0.032	1.16
Soil type										
Poor	−0.362	−2.95	−0.365	−2.81	−0.334	−2.07	−0.500	−2.91	−0.059	−3.10
Average	−0.057	−1.10	−0.050	−0.92	−0.081	−1.19	−0.116	−1.40	−0.092	−1.09
Toposequence										
topo1	0.005	0.03	0.071	0.37	−0.070	−0.34	−0.088	−0.40	−0.026	−0.12
topo2	0.061	0.35	0.114	0.61	−0.084	−0.43	−0.072	−0.34	−0.010	−0.04
topo3	0.057	0.30	0.077	0.38	−0.043	−0.18	0.067	0.25	0.095	0.35
topo4	0.014	0.06	0.021	0.08	−0.081	−0.28	−0.091	−0.30	−0.014	−0.04

Notes: The dependent variable is the log of the value of all plot output per hectare. These are OLS estimates with village-year-crop fixed effects (in the first four columns) or with village-year-crop and household fixed effects (in the final column). The t-ratios are calculated from heteroskedasticity-consistent standard errors.

amounts of land have particularly good land (as a consequence, perhaps, of a political process of land allocation), and thus are optimally farming their plots particularly intensively. Fortunately, it seems possible to address this issue decisively with these data, because many plots can be traced over time. Unobserved characteristics of the plot which are constant over the two seasons can be differenced away, so any correlation between such characteristics and household characteristics no longer influences the results.

Table 16.6 presents a series of simple tests of the exclusion restrictions with plot and crop-season fixed effects. The tests are cross tabulations of changes in plot output (and plot labor use) against changes in household size, area on other plots, and inflows of gifts and nonfarm income (net of crop-season effects). Row 1 indicates that output increases by more on plots controlled by households which grow in size than on plots controlled by households which shrink or remain the same size. The difference is significant only at the 10 percent level. Row 2 provides interesting, but statistically insignificant, evidence that output increases more on plots cultivated by households which increase their cultivated area than on plots controlled by other households. This would complement the similar striking finding in Burkina Faso, but the evidence in Kenya is quite weak. The third row provides weak evidence that output on plots controlled by households which experience increases in financial inflows declines by more than on plots controlled by other households, but the difference is not significantly different from zero at conventional levels.

Table 16.6: Tests of separation with plot and crop-season fixed effects

	Change in plot output	Mean change in output	$t: H_0$ is equal means
1	Δhouseholdsize > 0	227.1	–1.78
	Δhouseholdsize ≤ 0	–63.7	
2	Δarea on other plots > 0	109.3	–1.24
	Δarea on other plots ≤ 0	–84.2	
3	Δgifts and nonfarm income > 0	–54.5	1.19
	Δgifts and nonfarm income ≤ 0	119.7	
	Change in plot labor use	**Mean change in labor use/hectare**	**$t: H_0$ is equal means**
4	Δhouseholdsize > 0	7.27	–2.09
	Δhouseholdsize ≤ 0	–2.81	
5	Δarea on other plots > 0	–5.69	2.35
	Δarea on other plots ≤ 0	3.52	
6	Δgifts and nonfarm income > 0	0.89	–0.76
	Δgifts and nonfarm income ≤ 0	–2.10	

Note: Each variable is transformed by subtracting its crop-year mean, and then differencing over time for each plot. That is, for each variable A, the transformed variable $\tilde{A} = (A_{pct} - \bar{A}_{ct}) - (A_{pc,t-1}, \bar{A}_{c,t-1})$, where p is the plot, c is the crop, and t is the survey round. The cross-tabulation is calculated over these transformed variables.

Rows 4–6 examine labor inputs, and provide striking evidence against separation in the Kenyan data. Labor use grows significantly more on plots controlled by households which grow than on plots controlled by other households. Conversely, labor use diminishes significantly more on plots controlled by households which increase their cultivated area than on plots controlled by other households. Both results accord with conventional models of labor and land market imperfections.

Table 16.7 presents a multiple regression analysis of the relationship between the growth in plot output and the growth in household size, cultivated area, and financial inflows. The precision of the estimates is quite low, so no conclusions can be drawn from this set of results.

Table 16.7: LAD estimates of the determinants of plot yield, plot and crop-year fixed effects, IFPRI Kenya data

	Estimate	t
Household size	22.320	1.23
Area on other plots	–10.832	–0.58
(Area on other plots)^2	2.994	0.69
Gifts	–0.016	–1.05
Nonfarm income	0.001	0.68

The plot fixed effect analysis presented in this section is a useful response to the most important econometric worry associated with this research program — the possibility that unobserved characteristics of plots are correlated with the household characteristics, thus causing false rejections of the profit-maximization hypothesis. The procedure, however, does have significant weaknesses. Eliminating the plot fixed effect also eliminates much potentially useful information, and if there is measurement error in the household characteristics of interest, the impact of that measurement error is magnified relative to analysis without fixed effects. In this particular example, the short time period (6 months) between the two rounds is associated with only small changes in the relevant household characteristics and thus tests with relatively low power to detect violations of separation.

Moreover, the plot fixed effect does not eliminate the effect of unobserved shocks to plot productivity caused, for example, by rainfall variation. One interpretation of these results is that certain plots are subject to shocks which increase output and the marginal product of labor, and that to meet the increased demand for labor on these plots, cultivating households expand (Guyer [22] emphasizes the flexibility of the size and composition of many African households in response to changing labor demand conditions). This is a form of non-separation, but the analysis underscores the need to question any assumption that household size can be treated as exogenous.

Nevertheless, certain patterns do emerge from the data. The intensity of labor use on plots is positively correlated with household size, and decreases with the area cultivated on other plots by the household. The overall pattern, therefore, is that there is strong evidence that the separation hypothesis is violated in Kenya; the pattern (with one exception) corresponds to the relatively simple situation of absent labor and land markets. The exception is that plot output may be increasing in the area cultivated on other plots, which suggests an interesting relationship between nonlabor inputs and household resources. This result, however, is statistically significant only at the 20 percent level, and thus must be treated with extreme caution.

16.5 Profit Maximization in Africa

This paper has provided evidence against the hypothesis that farmers maximize profits in two African settings. In each case, it was shown that plot level production decisions are affected by factors other than prices and the characteristics of the plot itself, and thus that the hypothesis that production decisions are separable from the rest of the households' allocation decisions is not correct.

In Burkina Faso, violations of the separation hypothesis are indicated by correlations between the intensity of cultivation and short-term resource inflows (nonfarm income and gifts) over the months preceding planting and between the intensity of cultivation and farm size. The estimates indicate that plot yield increases by 11 percent when nonfarm income and gifts double. Plot yield declines by about 15 percent when the area cultivated on other plots doubles. There are strikingly large changes in the allocation of factors of production on a particular plot associated with increases in the area cultivated by the household on other plots. A doubling of farm size leads to a decline of over 15 percent in plot labor use per hectare, while a one standard deviation increase in farm size is associated with a decline in manure use per hectare of over 1,000 kilograms (the average manure application per hectare is about 3,250 kg).

In Kenya, the allocation of labor to plots is strongly correlated with household size and farm area. On average, 20 days of labor (the sum of male, female, and hired labor) is used on plots in the sample. On plots controlled by households which gained members, labor use increased by about 10 days more than on plots controlled by households which lost members. Similarly, on plots controlled by households whose overall farm size fell, labor use increased by about 9 days more than on plots controlled by households whose overall farm size grew.

In both Kenya and Burkina Faso, therefore, there are economically large and statistically significant divergences between actual input allocation decisions and those which would be predicted in a simple model of a profit-maximizing farmer. In neither case can the cause of the divergence be determined with certainty, using the broad-brush methods of this paper. A rejection of the separation hypothesis can be caused by any number of possible market failures. However, the pattern of results is suggestive: in Burkina Faso, the fact that plot output is correlated with nonfarm income and gifts as well as with farm size suggests that further investigation include a close look at capital and insurance markets; in Kenya, the fact that plot labor demand is correlated with household and farm size rather than with nonfarm income suggests that future work focus first on possible imperfections in land and labor markets.

References

1 Aryeetey, Ernest, Financial integration and development in sub-saharan Africa: A study of informal finance in Ghana, unpublished manuscript, ISSER/Legon, 1993.

2 Ashenfelter, O., A. Deaton and G. Solon, Collecting panel data in developing countries: Does it make sense?, Living Standards Measurement Study Working Paper No. 23, World Bank, 1986.

3 Banerjee, A. and A. Newman, Occupational choice and the process of development, Journal of Political Economy, 1993.

4 Bardhan, P., Size, productivity and returns to scale: An analysis of farm-level data in Indian agriculture, Journal of Political Economy (1973) 81.

5 Bardhan, P., Land, Labor, and Rural Poverty: Essays in Development Economics, Columbia University, New York, 1984.

6 Barrett, Christopher, On price risk and the inverse farm size-productivity relationship, Journal of Development Economics 51 (1996).

7 Benjamin, Dwayne, Household composition, labor markets, and labor demand: Testing for separation in agricultural household models, Econometrica (1992).

8 Benjamin, D., Can unobserved land quality explain the inverse productivity relationship?, Journal of Development Economics (1994).

9 Bhalla, Surjit, Does land quality matter? Theory and measurement, Journal of Development Economics, 29 (1988).

10 Binswanger, H. and M. Rosenzweig, Contractual Arrangements, Employment, and Wages in Rural Labor Markets, Agricultural Development Council, New York, 1981.

11 Blarel, B., P. Hazell, F. Place and J. Quiggin, The economics of farm fragmentation: Evidence from Ghana and Rwanda, World Bank Economic Review (1992).

12 Carter, Michael, Environment, technology and the social articulation of risk in West African agriculture, unpublished manuscript, Wisconsin, 1994.

13 Carter, Michael and Kieth Wiebe, Access to capital and its impact on agrarian structure and productivity in Kenya, American Journal of Agricultural Economics (1990).

14 Chaudhuri, Shubham, Crop choice, fertilizer use and credit constraints: An empirical analysis, unpublished manuscript, Princeton, 1993.

15 Cogill, Bruce and Eileen Kennedy, Income and nutritional effects of the commercialization of agriculture in southwestern Kenya, Research Report Series, No. 63, International Food Policy Research Institute, Washington, DC, 1987.

16 Collier, P., Malfunctioning of African rural factor markets: Theory and a Kenyan example, Oxford Bulletin of Economics and Statistics, 45/2 (1983).

17 Deaton, A., Saving and income smoothing in the Côte d'Ivoire, Journal of African Economies, 1(1) (1992) 1–24.

18 Eswaran, M. and A. Kotwal, Access to capital and agrarian production organization, Economic Journal, 96 (1986).

19 Fafchamps, Marcel, Sequential labor decisions under uncertainty: An estimable household model of West African farmers, Econometrica, 61 (1993) 1173–98.

20 Feder, G., The relation between farm size and farm productivity: The role of family labor, Supervision, and Credit Constraints, Journal of Development Economics (1985).

21 Gavian, Sara and Marcel Fafchamps, Land tenure and allocative efficiency in Niger, unpublished manuscript, ILRI, Addis Ababa, 1995.

22 Guyer, J., Household and community in African Studies, African Studies Review, 24 (1981).

23 Hays, H., The marketing and storage of food grains in Northern Nigeria, Samaru Miscellaneous Papers No. 50, Ahmadu Bello University, Zaria, Nigeria, 1975.

24 Hill, P., Rural Hausa, Cambridge University. Cambridge, 1972.

25 Hill, P., Population, Prosperity and Poverty: Rural Kano 1900 and 1972, Cambridge University, Cambridge, 1977.

26 Hill, P., Dry Grain Farming Families, Cambridge: Cambridge University, Cambridge, 1982.

27 Jones, W., Agricultural trade within tropical Africa: Achievements and difficulties, in: W. Jones, R. Bates and M. Lofchie (eds.), Agricultural Development in Africa: Issues of Public Policy, Praeger, New York, 1980, pp. 311–48.

28 Kennedy, Eileen, Effects of sugarcane production in southwestern Kenya on income and nutrition, in: J. von Braun and E. Kennedy (eds.), Agricultural commercialization, economic development, and nutrition, Johns Hopkins University Press for the International Food Policy Research Institute, Baltimore and London, 1989.

29 Kevane, Michael, Agrarian structure and agricultural practice: Typology and application to Western Sudan, unpublished manuscript, Harvard University, 1994.

30 Krishna, Raj, Theory of the firm: Rapporteur's report, Indian Economic Journal, 11 (1964).

31 Matlon, P., The ICRISAT Burkina Faso farm level studies: Survey methods and data files, Economics Group, VLS and Miscellaneous Papers Series, ICRISAT, Hyderabad, India, 1988.

32 Morduch, J., Risk, production and saving: Theory and evidence from Indian households, unpublished manuscript, Harvard University, 1990.

33 Pingali, P., Y. Bigot and H. Binswanger, Agricultural Mechanization and the Evolution of Farming Systems in Sub-Saharan Africa, Johns Hopkins, Baltimore, 1987.

34 Pitt, Mark and Mark Rosenzweig, Agricultural prices, food consumption and the health and productivity of Indonesian farmers, in: I. Singh, L. Squire and J. Strauss, Agricultural Household Models: Extensions, Applications, and Policy, Johns Hopkins, Baltimore, 1986.

35 Reardon, Thomas, Eric Crawford and Valerie Kelley, Links between nonfarm income and farm investment in African households: Adding the capital market perspective, American Journal of Agricultural Economics. 76 (1994) 1172–76.

36 Rosenzweig, Mark, Labor markets in low-income countries, in: Chenery and T.N. Srinivasan, eds. Handbook of Development Economics, North-Holland, Amsterdam, 1988.

37 Rosenzweig, M. and H. Binswanger, Wealth, weather risk and the composition and profitability of agricultural investments, Economic Journal (1993).

38 Savadogo, K., T. Reardon and K. Pietola, Farm productivity in Burkina Faso: Effects of animal traction and nonfarm income, American Journal of Agricultural Economics, 76 (1994).

39 Sen, A.K., Peasants and dualism with or without surplus labor, Journal of Political Economy (1966).

40 Singh, I., L. Squire and J. Strauss, Agricultural Household Models: Extensions, Applications, and Policy, Johns Hopkins, Baltimore, 1986.

41 Srinivasan, T.N., Farm size and productivity: Implications of choice under uncertainty, Sankhya: The Indian Journal of Statistics, Series B, 34(4) (1972).

42 Swamy, Anand, An economic analysis of rural labor and capital markets in the Punjab, c. 1920–40, Ph.D. dissertation, Northwestern University, 1993.

43 Townsend, R., Risk and insurance in village India, Econometrica, 62 (1994) 539–92.

44 Udry, C., Credit markets in Northern Nigeria: Credit as insurance in a rural economy, World Bank Economic Review, 4/3 (1990).

45 Udry, C., Gender, agricultural productivity and the theory of the household, Journal of Political Economy (1996).

Trade, Growth, and Development
G. Ranis and L.K. Raut

CHAPTER 17

Explaining Interlinking[1]

Clive Bell

Süd-Asien Institut der Universität Heidelberg, Im Neuenheimer Feld 330, D-69120, Heidelberg, Germany

17.1 Introduction

The topic of interlinking kept many economists of various stripes and persuasions busy throughout the 1980s.[2] Thereafter, the stream of theory began to abate somewhat, but empirical contributions have gathered to scarcely more than a trickle. As there are already extensive surveys of the literature up to the early 1990s (see, for example, Bell [5] and Otsuka, Chuma and Hayami [16]), it follows that one needs more compelling reasons to revisit this territory than the occasion of a *Festschrift*. The one pursued in this essay is that extant theory still has some weaknesses, both in its power to explain the existence of interlinking and in its undue emphasis on tenancy and credit. Given the accumulating evidence that interlinking is a widespread phenomenon, it is also natural to ask whether there is a case for public intervention to regulate the practice.

[1] This paper is based on the first two of three public lectures given at the 1998 Annual Winter Workshop at the Delhi School's Center of Development Economics, and subsequently at the Indira Gandhi Institute of Development Research, Mumbai. I am very indebted to the members of both faculties and to the students for their warm hospitality during my visit, and to the participants for valuable comments. My thanks go to Pinaki Bose and Lakshmi Raut for their careful and perceptive reading of an earlier draft, and to the *Deutsche Forschungsgemeinschaft* for research support. None of them, of course, is responsible for the errors that remain, nor for those that have crept in since.

[2] Srinivasan himself made an early theoretical contribution on tenancy and credit (Braverman and Srinivasan [10]) and was the leading architect of a large-scale household survey of three Indian States to ascertain how extensive interlinking is in practice, what forms it takes, and how it affects efficiency and welfare. The project was financed by the World Bank as RPO 671–89. The incidence of interlinking in its various forms is reported in Bell and Srinivasan [6], and the interplay of formal credit and informal credit-cum-marketing transactions in Punjab is analyzed in Bell, Srinivasan and Udry [8].

None of the theoretical issues considered here, it will be argued, has received quite the emphasis that it deserves. First, insufficient attention is sometimes paid to the possibility of redundancy among instruments in the presence of agency problems. When the principal's prime objective in imposing interlinking is to gain closer (indirect) control over the agent's doings by adding to the array of contractual instruments, such inattention can be fatal. Second, limited liability has emerged as a potentially important influence on the choice of contract and on the form of rural economic organization (Shetty [19], Basu [3]). In a fair part of the literature on interlinking, the question of limited liability is not even discussed, and in much of the rest, it tends to be swept under the carpet by invoking assumptions that do away with insolvency when contractual obligations fall due. Third, when comparing allocations chosen under interlinking with those under separate contracts, there has been little systematic analysis of the effects of changes in the order in which the principals move.

Fourth, interlinking between trading and finance continues to be neglected in theoretical work, despite substantial evidence that it is the most extensively practiced and, in terms of the associated volume of finance, important form. Supplier's credit is ubiquitous, and traders and middlemen play an important role in the financing of agriculture and fishing in many countries — the amount of finance involved being large. In the case of south India, for example, this claim is borne out by Harriss [12] and Janakarajan [13]. The comparative survey of three Indian States referred to above yielded the findings that the volume of finance associated with marketing dwarfed that associated with tenancy and labor, and that the incidence of the former sort of interlinking was highest in the most commercialized areas (Bell and Srinivasan [6]). There is, moreover, at least one further empirical regularity with which any serious theoretical analysis must come to grips: namely, that it is common for the producer to pay interest on the loan received from the trader. This was a virtually universal practice in the cotton-growing tract of Andhra Pradesh covered by the above-mentioned survey (Steinmetz [21]) and the general rule in the Punjabi villages (Bell, Srinivasan and Udry [8]). Platteau and Abraham [17] report the same for a commercialized fishing village in Kerala, although in two traditional villages, a flat commission fee of five per cent of the value of the catch was the norm.

With this general aim in view, it will be helpful to begin by recalling two definitions of interlinking: "An interlinked deal is one in which two or more independent exchanges are simultaneously agreed upon" (Basu [2]); and "An interlinked transaction is one in which the two parties trade in at least two markets on the condition that the terms of all such trades are jointly determined" (Bell [5]). Upon brief reflection, certain features of the Arrow–Debreu framework come to mind. In that setting, first, agents might as well be anonymous. Second, given that there is a full set of contingent (or insurance) markets, and that other well-known conditions are fulfilled, then the allocation of resources will be Pareto-efficient. Third, agents are implicitly subject to unlimited liability, there being an underlying mechanism that costlessly enforces the contracts associated with trading. In light of these features, it is very hard to see why interlinked transactions, with their inherently personal nature, would arise either in a perfectly certain world in which contractual obligations are always fulfilled, or in the presence of a complete set of markets when there is natural uncertainty and costless enforcement.

This conclusion provides the guiding, taxonomic principle that structures the paper. Section 17.2 deals with the canonical case in the theoretical literature, in which tenancy is interlinked with credit. It is divided into five subsections, beginning with a perfectly certain world, and then proceeding through a string of variations involving uncertainty, moral hazard, risk aversion, and limited liability. Section 17.3 is devoted to trading and credit, from which it emerges that theory in this area remains in a rather unsatisfactory state. Public policy is addressed briefly in Section 17.4, in which the main point is made by drawing upon an empirical study of trading and credit. The paper concludes with some brief remarks on what seem to be promising avenues for future research.

17.2 Tendency and Credit: A Medley of Models

By way of preliminary, we describe the main features of the setting and introduce much of the notation. To start with, it is assumed that output, Q, is produced by means of labor, L, and a fixed factor, which may be thought of as land

$$Q = \theta f(L), \tag{17.1}$$

where the random variable θ represents the state of nature and has the distribution function $F(\theta)$ with support $[\theta_0, \infty)$ and the normalization $E\theta = 1$. The function $f(\cdot)$ is smoothly increasing and strictly concave in L, and it satisfies the lower Inada condition. The agricultural cycle comprises two subperiods, which can be thought of as the "cultivation" and "harvest" seasons, respectively. For simplicity, it is assumed that labor inputs are committed only in the first season, a commitment that must be made before the state of nature is revealed.

A peasant's sole endowment is one unit of labor, and he has access to employment at the parametric wage rate w, which is his alternative to working as a tenant. Lenders, who also may be landlords and traders, have access to a safe placement for their funds at the parametric rate of interest r_0. The peasant does not, but he can put any cash obtained in the first subperiod into a secure hiding place for subsequent use in the second.

17.2.1 No uncertainty and unlimited liability

It should be clear from the foregoing remarks about the central role of uncertainty and limited liability that this subsection deals with a most unpromising setting for a discussion of interlinking; but these are precisely the salient assumptions in Bhadhuri [9], a paper that did much to spawn the vast literature on this topic. The point of subjecting it to rigorous scrutiny here is both to provide a benchmark for other cases and to reveal how matters can go wrong when assumptions are ill-chosen.

We begin with the peasant's reservation alternative, namely, to work for wages. Let his preferences over consumption in the two seasons be representable by the smooth, quasi-concave function $U(c_1, c_2)$, where c_i $(i = 1, 2)$ denotes his consumption in subperiod i. Given the opportunity to save at no interest, he must decide how much of his earnings w to consume in the cultivation season. Let

$$\bar{U} \equiv \max_{c_1}[U(c_1, w - c_1)]. \tag{17.2}$$

Now suppose, instead, that the peasant leases a plot of land. The rent takes the form of an output-share, $(1 - \alpha)$, plus a fixed payment, β, out of the harvest, where, in the absence of uncertainty, $Q = f(L)$. He also obtains a loan in the amount B at the rate r in the cultivation season to help finance consumption until the harvest is in.[3] His budget constraints in the two seasons may be written as

$$c_1 = B + w(1 - L) - s \tag{17.3}$$

and

$$c_2 = \alpha Q - \beta - RB + s, \tag{17.4}$$

where r is the interest rate payable on B, $R \equiv (1 + r)$, and s (≥ 0) is the amount put away in the first season to help finance expenditures in the second.

We begin with the case where these deals in the markets for tenancy and credit are interlinked. The landlord–moneylender's combined income from rent and usury at the end of the cycle is

$$\Omega = (1 - \alpha)f(L) + (\beta + RB) - R_0 B. \tag{17.5}$$

In the absence of uncertainty, there is nothing to prevent him from specifying a particular level of labor input. For the peasant's actual input can be perfectly

[3] Observe that the contracts are constrained to be linear, a restriction that may be justified by an appeal to the transactions costs of implementing arbitrarily nonlinear forms. If the latter are indeed feasible, then interlinking will offer no advantages over suitably formulated separate contracts and certain conditions concerning observability and the terms on which principals can trade are fulfilled (Ray and Sengupta [18]), and the superiority of interlinking over separate contracts under the restriction to linearity, should it exist, will be irrelevant.

inferred from the level of output and, since the penalties for shirking in the present can be extracted out of future income when there is unlimited liability, the input stipulation is perfectly credible. Bhadhuri [9] simply assumes that the level of output is independent of what the peasant does. Both possibilities are therefore covered by

Assumption 1. *The peasant has no influence over L, which may be taken as fixed.*

Inspection of (17.5) reveals that Assumption 1 renders one of the instruments, R or β redundant. That output is independent of the contractual parameters is crucial to an additional redundancy among instruments which is about to emerge.

Given the terms of the combined contract (α, β, B, R), the peasant's only choice is whether to accept it, and if so, how much to put by. Let

$$U^0 \equiv \max_s U(c_1, c_2) \tag{17.6}$$

The landlord–moneylender's optimum interlinked contract is the solution to the problem

$$\max_{(\alpha,\beta,B,R)} \Omega \qquad \text{subject to: } U^0 \geq \bar{U}. \tag{17.7}$$

By writing out the associated Lagrangian and the first-order conditions, and then appealing to the envelope theorem, it readily verified that the peasant's participation constraint binds at the optimum and that

$$U_0^1 / U_2^0 = R_0. \tag{17.8}$$

It should come as no surprise that, in a first-best setting, the peasant's marginal rate of substitution is equal to the interest factor of the safe placement, even though he has no direct access to it.

Let the value of Ω at the optimum be denoted by Ω^0. In view of the redundancy between β and R, the landlord–moneylender can attain the optimum by choosing $R = R_0$, for example, together with an appropriate choice of α, β and B to satisfy the participation constraint with equality and condition (17.8). In other words, he can reach the optimum without recourse to usury, a point made by Newbery [15] in a somewhat simpler formulation. It is clear from (17.3)–(17.5) and Assumption 1, moreover, that one of the rental parameters α and β is also redundant. Let $\Gamma^0 \equiv (\alpha^0, \beta^0, B^0, R_0)$ denote any non-usury, interlinked contract that attains the optimum.

This redundancy among instruments includes other special cases that yield Ω^0. All will be useful for future reference.

(i) No rent: $\alpha = 1$, $\beta = 0$, $R > R_0$.

Equivalent to (i) is the case of a pure fixed rent, interpreted as the peasant making a fixed payment β in the second period in exchange for the use of the land and the receipt of the sum B in the first.

(ii) Fixed rent: $\alpha = 1$, $\beta > 0$, $R = 0$.

With the optimum interlinked contract, as viewed by the landlord–moneylender, completely characterized, we turn to the case where a separate moneylender enters into the picture. This would arise, for example, if the landlord had moral scruples against lending, or simply no access to funds. With an optimal interlinked contract as a benchmark, will the landlord fare worse if he relinquishes, or has no control over, (B, R)? It is natural to treat this situation as a Stackelberg game in which the landlord moves first, choosing (α, β), and the moneylender second, choosing (B, R); for in practice, a peasant will find it very hard to obtain a loan if he does not have a tenancy. For logical completeness, however, we shall also deal with the case in which the lender moves first.

To be precise, the game proceeds as follows: At the first stage, the landlord offers the peasant the contract (α, β), with the provision that it becomes void in the event that the peasant is unable to find finance on what are to him satisfactory terms. If he accepts, the peasant then approaches the lender and divulges the terms of the tenancy contract. At this stage, the lender decides whether to offer the peasant a credit contract. If the latter accepts this offer, then he must return to the landlord to confirm the fact that he has secured finance in order to obtain the lease; indeed, it would be irrational of the peasant to do otherwise, given that he has done his calculations correctly all the way through. The sequence is the same, *mutatis mutandis*, if the lender moves first.

Proposition 1. *Under Assumption* 1, *the party who moves first in the Stackelberg game, in which the landlord chooses* (α, β) *and the lender* (B, R), *obtains the optimal payoff under interlinking, whereas the one who moves second obtains just his reservation payoff.*

Proof. (a) We begin with the case where the landlord moves first. The lender's income is

$$\Omega_m \equiv (R - R_0)B. \tag{17.9}$$

His problem is to

$$\max_{(B,R|\alpha,\beta)} \Omega_m \qquad \text{subject to: } U^0 \geq \bar{U}. \tag{17.10}$$

By writing down the associated Lagrangian and proceeding as above, we obtain, once more, (17.8) and $U^0 = \bar{U}$, which implicitly define the lender's reaction functions, given (α, β). Let the landlord choose the rental component (α^0, β^0) of the optimal interlinked contract Γ^0. By Assumption 1, this will yield him exactly Ω^0 as before, whatever the moneylender does — provided, of course, the latter finds it worthwhile to offer the peasant a credit contract. It is clear that if the moneylender were to reply with (B^0, R_0), then (17.8) and $U^0 = \bar{U}$ would be satisfied. It is also clear that he cannot improve thereon without violating the peasant's participation constraint; for Ω^0 is also the maximal surplus that can be extracted from the peasant under a first-best interlinked contract. Since the reply (B^0, R_0) yields the lender a net income of zero, his participation is just ensured. Thus, we have a subgame perfect equilibrium in which the landlord can do without lending and still obtain the net payoff Ω^0.

(b) Now let the moneylender move first, and note that the landlord has no alternative use for his land. Recall the no-rent interlinked contract $\alpha = 1$, $\beta = 0$, $R > R_0$. This yields Ω^0 by choosing $B = B^0$ and R such that $U^0 = \bar{U}$. If the lender makes this choice, the landlord will possess no reply that yields him a strictly positive payoff without violating the peasant's participation constraint. The reply $(\alpha = 1, \beta = 0)$, however, ensures that both he and the peasant will find participation (barely) attractive. ■

It should be noted that if the landlord does have access to funds, but at a rate that exceeds the lender's opportunity cost of funds, and if he enjoys the first move, then he will actually welcome the moneylender's appearance on the scene. For the above strategy will, in effect, give him access to funds at the lender's rate, and so improve on the net payoff from interlinking at the higher rate, namely, $\Omega^0(r > r_0)$.

At first glance, it might be thought that the failure of interlinking to be strictly superior to separate contracts in this setting can be laid at the door of Assumption 1, and that relaxing it to give the peasant some freedom of choice in how much labor to use on the tenancy would confer strict advantages on interlinking. As will now be demonstrated, however, this is not the case.

17.2.2 Moral hazard: Risk neutrality and unlimited liability

Relaxing Assumption 1 involves introducing both uncertainty about the state of nature and prohibitive costs of monitoring. Moral hazard then arises, for the peasant can choose to spend some or all of his time in wage employment, instead of applying himself diligently to cultivation. In order to make a sensible comparison with the result just derived in Section 17.2.1, one must assume that all parties are risk neutral.

Assumption 2. *The peasant's preferences may be represented as*

$$EU(c_1, c_2) = c_1 + \delta E c_2 \tag{17.11}$$

where δ (≤ 1) *is his subjective discount factor.*

It is clear that Assumption 2 yields the following convenient specialization of (17.2):

$$\bar{U} = w. \tag{17.2$'$}$$

Since the peasant will choose L and s only after the terms of the rental and loan contract have been settled, his problem is to

$$\max_{(c_1, c_2, s, L)} EU(c_1, c_2) \tag{17.12}$$

Substituting for c_1 and c_2 using (17.3) and (17.4), the first-order condition with respect to s yields the optimal choice $s^0 = 0$. Assuming that $L \leq 1 + B/w$, we also have

$$\partial EU/\partial L = -w + \delta\alpha f'(L) = 0, \tag{17.13}$$

which implicitly defines the peasant's reaction function, and hence the incentive constraint in the principal's problem. Observe that the lower Inada condition rules out $L = 0$, and that the only contractual parameter which has an influence on L is α.

Under an interlinked contract (α, β, B, R), the landlord–moneylender's expected net income will be given by (17.5), with $E\Omega$ replacing Ω. Employing the first-order approach to the incentive constraint, his problem is to

$$\max_{(\alpha, \beta, B, R)} E\Omega \qquad \text{subject to: (17.13) and } EU \geq w. \tag{17.14}$$

The solution is well known: He chooses $\alpha = 1$, thereby making his own payoff independent of the peasant's actions and the state of nature; he sets $\beta = 0$ for convenience; and he selects a pair (B, R) that yields the peasant just his reservation utility. From (17.13) and $EU \geq w$, this pair is

$$B^0 = w f'^{-1}(w/\delta) \text{ and } R^0 = f\,[f'^{-1}(w/\delta)]/B^0,$$

where B^0 is chosen to bring about $c_1 = w$, and R^0 to bring about $Ec_2 = 0$. This strategy yields

$$E\Omega^0 f\,[f'^{-1}(w/\delta)] - R_0 B^0.$$

As in Section 17.2.1, a separate moneylender now appears on the scene. In view of Assumption 2, the associated game has three stages. Either the landlord moves first, choosing (α, β), and the moneylender then replies by choosing (B, R), or conversely. The peasant moves last, choosing (L, s) in the light of (α, β, B, R).

Proposition 2. *If all parties are risk neutral and there is moral hazard, then in the three-stage game, the party to move first will do as well as under interlinking, while the other two will obtain just their reservation utilities.*

Proof. Suppose the landlord moves first. If he chooses $\alpha = 1$ and $\beta = f\,[f'^{-1}(w/\delta)]$, he will obtain $E\Omega^0$ whatever the lender does — provided the lender participates. The latter's best reply is (B^0, R_0), which yields his reservation payoff, and the peasant will choose $L = f\,[f\,'^{-1}(w/\delta)]$. These choices constitute a subgame perfect equilibrium in which the payoffs are $E\Omega_l = E\Omega^0$, $E\Omega_m = 0$ and $EU = w$. Note that the instrument β is now essential to the landlord.

If, instead, the moneylender moves first, let him choose the pair (B^0, R^0) derived above; for this will yield him $E\Omega^0$, provided the landlord's best reply is $\alpha = 1$ and $\beta = 0$. The latter's problem is to

$$\max_{(\alpha,\beta|B,R)} [(1-\alpha)\,f\,(L) + \beta] \qquad \text{subject to: (17.13) and } EU \geq w. \qquad (17.15)$$

It is readily checked from the first-order conditions that the optimum is indeed $\alpha = 1$, as intuition suggests in the presence of the lump-sum instrument β. Since, by hypothesis, the lender has chosen (B^0, R^0), it follows at once that the landlord must choose $\beta = 0$ so as not to violate the peasant's participation constraint. Hence the said choices constitute a subgame perfect equilibrium, in which the parties' payoffs are $E\Omega_\ell = 0$, $E\Omega_m = E\Omega^0$, and $EU = w$. ■

It should be remarked that the key point in these first two sections is that, in equilibrium, the financing of cultivation is such that L is independent of B and R, the lender's instruments. We now explore two possible routes out of this impasse, namely, risk aversion and limited liability.

17.2.3 Risk aversion and unlimited liability

We replace Assumption 2 with

Assumption 3. *The peasant's preferences over lotteries may be represented by*

$$EU = u(c_1) + \delta E u(c_2), \qquad (17.16)$$

where $u(\cdot)$ is a smoothly increasing and strictly concave function.

Replacing (17.11) with (17.16) in problem (17.12), the peasant's first-order conditions become

$$\partial EU/\partial L = -u_1' \cdot w + \delta E u_2' \cdot \alpha\theta f' = 0 \tag{17.17}$$

and

$$\partial EU/\partial s = -u_1' + \delta E u_2' \leq 0, \; s \geq 0 \text{ complementarily.} \tag{17.18}$$

Let $L^0 = L^0(\alpha, \beta, B, R)$ denote the peasant's optimal choice of labor input. We need to establish how L^0 responds to changes in B and R.

Given the strict concavity of $u(\cdot)$, it is plausible that the peasant will choose $s > 0$. In that case, (17.17) and (17.18) yield

$$Eu_2' \cdot (\alpha\theta f' - w) \geq 0,$$

Using this condition, Srinivasan [20] has shown that if absolute risk aversion is nonincreasing, then

$$Eu_2'' \cdot (\alpha\theta f' - w) \geq 0,$$

with strict equality if and only if there is constant absolute risk aversion (CARA). After differentiating (17.17) and (17.18) totally, some tedious algebra yields the following results:

$$\text{sgn}(\partial L^0/\partial B) = \text{sgn}\{-\delta u_1''[Eu_2'' \cdot (\alpha\theta f' - w)](R-1)\}$$

and

$$\text{sgn}(\partial L^0/\partial B) = \text{sgn}\{-\delta u_1''[Eu_2'' \cdot (\alpha\theta f' - w)]\}.$$

Hence, L^0 is insensitive to R if and only if there is CARA, and to B if and only if there is either CARA or credit without interest ($R = 1$). The intuition here is that an increase in either B or R will reduce c_2 in the absence of any other changes. Since $u(\cdot)$ is strictly concave, the adverse consequences thereof can be partly offset by increasing L or s, either of which will reduce c_1 in favor of c_2. If, by chance, $R = 1$, then changes in B can be costlessly neutralized by offsetting changes in s. That L^0 is otherwise sensitive to (B, R) means that the door is now open to coordination problems when there is a separate lender.[4]

If $s^0 = 0$, then even these very weak restrictions can be relaxed. Total differentiation of (17.17) yields

$$\frac{\partial L^0}{\partial R} = \frac{\delta E(u_2'' \cdot \alpha\theta f')B}{u_2'' \cdot w^2 + \delta E[u_2'' \cdot (\alpha\theta f')^2 + u_2' \cdot \alpha\theta f'']} \geq 0$$

[4] Subramanian [22] deals with a variant of this model, in which labor is employed only in the harvest season. The comparative static results are the opposite of those derived above if the peasant' s preferences exhibit decreasing absolute risk aversion.

and

$$\frac{\partial L^0}{\partial B} = \frac{u_1'' \cdot w + \delta E(u_2'' \cdot \alpha\theta f')R}{u_2'' \cdot w^2 + \delta E[u_2'' \cdot (\alpha\theta f')^2 + u_2' \cdot \alpha\theta f'']} \geq 0,$$

both of which will hold as strict equalities for $(B, R) \gg 0$ if and only if the peasant is risk neutral.

Inspection of (17.17) and (17.18) reveals that for any B, all pairs (β, R) such that $(\beta + RB)$ is constant will induce the peasant to choose the same value of L. It follows from (17.5) that one of the instruments β and R is superfluous under interlinking. In particular, $R = R_0$ is compatible with attaining the optimum when β is suitably chosen. Observe that if the optimum involves $\alpha < 1$, then $E\Omega^0$ will depend on the peasant's choice of L, which depends, in turn, on β, B and R. Whether interlinking is strictly superior, therefore, hinges on whether the optimum interlinked contract is characterized by output-sharing ($\alpha^0 < 1$). For if it is, the landlord might suffer losses from a contractual externality should he persist with output-sharing when there are separate contracts. In any event, let $(\alpha^0, \beta^0(R_0), B^0, R_0)$ be an optimal interlinked contract, that is, let it be a solution of problem (17.14) with (17.17) replacing (17.13). Then the set of all solutions is

$$S_1 = \{\alpha^0, \beta^0, B^0, R^0 : \beta^0 + R^0 B^0 = \beta^0(R_0) + R_0 B^0\} \tag{17.19}$$

This consideration motivates the following claim, which will also serve as a lemma:

Proposition 3. *If the peasant is risk averse and subject to unlimited liability, then the optimum interlinked contract is characterized by* $\alpha^0 < 1$.

Proof. The landlord–moneylender's problem is (17.14), with (17.17) replacing (17.13). Writing the Lagrangian as

$$\Phi = E\Omega + \lambda(EU - \bar{U}), \tag{17.20}$$

noting that the incentive constraint is $L^0 = L^0(\cdot)$, and employing the envelope theorem, we obtain the following first-order conditions

$$\begin{aligned}
\partial\Phi/\partial\alpha &= -f(L^0) + (1-\alpha)f'(\partial L^0/\partial\alpha) + \lambda\delta E u_2' \cdot \theta f(L^0) \geq 0, & (17.21)\\
&\quad \alpha \leq 1 \text{ complementarily} & \\
\partial\Phi/\partial\beta &= (1-\alpha)f'(\partial L^0/\partial\beta) + [1 - \lambda\delta E u_2'] = 0 & (17.22)\\
\partial\Phi/\partial B &= (1-\alpha)f'(\partial L^0/\partial B) + (R - R_0) + \lambda[u_1' - \delta E u_2' \cdot R] \leq 0, & (17.23)\\
&\quad B \geq 0 \text{ complementarily} & \\
\partial\Phi/\partial R &= (1-\alpha)f'(\partial L^0/\partial R) + B[1 - \lambda\delta E u_2'] = 0. & (17.24)
\end{aligned}$$

Suppose, contrary to the proposition, that $\alpha = 1$. Then (17.21) and (17.22) yield, respectively,

$$-f(L^0)[1 - \lambda\delta Eu_2' \cdot \theta] > 0$$

and

$$[1 - \lambda\delta Eu_2'] = 0.$$

Since, by virtue of the strict concavity of $u(\cdot)$, $Eu_2' \cdot \theta = Eu_2' \cdot E\theta + \text{cov}(u_2', \theta) < Eu_2'$ whenever $\alpha > 0$, we have an immediate contradiction. ■

The intuition for this result is that in the absence of the relevant insurance markets, the contract must strike a balance between covering the peasant against fluctuations in output and giving him incentives to produce output. If $\alpha = 1$, both incentives and the riskiness of c_2 are maximized, which is not optimal — unless, of course, he is risk neutral. It should be noted also that the result holds for all pairs (B, R), and hence for the three-stage game.

The landlord now eschews moneylending, and an independent financier appears on the scene. If the landlord moves first, the lender's problem is to

$$\max_{(B,R|\alpha,\beta)} (R - R_0)B \qquad \text{subject to: } EU \geq \bar{U}. \tag{17.25}$$

The associated first-order conditions are, abusing notation slightly, (17.23) and (17.24). If, instead, the moneylender moves first, then he chooses (B, R) subject only to the landlord's participation constraint; for with the peasant being subject to unlimited liability, the amount RB will be repaid with certainty whatever the other parties do — provided they participate. Together with Proposition 3, this fact suggests that whether interlinking confers advantages in the present setting depends on who moves first, and so signals a departure from Propositions 1 and 2.

Proposition 4. *If the peasant is risk averse and subject to unlimited liability, then in the three-stage game, interlinking is strictly superior if the landlord moves first and the peasant's preferences exhibit* DARA. *It offers no advantages, however, if the moneylender moves first.*

Proof. Suppose the landlord moves first. Since the expected combined profits from tenancy and moneylending are maximized only by pairs of contracts belonging to S_1, the landlord must choose a tenancy component (α^0, β^0) therefrom if he is not to do worse than under interlinking. Observe also that if the lender can find an $R > R_0$ and a B such that the peasant's participation constraint is satisfied, then he will make strictly positive profits. This suggests that the contracts comprising S_1 be partitioned as follows:

(i) $R^0 < R_0$. If the landlord chooses the associated (α^0, β^0) in the Stackelberg game, then the lender will not find it profitable to offer (B^0, R^0); the lender may, indeed, find participation unprofitable. By the definition of S_1, any deviation from (B^0, R^0) that also satisfies the peasant's participation constraint will affect the landlord adversely. Hence, the choice (α^0, β^0) will yield him an expected payoff inferior to that under interlinking.

(ii) $R^0 > R_0$. In this case, the landlord will lose the profits from usury if he gets out of the lending business. It follows that the only strategy that could yield him $E\Omega^0$ is $[\alpha^0, \beta^0(R_0)]$.

(iii) $R^0 = R_0$. The final step, then, is to prove that (B^0, R_0) does not solve problem (17.25) if the landlord chooses $[\alpha^0, \beta^0(R_0)]$. By the envelope theorem,

$$\partial EU^0/\partial B = u_1' - \delta E u_2' \cdot R.$$

Since $R_0 > 1$ and there is DARA, $\partial L^0/\partial B > 0$. Recalling that $\alpha^0 < 1$, it then follows from (17.23) that if $R = R_0$, then $\partial EU^0/\partial B < 0$ under the interlinked contract $[\alpha^0, \beta^0(R_0)]$, that is, the peasant is being force-fed with credit. Hence, the lender can make a positive profit by charging a rate somewhat higher than r_0, while offering a correspondingly smaller loan than B^0 so as to satisfy the peasant's participation constraint. It follows not only that the landlord will be worse off if he shuns, or cannot practice, lending, but also that the sum of his and the moneylender's expected net incomes will be smaller than under any (interlinked) contract in S_1.

If, instead, the lender moves first, let him choose the pair (B^0, R^0) from the interlinked contract in S_1 that yields no expected rents, that is, the associated pair (α^0, β^0) satisfies

$$(1 - \alpha^0) f(L^0) + \beta^0 = 0.$$

This implies that

$$R^0 = R_0 + [\beta^0(R_0) + (1 - \alpha^0) f(L^0)]/B^0.$$

If the lender makes this choice, the landlord cannot do better than offer the associated pair (α^0, β^0); for under unlimited liability, the peasant must pay the lender $R^0 B^0$ whatever the landlord does, provided of course, the landlord finds it attractive to participate, a condition that is just satisfied. ■

17.2.4 Risk neutrality and limited liability

The peasant's risk aversion introduced a particular form of nonlinearity into the previous subsection, as a result of which output-sharing between him and the landlord emerged in equilibrium, whatever the terms of the credit contract. With such output-sharing, moreover, the landlord's expected income depends on how diligently the peasant performs his tasks. Since limited liability also introduces a nonlinearity into the agents' payoff functions that arises from how much is produced, it is natural to ring the changes by returning to risk neutrality, that is, to Assumption 2. Observe at the outset that, in contrast to the foregoing sections, the lender is now no longer able to obtain a fixed payment with certainty.

To keep the analysis as simple as possible, we deny the peasant the opportunity to save, so that his liability at harvest time will be limited to his share of output. Put precisely, we have

Assumption 4. *The peasant's consumption levels in subperiods* 1 *and* 2 *are, respectively,*

$$c_1 = B + w(1 - L) \quad \text{and} \tag{17.3'}$$
$$c_2 = \max[\alpha Q - (\beta + RB), 0]. \tag{17.4'}$$

The limited liability clause expressed in (17.4′) implies that the peasant's consumption in the second period is bounded below by zero, whatever he does in the first period. In particular, he can always guarantee himself the consumption vector $(B+w, 0)$ by neglecting the tenancy completely in favor of wage labor. With this alternative of "strategic default" available to him, the contractual terms, whether interlinked or not, must be sufficiently attractive to induce diligent cultivation. For the principal, this means that (17.2′) must be modified to

$$\bar{U} = B + w, \tag{17.2''}$$

and the tenant's participation constraint becomes

$$EU \geq B + w,$$

from which it is clear that control over the amount lent assumes even stronger importance in the presence of unlimited liability.

We turn therefore to the peasant's choice of labor input, conditional on cultivation being attractive. Default may still occur in sufficiently adverse states of nature, but when it does so, it will be involuntary. Let the critical value of θ be denoted by

$$\theta_1 \equiv (\beta + RB)/\alpha f(L), \tag{17.26}$$

so that the peasant is forced into default for all $\theta < \theta_1$. His problem is to

$$\max_{c_1, L|(\cdot)} EU \quad \text{subject to: (17.3') and (17.4')} \tag{17.27}$$

Substituting for c_1 using (17.3′), recalling (17.11), and assuming that $L \leq 1 + B/w$, the first-order condition with respect to L is

$$\partial EU/\partial L = -w + \delta \int_{\theta_1}^{\infty} \alpha f'(L) \cdot \theta dF(\theta) \leq 0, \ L \geq 0 \text{ complementarily.} \tag{17.28}$$

If the vector (α, β, B, R) is such that $\theta_1 > \theta_0$, it follows at once that the peasant's choice of L will depend on (B, R), in contrast to the case of unlimited liability in Section 17.2.2. Since crop failures do occur in practice, the case where $\theta_1 > \theta_0$ is likely to be the rule.

We begin with interlinked contracts. The landlord–moneylender's net income is

$$\Omega = Q - R_0 B - \max[\alpha Q - (\beta + RB), 0],$$

so that (17.5) is replaced by

$$E\Omega = f(L) - R_0 B - \int_{\theta_1}^{\infty} [\alpha \theta f(L) - (\beta + RB)] dF(\theta). \tag{17.29}$$

It is clear that if $\theta_1 > \theta_0$, then setting $\alpha = 1$ will not sever the dependence of $\mathrm{E}\Omega$ on L, as was the case in Section 17.2.2. Recall also that $EU \geq B + w$ if the peasant is to put any effort into cultivation. Hence, the landlord is almost certain to be worse off if he does not control (B, R). Under interlinking, the redundancy between β and R remains as before, provided the condition $EU \geq B + w$ is satisfied.

When there is a separate moneylender, the peasant's limited liability requires a rule whereby the two principals' contractual claims on output are settled in the event that output is insufficient to cover both in full. It is plausible that the principal who moves first also enjoys seniority, for he is in a position to stipulate as much in the contract.

Assumption 5. *The first party to move also has the first claim on output.*

Proposition 5. *If the peasant is risk neutral and his liability is limited to his share of output, then under Assumption* 5, *interlinking is strictly superior to any outcome of the three-stage Stackelberg game, whoever moves first.*

Proof. Suppose the landlord moves first. Then (17.29) specializes to

$$E\Omega_\ell = f(L) - \int_{\theta_1}^{\infty} [\alpha \theta f(L) - \beta] dF(\theta),$$

and the dependence of L on B and R remains if $\theta_1 > \theta_0$. Loss of control over B is evidently damaging. Under this seniority rule, the moneylender's net income is

$$\Omega_m = \min[RB, \max(\alpha Q - \beta, 0)] - R_0 B.$$

We define the set S_2 analogously to S_1, that is, S_2 is the set of all contracts that yield an expected net profit of zero from the lending side of the business:

$$S_2 = \left\{ \alpha^0, \beta^0, B^0, R^0 : R^0 \int_{\theta_1}^{\infty} dF(\theta) = R_0 \right\}. \tag{17.30}$$

By an argument analogous to that used in the proof of Proposition 4, it is easily shown that in the three-stage Stackelberg game, the only strategy that could yield the landlord his expected net income from an optimal interlinked contract is the one in which he chooses the pair (α^0, β^0) associated with any contract in S_2. Should the landlord make such a choice, however, it can be shown that the lender can always find a strictly profitable reply that also respects the peasant's participation constraint.

If the lender moves first, his only hope of doing as well as under an interlinked contract is to choose a pair (B^0, R^0) from the set thereof that yield zero expected rents:

$$f(L) - \int_{\theta_1}^{\infty} [\alpha\theta f(L) - \beta] dF(\theta) = 0.$$

His expected profit therefrom is

$$E\Omega_m = \left[R^0 \int_{\theta_3}^{\infty} dF(\theta) - R_0 \right] B^0,$$

where $\theta_3 = RB/\alpha f(L)$, so that θ_3 depends on the landlord's choice of α directly and on (α, β) indirectly through L^0. It is clear, however, that given (B^0, R^0), the landlord can profitably reduce α and, in compensation, β, in order to keep cultivation attractive to the peasant. In doing so, he will also reduce $E\Omega_m$ below the level $E\Omega^0$ yielded by interlinking. ■

17.2.5 Risk neutrality, limited liability and full information

The assumption of limited liability turns out to be even more powerful than the analysis of the preceding subsection, in which moral hazard played a central role, might suggest. This claim is established in Basu, Bell and Bose [4], employing a simple, two-state model from which moral hazard has been banished.

Returning to Assumption 1, the peasant's labor input is fixed at unity. There are just two possible outcomes where production is concerned, namely,

$$Q = \{Q_1, Q_2\} \text{ with } Q_1 < Q_2.$$

The endogenous element in production is the probability that the harvest will be good, which is assumed to depend on the amount of working capital employed. Since the peasant's input of labor is fixed at unity, such working capital may be thought of as outlays on additional, complementary inputs. Let

$$\pi(B) \equiv \text{prob}(Q = Q_2), \tag{17.31}$$

where $\pi(\cdot)$ is a smoothly increasing, strictly concave function of B, which also satisfies the lower Inada condition and $\lim_{B\to\infty} \pi = 1$. The principals are assumed to possess a flawless monitoring and enforcement technology:

Assumption 6. *The use of labor in cultivation and of the loan B solely as working capital is costlessly enforceable. The landlord can observe the state, and so condition β thereon.*

Taken together, the assumption that $L = 1$ and Assumption 6 imply that $c_1 = 0$. That B is used only as working capital yields a direct dependence of output on the terms of the credit contract.

The parties' reservation utilities are, respectively, $\bar{\Omega}_\ell = \bar{\beta}$, $\bar{\Omega}_m = R_0 B$, and $\bar{U} = w$. The joint (expected) surplus yielded by a loan of size B is

$$\Sigma(B) = [(1-\pi)Q_1 + \pi Q_2] - (w + \bar{\beta} + R_0 B), \tag{17.32}$$

which is distributed among the parties in the proportions $(\lambda_\ell, \lambda_m, \lambda_p)$, depending on their bargaining strengths and the structure of the game.

We begin with an interlinked contract. In view of the redundancy of α and R when β can be specified contingently on the state, the contract is specified simply as (β_1, β_2, B), where the peasant's limited liability implies $\beta_i \leq Q_i$ $(i = 1, 2)$. The parties' expected utilities from such a contract are, respectively,

$$EU = (1-\pi)(Q_1 - \beta_1) + \pi(Q_2 - \beta_2) \tag{17.33}$$

and

$$E\Omega = (1-\pi)\beta_1 + \pi\beta_2 - R_0 B. \tag{17.34}$$

It is clear from (17.33) and (17.34) that both β_1 and β_2 effect lump-sum transfers, and that they can be varied so as to effect any distribution of the net surplus. Hence, B will be chosen so as to maximize $\Sigma(B)$. By the strict concavity of $\pi(B)$, $\Sigma(B)$ will have a unique maximum, at B^0 say, where B^0 satisfies

$$\pi'(B) \cdot (Q_2 - Q_1) - R_0 = 0. \tag{17.35}$$

At the optimum, the parties' expected utilities will be

$$EU = w + \lambda_p \Sigma(B^0)$$

and

$$E\Omega = \bar{\beta} + R_0 B^0 + \lambda_\ell \Sigma(B^0).$$

The landlord now gives up lending and a financier makes his entrance. In the ensuing two-stage Stackelberg game, the landlord first chooses (β_1, β_2)

and the lender follows with (B, R), the peasant being completely passive apart from the decision of whether to participate. Let

$$z \equiv \pi \cdot \min(RB, Q_2 - \beta_2) + (1 - \pi) \cdot \min(RB, Q_1 - \beta_1). \tag{17.36}$$

Then the parties' respective expected utilities may be written as, respectively,

$$EU = (1 - \pi)(Q_1 - \beta_1) + \pi(Q_2 - \beta_2) - z, \tag{17.37}$$

$$E\Omega_\ell = (1 - \pi)\beta_1 + \pi\beta_2, \tag{17.38}$$

and

$$E\Omega_m = z - R_0 B. \tag{17.39}$$

Observe that for any (β_1, β_2) such that $EU > 0$, R effects lump-sum transfers between the moneylender and the peasant. Hence, given (β_1, β_2), it follows from (17.37) and (17.38) that the choice of B, B^s say, is the (unique) solution to

$$\pi'(B)[(Q_2 - Q_1) - (\beta_2 - \beta_1)] - R_0 = 0. \tag{17.40}$$

Comparing (17.35) and (17.40), it follows that $B^s = B^0$ if and only if $\beta_1 = \beta_2$.

Let the "power structure" in the interlinked setting be such that $\lambda_\ell = 1$. If there exists a β^0 such that

$$\beta^0 = \bar{\beta} + \Sigma(B^0) < Q_1, \tag{17.41}$$

then the landlord can set $\beta^0 = \beta_1 = \beta_2$ without violating the others' participation constraints or the peasant's limited liability. In this case, therefore, interlinking offers no advantages. As noted in Section 17.2.4, however, harvest failures are not unknown, and if the low output in this two-state setting is sufficiently miserable, then (17.41) will be violated, and the landlord will profit from interlinking. Since he will choose $\beta_1 < \beta_2$ in the Stackelberg game, it also follows from a comparison of (17.35) and (17.40) that $B^0 > B^s$, so that the overall allocation therein will be inefficient.

17.3 Credit and Trading

As noted in the Introduction, on the available evidence, the bundling of credit and trade is the most frequently occurring form of interlinking, though this finds little, if any reflection in the literature up to the mid 1980s, with only a smattering of contributions thereafter. This would not matter much if the marketing side of the story could be accommodated without undue violence into the ubiquitous framework of tenancy, with some minor changes in notation and interpretation. In the case of trading, however, price as well as output is risky, a fact that influences marketing arrangements in practice, but does not always find a home in models. The examples that follow illustrate the difficulties involved in attempts to explain what happens in practice and indicate that the current state of theory is not fully satisfactory.

Gangopadhyay and Sengupta [11] postulate a perfectly certain world, which, it is now clear, is a dangerous thing to do. Output depends only on the amount of outside finance, the use of which can therefore be inferred from the level of output. Hence, (17.1) is replaced by

$$Q = f(B). \tag{17.1$'$}$$

Peasants have the option of borrowing at the parametric rate $\bar{r}$ and of selling their output at the parametric price p. They consume nothing in the first subperiod and, in view of unlimited liability,

$$c_2 = pQ - \bar{R}B$$

in the second. Let $B(\bar{R})$ maximize c_2. Then a peasant's reservation utility may be written as

$$\bar{U} = pf[B(\bar{R})] - \bar{R} \cdot B(\bar{R}). \tag{17.42}$$

A trader–moneylender, who has access to funds at the parametric rate $r_0 < \bar{r}$, enters the scene and offers the interlinked contract (q, B, R), under which he will purchase the entire output at price q. This contract yields him

$$\Omega = (p - q)f(B) + (R - R_0)B$$

and the peasant

$$c_2 = qf(B) - rB.$$

By writing $\alpha \equiv q/p$, we obtain

$$\Omega = (1 - \alpha)[pf(B)] + (R - R_0)B \tag{17.43}$$

and

$$c_2 = \alpha[pf(B)] - RB. \tag{17.44}$$

Comparing (17.43) and (17.44) with their counterparts in Section 17.2.1, namely, (17.5) and (17.4), respectively, we have Bhadhuri's model once more, with $\beta = s = 0$, and where $(1 - \alpha)$ may be interpreted either as the discount on the ruling spot price, or as the commission on the proceeds of sale at that price. Given that α and R operate as lump-sum transfer instruments, the trader's optimal choice of B, namely, B^0, maximizes $(\Omega + c_2)$ and thus is the solution to

$$pf'(B) - R_0 = 0. \tag{17.45}$$

No doubt aware of this awkward kinship, Gangopadhyay and Sengupta introduce a curious assumption:

Assumption 7. In *the context of an interlinked contract with a trader, the peasant chooses B, given the trader's choice of* (α, R).

This is very hard to justify given their implicit assumption that such contracts are exclusive. That they cite empirical evidence to the effect that peasants make choices over B serves only to underline a logical inconsistency in the structure of their model.

Be that as it may, under Assumption 7, the peasant's choice of B, namely, $B(\alpha, R)$, satisfies

$$\alpha p f'(B) - R = 0. \tag{17.46}$$

Comparing (17.45) with (17.46), it is seen that $B(\alpha, R) = B^0$ if and only if $\alpha = R/R_0$. To extract the entire surplus under Assumption 7, however, the trader must also choose (α, R) such that

$$\alpha p f(B^0) - RB^0 = \bar{U}.$$

Solving for R, we obtain

$$\alpha^0 = \bar{U} R_0 / [pf(B^0) - R_0 B^0],$$

and hence

$$\alpha^0 = \bar{U} / [pf(B^0) - R_0 B^0].$$

Assumption 7 therefore pulls off the trick of yielding an efficient contract, together with a definite selection (α^0, R^0) from the set of all pairs that arise under the foregoing redundancy of instruments. As argued above, it is not especially convincing, and the finding that the outcome is allocatively superior to that under the peasant's reservation alternative is a simple consequence of the assumption that the trader's opportunity cost of funds is lower than the parametric rate ruling in the market.

A quite different tack is taken by Bell and Srinivasan [7], who deal with a variant of the situation analyzed in Section 17.2.4, in which the peasant's limited liability is replaced by a penalty in the case of default. In their two-state world, output depends on the input of labor and the state of nature as follows:

$$\begin{cases} \text{(i)} & \text{if } L = 0, \text{ then } Q = 0; \text{ and} \\ \text{(ii)} & \text{if } L = 1, \text{ then } Q \in \{0, Q_2\}. \end{cases} \tag{17.47}$$

where the probability of producing Q_2 is fixed, at π. Moral hazard then arises through.

Assumption 8. *The peasant's unit of labor is indivisible, but its use is prohibitively costly to monitor.*

It is clear from (17.47) that the peasant is then free to choose between diligent cultivation ($L = 1$) and complete shirking by working for wages ($L = 0$).

Ruling out onlending at interest or storage between subperiods, the whole of any loan B will be consumed in subperiod 1. Hence, if a default does occur, it will be total. The peasant's dated consumption vector is:

(i) if $L = 0$, then $c_1 = B + w$ and $c_2 = -\Psi(RB)$; and

(ii) if $L = 1$, then $c_1 = B$ and $c_2 = \begin{cases} \alpha p Q_2 - RB & \text{with probablity } \pi \\ -\Psi(RB) & \text{with probability } 1 - \pi \end{cases}$

where $\Psi(RB)$ is the cash value of the damage suffered by the peasant in his future dealings if he defaults on the amount RB. Observe that in this setting, none of the contractual parameters can effect lump-sum transfers.

In a strong departure from the standard assumption that the peasant faces a monopolistic principal, and in recognition of the fact that there are commonly well over a hundred active traders and commission agents in most marketing centers of any size in South Asia, Bell and Srinivasan choose a radically different market structure:

Assumption 9. *There is free entry into trading and moneylending, so that the expected profits therefrom under diligent cultivation are zero.*

The trader–moneylender's participation constraint must be augmented by the incentive-compatibility condition that, confronted with any exclusive contract (α, B, R), the peasant prefer $L = 1$ to $L = 0$. Let

$$\begin{aligned} S_3 = \{&\alpha, B, R : \alpha \in [0,1],\ B > 0,\ R \geq 0;\ E\Omega(L = 1) \geq 0; \\ &EU(L = 1) \geq EU(L = 0)\}. \end{aligned} \qquad (17.48)$$

Thus, S_3 defines the menu of all exclusive interlinked contracts that traders will offer under Assumption 9. Assuming that S_3 is non-empty, the peasant's problem is to

$$\max_{(\alpha,B,R)} EU \qquad \text{subject to } (\alpha, B, R) \in S_3. \qquad (17.49)$$

The main result is that the solution to problem (17.49) is always such that the contract involves a discount on the ruling spot price of output and a loan without interest. The first finding is pleasing; but in view of the extensive evidence that such contracts very frequently specify a positive rate of interest, the second is decidedly not so. As the authors themselves note, the model is therefore wanting in an important way.

17.4 Public Policy

What lessons, if any, does this exploration of theoretical issues hold for the design of public policy? One school of thought would argue that if interlinking is chosen by individually rational agents, then it must yield net advantages, so that governments should not meddle in matters. The flaw in such reasoning is that in settings where markets are not complete and information is not symmetrically held, individually rational choices in drawing up contracts may lead to collectively inefficient outcomes. Since no general results beyond the so-called second-best theorem are available, however, each case for intervention must be constructed on its merits in the circumstances in question, including the associated administrative costs. Given the theoretical difficulties involved and the burdensome nature of the complementary empirical work, it is scarcely surprising that rigorous attempts to evaluate proposals for intervention are extremely rare.

The following example illustrates what is involved. In 1973, the Indian government eliminated the wholesale trade in wheat as the first step in a plan to nationalize the grain trade. This was a dismal failure, and private wholesale trading was reinstated in the following year on the condition that traders "would transfer 50 percent of their purchases from farmers to the government at the procurement price. Monitoring and enforcement of this proved to be impossible" (Joshi and Little [14, p. 110]). The connection with interlinking turns out to have been important; for the main source of wheat surpluses was (and remains) the Punjab, and traders and commission agents were a vital source of credit to Punjabi farmers — according to the RPO 671–89 study, on terms that involved interlinkage in the great majority of cases. Bell, Srinivasan and Udry [8, p. 582] estimate that a ban on private trade would have reduced the supply of private credit to agriculture by 70 percent and the supply of all credit to that sector by somewhat over 35 percent.

Suppose now that there were a proposal to ban the interlinking of trade and credit. Given the monitoring and enforcement problems related to procurement, it appears that the only way to effect such a ban would be to ban private trading altogether. Seen in this light, the above estimates of the associated effects on the financing of Punjabi agriculture are sobering, even if the state trading corporation were just as efficient in serving farmers as private traders. These estimates do not clinch the argument, however, because they take no account of other effects that would ensue from the general reallocation of resources attendant upon the suppression of interlinking as a contractual device. While simple intuition might suggest that the latter cannot plausibly be so large as to undo the damage wrought by the contraction of credit, the case is still not rigorously sewn up.

17.5 Concluding Remarks

Two lessons can be drawn from Section 17.2. First, even in such apparently unpromising circumstances as risk neutrality and complete and symmetrically held information, the tenant's limited liability may make interlinking advantageous. The same holds under moral hazard if the tenant is risk averse, but is subject to unlimited liability — though in both cases, the advantage arises only if the landlord moves first in the sequential game with two separate principals, the size of the plot being fixed. In the models considered above, therefore, limited liability and the order of moves matter. The second lesson is more general: it is hard to resist the conclusion that there is not much left to say about the interlinking of tenancy and credit contracts, at least in a partial equilibrium setting. Much more promising territory, for newcomers and old hands alike, is the relation between trading and credit, precisely because it has features that are not well captured by models of tenancy and credit, and yet has been rather neglected in the literature.

The other weakness in our knowledge is a shaky grasp of the nature and importance of the general equilibrium effects of policies designed to regulate or promote interlinking. Since interlinking is itself a response to imperfections in, or outright failures of, particular markets, including those for risk, a thorough welfare analysis of interventions poses no mean intellectual challenge. In view of Srinivasan's own exemplary record in furthering able young scholars, it seems fitting to have concluded by pointing out a collection of unresolved puzzles worthy of their attention.

References

1 Bardhan, P. (ed.), The Economic Theory of Agrarian Institutions, Clarendon Press, Oxford, 1989.

2 Basu, K., The emergence of isolation and interlinkage in rural markets, Oxford Economic Papers, 35 (1983) 262–80.

3 Basu, K., Limited liability and the existence of share tenancy, Journal of Development Economics, 38 (1992) 203–20.

4 Basu, K., C. Bell and P. Bose, Interlinkage, Limited Liability and Strategic Interaction, mimeo, University of Memphis, 1998.

5 Bell, C., Credit markets and interlinked transactions, in: H.B. Chenery and T.N. Srinivasan (eds.), Handbook of Development Economics, North-Holland, Amsterdam, 1988.

6 Bell, C. and T.N. Srinivasan, Interlinked transactions in rural markets: An empirical study of Andhra Pradesh, Bihar and Punjab, Oxford Bulletin of Economics and Statistics, 51 (1989) 73–83.

7 Bell, C. and T.N. Srinivasan, Some aspects of linked product and credit contracts among risk-neutral agents, in: P. Bardhan (ed.), The Economic Theory of Agrarian Institutions, Clarendon Press, Oxford, 1989.
8 Bell, C., T.N. Srinivasan and C. Udry, Rationing, spillover, and interlinking in credit markets: The case of Rural Punjab, Oxford Economic Papers, 49 (1997) 557–85.
9 Bhadhuri, A., Agricultural backwardness under semi-feudalism, Economic Journal, 83 (1973) 120–37.
10 Braverman, A. and T.N. Srinivasan, Credit and sharecropping in agrarian societies, Journal of Development Economics, 9 (1981) 289–312.
11 Gangopadhyay, S. and K. Sengupta, Small Farmers, moneylenders and trading activity, Oxford Economic Papers, 39 (1987) 333–42.
12 Harriss, B., Paddy and Rice Marketing in Northern Tamilnadu, Sangam Publishers, Madras, 1979.
13 Janakarajan, S., Interlinked transactions and the market for water in the agrarian economy of a Tamilnadu Village, 1992.
14 Joshi, V. and I.M.D. Little, India: Macroeconomics and Political Economy 1964–1991, Oxford University Press, New York, 1994.
15 Newbery, D.M.G., Tenurial obstacles to innovation, Journal of Development Studies, 11 (1975) 263–77.
16 Otsuka, K., H. Chuma and Y. Hayami, Labor and land contracts in agrarian economies, Journal of Economic Literature, 30 (1992).
17 Platteau, J.-P. and A. Abraham, An inquiry into quasi-credit contracts: The role of reciprocal credit and interlinked deals in small-scale fishing communities, Journal of Development Studies, 23 (1987), 461–90.
18 Ray, D. and K. Sengupta, Interlinkages and the pattern of competition, in: P. Bardhan (ed.), The Economic Theory of Agrarian Institutions, Clarendon Press, Oxford, 1989.
19 Shetty, S., Limited liability, wealth differences and tenancy contracts in agrarian economies, Journal of Development Economics, 29 (1988) 1–22.
20 Srinivasan, T.N., Farm size and productivity: Implications of choice under uncertainty, Sankhya: The Indian Journal of Statistics, Series B, 34 (1972) 409–20.
21 Steinmetz, A.M., Vermarktung, Kredit und Interlinking in Ländlichen Gebieten, Diplomarbeit, Heidelberg University, 1997.
22 Subramanian, S., Wage labor, sharecropping, and credit transactions, Oxford Economic Papers, 47 (1995) 326–56.
23 Zusman, P., Peasants' risk aversion and the choice of marketing intermediaries and contracts: A bargaining theory of equilibrium marketing contracts, in: P. Bardhan (ed.), The Economic Theory of Agrarian Institutions, Clarendon Press, Oxford, 1989.

Trade, Growth, and Development
G. Ranis and L.K. Raut

CHAPTER 18

The Kibbutz as a Labor-Managed Club: Public and Private Goods, Incentives and Social Control[1]

Michael Byalsky,[a] Michael Keren[b] and David Levhari[c]

[a]Falk Foundation for Economic Research in Israel, Hebrew University, Jerusalem 91905 Israel

[b,c]Department of Economics, Hebrew University, Jerusalem 91905 Israel

18.1 Introduction

The 1980s was a decade of soul-searching for the kibbutz movement (see Barkai [1] for a classic analysis of the kibbutz and its economic success in the 1970s). Before the credit squeeze of the mid-decade, both the kibbutz movement and its creditors believed in the implicit government guarantee of the debts of the movement, and the softening of the budget constraint of the movement and of each individual kibbutz was a direct result. The hardened budget constraint, ushered in with the stabilization program of 1985, brought the day of reckoning. The movement that seemed to have emerged economically strengthened in the 1970s was suddenly shown to be quite fragile, with many of its member-kibbutzim unable to cover their living expenses and even more of them deep in debt, with quite a few hopelessly in the red. The movement, which has never been short of soul-searching, came up with quite a few ideas for reform. For instance, to increase efficiency it was suggested that the production and the consumption sides of the kibbutz be separated. This idea strongly affects the "self labor" principle of the kibbutz, the principle that makes the kibbutz not employ non-members and that makes an LMF (a labor-managed firm). Another idea has been to tie some of the remuneration of kibbutz members to their contribution to production, a direct attack on the dogma of equality (Helman [11]).

[1]We are grateful to Michael Ferrantino who, in our absence, presented the paper for us, discussed it, and sent us his perceptive and fruitful comments. We also wish to thank the Israeli Ministry for Science and the Falk Foundation for financial support.

To be able to examine these reform ideas and anticipate their probable effects, we need some credible model of the kibbutz. No such model exists at present. The model of the LMF[2] has usually been applied uncritically to the kibbutz, but as Satt [21] and Ben-Ner [2, 3] have pointed out such applications lead to conclusions that run counter to observed facts. The reason is very simple: the model is concerned only with the productive side of the kibbutz, and disregards the common consumption and communal life in general, an integral if declining part of the very essence of the kibbutz. One of our aims below is to integrate this aspect into the LMF model, by adding a public good, produced and consumed by kibbutz members. This, in fact, makes the kibbutz a labor-managed club (see Section 18.2). It does, however, raise a problem: it imparts a degree of increasing returns to the kibbutz. Now, although one tends to hear more complaints of kibbutzim that are too small and therefore not self-sufficient—and the desire for size fits in well with the dominant strands of kibbutz dogma[3] — in really large kibbutzim one also hears complaints that the farm (and we shall use this term to translate the Hebrew meshek) has become too large. A clear disadvantage of size is the reduced relation between members' efforts and their individual well being, a relation analyzed by Holmström [13] in his "Moral Hazard in Teams." We therefore add this aspect in Section 18.3. Section 18.4 examines the effect of social control and social pressure on the behavior of kibbutz members and their welfare, in a manner suggested by Kandel and Lazear [15]. Section 18.5 confronts our model with the problems raised by Satt [21] and Ben-Ner [2, 3] and concludes this paper with a brief discussion of the proposed reforms.

18.2 The Kibbutz as a Club

Suppose first that the kibbutz has L members, each of whom supply a single unit of labor. The kibbutz hires capital at a given interest rate, r, and finances its interest payments out of the partial proceeds of its marketable private product, X; the remainder is consumed in equal shares by the members. It also produces a public good, G, all of which is used exclusively by the members. The inputs, L and K, are used in the production of both goods. Simple Cobb–Douglas functions are assumed:

$$X = X(L_1, K_1) = L_1^{\beta_1} K_1^{1-\beta_1}, \tag{18.1}$$

[2] See Ward's pioneering [28] paper and the relevant parts of his [29] *Socialist Economy*. See also Domar [10] and Vanek [26, 27]. Meade's [18] analysis is a seminal generalization.

[3] Though not with all strands: one of the movements, which merged into the Union of Kvutzot and Kibbutzim and then became part of the United Kibbutz Movement (Takam), believed in small, face-to-face communities. But kibbutzim belonging to this movement, called the Association of Kvutzot (Hever haKvutzot), also grew, probably proving that the club aspect of the kibbutz is stronger than their dogma of smallness.

$$G = G(L_2, K_2) = L_2^{\beta_2} K_2^{1-\beta_2}, \tag{18.2}$$

where L_j, K_j ($j = 1, 2$) are the labor and the capital which are used in the production of the private ($j = 1$) and the public ($j = 2$) goods respectively. Clearly,

$$L_1 + L_2 \leq L, \tag{18.3}$$

and

$$K = K_1 + K_2. \tag{18.4}$$

The utility function of the ith kibbutz member, $i = 1, ..., L$, is also assumed to be of the Cobb–Douglas family:

$$U^i(y, G) = y^{\alpha} G^{1-\alpha}, \tag{18.5}$$

where

$$y = \frac{X(L_1, K_1) - \frac{r}{p}K}{L} \tag{18.6}$$

is the amount of the private good X left over for consumption by each kibbutz member, p being the relative price of the marketable good X. Since all members are assumed for now to be identical, the superscript i is henceforth omitted.

The problem that the economic leadership of the kibbutz faces is to maximize the utility of each kibbutz member, (18.5), subject to the definitions (18.1), (18.2), (18.4), (18.6), and constraint (18.3). In other words, they have to determine the division of labor into L_1 and L_2, hire capital at the given interest rate of r, and determine the output of the marketable good X and the public good G. From the first order conditions of their problem with respect to L_i, the requirement that the marginal additions to utility through L_1 and L_2 be equal (i.e., by the equality of $\partial U/\partial L_j$, $j = 1, 2$, to the shadow price of labor),[4] we obtain,

$$\frac{\alpha\beta_1 X}{yLL_1} = \frac{(1-\alpha)\beta_2}{L_2},$$

or

$$\frac{L_2}{L_1} = \frac{(1-\alpha)\beta_2}{\alpha\beta_1 X} yL. \tag{18.7}$$

By $\partial U/\partial K_1 = 0$,

$$\frac{(1-\beta_1)X}{K_1} = \frac{r}{p} \equiv \rho,$$

[4] We assume here an efficient allocation of members' labor between the two branches of production. That this was the case in the pre-1985 kibbutz is disputed by our sociologist colleague, Chezi Dar. Dar claims that there was a wasteful use of labor in the G production. Only the post-1985 hardening of the budget constraints seems to have led to a more economic employment policy in kibbutzim.

and the notation of $r/p = \rho$ is used henceforth. This, upon rearrangement, leads to

$$X = \frac{\rho}{1-\beta_1} K_1 \equiv B^{-1} K_1, \tag{18.8}$$

and, when (18.1) is substituted for X,

$$K_1 = \left[\frac{1-\beta_1}{\rho}\right]^{1/\beta_1} L_1 \equiv B^{1/\beta_1} L_1, \tag{18.9}$$

or, when (18.9) is substituted for (18.8)

$$X = \frac{\rho}{1-\beta_1} \left[\frac{1-\beta_1}{\rho}\right]^{1/\beta_1} L_1 \equiv B^{(1-\beta_1)/\beta_1} L_1. \tag{18.10}$$

$\partial U/\partial K_2 = 0$ leads to

$$\frac{(1-\alpha)(1-\beta_2)}{\alpha K_2} = \frac{\rho}{yL},$$

hence to

$$K_2 = \frac{(1-\alpha)(1-\beta_2)}{\alpha\rho} yL \equiv AyL. \tag{18.11}$$

Substituting (18.10) in (18.7),

$$L_2 = \frac{(1-\alpha)\beta_2}{\alpha\beta_1} B^{(\beta_1-1)/\beta_1} yL \equiv DyL. \tag{18.12}$$

Finally, plugging in (18.10) for X, (18.9) for K_1, and (18.11) for K_2, in the definition of y, (18.6), and rearranging, we get

$$\left(B^{(1-\beta_1)/\beta_1} - \rho B^{1/\beta_1}\right) L_1 = (1+\rho A) yL,$$

or

$$yL = \frac{(B^{-1}-\rho) B^{1/\beta_1}}{1+\rho A} L_1 \equiv CL_1. \tag{18.13}$$

From (18.12) and (18.13) we get $L_2 = DCL_1$, hence

$$L = L_1 + L_2 = (1+DC) L_1, \tag{18.14}$$

from which L_1 is derived below, in Lemma 1:

Lemma 1 *The utility optimizing labor and capital inputs to the X and G branches of the kibbutz are proportional to the membership L, where the proportions depend on the parameters of the production and utility functions and on the prices r and p.*

Proof L_1 is obtained from (18.14):

$$L_1 = \frac{L}{1+DC} = [1-(1-\alpha)\beta_2]L \equiv \lambda_1 L.$$

The values for the remaining inputs are all proportional to L_1:

$$\begin{aligned} L_2 &= DCL_1 = (1-\lambda_1)L = (1-\alpha)\beta_2 L \equiv \lambda_2 L, \\ K_1 &= B^{1/\beta_1}L_1 = [1-(1-\alpha)\beta_2]B^{1/\beta_1}L \equiv \kappa_1 L, \\ K_2 &= ACL_1 = \frac{(1-\alpha)\beta_1(1-\beta_2)}{1-\beta_1}B^{1/\beta_1}L \equiv \kappa_2 L, \end{aligned}$$

which establishes the lemma. ■

Observe that the usual result, by which logarithmic (or Cobb–Douglas) utilities entail a fixed-proportional division of expenditures, translates here into a fixed proportional distribution of the constraining resource, labor or membership. The extension that outputs X and G are also proportional, is immediate:

Lemma 2 *The outputs of X and G are proportional to L.*

Proof Let

$$\log \xi = \beta_1 \log \lambda_1 + (1-\beta_1)\log \kappa_1,$$

and

$$\log \gamma = \beta_2 \log \lambda_2 + (1-\beta_2)\log \kappa_2.$$

Then $X = \xi L$ and $G = \gamma L$. ■

This leads us to the fundamental proposition on the kibbutz:

Proposition 1 *If the labor input of the kibbutz member is constant and independent of kibbutz size, then the optimal consumption y of the private good X by kibbutz members is independent of kibbutz size, but the consumption of the public good is increasing with size, and the kibbutz member's utility is increasing with membership.*

Proof It is simple to compute and find that y of (18.6) becomes

$$y^* = \frac{\xi L - \rho(\kappa_1+\kappa_2)L}{L} = \xi - \rho(\kappa_1+\kappa_2) \equiv \chi, \tag{18.15}$$

and, as Lemma 1 shows,

$$G* = \gamma L.$$

Consequently

$$U^i = (\chi^\alpha \gamma^{1-\alpha}) L^{1-\alpha} \equiv \psi L^{1-\alpha}, \tag{18.16}$$

where ψ is a constant, but $L^{1-\alpha}$ is increasing in L for $\alpha < 1$. ■

Proposition 1 is both illuminating and worrying. It is illuminating, because it shows the potential relevance of the kibbutz: it is public consumption, when $1-\alpha > 0$, which gives this socio-economic organization its stability, its desire to grow, and its quality of life that defends it against the lures of the outside world. When $\alpha = 1$, i.e., when we have no sharing in public consumption, the organization becomes very brittle. For when y lies below incomes available in the private economy, it is clearly hard to make members stay in the kibbutz. But when it exceeds the comparable income in the surrounding capitalist society, it can usually be found that a reduction in L, the attrition of membership, will benefit the remainder, and the kibbutz will tend to shrink to a single member. This is the case analyzed by Satt [21] and Ben-Ner [2, 3]. What is unsatisfactory, though, is that the kibbutz of this section will tend to grow to infinity.[5] The following section deals with this problem.

18.3 Incentives in the Kibbutz

Holmström's [13] seminal paper provides us with a natural way of examining incentives in the kibbutz. We maintain our assumption that all L kibbutz members are identical agents, but now each of them may exert himself, and his effort is an input into production. It will now prove necessary to distinguish clearly between each member's own effort, e_i, $i = 1, ..., L$, and the L-dimensional vector of all members' effort levels, the bold-faced $\mathbf{e} = (e_1, ..., e_L)$. It will also be useful to denote by

$$\mathbf{e}_{-i} = (e_1, ..., e_{i-1}, e_{i+1}, ..., e_L)$$

the $(L-1)$-dimensional vector of the effort levels of all other members but i. Each member's own effort is a utility-reducing input, and we denote by $V^i(e_i)$, $i = 1, ..., L$, i's disutility of effort. As in Section 18.2, U^i denotes i's utility from private and public consumption, and W^i is the sum of the utility and disutility of effort. Thus

$$W^i = W^i(y, G, \mathbf{e}_{-i}, e_i) = U^i(y^i(\mathbf{e}), G(\mathbf{e})) - V^i(e_i). \tag{18.17}$$

[5] Some of the originators of the first kibbutz movements dreamed of a kibbutz which would comprise the whole of the country. It is this model they were possibly viewing, because they had little idea of the problems of incentives, the subject of the following section.

The i superscript is omitted whenever there is no danger of confusion. As is usual, $V' > 0$, $V'' > 0$, with $V(0) = 0$.

We also need to introduce the effect of members' effort on production. Let $\overline{e}$ denote the mean effort level of all kibbutz members, $\overline{e} = (e_1 + \cdots + e_L)/L$. The simplest assumption is that the effective input of labor is just the sum of the effort levels in each branch of production, i.e.,

$$X = X(L_1, K_1, \mathbf{e}) = (\bar{e}L_1)^{\beta_1} K_1^{1-\beta_1}, \tag{18.18}$$

and

$$G = G(L_2, K_2, \mathbf{e}) = (\bar{e}L_2)^{\beta_2} K_2^{1-\beta_2}. \tag{18.19}$$

The building blocks of the first order conditions for both a Pareto optimal and a Nash equilibrium allocation will now be prepared. Let U_e, G_e, u_e, etc., denote the partial derivative of U^i, G^i, y^i, etc., with respect to e_j: due to the full symmetry assumed, the omission of both subscript and superscript in $\partial U_i / \partial e_j$ and the like need cause no confusion:

$$U_e(y, G) = U_y y_e(\mathbf{e}) + U_G G_e(\mathbf{e}) = U \left[\frac{\alpha}{y} y_e(\mathbf{e}) + \frac{1-\alpha}{G} G_e(\mathbf{e}) \right]. \tag{18.20}$$

By the definition of y, (18.6), and the new expressions for X and G, (18.18) and (18.19),

$$y_e = \frac{1}{L} X_e = \frac{\beta_1 X}{\bar{e}L^2},$$

and

$$G_e = \frac{\beta_2 G}{\bar{e}L}.$$

Substituting these two expressions into (18.20) and remembering the definition of y in (18.6), we get

$$U_e(y, G) = U \left[\frac{\beta_1 \alpha X}{\bar{e}L^2 y} + \frac{\beta_2(1-\alpha)G}{\bar{e}LG} \right] = \frac{U}{\bar{e}L} \left[\frac{\beta_1 \alpha}{1 - \rho\frac{K}{X}} + \beta_2(1-\alpha) \right]. \tag{18.21}$$

These building blocks are used in the construction of the first-order conditions of the Pareto allocation in the next section, and of the Nash equilibrium in the following one.

18.3.1 The Pareto optimal allocation

As a benchmark, we first calculate the Pareto optimal allocation of effort in the kibbutz. We maximize member i's utility, subject to his fellow members' welfare not declining below $\underline{W}^j$, $\forall j$, where $\underline{W}^j$ is an exogenously given utility level:

$$e_i^P = \operatorname{argmax}\left\{W^i(y, G, \mathbf{e}_{-i}, e_i) + \sum_{j \neq i} \mu^j [W^j(y, G, \mathbf{e}_{-j}, e_j) - \underline{W}^j]\right\}, \ \forall i. \tag{18.22}$$

Using (18.17), the values of the j can be calculated by perturbing $\mathbf{e}$. The differentiation of (18.22) with respect to the elements of $\mathbf{e}$ provides us with the following set of L equations:

$$U_e^i - V^{i\prime} + \sum_{j \neq i} \mu^j U_e^j = 0,$$

$$U_e^i + \sum_{j \neq i} \mu^j (U_e^j - V^{j\prime}) = 0, \quad j \neq i.$$

Subtracting the ith equation from the jth, and using the assumption of the homogeneity of the members, we obtain

$$\mu^j V^{j\prime} = V^{i\prime},$$

from which we learn that $\mu^j = 1$, $j = 1, .., L$, hence that equation (18.22) can be written

$$e_i^P = \operatorname{argmax}\left\{W^i(y, G, \mathbf{e}_{-i}, e_i) + \sum_{j \neq i} [W^j(y, G, \mathbf{e}_{-j}, e_j) - \underline{W}^j]\right\}, \ \forall i. \tag{18.23}$$

The first-order conditions therefore become

$$V'(e_i) = LU_e^i(y, G), \forall i,$$

or, using (18.21),

$$V'(e_i) = \frac{U}{\overline{e}}\left[\frac{\beta_1 \alpha}{1 - \rho \frac{K}{X}} + \beta_2(1 - \alpha)\right], \ \forall i. \tag{18.24}$$

The second-order conditions for a maximum can be checked to establish that this is a maximum.

A comment: as equation (18.22) shows, the Pareto optimal effort decision is, in effect, an altruistic decision, because it maximizes the sum of the welfare of all kibbutz members.

18.3.2 The Nash equilibrium

Suppose that each member is selfish and considers only his own welfare. The problem is now turned into a noncooperative game with the payoffs $W^i(y, G, \mathbf{e})$ of (18.17) for the member (player) i, $i = 1, ..., L$. A Nash equilibrium is assumed: each member i maximizes his payoff W^i, assuming as given the other actors' effort, $\mathbf{e}_{-i}$, and the equilibrium condition is

$$e_i^N = \text{argmax}[U^i(y(\mathbf{e}_{-i}^N, e_i), G(\mathbf{e}_{-i}^N, e_i)) - V(e_i)], \quad \forall i. \tag{18.25}$$

The first-order conditions for (18.25) are

$$V'(e_i) = U_e^i(y, G), \quad \forall i, \tag{18.26}$$

or[6]

$$V'(e_i) = \frac{U}{\bar{e}L}\left[\frac{\beta_1 \alpha}{1 - \rho\frac{K}{X}} + \beta_2(1 - \alpha)\right].$$

Figure 18.1 illustrates the difference between the Pareto and Nash allocations of effort, conditions (18.22) and (18.25). The $MU(h)$ curves denote effort's marginal utility from increased consumption, for the Pareto Optimum ($h = PO$) and the Nash Equilibrium ($h = NE$), and $MdUe$ stands for the marginal disutility of effort. It clarifies why $e^N < e^P$:

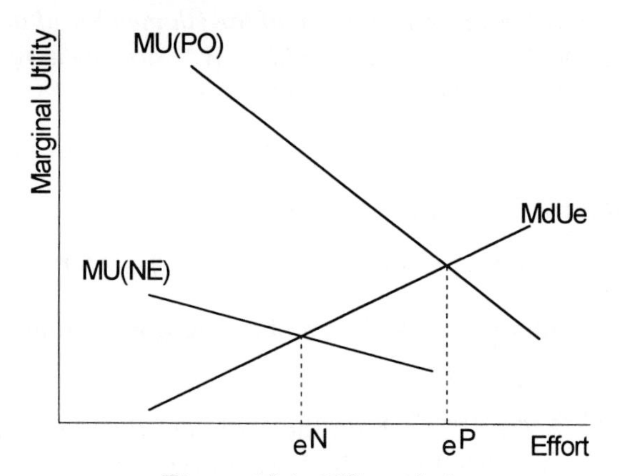

Figure 18.1: Effort choice

Proposition 2 (Holmström) *The Nash equilibrium levels of effort in a kibbutz are always lower than the Pareto optimal ones, as long as $L > 1$.*

[6] As for (18.22) and e^P, the second-order conditions for (18.25) show that e^N is indeed a maximizer.

The proof follows directly from the convexity of the effort disutility function V.

It is quite straightforward to show that Lemmas 1 and 2 apply here with obvious adjustments:

Lemma 3 *The optimal inputs of labor to the X and G branches of the kibbutz with endogenous effort inputs are, as in Lemma 1, proportional to the membership of the kibbutz, and the optimal inputs of capital and outputs are proportional to their effective labor.*

Proof is identical to that of Lemmas 1 and 2, except that here $K_j = \kappa_j \overline{e} L$, $j = 1, 2$, $X = \xi \overline{e} L$, and $G = \gamma \overline{e} L$.

Let y^P and y^N denote the Pareto optimal and Nash levels of y_i, respectively.

Lemma 4 *The optimal level of private consumption, y^h, $h = P, N$, of kibbutz members in the kibbutz with discretionary effort, is proportional to the average kibbutz member's effort $\overline{e}$. More precisely, $y^h = \overline{e}\chi$, where χ has been defined in* (18.15).

Proof is immediate.

Lemma 5 *The utility from consumption of the ith member of the kibbutz with discretionary effort, U^i, and his total utility W^i, are given by the following functions of the vectors of labor L and efforts $\mathbf{e}$:*

$$U^i[y(\overline{e}^h), G(\overline{e}^h), L] = \psi \overline{e} L^{1-\alpha}, \quad h = N, P, \tag{18.27}$$

where the constant ψ is as in (18.16), *and*

$$W^i[y(\overline{e}^h), G(\overline{e}^h), L] = \psi \overline{e} L^{1-\alpha} - V(e_i), \quad h = N, P. \tag{18.28}$$

Proof As with Lemmas 3 and 4, the proof is analogous to that of Proposition 1. ■

Observe that by (18.27), U^i, for both the Nash and the Pareto allocations, is linear in $\overline{e}$ (and e_i, $\forall i$). See Figure 18.2, where the notation is identical with that of Figure 18.1. U_e of (18.21) becomes

$$\begin{aligned} U_e(y, G) &= \frac{U}{\overline{e}L}\left[\frac{\beta_1 \alpha X}{X - \rho K} + \beta_2(1-\alpha)\right] \\ &= \psi\left[\frac{\beta_1 \alpha \xi}{\xi - \rho(\kappa_1 + \kappa_2)} + \beta_2(1-\alpha)\right] L^{-\alpha} \equiv \psi L^{-\alpha}, \end{aligned} \tag{18.29}$$

since the square brackets equal $\lambda_1 + \lambda_2 = 1$.

The main question of interest has not been clarified yet: does W increase in L without bound, as in the kibbutz with a fixed level of effort of Section 18.2, or can we find conditions under which there will be an internal optimum membership of the kibbutz? We therefore try in the following section to examine the relation between the size and the welfare level of the kibbutz under a very simple effort disutility function, the quadratic function.

18.3.3 Size effects: Quadratic disutility

Since $U(\cdot, \mathbf{e})$ is linear in $\overline{e}$ (and e_i, $\forall i$), a quadratic disutility function is the simplest we can use to find an explicit solution and show by example that the quality of life in a Nash-equilibrating kibbutz of our construction need not improve indefinitely as its size increases. Thus, let

$$V(e_i) = \tfrac{1}{2} b e_i^2, \quad V'(e_i) = b e_i, \tag{18.30}$$

where $b > 0$. Setting (18.30) and (18.29) in (18.26),

$$e_i^N = \tfrac{\psi}{b} L^{-\alpha}, \quad \forall i,$$

i.e., the optimal level of effort e_i is a declining, convex hyperbola-like power function in the membership L. For the sake of completeness, the Pareto optimal level of effort is similarly obtained as

$$e_i^P = \tfrac{\psi}{b} L^{1-\alpha}, \quad \forall i, \tag{18.31}$$

i.e., an increasing concave parabola-like power function in L, and

$$e_i^P = e_i^N L, \quad \forall i.$$

See Figure 18.2. Substituting these effort levels into the members' utility function, U of (18.27), we obtain

$$U_i^N = \frac{\psi^2}{b} L^{1-2\alpha}, \quad U_i^P = \frac{\psi^2}{b} L^{2(1-\alpha)},$$

i.e., U^N is increasing in L only if $\alpha < 1/2$. Also,

$$U^P = U^N L,$$

and U^P is always increasing in L. But it is W^N we are after:

$$W^N = \frac{\psi^2}{b} L^{1-2\alpha} - \frac{\psi^2}{2b} L^{-2\alpha} = \frac{\psi^2}{2b} \cdot \frac{2L-1}{L^{2\alpha}}. \tag{18.32}$$

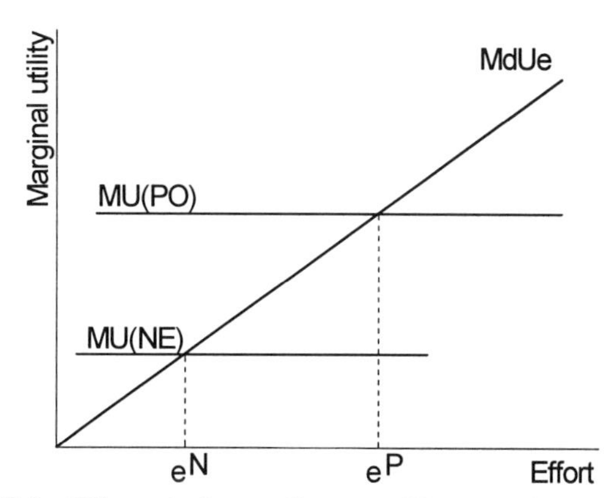

Figure 18.2: Effort choice — linear utility, quadratic disutility

When $\alpha \leq 1/2$, W^N is increasing in L throughout, but when $\alpha > 1/2$, there exists a welfare maximizing membership. Differentiating W^N with respect to L, we find that for $\alpha > 1/2$ the welfare maximizing membership is

$$L^* = \frac{\alpha}{2\alpha - 1}. \tag{18.33}$$

The second derivative of W^N shows that W^N also has an inflection point to the right of L^*, at

$$L = \frac{\alpha + \frac{1}{2}}{2\alpha - 1} > L^*.$$

As for W^P, the member's welfare at the Pareto Optimum is increasing without bound:

$$W^P = \frac{\psi^2}{2b} L^{2(1-\alpha)} = W^N \cdot \frac{L^2}{2L - 1} > W^N, \tag{18.34}$$

for $L > 1$. The two panels of Figure 18.3 illustrate W^N and W^P for $\alpha < 1/2$ (panel a) and $\alpha > 1/2$ (panel b). The conclusions of this discussion are summarized in Proposition 3.

Proposition 3 *When the disutility of effort, V, is quadratic as in equation (18.30), W^N, the kibbutz member's welfare under a Nash equilibrium, depends on α:*

(i) *When $\alpha < 1/2$, W^N is increasing in L and converges to $cL^{1-2\alpha}$, where c is a positive constant.*

(ii) *When $\alpha = 1/2$, it is an increasing concave function which converges to c.*

(iii) *When $\alpha > 1/2$, it has a finite maximum, and converges to the decreasing $cL^{1-2\alpha}$.*

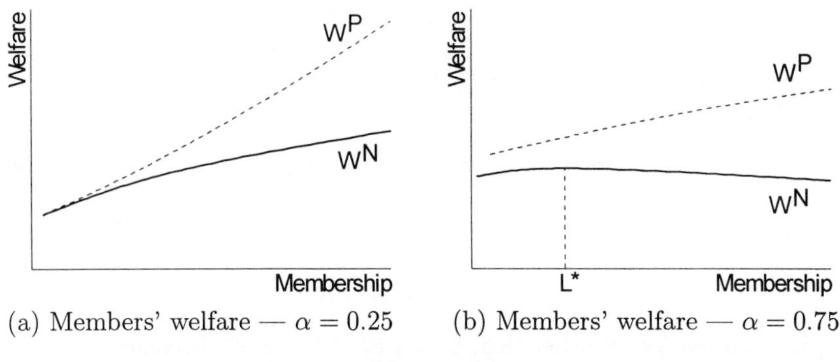

(a) Members' welfare — $\alpha = 0.25$ (b) Members' welfare — $\alpha = 0.75$

Figure 18.3

It is this latter condition which is of interest: a high α implies a great weight to private rather than public consumption. It is when private consumption becomes important that there is a utility-punishment for letting the kibbutz grow too large. When α is very low and public consumption has an overarching role, then the kibbutz should grow as large as it possibly can. We believe that this distinction makes sense, and that quite a few developments in the kibbutz can be explained with its aid.

18.4 Peer Pressure

Anyone familiar with life in the kibbutz knows that peer pressure is an important element in members' choices of their effort level. Thus, the Pareto optimal effort level will not be forthcoming, because it assumes that members choose to exert themselves without regard to their own interests. Neither will the Nash equilibrium of the previous sections, because it assumes that members can make their choices with no regard to their comrades' opinions. Kandel and Lazear [15] have shown a simple way of introducing peers' opinions into the picture. They posit a peer pressure function, $P^i(e_i, \mathbf{e}_{-i})$, which denotes the pressure applied against each member as a function of his own and his comrades' effort levels. The ith member's welfare level now becomes

$$W^i = W^i(y, G, e_i, \mathbf{e}_{-i}) = U^i(y, G) - V^i(e_i) - P^i(e_i, \cdot), \qquad (18.35)$$

where P^i depends on other variables and parameters. $P^i(\cdot) > 0$ denotes the ith member's utility loss through social pressure applied by his peers. $P^i_{e_i}(\cdot) < 0$, i.e., P is decreasing in the member's own effort, and may be increasing in others' monitoring efforts. We deal with the very simplest case, in which P is quadratic in the difference $a - e_i$, where a is an effort norm imposed by the kibbutz "social planner" (see Section 18.4.1 for the choice of a). Thus, $P(e_i) = \frac{\gamma}{2}(a - e_i)^2$, where (following Kandel and Lazear [15]) the parameter

$\gamma > 0$ denotes the punishment applied when the member reduces his effort e_i by a single unit. As before (see Section 18.3.3), we assume that V, effort disutility, is quadratic: $V(e_i) = \frac{1}{2}be_i^2$. The welfare function of kibbutz member i now takes the form

$$W^i = U^i - \tfrac{1}{2}be_i^2 - \tfrac{\gamma}{2}(a - e_i)^2. \tag{18.36}$$

The first-order condition of (18.36), with respect to e_i, provides us with the Nash equilibrium

$$U_{e_i} - V'(e_i) - P'(e_i) = 0, \quad \forall i. \tag{18.37}$$

Applying Lemma 5 of Section 18.3, $U = \psi\, \bar{e}L^{1-\alpha}$, (18.37) becomes

$$\frac{\psi}{L^\alpha} - be_i + \gamma(a - e_i) = 0, \quad \forall i.$$

And e^S, the Nash optimal level of each kibbutz member's effort when social pressure is applied, is

$$e_i^S = e^S = \frac{\psi L^{-\alpha} + \gamma a}{b + \gamma}. \tag{18.38}$$

This solution is illustrated in Figure 18.4. As the figure makes abundantly clear, social pressure will increase exertion, but not necessarily to the Pareto optimal level.

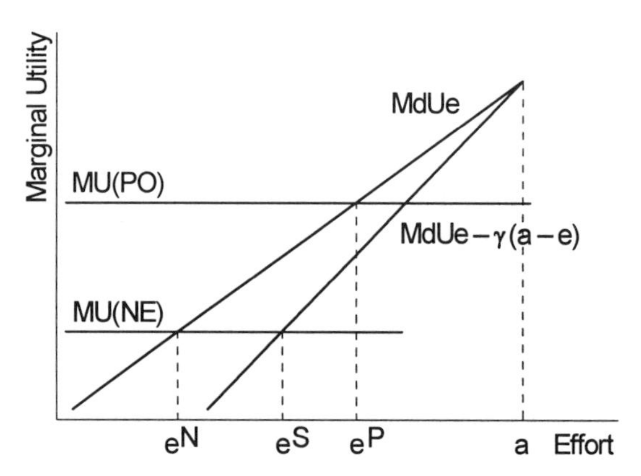

Figure 18.4: Effort choice with social pressure

When (18.38) is set into (18.36), the optimal level of each kibbutz member's welfare can be computed:

$$\begin{aligned} W^S(L) &= \psi e^S L^{1-\alpha} - \frac{b \cdot (e^S)^2}{2} - \frac{\gamma(a - e^S)^2}{2} \\ &= \frac{1}{b+\gamma}\left[\frac{\psi^2}{2} \cdot \frac{2L - 1}{L^{2\alpha}} + \gamma\psi a L^{1-\alpha} - \frac{\gamma a^2 b}{2}\right]. \end{aligned} \tag{18.39}$$

The first-order condition for an extreme point for $W^S(L)$ is

$$W^{S\prime}(L) = \frac{\psi}{(b+\gamma)L^{2\alpha+1}}\left[\psi(1-2\alpha)L + \psi\alpha + \gamma a\,(1-\alpha)L^{\alpha+1}\right] = 0, \quad (18.40)$$

and the internal extreme points, if they exist, solve

$$\Omega(L) \equiv \psi(1-2\alpha)L + \psi\alpha + \gamma a(1-\alpha)L^{\alpha+1} = 0. \qquad (18.41)$$

As in Section 18.3.3, the behavior of W^S depends on α:

(1) When $\alpha \leq 0.5$, i.e., when tastes are communalistic and there is a high preference for the public good, (18.40) has no solution for a positive L and $W^S(L)$ is increasing throughout.

(2) When $\alpha > 0.5$, that is, when members are individualistic and place a high value on the consumption of the private good, an internal solution to (18.41) may exist, and W^S may have both increasing and decreasing segments. To analyze the behavior of $\Omega(L)$ we take a look at Ω':

$$\Omega'(L) \equiv \psi(1-2\alpha) + \gamma a(1-\alpha^2)L^{\alpha} = 0.$$

$\Omega' = 0$ when

$$L = \hat{L} \equiv \left[\frac{\psi(2\alpha-1)}{\gamma a(1-\alpha^2)}\right]^{1/\alpha}, \qquad (18.42)$$

and it is easy to show that

$$\Omega'(L) \gtreqless 0 \Leftrightarrow L \gtreqless \hat{L}.$$

Ω is declining in the interval $L < \hat{L}$ and increasing in $L > \hat{L}$, and $\hat{L}$ in (18.42) provides a minimum to $\Omega(L)$ (see Figure 18.5).

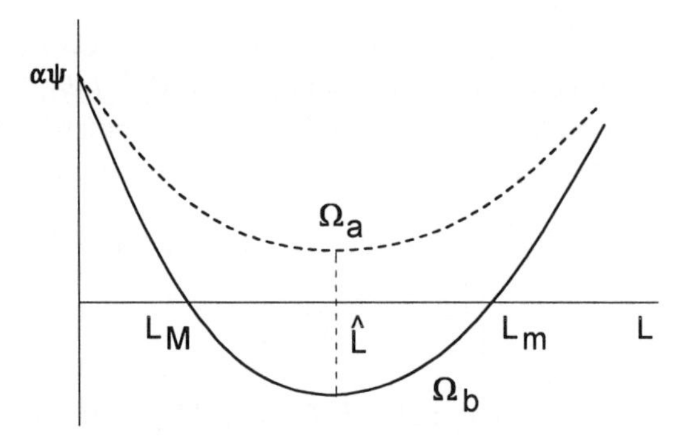

Figure 18.5: $\Omega(L)$

Since $\text{sgn}[W^{S\prime}(L)] = \text{sgn}[\Omega(L)]$, the existence of an internal maximizer, when $\alpha > 0.5$, depends on $\Omega(L)$. $\Omega_a(L)$ and $\Omega_b(L)$ in Figure 18.5 illustrate the two possibilities:

(a) $\Omega(L) \geq 0$ throughout, and W^S is increasing unboundedly in kibbutz membership L (Ω_a in Figure 18.5). In this case an extreme point of $W^S(L)$ does not exist.

(b) $\Omega(L) < 0$ for some L, as in Ω_b in Figure 18.5. Hence, there exist values of membership L_M and L_m, $L_M < L_m$, such that $\Omega(L_M) = \Omega(L_m) = 0$. But this is equivalent to $W^{S\prime}(L_M) = W^{S\prime}(L_m) = 0$, so that for $L < L_M$ and $L > L_m$, W^S is increasing, and for $L \in (L_M, L_m)$, W^S is declining (see Ω_b in Figure 18.5).

In other words, W^S has two extreme points in this case: the local maximum L_M, followed by the local minimum L_m, as illustrated in $W^S(L|b)$ in Figure 18.6.

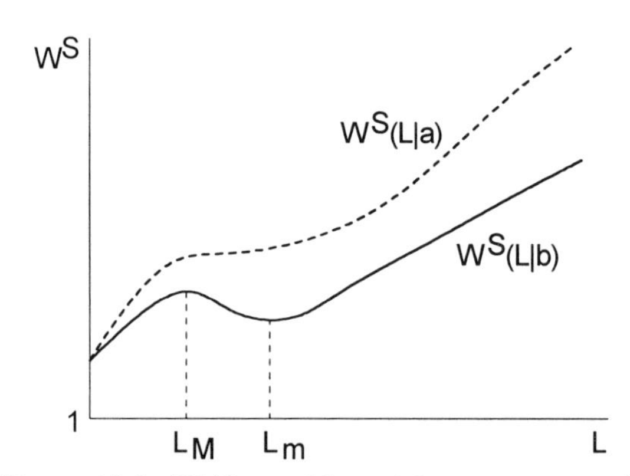

Figure 18.6: Welfare with social pressure, $\alpha > 0.5$

We can go a step further, and given that $\alpha > 0.5$, we can compute the values of the parameters at which W^S behaves like $W^S(L|b)$, i.e., for which it has internal extreme points. For this we have to find out whether $\Omega(\hat{L}) < 0$, where $\hat{L}$ has been defined in (18.42). Plugging (18.42) in (18.41),

$$\begin{aligned}\min \Omega(L) &= \Omega(\hat{L}) = \hat{L} \cdot [\psi(1-\alpha) + \gamma a(1-\alpha)\hat{L}^{\alpha}] + \alpha\psi \\ &= \alpha\psi \cdot \left[1 - \hat{L} \cdot \frac{2\alpha - 1}{1+\alpha}\right].\end{aligned}$$

Hence, $\min \Omega = 0$ if and only if $\hat{L} = \frac{\alpha+1}{2\alpha-1}$, that is, given (18.42), $\frac{\psi(2\alpha-1)}{\gamma a(1-\alpha^2)} = \left(\frac{\alpha+1}{2\alpha-1}\right)^{\alpha}$. We can use this condition to define a boundary value of the punishment parameter γ:

$$\gamma_0 \equiv \frac{\psi(2\alpha-1)}{a(1-\alpha^2)} \cdot \left(\frac{2\alpha-1}{\alpha+1}\right)^{\alpha} = \frac{\psi}{a(1-\alpha)} \cdot \left(\frac{2\alpha-1}{\alpha+1}\right)^{\alpha+1}.$$

Now, for $\alpha > 0.5$ and values of $\gamma \geq \gamma_0$, $\Omega'(L) \geq 0$ throughout, and for $\gamma < \gamma_0$, $\Omega'(L) < 0$ for $L_M < L < L_m$. This leads us to the following proposition:

Proposition 4 *In a kibbutz with social pressure:*

(i) *welfare increases without bound when either $\alpha \leq 0.5$, or $\alpha > 0.5$ and $\gamma \geq \gamma_0$.*

(ii) *when $\alpha > 0.5$ and $\gamma < \gamma_0$, then $W^S(L)$ possesses an internal local minimum L_m and a local maximum L_M.*

We also learn from (18.42), that when $\gamma \to 0$, $a \to 0$, or $\alpha \to 1$, then $\hat{L} \to \infty$, and since $L_m > \hat{L}$, $L_m \to \infty$, and we are back in the previous case without social pressure: $W^S(L)$ possesses an internal maximum. That is, a declining γ or a, or an increasing α, increases the range where $\Omega(L)$ is declining. When, for any given α, γ or a increase, then $\hat{L}$, L_M, and L_m decline. The same happens when, for given γ and a, α converges to 0.5. The region of L where $W^S(L)$ first rises and then declines shrinks and then disappears, and W^S becomes a monotonically increasing function.

18.4.1 Effort norms enforced by social pressure

The previous section assumed that a closely-knit society such as the historic kibbutz tries to impose effort norms, a, on its members, and uses social pressure for this purpose. We now inquire about the impact of the choice of these norms. It seems natural to us that the Pareto optimal effort level be chosen as the norm to be imposed. The latter, e^P, was computed in equation (18.31) in Section 18.3.3. Thus,

$$a = e^P = \frac{\psi}{b}L^{1-\alpha}.$$

Plugging this effort norm into the welfare function of Section 18.4, equation (18.39), we obtain

$$\begin{aligned} W^S(L) &= \frac{\psi^2}{2(b+\gamma)} \cdot \frac{2L-1}{L^{2\alpha}} + \frac{a\gamma}{b+\gamma}\left(\psi L^{1-\alpha} - \frac{ab}{2}\right) \\ &= \frac{b}{b+\gamma}W^N(L) + \frac{\gamma}{b+\gamma}W^P(L), \end{aligned}$$

where W^N and W^P are the Nash and the Pareto optimal level of the kibbutz members' welfare, (18.32) and (18.34) respectively, of Section 18.3.3 without peer pressure. Thus, in this case we get

Proposition 5 W^S, *the Nash-optimal level of the kibbutz members' welfare in the kibbutz with a quadratic cost of peer pressure, is the weighted average of the Nash-equilibrium and the Pareto optimal welfare level without peer pressure, where the weights are proportional to* b *and* γ, *respectively.*

Proposition 5 holds for a wider class of costs of peer pressure. For instance, it holds for linear costs. We view this as one of the main findings of our model, and we use it below to draw some empirical conclusions about recent developments in the relative strength of the kibbutz as an institution.

The main contribution of the present section to our understanding of the institution of the kibbutz is in the closer relation to reality that it provides. Even our simple model shows that an internal maximizer does sometimes exist, though this is not the global maximizer. Now, the global maximum at $L \to \infty$ should not be taken too literally: it assumes that the kibbutz should be able to exercise an unchanged degree of social control, regardless of size. Clearly, γ is size-dependent, declining as L increases: in a very large kibbutz, it would be easy for any member to find a niche in which to hide from their comrades. Thus the local maximum is likely to be close to the global one. And this ideal size is attained with a larger number of members than the kibbutz with no social pressure and control of Section 18.3.3; compare $\hat{L}$ of (18.42) to L^* of (18.33).[7]

18.5 The Relevance of the Model

The public good G, in the real kibbutz, stands for many different phenomena. It represents the consumption of public services and amenities, such as the swimming pool; and some of the services of the common dining hall, such as cultural offerings and the like. But it also stands for the social benefits of a large membership, where size imparts the ability to have a wider choice in the selection of one's social environment and one's friends. Furthermore, it represents all the elements of increasing returns in production, because once such returns to scale enter through consumption there is little sense in duplicating them through similar effects in the production of the marketable good X.

Consider $1 - \alpha$, the weight of the public good in the member's utility: the importance of communal consumption and of contacts with one's comrades

[7] The comparison obviously depends on many parameters, and $L_M < \hat{L}$. But L^* of (18.33) whereas $\hat{L}$ of (18.42) can be quite large.

is also related to the ideological fervor of kibbutz members. An increasing interest in private consumption, a growth of α, would mean that members have become preoccupied with their own private life and less concerned with communal affairs, as has happened over time. Such a change would lead to a reduced cohesion of the commune, and a decreased advantage of kibbutz membership over private and individualistic existence. It also would imply a reduction in the optimal size of the kibbutz, as shown by (18.33).

In what sense does the addition of G make the LMF model more relevant? Consider first the criticism of Satt [21] and Ben-Ner [2, 3]. It can be nested in our model by assuming that $\alpha = 1$, i.e., by assuming that there is no public good in the system. The essence of their criticism is very simple: now (18.5) becomes

$$U^i = y,$$

i.e., the kibbutz member's utility is linear in his income and becomes easily comparable to that of a worker in the private sector, whose real income is w. The alternative to membership is now this wage, and if

$$w > y,$$

there is nothing that attracts the worker to kibbutz membership. But w is also the cost of employing a non-member in the kibbutz. Consider the case of

$$w < y,$$

in which the kibbutz happens to be more efficient and profitable than the outside economy. In this case, if an outsider replaces a member, all other members can divide among themselves the difference saved between the member's consumption y and the outsider's wage, $y - w$. This advantage will lead to the replacement of members by hired labor, against the principles of the kibbutz movement. Outsiders working in the kibbutz are quite prevalent, but the tendency to get rid of members has not been documented.

Once $\alpha < 1$, and the kibbutz produces a communal good, the simple comparability of income in the kibbutz and in the private economy becomes irrelevant. If we posit that individuals may differ as to their liking for the public good, then a ground for separation into kibbutz-preferring and kibbutz-detesting individuals suggests itself: high-α individuals, whom we may call individualists, will choose to live in the private economy, while low-α individuals will be those who prefer kibbutz life.

Yet this is not the end of the model's contribution. Changes over time in a couple of variables can explain some of the present difficulties of the kibbutz. The affinity to public consumption, in society in general and the kibbutz in particular, has been declining over time. Services which used to be subject to indivisibilities and increasing returns, and which therefore favored public consumption, have become privatized once the indivisibilities became insignificant. This is particularly true for cultural services, luxuries whose weight in consumption has been growing. Consider television, which has replaced the local low-fidelity cinema (viewed on the creaking chairs of the communal dining hall); tourism; and even education: the increasing weight of higher and specialized education, which cannot be provided by the kibbutz, has turned it from a public to a private service. The same is true for the consumption of food: as long as income was low, all that mattered was the proper intake of calories and vitamins, and this could be supplied by the communal kitchen and gobbled down in the communal dining hall. This saved the need of investing in separate kitchens in the austere housing of the individual members. With the increase in the standard of living came the preference for tastier food and choice, and the improved housing of the members permitted the construction of individual kitchens. The public good that is the communal meal in the dining hall has been dethroned.

In our model, this means that α has been increasing over time. This has had a double effect: the optimum kibbutz-size has been declining, and the potential preference of the kibbutz over its alternatives has declined. And this is related to another parameter, γ. But first, a brief discussion of γ's background.

The social pressure exercised by members of the community depends closely on the cohesion of the kibbutz. A small, family-like community can exert very strong pressure, but a large amorphous body, which is in effect subdivided into smaller face-to-face groupings, is much more limited in the sanctions it can employ. Therefore, γ is not independent of the size of the kibbutz: $\gamma(L)$, $\gamma' < 0$. It would be easy but messy to add this to the model, and the outcome is straightforward: as L grows, the weight of the selfish Nash equilibrium grows. This means that very large kibbutzim would gain little from social pressure, and the optimum kibbutz scale would therefore be finite and not too large. See the discussion at the end of Section 18.4.1.

This is not all. The same historical changes which have affected tastes for public vs. private goods have also had an impact on γ. Social sanctions were often imposed through the consumption of public goods. For instance, those who were considered shirkers were ostracized, and nobody would join them at the communal meals. With the reduced use of the communal dining hall, the efficacy of such sanctions was blunted. Thus, the increase of α carries with it a reduction of γ. With an increase in individualism came a decline in the desire and ability of kibbutz members to impose social norms on their mates. In other words, the kibbutz has become more of a collection of selfish individuals than a community imbued with common values, and γ has declined. As a result, optimum size has declined, and more importantly, the comparative advantage of the kibbutz for those whose tastes are for public goods has declined. Thus, both the decline in γ and the increase of α bring with them the weakening of the kibbutz. Are the present attempts of reform relevant to the disease we describe?

The main reform that is being introduced in many kibbutzim is to introduce contribution dependent rewards; i.e., to abolish the equality in consumption, the mainspring of kibbutz ideology. It is quite easy to see that this modifies somewhat the analysis of Section 18.3.2, but does not change it sufficiently. Holmström's [13] curse still afflicts the kibbutz. But this change does affect social cohesion, and reduces still further γ. The net effect may be detrimental rather than beneficial. Other reforms are organizational: a clearer separation between the production and the consumption spheres in the kibbutz is suggested. It would involve the introduction of capitalist business administration techniques into the former, along with outside directors and more professional management. Here again, γ may be affected, but to a lesser extent, and the gains may be more significant. But neither reform comes to grips with the main historical trends which weaken the kibbutz. Will it survive?

References

1 Barkai, Haim, Growth Patterns of the Kibbutz Economy, North-Holland, Amsterdam, New York and Toronto, 1977.

2 Ben-Ner, Avner, On the stability of the cooperative type of organization, Journal Comparative Economics 8(3) (1984) 247–60.

3 Ben-Ner, Avner, Preferences in a communal economic system, Economica, 54 (1984) 207–21.

4 Berglas, Eitan, On the theory of clubs, American Economic Review 66 (1976) 116–21.

5 Berglas, Eitan and David Pines, Clubs, local public goods and transportation models, Journal Public Economics 15 (1981) 141–62.

6 Berglas, Eitan, Elhanan Helpman and David Pines, The economic theory of clubs: Some clarifications, Economics Letters 10 (1982) 343–48.

7 Bonin, John P. and Louis Putterman, Economics of cooperation and the labor-managed economy, in: Fundamentals of Pure and Applied Economics, Vol. 14, Harwood Academic Publishers, London, 1987.

8 Buchanan, J.M., An economic theory of clubs, Economica 32 (1965) 1–14.

9 Dar, Chezi, The firms and the families in the economic crisis of the kibbutz, unpublished typescript, 1989.

10 Domar, Evsey D., The Soviet collective farm as a producer cooperative, American Economic Review 56(4) (1986) 734–57.

11 Helman, Amir, Kibbutz crisis as incentive to radical changes. Presented at the Bar-Ilan International Conference on Theoretical and Applied Aspects of Labor Managed Firms, 25–28 May, 1992.

12 Hillman, A.L. and P.L. Swan, Club participation rules for Pareto-optimal clubs, Journal of Public Economics 20 (1983) 55–76.

13 Holmström, Bengt, Moral hazard in teams, Bell Journal of Economics 13(2) (1982) 324–40.

14 Ireland, Norman J. and Peter J. Law, The Economics of Labor-Managed Enterprises, St. Martin's Press, New York, 1982.

15 Kandel, Eugene and Edward P. Lazear, Peer pressure and partnership, Journal of Political Economy 100(4) (1992) 801–17.

16 Keren, Michael and David Levhari, Some capital market failures in the socialist labor managed economy, Journal of Comparative Economics 16(4) (1992) 655–69.

17 Keren, Michael and David Levhari, The Israeli Moshav as a credit union: Adverse selection and moral hazard, in: C. Csaki and Y. Kislev (eds.), Agricultural Cooperatives in Transition. Westview Press, Boulder, CO, 1993.

18 Meade, James E., The theory of labour-managed firms and of profit sharing, Economic Journal 82 (1972) 402–28.

19 Meade, James E., Agathotopia. The Economics of Partnership, Pergamon and Aberdeen University Press, London, 1989.

20 Putterman, Louis, Incentives and the kibbutz: Toward an economics of communal work motivation, Zeitschrift für die Natiaonalökonomie 43(2) (1983) 157–88.

21 Satt, Ehud, Problems in the kibbutz economy, Ph.D. Thesis, Stanford University, 1979.

22 Satt, Ehud and Haim Ginzburg, On the dynamic effects of using hired-labor in kibbutz: Theory and case studies, Journal of Comparative Economics 16(4) (1992) 688–700.

23 Satt, Ehud, Relative deprivation in the kibbutz economy: An exploration of the concepts of equality and equity, Economica 63 (1996), S87–101.

24 Shachmurove, Yochanan and Uriel Spiegel, Effects of coordination and transition costs on LMFs size. Presented at the Bar-Ilan International Conference on Theoretical and Applied Aspects of Labor Managed Firms, 25–28 May, 1992.

25 Tiebout, C.M., A pure theory of local expenditure, Journal of Political Economy 64 (1956) 416–24.

26 Vanek, Jaroslav, Decentralization under workers' management: A theoretical appraisal, American Economic Review 59 (1969) 1006–14.

27 Vanek, Jaroslav, The General Theory of Labor-Managed Market Economics, Cornell University Press, Ithaca, New York, 1970.

28 Ward, Benjamin, The firm in Illyria: Market syndicalism, American Economic Review 48(4) (1958) 566–89.

29 Ward, Benjamin, The Socialist Economy: A Study of Organizational Alternatives, Random House, New York, 1967.

Trade, Growth, and Development
G. Ranis and L.K. Raut

CHAPTER 19

What Do We Know About Production Sets Arising from an Activity Analysis Model with Integral Activity Levels?

Herbert E. Scarf

Cowles Foundation for Research in Economics, 30 Hillhouse Avenue,
Yale University, New Haven, CT 06520

19.1 Introduction

It gives me great pleasure to present these remarks on the occasion of T.N. Srinivasan's 65th birthday celebration. I've known T.N. for almost four decades, as a friend and admired colleague and for the last several months as the Chairman of our department. We have been daily companions at the Cowles coffee hour for many years; we have endured more Departmental meetings than I care to remember; we have enjoyed many lunches, attended conferences and we have actually taught a course together. I consider myself extremely fortunate to have shared this extended period of time with a comrade whom I have found to be so consistently stimulating and impressive.

The course that we taught together was on Applied General Equilibrium Analysis. My responsibility was to discuss the mathematical formulation of the Walrasian model in order to demonstrate the existence of an equilibrium and to provide some elementary algorithms for its solution. This material took a few weeks at best, and then T.N. took over, presenting to the students — and to me as well — a beautiful and erudite display of the many applications of equilibrium theory to problems of international trade and public finance. T.N. is one of those scholars who has read everything and never forgets what he has read. His taste, sense of proportion, and devotion to our subject are a delight for me. I marveled at his lectures and I look forward to the next time that we will offer this course together.

I cannot remember whether we actually discussed general equilibrium models in which the production possibility set exhibited increasing returns to scale. The classical Walrasian model requires, in the most basic way, that the production set be convex, and this rules out the introduction of economies of scale. If economies of scale are immediately available to producers, profit maximizing at fixed competitive prices will lead to larger and larger scales of operation, eliminating any possibility of equilibrium between supply and demand. For an equilibrium model with economies of scale to be manageable, the relation between prices and supply decisions cannot be based on optimization with respect to fixed prices; other assumptions are necessary.

Of course, we have seen many theoretical and applied discussions involving average cost pricing, marginal cost pricing, and other variations as well. These alternatives to perfect competition may be perfectly accurate descriptions of economic reality; or if not fully accurate, at least useful for the illustration of economic issues. But for me these non-competitive pricing rules are frustrating because they lead to outcomes which no longer satisfy the optimality theorems of welfare economics.

I have been preoccupied, for many years, with the issue of optimality without the assumption of convexity. My first attempt in this direction was to study the *core* of an economy with non-convex production possibility sets. Production and distribution decisions in the core of an economy are automatically Pareto Optimal, and it seemed likely to me, when I first came upon this subject, that the core would provide a replacement for the competitive equilibrium, allowing for optimality while still relating individual consumption to individual ownership of goods and factors of production.

But it turns out that the core is not a promising approach to economies of scale in production. If the production possibility set is not convex, then one can always find a perfectly ordinary economy — in terms of utilities and initial holdings — in which all coalitions of consumers have access to this common production set, but for which the core is empty: any production and distribution plan proposed by the economic community will be overturned by some coalition of consumers.

How to proceed? What I've done is to spend a lot of time thinking about the detection of optimality in the absence of convexity. In particular, I've looked at the most extreme form of non-convexity: indivisibilities in production. What I'd like to do is to share with you today some insights into the structure of production possibility sets that arise from activity analysis models in which the activity levels are required to assume integral values, and to suggest how some knowledge of this structure might be useful in verifying optimality.

19.2 An Innocent Question

Suppose that we are given three coins of denomination a_1, a_2, a_3. For a specific integer b, we can make change using these three numbers if b can be written as a non-negative integer combination of the three coins: $b = a_1h_1 + a_2h_2 + a_3h_3$, with h_i non-negative integers. For example, if $a_1, a_2, a_3 = (5, 7, 11)$, we can make change with $b = 5, 7, 10, 11, 12$, and every $b \geq 14$. If $a_1, a_2, a_3 = (5, 9, 13)$, we can make change with $b = 5, 9, 10, 13, 14, 15, 18, 19, 20$, and every $b \geq 22$.

If the three numbers a_1, a_2, a_3 have a common factor, say 5, then one can only make change with numbers b which are themselves multiples of 5. It is intuitively clear, and not at all hard to show, that if a_1, a_2, a_3 have no common factor bigger than 1, then one can make change for all sufficiently large b. In other words there is some largest integer, say $b(a_1, a_2, a_3)$, for which one cannot make change. In our examples $b(5, 7, 11) = 13$ and $b(5, 9, 13) = 21$.

How can one find this largest non-changeable integer for an arbitrary triple a_1, a_2, a_3? One's first thought is to use some form of enumeration. It is easy to obtain an upper bound for this integer (the product a_1, a_2, a_3 will do), and then check each integer below this upper bound. This certainly works, but it is a terribly slow algorithm for large values of the triple.

I have written a simple computer routine, with no more than 30 lines of code, that tells me, instantaneously, that

$$\mathbf{b(1110769831, 8779066531, 677533209)}$$
$$= \mathbf{393324634878659}.$$

Take a moment (or more) to speculate how you might carry out this calculation rapidly. It is easy, but not at all obvious (Scarf and Shallcross [4]).

What type of problem is illustrated by this question? I believe that it is an economically meaningful question that arises in the study of production sets based on an activity analysis model with integer activity levels; in other words, in the analysis of indivisibilities in production. If we are given positive integers $a_1, a_2, ..., a_n$, the question of whether a particular integer b can be written as $a_1h_1 + a_2h_2 + \cdots + a_nh_n$, with $h_1, h_2, ..., h_n$ non-negative integers, is a simple example of an *integer programming problem.* In a more familiar form, the question can be phrased by asking whether the optimal value of the integer program

$$\begin{aligned} \max \quad & a_1h_1 + a_2h_2 + \cdots a_nh_n \qquad \text{subject to} \\ & a_1h_1 + a_2h_2 + \cdots a_nh_n \leq b, \\ & h_1, h_2, ..., h_n \geq 0 \text{ and integral,} \end{aligned}$$

is, in fact, equal to b.

Linear programs arise from an activity analysis model of production with **continuous** activity levels by selecting a row of the activity analysis matrix as the objective function, and by specifying a particular endowment of factors. In an integer program the activity levels are required to assume integral values. The value of the optimum, as a function of the factor endowment, is one of many production functions associated with the production possibility set generated by the activity analysis model.

Solving an integer program for a particular right hand side yields an evaluation of the production function at a specified point. Much is known about the intrinsic difficulty of solving integer programs from the point of view of complexity theory. Integer programming problems are what is known as NP complete problems; if they can be solved in polynomial time, then virtually all programming problems are easy. But it is important to realize that our problem of finding the largest non-changeable integer, $b(a_1, ..., a_n)$, is not a question about a particular value of the production function; it is a global question about the production function as a whole. In our case, we are asked to find that endowment level where the production function becomes linear: the particular point after which indivisibilities no longer matter. Questions about the entire production function are clearly more difficult to answer than questions about a particular value.

How can we solve integer programs in such a way that questions about the entire production set can be addressed? In the next section I will provide a brief review of the continuous problem, and then turn to discrete production possibility sets.

19.3 A Brief Review

A general activity analysis model with continuous activity levels is specified by a matrix, say

$$A = \begin{bmatrix} c_1 & c_2 & \cdots & c_n \\ -a_{11} & -a_{12} & \cdots & -a_{1n} \\ -a_{21} & -a_{22} & \cdots & -a_{2n} \\ \vdots & \vdots & \ddots & \vdots \\ -a_{m1} & -a_{m2} & \cdots & -a_{mn} \end{bmatrix}.$$

The production possibility set Y associated with A consists of those production plans y with $y \leq Ax$ for some non-negative activity vector x. The coordinates of y with negative entries are inputs into production, and those with positive entries are outputs. Of course, a particular commodity can be an input in one plan and an output of another. The inequalities are introduced in order to allow for the possibilities of free disposal.

The following figure represents the production possibility set associated with the matrix

$$\begin{matrix} O \\ L \\ K \end{matrix} \begin{bmatrix} 2 & 3 & 3 \\ -1 & -3 & -6 \\ -3 & -2 & -1 \end{bmatrix},$$

augmented by 3 disposal activities. The figure illustrates the general fact that the production possibility set associated with a continuous activity analysis model is a closed, convex cone.

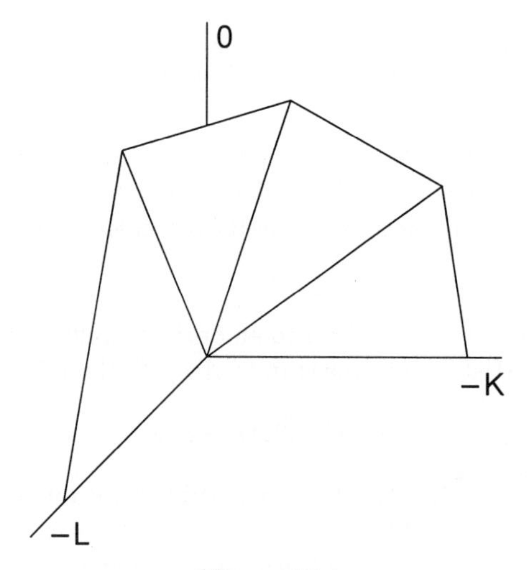

Figure 19:1

Remark. Assume that there is a non-zero price vector $\pi = (\pi_i, ..., \pi_m) \geq 0$ such that each activity makes a non-positive profit, i.e., $\Sigma\pi_i a_{ij} \geq c$ for all j. Then the production set Y is a closed convex cone with a well-defined, finite upper boundary.

In a linear program we consider those production plans that use no more than a specified factor endowment $b = (b_1, ..., b_m)$ and select from those plans the particular one that maximizes output. In formal notation the linear program is given by

$$\begin{array}{lll} \max & c_1\ x_1 + \cdots + c_n\ x_n & \text{subject to} \\ & a_{11}x_1 + \cdots + a_{1n}x_n \leq b_1 & \\ & a_{21}x_1 + \cdots + a_{2n}x_n \leq b_2 & \\ & \quad \vdots & \\ & a_{m1}x_1 + \cdots + a_{mn}x_n \leq b_m & \\ & x \geq 0 & \end{array}$$

In my opinion, the most important economic feature of linear programs is that optimality can be recognized using competitive prices. We have the famous

Pricing Test for Optimality. A vector of activity levels $x^* = (x_1^*, ..., x_n^*)$ which satisfies the constraints will be optimal if we can find a price vector

$\pi^* = (\pi_1^*, ..., \pi_m^*) \geq 0$ such that

$$\sum_j a_{ij} x_j^* \leq b_i \text{ with } = \text{ if } \pi_i^* \geq 0,$$

$$c_j \leq \sum_{i=1,m} \pi_i^* a_{ij} \text{ with } = \text{ if } x_i^* > 0.$$

The first condition states that the price of a factor in excess supply is 0; the second condition that each activity make a non-positive profit, and that this profit be equal to 0 if the activity is used at a positive level in the proposed feasible solution.

The proof that the test does lead to an optimal solution is straightforward. If $x = (x_1, ..., x_n)$ satisfies the constraints $Ax \leq b$, then

$$\pi^* b \geq \pi^* Ax \geq cx,$$

so that the value of the objective for any feasible solution is bounded above by $\pi^* b$, the value of the stock of factors at prices π^*. But if x^* and π^{**} are related as in the pricing test, these inequalities become equalities and $\pi^* b = c \cdot x^*$.

The fact that the pricing test implies optimality makes no use of convexity arguments at all. But the convexity properties of linear programming with continuous activity levels are used in a crucial way in order to deduce the existence of such prices. The argument for existence begins by observing that the production plan

$$y^* = (z^*, -b_1, ..., -b_m),$$

where $z^* = \Sigma c_j x_j^*$, is an efficient plan in Y; it lies on the upper boundary. The classical theorem about separating boundary points of a convex set by a hyperplane asserts the existence of a price vector

$$\pi^* = (1, \pi_1^*, ..., \pi_m^*)$$

such that

$$\begin{aligned} \pi^* y^* &= 0 \text{ and} \\ \pi^* y &\leq 0 \text{ for all } y \in Y. \end{aligned}$$

But this is precisely the price vector we are looking for in the test of optimality. To see this we make the observations:

1. The equality $\pi^* y^* = 0$ states that $z^* = \Sigma c_j x_j^* = \pi^* b^*$.

2. If we select y to be the production plan $(c_j, -a_{1j} - \cdots - a_{mj})'$ given by the jth column of the matrix A, we see that $c_j < \Sigma_{i=1,m} \pi_i^* a_{ij}$.

We have the following sequence of inequalities:

$$\begin{aligned}\pi^* b^* &= z^* = \sum_j c_j x_j^* \leq \sum_j x_j^* \sum_i \pi_i^* a_{ij} \\ &= \sum_i \pi_i^* \sum_j a_{ij} x_j^* \leq \pi^* b^*.\end{aligned}$$

Since the first and last terms in the sequence are equal, so must all of the potential inequalities be equalities. It follows that $x_j^* = 0$ if $c_j < \Sigma_{i=1,m} \pi_i^* a_{ij}$ and $\pi_i^* = 0$ if $\Sigma_j a_{ij} x_j^* < b_i$. In other words, these prices and the optimal activity levels will satisfy the complimentary conditions of the pricing test.

Prices are used to test the optimality of a proposed feasible solution to the linear program specified by a particular factor endowment b. But prices have an addition feature: they permit us to describe the production function in the vicinity of b.

Prices and Marginal Productivity. The prices π^* can be interpreted as *marginal productivities*: For small δ the maximum output obtainable from the factor endowment $(b_1, ..., b_i + \delta, ..., b_m) = z^* + \pi_i^* \delta$.

19.4 The Failure of the Pricing Test for Integer Programs

The most elementary form of non-convexity in a production possibility set arises from an activity analysis model in which some, or all, of the activity levels are required to assume integral values. As we have mentioned, the mathematical programming problems encountered under this additional requirement are known as integer programs:

$$\begin{aligned}\max \quad & c_1\, h_1 + \cdots + c_n\, h_n \qquad \text{subject to} \\ & a_{11} h_1 + \cdots + a_{1n} h_n \leq b_1 \\ & a_{21} h_1 + \cdots + a_{2n} h_n \leq b_2 \\ & \quad \vdots \\ & a_{m1} h_1 + \cdots + a_{mn} h_n \leq b_m \\ & h \geq 0 \textbf{ and h integral.}\end{aligned}$$

The most important distinction between linear and integer programming problems is that optimality is typically not verifiable by means of competitive prices. The pricing test described above for linear programs is certainly applicable for integer programs in the sense that if we are given a feasible **integral** solution h^* to the constraints of the integer program, a price vector π^* which yields a zero profit for the activities in use, and a non-positive profit for the remaining activities, then h^* is certainly the optimal solution to the integer program. And there are interesting classes of integer programs for which such prices do exist. Transportation problems and network flow problems do have the feature that if demands are integral, the optimal solution to the linear program will turn out to be integral. The requirement of integrality is automatically met, and the linear programming prices will do just as well for the corresponding integer program.

But unfortunately, this is not a general phenomenon. The optimal solution to the integer program

$$\begin{aligned} \max \quad & 3h_1 + h_2 \qquad \text{subject to} \\ & 2h_1 + h_n \leq 3 \\ & h_1, h_2 \geq 0 \textbf{ and integral.} \end{aligned}$$

is given by $h^* = (1, 1)$. Both activities are used at the optimum, but there is no price π for the single factor yielding a profit of zero for both activities. At the price $\pi = 3/2$ the first activity will make a profit of zero, but the second has a negative profit, and should not be used at all. At $\pi = 1$ the second activity makes a profit of zero, but the first will make a positive profit, and should therefore be used at higher and higher levels. There is no price that picks out the set of activities in use at the optimal solution of a general integer program. This is a terrible problem for economic theory if indivisibilities are considered to be an important source of economies of scale in production. Indivisibilities inevitably lead to centralized, non-competitive behavior at some scale.

The mathematical reason that decentralized prices do not exist for integer programs is that the production possibility set associated with an activity analysis model with integral activity levels is not a convex set, and the standard arguments of convex analysis are not available. Let us take a look at these sets.

19.5 Discrete Production Sets

The coin problem leads to a family of integer programs

$$\begin{aligned} \max \quad & a_1h_1 + a_2h_2 + \cdots + a_nh_n \qquad \text{subject to} \\ & a_1h_1 + a_2h_2 + \cdots + a_nh_n \leq b, \\ & h_1, h_2, ..., h_n \geq 0 \text{ and integral,} \end{aligned}$$

which can be viewed as arising from a particular discrete activity analysis model. It is a very special model with a single output and a single input, which are, in fact, equal. The activity analysis matrix is given by

$$A = \begin{bmatrix} a_1 & a_2 & \cdots & a_n \\ -a_1 & -a_2 & \cdots & -a_n \end{bmatrix}$$

and the production possibility set consists of all pairs (y_0, y_1) with $y \leq Ah$ for some non-negative, integral h. Figure 19.2 represents this production possibility set for the problem of three coins with $(a_1, a_2, a_3) = (5, 7, 11)$. In order to illustrate the possibilities of free disposal, I have appended a negative orthant to each of the production plans generated by the activity analysis matrix. We notice that there is a well defined upper boundary of the production set, in spite of the fact that the basic set of possibilities is discrete. This figure records the fact that $b(5, 7, 11) = 13$, but it is not rich enough for us to calculate precisely what this value is without an exhaustive search.

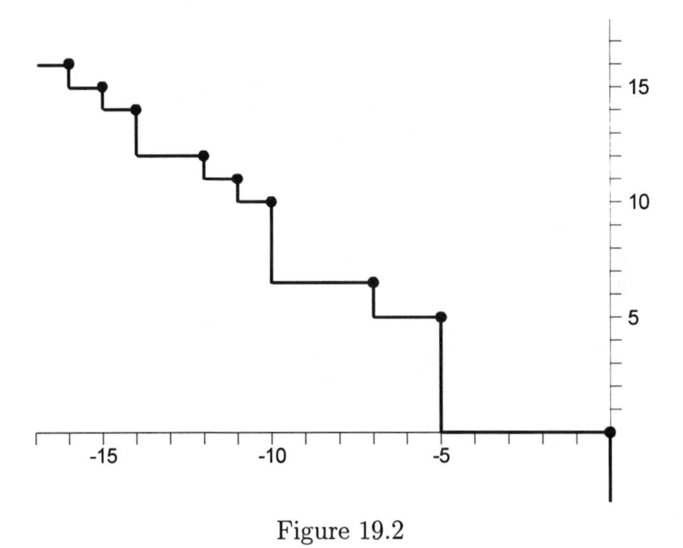

Figure 19.2

Let us leave the coin problem for a moment and take a look at discrete production sets with three or more commodities. In the following figure, I've drawn a portion of the production set based on the activity analysis model

$$\begin{matrix} -L \\ O_1 \\ O_2 \end{matrix}\begin{bmatrix} -2 & -1 \\ 4 & -4 \\ -1 & -3 \end{bmatrix},$$

where the activity levels are arbitrary integers. At any point in the horizontal plane, the depth is the labor cost of producing the output combination O_1, O_2. A negative orthant is appended to each plan generated by the activity analysis model, giving rise to a full dimensional upper surface for the production set. The example does seem to have a vague connection with our coin problem in the sense that $(5, 7, 11)A = 0$.

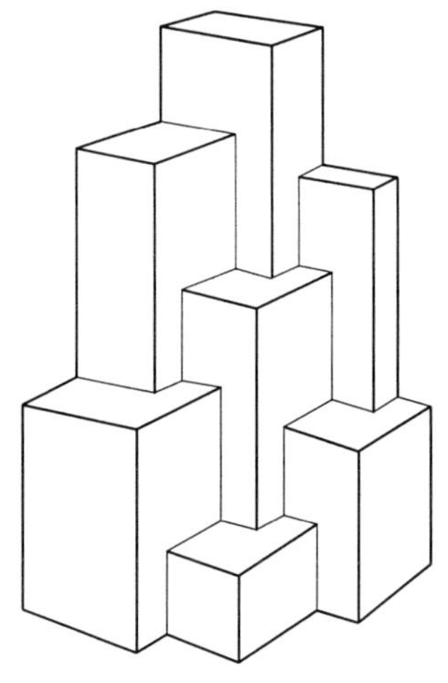

Figure 19.3

The typical integer program arising from this model is to minimize the labor input required to produce a particular net output of the two produced goods

$$
\begin{aligned}
\max \quad & 2h_1 + \ \ h_2 \qquad \text{subject to} \\
& 3h_1 - 4h_3 \geq r_1 \\
& -h_1 + 3h_2 \geq r_2
\end{aligned}
$$

Let us consider the instance in which $(r_1, r_2) = (1, 3)$. The plan $(-13, 3, 4) = A(5, 3)$ satisfies these constraints. Is it the optimal solution for this pair of requirements?

A pricing test is not conclusive. It is true that the price vector $\pi = (5, 7, 11)$ will yield a zero profit for both activities. But given the simplicity of this particular activity analysis model (an input/output table with integral activity levels), these prices yield a zero profit for all production plans not involving free disposal, and don't discriminate one such plan from another.

If a pricing test is not sufficient to detect optimality, will a quantity test work? In other words, take the activity vector (5,3) and compare it with alternative activity vectors which are close, and which decrease labor cost. The most naive comparisons are with those activity vectors in which one or both of the activity levels are decreased by a single unit. We have

$$
\begin{aligned}
A(4,3) &= (-11, 0, 5) \\
A(5,2) &= (-12, 7, 1) \\
A(4,2) &= (-10, 4, 2).
\end{aligned}
$$

Each of these neighboring comparisons generate production plans which lower cost, but none of them produce the required net outputs. None of them lie in the translate of the positive orthant with the origin located at $(-13, 1, 3)$.

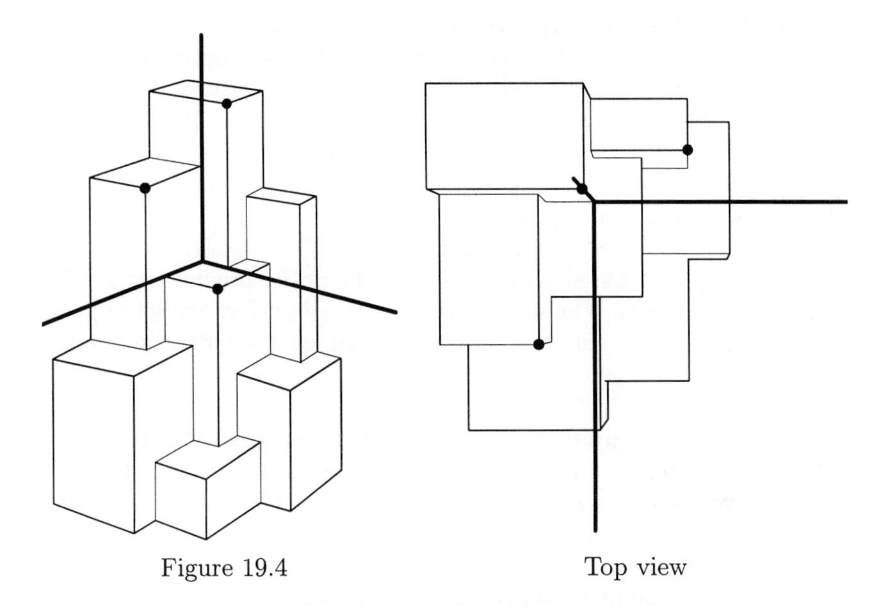

Figure 19.4 Top view

A possible conclusion is that the activity vector (5,3) is indeed optimal. But this is false: the vector (3,2) generates the plan $A(3,2) = (-8,1,3)$ that satisfies the constraints of our problem with a decrease in cost. As we see, there is a way of detecting the non-optimality of the previous proposal, but it is more sophisticated than the simple notion of considering only those adjacent vectors of activity levels which differ from the proposed solution by no more than a single unit.

19.6 The Quantity Test

Associated with an activity analysis matrix A (Scarf [2, 3]), in general position, is a finite set of integer vectors, symmetric about the origin, $N(A)$, which forms a *test set* for the family of integer programs obtained by selecting a row of A as the objective function and imposing resource constraints on the remaining factors:

$$\begin{aligned} \max \quad & a_0 h \qquad \text{subject to} \\ & a_1 h \geq b_1 \\ & \vdots \\ & a_m h \geq b_m \\ & h \text{ integral.} \end{aligned}$$

The set $N(A)$, whose members are called the **neighbors of the origin**, is a test set in the following sense:

The Quantity Test for Optimality. Let k satisfy the constraints for a particular selection of the factor endowment b. Then k is optimal if for every $h \in N(A)$ the vector $k + h$ either violates one of the constraints, or yields a lower value of the objective function. The set of neighbors, $N(A)$, is *minimal* in the sense that if a single member of the set is eliminated, a particular instance can be found in which testing with the smaller set of neighbors will not be sufficient to reveal optimality.

The quantity test based on neighbors replaces the pricing test for linear programming problems. The set of neighbors also describes the set of possible changes in the optimal solution that arise from small changes in the factor endowment.

Neighbors and Marginal Analysis for Discrete Production Sets. Let h be the optimal solution for an integer program with a given factor endowment. If the factor endowment is gradually increased, the first change in the optimal solution is obtained by adding a neighbor $k \in N(A)$ to h.

There is a very simple construction which yields the set of neighbors for a general discrete production set based on an activity analysis model with integral activity levels. Let me describe the construction in the case that we

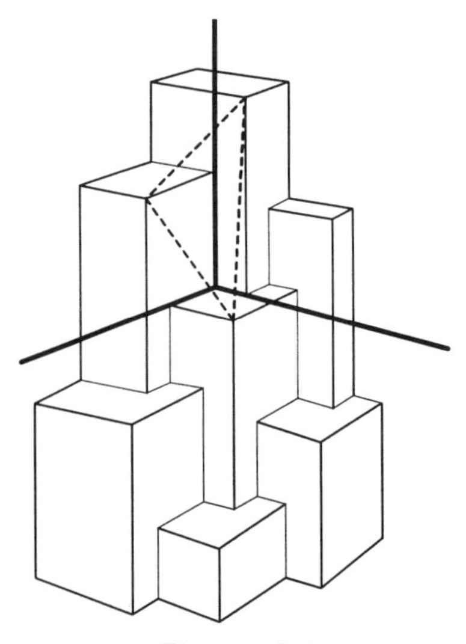

Figure 19.5

can easily visualize, a matrix A with 3 rows and 2 columns. The extension to higher dimensions is straight forward.

Let

$$A = \begin{bmatrix} a_{01} & a_{02} \\ a_{11} & a_{12} \\ a_{21} & a_{22} \end{bmatrix},$$

and let Y be the discrete production set obtained by depending a negative orthant to every $y = Ah$ as h ranges over all of the integers. Assume that there is a positive vector π such that $\pi A = 0$, so that all of the y vectors lie on the plane $\pi y = 0$. We shall construct a collection of triangles with vertices in the set of production plans $y = Ah$.

Triangulating the Production Possibility Set. Take the positive orthant and move it parallel to itself, so that it lies entirely above the plane $\pi y = 0$. Then lower the orthant gradually until no further movement is possible without passing through one of the points in the production set. The movement will be stopped by 3 plans, say

$$\begin{aligned} y^{j_0} &= Ah^{j_0} \\ y^{j_1} &= Ah^{j_1} \\ y^{j_2} &= Ah^{j_2}, \end{aligned}$$

with the origin of the translated orthant located at the point

$$\min[y^{j_0}, y^{j_1}, y^{j_2}].$$

The vectors$\{y^{j_0}, y^{j_1}, y^{j_2}\}$ will be the vertices of a simplex in our triangulation. The collection of triangles reached by all ways of lowering the translated positive orthant is the triangulation of the production possibility set.

The set of neighbors of a lattice point k is then defined to consist of those lattice points h such that Ak and Ah are *vertices of a common triangle* in this subdivision. The set $N(A)$ is precisely the set of lattice points that are *neighbors of the origin.*

If h^0, h^1, h^2 generate a triangle in this subdivision, then so does an arbitrary lattice translate $h^0 + k$, $h^1 + k$, $h^2 + k$. It can be shown that the collection of triangles arising from a matrix with 3 rows and 2 columns will consist of 2 specific triangles and all of their lattice translates. When the triangulation is drawn in the space of activity levels, the triangles will form a simplicial subdivision as in the following figure:

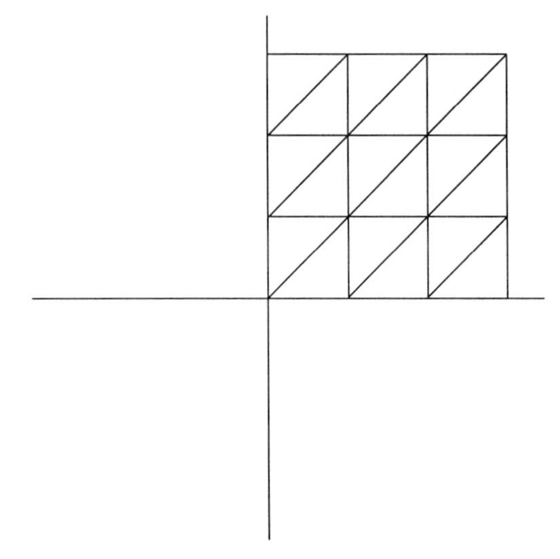

Figure 19.6

It follows that for a matrix of this size, there will be precisely 6 neighbors of the origin. In the example in which the activity analysis matrix is given by

$$\begin{bmatrix} -2 & -1 \\ 3 & -4 \\ -1 & 3 \end{bmatrix}$$

it can be shown that the 2 triangles in the simplicial subdivision are given — up to lattice translates — by the production plans

$$(0,0,0), (-2,3,-1), (-5,2,1) \text{ and}$$
$$(0,0,0), (-3,-1,2), (-5,2,1).$$

The neighbors of the origin are the 3 production plans

$$(-2,3,-1)$$
$$(-3,-1,2)$$
$$(-5,2,1)$$

and their negatives.

If the translated positive orthant is stopped in its downward movement at the first of these triangles, its origin will be located at the point

$$p^1 = \min[(0,0,0), (-2,3,-1), (-5,2,1)] = (-5,0,-1).$$

For the second triangle the origin will be at

$$p^2 = \min[(0,0,0), (-3,-1,2), (-5,2,1)] = (-5,-1,0).$$

If the activity analysis matrix has a larger number of rows or columns than this example, the set of neighbors can be quite large. The cardinality of $N(A)$ is not bounded by a function of the dimension of A alone; it depends on the complexity of the entries in the activity analysis matrix. There is much to be said about the structure of the set of neighbors and the simplicial complex that we have just defined. But that will take us quite far afield, and it is best to finish this discussion with a return to our inquiry about making change.

19.7 Back to the Coin Problem

Remember that the above activity analysis model was related to the coin problem given by $a = (5, 7, 11)$. What is surprising is that there is enough information in the two triangles calculated in the previous section in order to deduce the value of $b(5, 7, 11)$. It is simply obtained by calculating $-ap^i$, finding the larger of these two numbers and subtracting $\sum a_i$. In our example, $-ap^1 = 36$, $-ap^2 = 32$ and

$$36 - \sum a_i = 36 - 23 = 13.$$

This same algorithm works for any problem of 3 coins. It is the algorithm that was used to calculate the example at the beginning of this paper. I have the code with me and would be delighted to run some other examples for anyone who might be interested.

The coin problem with n coins $(a_1, a_2, ..., a_n)$ can be answered readily if we can decide, for a given integer b^0, whether the system

$$a_1h_1 + a_2h_2 + \cdots + a_nh_n = b$$

has a solution in non-negative integers $\{h\}$, for **all** $b \geq b^0$. If we have a good upper bound for $b(a_1, a_2, ..., a_n)$ and we can answer this question for arbitrary b^0, then repeated bisection will lead us to the precise value of $b(a_1, a_2, ..., a_n)$ rapidly.

Our colleague Ravi Kannan [1] has raised, and answered, a much deeper question that bears on general production sets arising from an activity analysis model with discrete activity levels.

Theorem 1. *Let A be a matrix with m rows and n columns, and let P be a convex polyhedron in m space. The question of whether the system $Ah \leq b$ has a non-negative integral solution for every integral b in P can be resolved in polynomial time (for fixed n and m).*

In applying Kannan's theorem to the coin problem, the system of linear inequalities is a single equality $a_1h_1 + a_2h_2 + \cdots + a_nh_n = b$, and the convex polyhedron is the interval on the line: $\{b : b \geq b^0\}$. For a general discrete activity analysis model

$$A = \begin{bmatrix} c_1 & c_2 & \cdots & c_n \\ -a_{11} & -a_{12} & \cdots & -a_{1n} \\ -a_{21} & -a_{22} & \cdots & -a_{2n} \\ \vdots & \vdots & \ddots & \vdots \\ -a_{m1} & -a_{m2} & \cdots & -a_{mn} \end{bmatrix},$$

the theorem can be used to answer, for example, the question of whether for all integral factor endowments b in a convex polyhedron P, the maximum output is greater than or equal to any specified value; and other global questions about the production set as well.

References

1 Kannan, Ravindran, Test sets for integer programs, ∀∃ sentences, in: Dimacs Series on Discrete Mathematics and Theoretical Computer Science, Vol. 1, American Mathematical Society, 1990, pp. 39–47.

2 Scarf, Herbert E., Production sets with indivisibilities, Part I: Generalities, Econometrica, 49(1).

3 Scarf, Herbert E., Production sets with indivisibilities, Part II: The case of two activities, Econometrica, 49(2) (1981).

4 Scarf, Herbert E. and David F. Shallcross, The Frobenius problem and maximal lattice free bodies, Mathematics of Operations Research, 18(3) (1994).

Index

A

Ades, Alberto 219
agricultural products 162–3, 164
agriculture 287
- and trade, 162 178, 183
- in Indian economy 203, 284–6
anti-dumping protection 144, 151, 152
APEC 147–8, 156, 162, 183–5, 191
- and free trade agreements 163, 195, 196–7
- and the Japanese rice market 190, 192
Argentina 81, 84
Armington trade model 134–5, 173
Armington, P.S. 128, 133
Arrow–Debreu model 396
Asian crisis 71
Asian Pacific Economic Cooperation
–, *see* APEC
Association of South East Asian Nations (ASEAN) 162

B

Baby Boom generation 321–3
balance of payments 81, 85, 98–9, 104, 114–5
- and crises, 78, 86
–, deficit in 239, 242
Baldwin, Richard E. 140, 154
Bardhan, P.K. 14
Barrett, S. 359–60
Barro–Becker growth model 43
Bartel, A. 320, 342–3
Basic Agriculture Act (1961) 188
Beals, Richard 25
Becker, Gary S. 216
Bell, Clive 210, 395–418
Ben-Porath, Y. 295–6
Benassy, Jean-Pascal 212–4
Benjamin, Dwayne 369
Besley, T. 219
Bhagwati, Jagdish 57–70, 142, 148, 216, 276
Bhalla, Surjit 369
Bhattarai, Keshab 125–136
Blanchard, Olivier 25
bloc goods 174–5, 177
Bogor Declaration 162
Brander, J.A. 91, 108, 118
Brander–Spencer model 111–2
Brazil 81
Brecher, R.A. 216
bribes 218–9, 221–2
Burkina Faso 374–84
- and Kenya 384–5, 387, 389–90
Bush, George 153, 162
Bylasky, Michael A. 419–41

C

Cameron, S. 299
Canada 161
Canada–United States Free Trade Agreement (CUFTA) 161
capital 34, 44, 203–4, 254
–, accumulation of 64–6, 68
–, inflow of 82–6, 315
capital market imperfections 383
captial stock 33–4, 268, 342
–aggregated 302

capital–labor ratio 46, 48
Carraro, C. 359–60
cartels 100, 103–4
Caselli, F. 321
Chakravarty, S. 277
Chamberlin competition model 101–2, 106
Chander, Parkash 353–61
chemical-based industries 254
children 40
- and associated costs 42
- and rate of consumption 33–4
–, *see also* fertility
Chile 84
China 63
choice of technique theory, 4–6, 9–10, 12, 15
Clinton, Bill 153
closed economy 278, 293–4, 315, 320
- and earnings 325, 327, 340–3
coalition formation 96, 361
- and stability 360
Cobb–Douglas function 129, 131, 190
college attendance 327, 331–2, 340–1, 345
–, probability of 311
–, *see also* schooling
communal living 420
- and consumption 433, 436
competition 90, 112–3, 118
- Chamberlin model 101–2, 106
- Cournot model, 101–2, 109
- Stackelberg model, 101–2
computable general equilibrium (CGE) models 202, 210–1, 231
constant absolute risk aversion (CARA) 404
consumers 114
–, welfare level of 193–4
consumption 131, 174, 190, 234, 297–300
- private 428, 444
- public 245, 424, 437
consumption program 29–32
corruption 218–9
- investigations of 220
Cournot competition model 101–2, 109
credit 79, 398
- and contracts, 400–1

D
demand function 172–3
descendants
–, *see also* children
devaluation 245, 255, 258
- and overdevaluation 83
developing countries 4–6, 71–2, 80, 138–9, 231
–, corruption in 219
–, economic interventions in 145
–, trade liberalization in 142, 152, 156
development economics 363
Di Tella, Rafael 219
Directly Unproductive Profit Seeking (DUP) 216
disequilibrium economics 212, 215–6
Dixit, A.K. 105
Dixit–Stiglitz model 105–6
Dobb, M. 10–2
dollar standard 72
Domino Theory of Regionalism 140
dynamic time path dimension 353, 357, 360

E
Easterlin hypothesis 46–7
EEC
- *see* European Economic Community
Eckaus, Richard S. 205
economic integration 143
–, degree of 142, 148, 150, 155, 161
economy of scale 173
education
–, *see also* college
–, *see also* schooling

electricity, subsidy of 221
endogenous growth theory 108
endowments 97, 110
- fixed land and labor 373
environmental standards 147, 152–3
- and emission levels 356
- and global pollution 353–4
equilibrium models 108, 403, 444
European Economic Community 161
European Union 103
excess demand theorem 115
exchange rate policies 71–2, 82, 258
–, change of 72, 230
exchange rates 73–5, 77–8, 84–5, 125, 127
- fixed 71, 80, 82
- fixed-but-adjustable 72
- floating 79
–, in India 229
- nominal 78–9, 83–4
–, stability of 145
exclusion restrictions 369, 371, 373, 384
exports 63, 74, 236, 243, 276
- and tariffs 92
- trade blocs 177
–, demand for 126–7, 129, 269
–, supply of 72, 176

F
Factor Price Equalization theorem 61
farm size 364, 369, 390
- and plot yield 377
Feldman–Mahalanobis model 203
Fernando, A.C. 214
fertility 39, 40, 43–4
- endogenous 50, 52
- exogenous 46, 51–2
–, rate of 45
Findlay, Ronald 25
fiscal adjustment 229–30, 243, 245–7
fiscal balance 73, 239–40
- external 86
- internal 243, 245, 257–8
fiscal deficit 242–3
- and India 229, 236
Fischer, Stanley 25
Fisher, Irving 21–4, 26
flat tax 337, 340, 342
Food Control Act of 1947 186–7
food shortage 187
foreign investment 144, 258, 277
Free Trade Agreement of the Americas 153
Frisch, Ragnar 204, 208–9, 218
full employment 247

G
Galenson–Leibenstein suggestion 7
Gangopadhyay, S. 413
Garnaut, Ross 63
GATT 78, 138, 141–2, 146, 151–2
- and tariff negotiations 164
GATT/WTO 90, 111, 118
General Agreement on Tariffs and Trade –, *see* GATT
general equilibrium models 207–10, 222, 223, 291–3, 296, 303–4, 336–44
- and labor earnings 291–3, 296, 332–3
- and tariff rates 91, 95, 99
Ghosh, Madanmohan 125–36
Ginsburgh, V. 206
Go, Delfin S. 229–72
Goodwin 204
Goto, Junichi 161–98
governments 238
- and consumption 237, 247, 255
- and corruption 222
- and economic intervention 215
Greece 179
Greenwood, J. 314
growth model dynamics 39
Grubel, H.G. 169

H
Hahn, Frank 209

Hall, Kevin 149
Harcourt, G.C. 208
Harrod–Domar model 203–4, 222, 276
Hawkins–Simon condition 278
heavy industries 6, 12, 254
Heckman, James J. 291–349
Heckscher–Ohlin models 89, 113, 165
Heijdra, B.J. 105
Hirschman, A.O. 276
human capital 298–301, 313, 321, 337, 341–2
- and production 306–9
–, investment in 294, 303, 314, 316, 322
human capital stock 301–5, 316

I
IMF
–, *see* International Monetary Fund
imports 63, 74–5, 85, 92, 128, 179–81, 243
- and penetration ratios 67
- and prices 130, 235
- and tariffs 97
–, controls of 276
–, demand for 72
–, restriction of 216
–, supply of 125, 127, 129
income 50, 105
- and savings 269
- and welfare 291
–, growth rate of 65
–, inequality of 322
–, *see also* wages
India 71, 202, 212–3
–, economic development of 201, 207, 231, 244
–, foreign trade and 229
–, government of 217
–, infrastructure of 277
indifference curves 22
inflation 73–4
- and anti-inflation devices 84
–, rate of 75, 81–3
infrastructure 285–7
- in Indian economy 284–6
- of India 277
inner-oriented trade 78–9
interest rates 326, 343
interlinked contracts 401, 407, 409–11, 415
interlinked transactions 397
interlinking 395–6, 401, 406, 412
- of trade 416
International Monetary Fund 72, 76
- and stabilization programs 77–9, 81
intraindustry trade indices 169–70
investment 74, 204, 242, 256, 299–300, 333
- in low skill workers 319

J
Japan 63, 154
- and the rice market 163, 183–7
Johnson, Harry 91

K
Kaldor, Nicholas 207, 209, 211
Kanaan Oussama 205
Kandel, Eugene 431
Kannan's theorem 457
Kannan, Ravi 457
Kenya 375, 383–9
Keren, Michael 419–41
Khan, M. Ali 9–19
kibbutzim 419–41
- and peer pressure 438
–, as labor-managed club 420
–, side of 431
Koopmans, Tjalling C. 23, 27, 29, 30
Korea 79
Krueger, Anno O, 71–87, 143
Krugman–Armington model 173

L
labor 63, 318–9, 332, 423, 452
–, accumulation of 64–5
– and capital 422
– and land markets 372, 388
–, demand for 381
– low skilled 312, 318, 324–5
– skilled 66, 295, 312, 314–5
–, supply of 298
– unskilled 315
labor force distribution 269
labor markets 238–9, 315, 381
labor standards 147, 152–3
labor-intensive goods 58–9, 61–2, 66–9
–, prices of 65
ladder of competitive advantage 63
land quality 370–1
Lazear, Edward P. 431
leading sector 280
–, identification of 282–86
Leontief, Wassily 34
Levhari, David 419–41
Levy, Philip I. 137–59
limited liability 396, 398, 403, 407–12, 417
–, *see also* unlimited liability
Linear Expenditure System 269
linear programs 447–8, 450, 454
– and production models 445–6
Lipschitz condition 30
Lloyd, P.J. 169
Loasby, Brian J. 215
Lochner, Lance 291–349
log wages, 320, 322, 331
long-run optimal growth factor 274–5, 280–2
Lucas, Robert 23

M
machines, production of 4, 6, 7, 13, 15
Mahalanobis model 203–5
Mahalanobis, P.C. 6, 12, 203
Majumdar, Mukul 273–89
Manne, Alan S. 206
Mansur, A. 215
Mantel, Rolf R. 21–37
manufacturing 162, 166, 178–9, 182, 284–6
manufacturing sectors 257
marginal impatience 21
– decreasing 24
– increasing 23–5, 33
mark-ups 112, 116
Mauro, P. 219
Mayer, W. 216
Mclaren, J. 219
Meade, James 72
metal industries 251
Mexican economy 83
Mexico 80–1, 150–1, 162
Mincer "rate of return" measures 294, 309, 324, 329–32, 344
Mitra, Pradeep 229–72
modified value added tax (MODVAT) 236–7
monopolies 101–2, 106, 108, 111, 113
moral hazards 403, 410, 414
mortgage loans 329
multi-sector models 273
multilateral trading system 147, 154

N
NAFTA 57–8, 61, 149–50, 153, 163
Narayana, N.S.S. 201–28
Nash allocation 428
Nash equilibrium 95, 99, 103–4, 113–4, 134–5
– and kibbutzim 427, 429–30, 432, 438
Nash optimal level 436
Nash tariff game 109, 111, 129, 133
Nehru, Pandit 201–2, 223
new trade theory 90–1, 99, 104, 118
Nishimura, Kazuo, 39–53
nonfarm resources 380, 383, 390

North–South preferential trade arrangements 146–7, 149, 152
numeraire goods 105, 107, 112–3, 117
 –, consumption of 190

O
Obstfeld, Maurice 24–5
offer curves 66–8, 125, 129, 131–3
OLG
 –, *see* overlapping generations
oligopolies 91, 95, 99–101, 109, 113, 115
on the job training 297, 303, 309, 318
 – and human capital 298
 – and investment 339
open capital markets 327, 333
open economy 292–3, 326, 336, 341–4
open regionalism 148
optimal growth rate 275, 285
optimal tariffs 91–3, 95, 97–99
Ossella, Ilaria 273–89
overlapping generations (OLG) 41–2, 52

P
Pant, Pitamber 202
Pareto optimal level of effort 208, 359, 425–32, 435, 444
Parikh, Kirit S. 201–28, 205, 206, 207
partial equilibrium model 112, 293–4, 333, 337–42, 417
peer pressure
 – in kibbutzim 431, 435, 436, 438
Peron–Frobenius eigenvalue 281
physical capital 294, 298–301, 313, 336
 –, decline of 341
planning models 205, 206, 211–2
plot yield 377–80, 386, 388–9
 – and quality 371–2, 380, 383
 – and size 385
plots 367–8, 380, 384
 – and farming 366
 –, characteristics of 372, 375, 377, 387–8
 – multiple 375
 –, size of 370
populations, classification of 217
Portugal 183, 186
 – and imports 181–2
Prasannan, R. 220
preferential trade liberalization 139, 154–6
 – of APEC rice 193
preferential trading arrangements 137, 139, 142–3, 146–8, 156, 361
price differentials 317
price function, 97–8
prices 60, 97, 128, 452
 –, adjustment of 296
 –, decentralization of 450
 –, elasticity of 345
 –, normalization of 98–9, 107, 114
private trading 416
producers 107
 – strategic 109, 110
product differentiation 163, 166, 169–70, 177–8, 183, 193–5
 –, degree of 166, 173–4
production 132, 188, 203, 420, 425
 – in kibbutzim 419, 424, 436, 439, 444
production function 51, 173, 301–2, 312, 446, 449
production sets 451, 458
profit maximization 375–89
 – in African farms 389
profit taxes 96, 100–1, 107
protectionism 103
PTAs
 –, *see* preferential trading arrangements
public policy in India 416

Q
QRs
 –, *see* quantitative restrictions

quantitative restrictions 73–4, 77–9, 81

R
Ramsey model 39–42, 50, 51–2
Ramsey, Frank P. 7, 11
Ramsey–Fisher consideration 14
rationing 213
Raut, Lakshmi K. 39–53
real interest rate 343, 345
real wages 57–60, 67–9, 240, 255, 322
  –, decline of 61
  –, *see also* wages
regionlism 153–4
  – and economic integration 162, 174, 178, 183
rice
  – as differentiated 167
  – Basmati 213
  –, brands of 168
  – domestically produced 193
  –, exports of 191
  – import of foreign 191
  –, price of 188
  –, production of 191
  –, stockpile of 188
rice imports 165, 167
  – and Japan 183
risk aversion 403–7
risk neutrality 407–12
Robinson, Joan 14–5
Robinson, S. 215
Romer, Paul M. 205
Rudra Ahok 205–6, 212, 215, 217
rural economies 396
  – in Africa 363, 385
rural markets 364, 367

S
Salinas, Carlos 150, 162
savings 44, 84, 204, 300, 313
  – and investment 238, 244
Scarf, Herbert E. 210, 443–58
schooling 287, 292, 297, 323, 330
  – and earnings 333–4
  –, choice of 299
  –, level of 294, 297, 301, 304, 331
  –, *see also* college attendance
Scitovsky, T. 276
Sen, Amartya K. 10–1, 217, 373
Sengupta, K. 413
separation 363, 367–73, 376–7
  –, tests of 387
  –, violations of 364
separation hypothesis 374, 384, 389–90
setup costs 94–5, 108
Sicherman, N. 320, 342–3
simultaneous buy and sell (SBS) formula 189
Siniscalco, D. 359–60
skill investment 334, 342
skill prices 311, 315, 319, 332, 334
skilled labor 292, 296, 315
Solow model 40
Solow, Robert M. 3–8, 9–11
South–South preferential table arrangements 146, 149
Spain 183, 186
  – and imports 180–2
Spencer, B.J. 91, 108, 118
Srinivasan, T.N. 3–8, 9–19, 57, 125–7, 291–3
  – and Clive Bell 210, 41, 415
  – and investment models 4–6
  – and labor and environmental standards 147
  – and multilateral liberalization 154–5
  – and preferential trading arrangements 137–8
SS theorem
  –, *see* Stopler–Samuelson theorem
stabilization programs 77, 240, 242, 256, 258
Stackelberg competition model 101–2
Stackelberg game 109, 111, 400, 406,

409–12
Stiglitz, J.E. 13, 15, 105
Stokey, Nancy 23
stop–go pattern of growth 75–7
Stopler–Samuelson theorem (SS theorem) 59–62, 68
Streeten, P.P. 276

T
Taber, Christopher 291–349, 307
tariff games 91–2, 95, 100, 102–3
- and Nash equilibria 129, 135
tariff reduction 244, 247, 254, 258
tariffs 59, 90, 92–3, 98–101, 111-2, 141, 266
- and markups 115, 117
- and rice 189–90, 193
- fixed 103
- on imports 230
- protective 165, 237, 244
-, rates of 113, 164, 172, 191
tax proposals 338, 340
taxes 337
- excise 232
- progressive 340
Taylor, L. 205–6, 210–1
technical change 64–5, 282, 320–1, 332, 343
tenancy 397–8, 400, 408, 412
- and credit 395, 417
- and labor 396
The Theory of Interest (Fisher) 22
tomato industry
- in Mexico and the US, 151–2
total factor productivity 42, 232, 245, 247, 255
trade 61, 94–5, 110, 139, 257
-, balance of 127, 130, 135
- with poor countries 57
trade barriers 164, 183
trade closure 125–29, 135
trade diversion 139–40, 149
trade liberalization 58–60, 64–8, 137–52, 155, 184–5, 229–31, 245–6, 252–8
- and agricultural trade 163
- and rice imports 192–5
- discriminatory 142
- global 184
- internal and external 255
- multilateral 140–1, 143, 148, 152, 154
- partial 195
- unilateral 148
trade models 90, 108, 134
tuition policies 333–4, 336, 342, 344–5
- and subsidies 332–3, 335
Tulkens, H. 354–6
Turkey 79, 84
-, exchange rate of 76

U
Udry, Christopher 363–93
United States 154, 161–2
unlimited liability 399, 403–7, 413, 417
unskilled labor 57, 67, 292, 295–6, 315
Uruguay Round 144, 155, 167, 184, 189
Usawa, Hirofumi 24

V
Viner, Jacob 139–41

W
Waelbroeck, J. 206
wage inequality 292, 319–20, 323, 332, 340–1, 344
- increasing 295–6, 312, 314
wages 63, 293, 302, 307, 310
- nominal 239–40
-, progressivity of 339
-, *see also* real wages
Walrasian model 207, 209–11, 443
Weitzman, M.L. 14
welfare 28–30, 32, 211–2, 435
- of Japanese consumers 193–4

– of kibbutz members 429, 432
Whalley, John 125–36, 215
Willingness lines 22
Woods, Bretton 72
workers
–, *see also* human capital
–, *see also* labor
World Trade Organization 138, 142–4, 151, 155, 156
worldwide trade treaty 356, 359–60
– on global pollution 354
WTO
–, see World Trade Organization

Y
Yang, X. 105
Yorukoglu, M. 314

Z
Zedillo, Ernesto 150